Gary Gann

3M

APPLIED TECHNIQUES
IN STATISTICS FOR
SELECTED INDUSTRIES

APPLIED TECHNIQUES IN STATISTICS FOR SELECTED INDUSTRIES

Coatings, Paints, and Pigments

H. EARL HILL

Lord Corporation
Erie, Pennsylvania

JOSEPH W. PRANE

Industrial Consultant
Elkins Park, Pennsylvania

A Wiley-Interscience Publication

JOHN WILEY & SONS

New York   Chichester   Brisbane   Toronto   Singapore

Library of Congress Cataloging in Publication Data:

Main entry under title:

Applied techniques in statistics for selected industries.

    "A Wiley-Interscience publication."
    Includes index.
    1. Coatings—Statistical methods. 2. Paint—
Statistical methods. 3. Pigments—Statistical methods.
I. Hill, H. Earl. II. Prane, Joseph W.

TP156.C57A67 1984   667′.9′015195   83-21750
ISBN 0-471-03791-5

Printed in the United States of America

10 9 8 7 6 5 4 3 2 1

CONTRIBUTORS

JOHN BAX
Scott Bader (USA), Inc.
Richmond, California

JOHN COMPTON
College of Graphic Arts &
Photography
Rochester Institute
of Technology
Rochester, New York

FRANK CULLEN
Loyola College
Baltimore, Maryland

GEORGE C. DERRINGER
Battelle
Columbus Laboratories
Columbus, Ohio

ROGER E. ECKERT
School of Chemical Engineering
Purdue University
West Lafayette, Indiana

BHAGWATI P. GUPTA
Lord Corporation
Erie, Pennsylvania

GERALD J. HAHN
General Electric Company
Corporate Research &
Development
Schenectady, New York

H. EARL HILL
Lord Corporation
Erie, Pennsylvania

JAMES R. KING
TEAM
Tamworth, New Hampshire

JOHN V. KNOPP
1025 Reynolds Road
Johnson City, New York

GARY E. MEEK
College of Business Administration
University of Akron
Akron, Ohio

THOMAS D. MURPHY
American Cyanamid Company
Consumer Products Research
Division
Clifton, New Jersey

JOSEPH W. PRANE
213 Church Road
Elkins Park, Pennsylvania

ANTHONY A. SALVIA
Behrend College
Pennsylvania State University
Erie, Pennsylvania

ADAM ZANKER
Kiriat—Jam "G"
Israel

PREFACE

We have tried to bring together in this book an organized, self-contained, empirical collection of topics in mathematics and statistics for the industrial chemist and engineer. Because of the background and experience of the editors and several of the individual chapter authors, the thrust of the book is toward the solution of problems and the design of experiments in the coatings and allied industries. However, this book should be of value to individuals in other basic and applied polymer industries, for example, plastics, elastomers, adhesives, sealants, fibers, and films. The statistical methods discussed are invaluable tools and, when used in well-planned designs, are universal in their application.

We have attempted to present the material in this book at a level somewhere intermediate between that of standard statistical textbooks and other books that deal almost entirely with the application of statistical methods. Although there exist many published papers in the coatings industry with a statistical foundation, the emphasis is nearly always on the experiment first with the statistical analysis, frequently very brief, merely as support. This book follows the reverse format. The emphasis, where possible, is on practical statistical applications, where theoretical exposures are generally presented with minimum proofs or derivations, and the experiments are used as examples. The contributions, although self-contained, are interrelated. The plan of the book is given at the end of Chapter 1. In a few cases there will be some overlapping of information. This was inevitable in view of the independence of the chapters.

The coatings chemist or engineer is no stranger to statistics, but he or she is frequently faced with translating examples from unfamiliar fields such as agriculture or medicine. This book is a first attempt to give industrial chemists or engineers in the general field of polymer science a set of topics, in their language where practical, which are basic to their everyday use. Hopefully, as an extension of these applications, it will also be of value to the chemists and engineers working for the raw material suppliers to the coatings industry, those employed by the users of coatings, government coatings chemists, and so on. Examples have been taken from work in the coatings, chemical, and allied industries to illustrate the application of the methods.

In the early chapters, the necessary working foundations of mathematics and statistics are developed through introductory and descriptive material, followed in subsequent chapters by practical applications such as regression analysis. There are no exercises as such, since this book is primarily intended to be an orienting, general reference text for the chemist and engineer using the techniques, rather than a teaching textbook. Where at all possible, the examples and case histories have been selected from the coatings industry, high polymers, or from general examples that are not too far afield. In general, the only exceptions to this have been in discussions of, for example, probability where it is easier to develop first principles from familiar examples such as those concerning playing cards, dice rolling, and so on.

Although most practicing industrial chemists and engineers use computers of all types and capabilities in their work, no formal attempt has been made to discuss the use of computers in developing solutions to problems. Rather, the emphasis has been on developing the conceptual ideas of data analysis, with the opinion that with fundamental methods and understanding well in hand, interfacing with calculators and computers will follow naturally. However, this is not meant to minimize the use of computers since their wide availability has made traditional, formerly cumbersome analysis methods so much more convenient and accurate. For this same reason though, shortcut methods such as estimation from the range, have been deemphasized. However, unlike the problem of the "chicken and the egg," there is no question which comes first and the basis of written computer programs stems from the knowledge and understanding of first principles and concepts. Hopefully, this book may also help in particular cases to bridge the gap in theory and practice between the industrial coatings chemist and engineer, and the computer specialist.

The techniques that are presented represent only a fraction of those now available. We have tried to base the selection on those of fundamental importance and greatest utility. Consideration has also been given to the degree of correspondence between the practical situations from which data are obtained and the mathematical treatments on which analyses of the data are made. No attempt has been made to treat all methods exhaustively but rather the fundamentals of each are emphasized with supporting examples. Detailed references to original papers or more complete works are supplied for those desiring to dig deeper into any one subject area.

When this book was being planned, it was hoped, regardless of diversity of the topics, that a uniform nomenclature could be used throughout. As work progressed, it became evident that this was not completely possible. Two approaches have been taken to solve this problem. One, a general glossary of terms and symbols has been included in the Appendix. Two, where possible, terms have also been defined in the chapters where they first appear. Within the context of each chapter, this latter procedure should help resolve the problem of symbols with different meanings.

General tables of statistical functions have been included in the Appendix to enhance its value as a work of reference. However, other tables of a more specific nature will be found at the end of several of the chapters; see, for example, Chapter 4 on Sample Size and Chapter 16 on Control Charts for Variables. A comprehensive index has also been included.

We gratefully acknowledge permission from various authors and publishers to use or reproduce portions of several books and papers. Sources are given prior to the Appendix tables.

The authors are indebted to the Lord Corporation for generous support in terms of preparation of the manuscript and illustrations.

H. EARL HILL
JOSEPH W. PRANE

Erie, Pennsylvania
Elkins Park, Pennsylvania
February 1984

CONTENTS

APPLIED TECHNIQUES IN STATISTICS FOR SELECTED INDUSTRIES

BASIC CONCEPTS

JOSEPH W. PRANE

213 Church Road
Elkins Park, Pennsylvania

1.1. INTRODUCTION

The purpose of presenting this particular feature on statistical methods in the coatings industry is to acquaint you with the many benefits that can be gained from the application of statistical procedures in three important areas of paint technology—research and development, testing and evaluation, and production.

We are well aware that the study of surface coatings is attended by many difficulties stemming from the inability to obtain reproducible test results. Uncontrollable factors such as atmospheric conditions, thickness of paint films, the human element, and so on—all contribute to varying results from a particular test method. In addition, many physical properties of paint films must be approximately measured since, in many cases, there are no satisfactory quantitative test methods available.

It is in this particular phase of coating technology that statistical methods have much to offer. For in statistical techniques we have the tools and the power to design our test methods, analyze our test results, correlate our findings, and predict precision and specification compliance. But most important is the fact that statistical methods can provide reliable results for making decisions.

Many will ask what is required to utilize statistical techniques in the ordinary paint plant. First of all, laboratory individuals must be versed in statistical methods to plan experiments and analyze data. This may be accomplished by training key laboratory personnel in the various phases of statistical procedures, by holding scheduled seminars within the company or by having individuals attend a short course(s) in statistics. Such training should adequately prepare the individual to apply statistical methods to most of the problems he or she encounters. For advanced techniques, the services of a professional statistician are needed.

In the way of equipment, a calculator is a necessity and the type chosen will depend on the amount of computation involved. For greater in-depth statistical analysis, access to a modern computer is required.

In conclusion, we would like to emphasize that the cost of utilizing statistical procedures in your laboratories and plants is exceedingly small when compared with the many savings and advantages that can be derived from such techniques.

Background of Statistical Procedures

Many companies and organizations in the coatings industry are enjoying economies of operation and increased precision of results thought unattainable just 10–15 years ago. Did some miracle bring this about? Hardly. Rather, it was the slow but steady evolution of the use of statistical methods for the presentation and analysis of data and interpretation of results. This evolution has been particularly evident in the fields of quality control and experimental design and analysis.

Formerly, these methods were the province of mathematicians, biologists, and various groups in the social sciences, such as agronomists, psychologists, and actuarians. Chemists and other physical scientists used numbers and their interrelationships during their experimental work. However, many of them were reluctant to employ statistical analysis, even if they may have been familiar with the concepts, because the particular factor or property they were studying could usually be measured so closely; in addition, it usually would vary over such a small range that the true value could be established within narrow limits. Indeed, measurements of an absolute nature, such as atomic numbers or the speed of light, were entered into our record books without the use of statistics as we know it.

It was during World War II that the omnipotent statistical methods, borrowed from the social scientists and introduced into the war effort, made their mark and received their impetus. In the fields of quality control, sampling, inspection, fire control, and artillery and shell design, use of these methods resulted in a considerable reduction of critical manpower requirements with a usual gain in precision and accuracy of results.

During the years that statisticians were developing statistical techniques, marked changes were taking place in pure and applied research (1). The problems became more complex, and tolerances became tighter. Successful competition called for more than buying cheap and selling expensive. Successful competition called for making better products at lower costs. Large industrial research laboratories came into existence and great research institutions pooled the skills of many investigators to solve these problems. Research and development budgets were increased and became an important item in the cost of doing business. At this point there was good reason to examine both the planning of research projects and the laboratory execution of the work to see if any techniques were being overlooked that would increase the return obtained from the research and development dollar.

This examination led a number of industries and laboratories to look more closely at the hitherto ignored, and sometimes despised, techniques of statistics. It was found that statistics and particularly the statistical design of experiments had potentially impressive leverage effects on the amount of information obtained from experimental work. Many engineers and scientists began to study statistics. It speaks well for the subject that even the basic elements, easily learned in a few hours, yield impressive dividends.

Efforts toward improving research thinking and research methods must be continued vigorously. Two major hindrances to progress are still very much in evidence; hence, for the adherents of the statistical approach, this task still remains rather formidable. The first one regards the applicability of many of the statistical methods. There is sometimes an extremely wide gap between a statistical model and the composition of an experiment—assumptions that cannot be met, variables that are difficult to control, and procedures that may be difficult to execute. An effective compromise can be agreed upon, perhaps by both modifying the statistical model into a less stringent, more workable form, and also, at the same time, making minor changes in the manner of performing the experiment.

The second hindrance, and the one more difficult to resolve, is the sheer lack of recognition among research personnel that the statistical method is a truly effective and powerful tool in scientific research. The statisticians themselves are not free of responsibility for this unfortunate situation, for often their uncompromising insistence on unduly stringent statistical models or their overemphasis on mathematics have taken the place of a properly sympathetic interest in the field of application, for example, chemistry, engineering, or biology. Hence, a research worker who might otherwise be cooperative has been alienated. Compromise here would be important, since the research and development worker does not have to be a statistical purist to make use of these methods.

Most researchers have intuitively organized their work on a factorial basis even though they do not actually proceed to the final analysis of variance. Here, statistics might be termed "scientific intuition," because it goes one step further in helping a scientist draw more reasonable conclusions from the data obtained in an experiment set up under the scientific method.

Applications of statistical methods are universal; particular application to the coatings industry involves design of experiments in research and development, pigmentation studies, vehicle components, exposure testing, and pilot plant work. Examples of use in quality control and production include: purchasing and product specifications, sampling plans, control of pigments, oils, resins, thinners, production control, and correlation tests. These will be discussed in detail in later sections with particular reference to savings of material and manpower in testing and production.

These methods are not restricted to large companies, to be considered as a luxury to be enjoyed only by those with extensive technical budgets. The techniques are relatively uncomplicated (if one forgoes the derivations) and can be used by individuals and organizations of any size. All that is needed is the know-how, a calculator or a microcomputer, the data and situation to analyze, and a desire to get the best results possible in the most economical fashion.

Fundamentals and Tools

Chemists and engineers in our development and control laboratories are constantly making measurements of such properties as gloss, viscosity, fineness of grind, weight per gallon, acid value, nonvolatile content, tinting strength, and so on. Frequently, they make measurements in duplicate or triplicate, thinking that repetition of the measurement will lead to reproducibility of results. Intuitively, most chemists realize that measurements without "checks"—without a rough determination of precision or experimental error—are almost meaningless.

They realize that many factors, both known and unknown, will influence their tests. For example, some of the known factors are temperature, humidity, film thickness, and so on. These and many others are known to exist and have some effect upon the precision (reproducibility) and accuracy (relation to a standard) of these results. But over and above these known causes for error are the unnumbered unknown, chance, uncontrollable, randomly occurring causes for deviation—made up in part by interactions between many of the known causes.

If the chemist or operator gets different results when checking hiding power of a paint, what or who is to blame? Is it the test method, the operator, the instrument used, or perhaps a combination of all of these variables and many more, some completely unknown, which contribute to the makeup of a particular test result?

Also, by using the same paint and test method, will the operator get the same results tomorrow or next month? Will the operator get the same results as a different chemist working under the same conditions?

We all know that problems of this type come up continually in quality control, specifications, and referee work, and also in cooperative work and sponsored work such as that done by the technical committees of the ASTM or AOCS. What we may not realize is that in statistical techniques we have the tools and power to design our test methods, analyze our test results, correlate our findings, and predict precision and specification compliance to the extent that management wishes to pay for.

Concepts

Statistical methods involve the collection, presentation, analysis, and interpretation of numerical data. The structure of the whole statistical analysis depends on the trustworthiness of the first element, the raw data.

An illustration cited by Stamp is to the point (2): Harold Cox, when a young man in India, quoted some Indian statistics to a judge. The judge replied, "Cox, when you are a bit older, you will not quote Indian statistics with that assurance. The government are very keen on amassing statistics—they collect them, add them, raise them to the nth power, take the cube root and prepare wonderful diagrams. But what you must never forget is that every one of those figures comes in the first instance from the 'chowty dar' (village watchman), who just puts down what he damn pleases." It should be added that this story refers to the India of the days long past. Today India has many able statisticians and an active statistical society. Presumably the "chowty dar" no longer functions as the source of local statistical information.

Fortunately modern statistical methods are quite discriminating. False data usually become apparent early in the analysis, and warning signals are set up to guide the experimenter into more fruitful channels.

1.2.   FREQUENCY DISTRIBUTIONS

Let us start by looking at a historical but real set of data typical of what may have been found or compiled in the coatings industry. In Table 1.1 is shown the raw data for optical density ratios for 43 films of the same moisture curing clear urethane coating. The effects of UV absorbers were being studied; transmitted intensity was measured at 4150 Å before and after five months exterior exposure. Their intensities were compared to the incident intensity, the optical densities calculated, and the relative stability, S, computed and tabulated.

The following example shows the process whereby we arrive at the figures to be treated statistically:

O.D. = optical density

I_0 = incident intensity (e.g., 100%)

I_1 = transmitted intensity before exposure (e.g., 87%)

I_2 = transmitted intensity after exposure (e.g., 55%)

λ = wavelength of measurement = 4150 Å = 4.5 nm

$$O.D. = \log_{10}\left(\frac{I_0}{I}\right)$$

$$S = \text{relative stability} = \left(\frac{O.D._2}{O.D._1}\right)_{\lambda}$$

$$S = \frac{\log_{10}(100/55)}{\log_{10}(100/87)} = \frac{2.0000 - 1.7404}{2.0000 - 1.9395} = \frac{0.2596}{0.0605} = 4.29$$

Note that Table 1.1 lists the S values as X; this is a notation that we shall be using frequently. Observe also that other notations and computations are shown in this table for convenience; their purpose and meanings will be explained shortly.

It is apparent that very little information is forthcoming unless the figures are rearranged, since, when the data are in this form, it is a tedious job to find even the lowest and highest value. It is more difficult to ascertain around what value the figures tend to concentrate or if indeed

TABLE 1.1.   Raw Data—Relative Stability of 43 Urethane Films ($S = X$)[a]

X	X^2	X	X^2
2.30	5.290	2.26	5.108
2.40	5.760	2.15	4.623
2.84	8.066	2.22	4.928
2.76	7.618	2.22	4.928
2.03	4.121	2.49	6.200
1.99	3.960	1.74	3.028
2.74	7.508	1.55	2.403
3.32	11.022	1.25	1.563
1.92	3.686	1.14	1.300
1.70	2.890	1.22	1.488
1.87	3.497	1.58	2.496
1.89	3.572	2.93	8.585
2.74	7.508	1.03	1.061
2.57	6.605	0.70	0.490
3.57	12.745	5.01	25.100
3.61	13.032	4.14	17.140
3.91	15.288	0.94	0.884
3.91	15.288	3.91	15.288
1.40	1.960	2.29	5.244
4.24	17.978	1.61	2.592
2.87	8.237	1.53	2.341
2.78	7.728		

[a]Note: $\Sigma X = 103.27$, $\Sigma X^2 = 290.149$, $n = 43$, $\bar{X} = 2.402$.

TABLE 1.2. Array of Relative Stability Data for 43 Urethane Films

5.01	2.84	2.03	1.14
	2.78	1.99	
4.24	2.76	1.92	1.03
4.14	2.74	1.89	0.94
	2.74	1.87	0.70
3.91	2.57	1.74	
3.91	2.49	1.70	
3.91	2.40	1.61 ← (Q_1)	
3.61	2.30	1.58	
3.57	2.29	1.55	
3.32	2.26 ← Median	1.53	
	2.22 (Q_2)	1.40	
2.93	2.22	1.25	
2.87 ← (Q_3)	2.15	1.22	

they do show such a concentration. These and other steps in analysis are facilitated by rearranging and summarizing the data.

In Table 1.2, the data have been arranged in an array, in descending order of magnitude. This arrangement shows the range of the data to be from 0.70 (the lowest) to 5.01 (the highest); there is also an apparent concentration in the neighborhood of 2.1–3.0. (Other notations in this table are explained later.)

Table 1.3 shows the data of Table 1.2, summarized in a *frequency distribution*. It is obvious that the frequency distribution does not show the details of the array, but much is gained by the summary. The range and concentration of the values are still shown as before. However, the pertinent point of this table is to establish the concept of the frequency of occurrence of the data within nine classes or groups. The class interval i comprises 0.54 units of relative stability.

TABLE 1.3. Frequency Distribution of Relative Stability Data

S Data		f = Number of Data Points
Class	Midpoint	Within Indicated Limits
0.42–0.96	0.69	2
0.96–1.50	1.23	5
1.50–2.04	1.77	11
2.04–2.58	2.31	9
2.58–3.12	2.85	7
3.12–3.66	3.39	3
3.66–4.20	3.93	4
4.20–4.74	4.47	1
4.74–5.28	5.01	1

Having thus classified the data, rapid computations of many other properties of this distribution can be made (shown later), which will assist in describing and analyzing the data.

Figure 1.1 shows a plot of the data in Table 1.3 as a frequency distribution curve. Examination of the graph shows that the midpoints of the classes or groups chosen are plotted on the abscissa and the frequency of occurrence f is plotted on the ordinate. The number of classes is chosen so that the distribution will not show noticeable irregularities when plotted (too many classes), or that so many frequencies will be crowded into a class as to cause much information to be lost (too few classes). An excellent, in-depth discussion of frequency distribution data treatment and plotting can be found in Ref. 3.

Generally speaking, from 6 to 16 classes should be selected. An absolute minimum to be used may be calculated from the following equation:

$$i = \frac{\text{range}}{1 + 3.322 \log n} \tag{1.1}$$

where i = the class interval and n = the number of items in the distribution.

The class intervals and limits should be so selected that the midpoints are truly representative of each class. For the data in Table 1.2, with a range of 4.31 (5.01–0.70), the minimum class interval calculates to 0.67. As shown, we have chosen i = 0.54 for ease of calculation and representation, yielding nine classes.

An interesting and useful representation of such data can also be made in a *cumulative frequency distribution*. In Table 1.4, the data of Table 1.3 are rearranged to show how many and what proportion of the relative stability values were more than the stated figures. The absolute data are plotted in Figure 1.2 and the percent values in Figure 1.3 (on *probability paper*).

TABLE 1.4. Cumulative Frequency Distribution of Relative Stability Data

S Data	Number of Data Points	Percentage of Total
0.42 or more	43	100.00
0.96 or more	41	95.53
1.50 or more	36	83.72
2.04 or more	25	58.14
2.58 or more	16	37.21
3.12 or more	9	20.93
3.66 or more	6	13.95
4.20 or more	2	4.65
4.74 or more	1	2.33

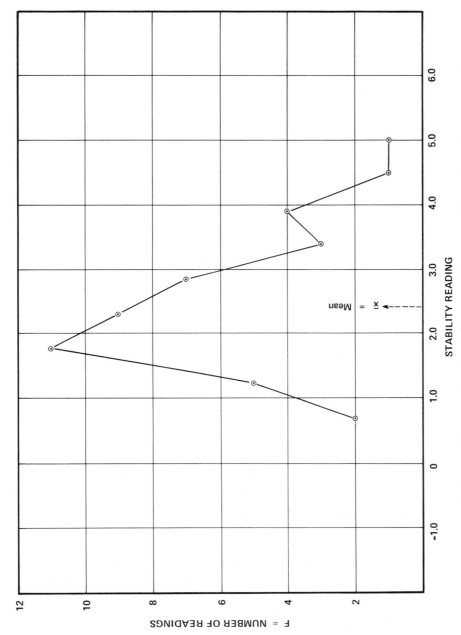

FIGURE 1.1. Frequency distribution curve of stability values of 43 urethane films.

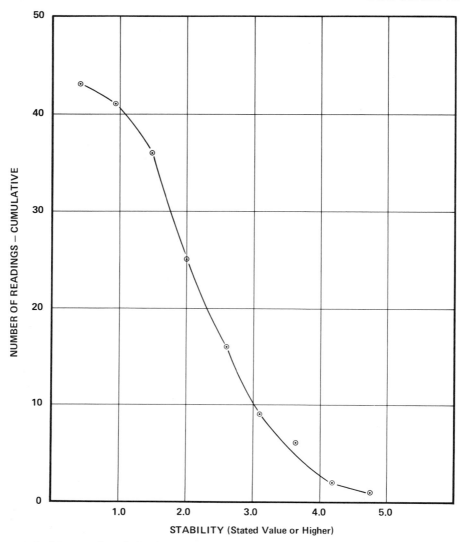

FIGURE 1.2. Cumulative frequency distribution plot of absolute data "ogive" curve.

These curves are called *ogives* (or *histograms*)and are helpful in visually classifying data as to quality or magnitude. (If the ogive of the data on the probability paper is a straight line, the so-called *normal curve* may be used to describe the distribution.)

1.3. CENTRAL TENDENCY

Now let us examine some of the properties of such data distributions: First we will consider the various measures of *central tendency*, that is, the

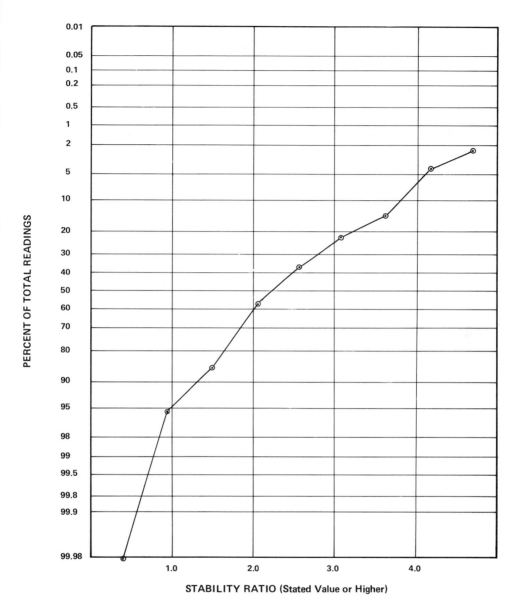

FIGURE 1.3. Cumulative frequency distribution plot of percent of total readings "ogive" curve.

characterization of the values that tend to concentrate in the center of the distribution.

Mean

Most familiar of these is the *arithmetic mean* or more commonly, the *mean* or the *average* or the *norm*. This of course is obtained by summing the values and dividing by the number of values.

$$\overline{X} = \frac{X_1 + X_2 + X_3 + \cdots + X_n}{n} = \frac{\Sigma X_i}{n} \tag{1.2}$$

For a frequency distribution, the arithmetic mean may be calculated as follows:

$$\overline{X} = \frac{f_1 X_1 + f_2 X_2 + f_3 X_3 + \cdots + f_n X_n}{f_1 + f_2 + f_3 + \cdots + f_n} = \frac{\Sigma fX}{n} \tag{1.3}$$

where X_1, X_2, X_3,... represent the midpoints and f_1, f_2, f_3 the frequencies.
For the distribution shown in Table 1.1, $\overline{X} = 2.402$ by equation (1.2). Equation (1.3) yields a value of $\overline{X} = 2.405$, which is very close to 2.402.

Several short cuts to the computation of $\overline{X}$ for grouped data exist. These depend on the assumption of a value $\overline{X}_d$ as the mean and computing the necessary correction to obtain $\overline{X}$. Here the symbol d signifies $X - \overline{X}_d$ and $\overline{X}_d$ may be the midpoint of any class. The resulting equation is

$$\overline{X} = \overline{X}_d + \frac{\Sigma fd}{n} \tag{1.4}$$

Further simplification is possible by expressing d as d', the deviation in terms of class intervals, with this result:

$$\overline{X} = \overline{X}_d + \left(\frac{\Sigma fd'}{n}\right)i \tag{1.5}$$

These will not be calculated here but may be obtained from a consideration of Table 1.3. $\overline{X}$ calculated by equation (1.4) and (1.5) will check the value $\overline{X}$ from equations (1.2) and (1.3).

Two important features of the arithmetic mean should be considered here: (a) $\Sigma(X - \overline{X})$ or Σx, the algebraic sum of the deviation of each individual value from the mean equal zero ($\Sigma x = 0$); (b) the sum of the squares of the deviations (Σx^2) is a minimum. This will be used later in the calculation of standard deviation.

Median

A second common measure of central tendency is the *median*, M_e, which is the value that divides a distribution of an odd number of units so that an equal number of items are on either side of it. It is apparent that the median is easily determined from an array of the data. In Table 1.2, the median of the distribution of stability data is 2.26.

For an ordered data set of an even number of units, the median is defined as the mean of the two middle units. From a histogram viewpoint, the median is that value of the data set which exactly divides the histogram into two equal parts.

Associated with the median are the *quartiles*, which divide the distribution into four equal parts. Q_1, the first or lower quartile, is the value located so that one-quarter of the items fall below it and three-quarters of the items exceed it. Q_2 is the median. Q_3, the third or upper quartile, is the value so located that three-quarters of the items fall below it and one-quarter exceed it. For Table 1.2, the measures are as follows; $Q_1 = 1.61$; $Q_2 = 2.26$; $Q_3 = 2.87$.

Mode

A third measure is the *mode*, which is the most frequent or typical value in a distribution, the point around which the items tend to be most heavily concentrated. In some cases, a mode may not exist, and even if it does, it may not be unique. The distribution of the individual stability data is unclear from Table 1.2. However, the modal group or class is 1.50–2.04, with a midpoint of 1.77 (Figure 1.1, Table 1.3).

For distributions of this type (which approach *normality* and are but slightly skewed), Karl Pearson, the eminent statistician, has expressed the relationship between the mean, median, and mode in this empirical equation:

$$\text{mode} = \text{mean} - 3(\text{mean} - \text{median}) \qquad (1.6)$$

Effect of Extreme Values

When *skewness* (lack of symmetry) is not general but due to a few items deviating a great deal from the mode, the median will be only slightly affected. The arithmetic mean, however, is affected by the value of every item in the series, and the presence of one or a few extremely large (or small) items in a series may result in a mean which is very misleading. As ordinarily computed, the mode is not at all influenced by the presence of a few unusually high (or low) extreme values.

Other Minor Means

Other measures of central tendency which are less frequently used are the *geometric* and *harmonic means* (G.M. and H.M.):

$$\text{G.M.} = \sqrt[n]{X_1 \cdot X_2 \cdot X_3 \cdots X_n} \tag{1.7}$$

The geometric mean is sometimes used to average ratios. It follows that the logarithm of the geometric mean is equal to the arithmetic mean of the logarithm of the observed values.

$$\text{H.M.} = \frac{1}{1/X_1 + 1/X_2 + 1/X_3 + \cdots + 1/X_n} = \frac{n}{\Sigma(1/X)} \tag{1.8}$$

This mean is sometimes used in averaging rates of change.

1.4.   SPREAD OF A DISTRIBUTION—DISPERSION

We have seen that distributions have a definite central tendency. However, it is also evident that the dispersion of the points or values around their mean can vary greatly. There are several types of measures of dispersions, which are absolute and measured in units of the problem.

Range

This is a rather crude, but useful and easy to understand measure, referring to the lowest and highest value of the distribution. For example, in Table 1.2, the range is 0.70 to 5.01 = 4.31. The range has one important disadvantage. Being based on extreme values, it is misleading if one or both of these values is an unusual occurrence.

Average Deviation (A.D.)

This is the sum of the deviation of the items from the arithmetic mean, without regard to sign, divided by the number of items. Thus,

$$\text{A.D.} = \frac{\Sigma|X|}{n} \tag{1.9}$$

or for a frequency distribution

$$\text{A.D.} = \frac{\Sigma f|X|}{n} \tag{1.10}$$

where $|\;|$ means the signs are neglected.

For a normal distribution (symmetrical) 57.7% of the items are included within the range $\overline{X} \pm \text{A.D.}$

Quartile Deviation

This measure of dispersion is based on the lower and upper quartiles Q_1 and Q_3 as is given by the following expression:

$$Q = \frac{Q_3 - Q_1}{2} \tag{1.11}$$

For a symmetrical series, or one nearly so, $\overline{X} \pm Q$, will include 50% of the values. Data from Table 1.2 shows

$$Q = \frac{2.87 - 1.61}{2} = \frac{1.26}{2} = 0.63$$

Also, from Table 1.2, $\overline{X} \pm Q = 2.40 \pm 0.63 = 3.03 - 1.77$. Of the 43 items, 21 are in this range, or 48.8%.

Standard Deviation

This is one of the most important and critical measures in all of statistics, and one which will enter into almost all of our considerations for the rest of this chapter. It is represented by the Greek letter σ for the population and as s for the sample of data points drawn from the population. The standard deviation (or root-mean-square deviation as it is sometimes known) for the sample is obtained as follows:

$$s = \sqrt{\frac{\Sigma(X_i - \overline{X})^2}{n-1}} = \sqrt{\frac{\Sigma x^2}{n-1}} \tag{1.12}$$

The steps involved in the actual calculation of s from raw data are as follows:

1. Determine the deviation x of each item from $\overline{X}$.
2. Square these deviations.

3. Total them.
4. Divide this sum by $n - 1$.
5. Take the square root.

For a large number of readings, this could become very laborious and tedious. A short cut computation for s is

$$s = \sqrt{\frac{\Sigma X_i^2}{n-1} - \left(\frac{\Sigma X_i}{n-1}\right)^2} \qquad (1.13)$$

This compilation has been performed in Table 1.1, with the following results:

$$s = \sqrt{\frac{290.149}{42} - \left(\frac{103.27}{42}\right)^2} = \sqrt{6.908 - 6.046} = \sqrt{0.862}$$

$$s = 0.93$$

For grouped data, in a frequency distribution,

$$s = \sqrt{\frac{\Sigma f(x)^2}{n-1}} \qquad (1.14)$$

Shortcut methods include the following:

$$s = \sqrt{\frac{\Sigma fd^2}{n-1} - \left(\frac{\Sigma fd}{n-1}\right)^2} \qquad (1.15)$$

or

$$s = i\sqrt{\frac{\Sigma f(d')^2}{n-1} - \left(\frac{\Sigma fd'}{n-1}\right)^2} \qquad (1.16)$$

These have not been computed here, but may be readily calculated from Table 1.3.

Several important properties and uses of the standard deviation are:

1. It is affected by all values.
2. A great importance is given to extreme values (since deviations are squared).
3. It is one of the factors involved in the equations for normal and skewed curves.
4. It is invaluable in reliability testing and correlation analysis.

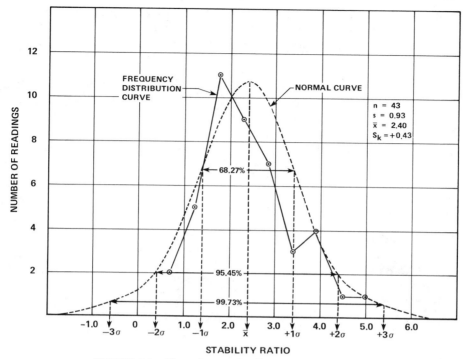

FIGURE 1.4. Normal curve fitted to stability value data.

The standard deviation is a much used measure of the spread of a series of data. For a normal distribution, 68.27% of the values are included within the range $\overline{X} \pm s$, 95.45% within $\overline{X} \pm 2s$, and 99.73% within $\overline{X} \pm 3s$. This is shown in Figure 1.4.

Referring to Tables 1.2 and 1.3, with $s = 0.93$, $\overline{X} \pm s = 2.40 \pm 0.93 = 3.33 - 1.47$. Within this range are 28 items or 65.1%. For $\overline{X} \pm 2s$, $2.40 \pm 1.86 = 4.26 - 0.54$, there are 42 items within this range, or 97.7%. Finally, for $\overline{X} \pm 3s$, $2.40 \pm 2.79 = 5.19 - 0$, there are 43 or 100% of the items within this range. This is only a fair to poor check of the expected values shown above, and indicates what has been evident all along—that our data are not normally or symmetrically distributed.

Coefficient of Variation

The actual variation or dispersion as determined from the standard deviation or other measure of dispersion is referred to as the absolute dispersion. However, a deviation of 1 oz in a batch of 3000 lb is quite different in effect from the same deviation of 1 oz in a laboratory batch of 500 g. A measure of this effect is provided by the relative dispersion, which is obtained by dividing the sample standard deviation by the mean, and this value, called the coefficient of variation, is given by

$$V = \frac{s}{\overline{X}}$$
(1.17)

and is generally expressed as a percentage.

The coefficient of variation is dimensionless. Because of this, it is useful in comparing distributions that are expressed in different units. A disadvantage of the coefficient of variation is that it lacks utility when $\overline{X}$ approaches zero.

1.5. TYPES OF DISTRIBUTIONS

Normal Curve

A frequency distribution usually represents a sample drawn from a much larger population or universe. Even though a sample is composed of a few hundred or frequently even fewer items, it may be a reasonable representation of the larger universe from which it was drawn. Since it is virtually never possible to measure all of the individuals or items comprising a universe, we must form our notion of the larger group from a study of a sample. We may therefore fit any one of a number of types of curves to a frequency distribution to attempt to describe what appears to be the general form of the curve for the entire population.

For the purposes of this chapter, we will discuss mainly the *normal curve*, and mention other types in passing (2).

The concept of the normal curve appears to have been originally developed by Abraham De Moivre and explained in a mathematical treatise which its author believed had no practical application other than as a solution of problems encountered in games of chance. Gauss later used the curve to describe the theory of accidental errors of measurements involved in the calculation of the orbits of heavenly bodies. Because of Gauss' work, this curve is sometimes referred to as the *Gaussian curve*.

In fitting a normal curve it is assumed that only chance errors are present and that the arithmetic mean represents the best approximation of the true value. It will be observed (Figure 1.4):

1. That small errors are more frequent than large ones.
2. That very large errors are unlikely to occur.
3. That positive and negative errors of the same numerical magnitude are equally likely to occur; in other words, the curve is symmetrical.

Because the fitted curve represents the relationship between the magnitude of an error and the probability of its occurrence in a given series of measurements, it is frequently termed the *normal probability curve* or simply the *normal curve*.

The equation of the normal curve is

$$Y_c = \frac{ni}{\sigma\sqrt{2\pi}} e^{-x^2/2\sigma^2} \tag{1.18}$$

where Y_c = the computed height of an ordinate at the distance x from the arithmetic mean. The other variables have been shown previously.

Figure 1.4 shows a normal curve fitted to the data of Table 1.3. The frequency distribution curve of Figure 1.1 is superimposed on Figure 1.4 to show their relationships.

It is obvious that the data of Tables 1.1 and 1.3 are not distributed according to the normal curve. This was brought out further by the nonlinear ogive of Figure 1.3.

Skewness

Figure 1.1, then, represents an asymmetrical or *skewed curve*. The skewness is referred to in the direction of the extreme values, or speaking in terms of the curve, in the direction of the longer tail. In this case, the data are skewed positively or to the right.

Pearson's quantitative measure of symmetry is

$$\text{skewness} = \frac{3(\overline{X} - \text{median})}{s} = s_k \tag{1.19}$$

For the curve in Figure 1.1,

$$s_k = \frac{3(2.40 - 2.26)}{0.98} = +0.43$$

Another measure of skewness is the so-called third moment about the mean, π_3 (s^2 is the second moment, π_2).

$$\pi_3 = \frac{\Sigma x^3}{n - 1} \tag{1.20}$$

Both s_k and π_3 are zero for normal curves.

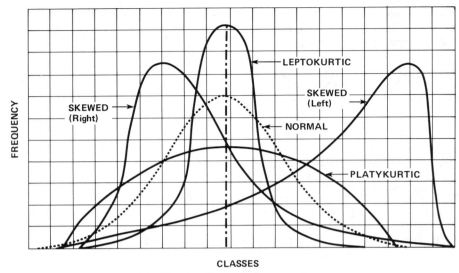

CLASSES

FIGURE 1.5.   Types of non-normal curves.

Kurtosis

Kurtosis is a measure of the peakedness or flatness of a frequency distribution curve. Curves of this type compared to the normal curve are shown in Figure 1.5. The more peaked curves are called *leptokurtic* and the flat-topped curves are called *platykurtic*.

An absolute measure of kurtosis is given by the fourth moment about the mean, π_4.

$$\pi_4 = \frac{\Sigma x^4}{n-1} \tag{1.21}$$

This may be put on a relative basis by dividing by s^4 (or π_2^2) thus:

$$a_4 = \frac{\pi_4}{s^4} = \frac{\pi_4}{\pi_2^2} \tag{1.22}$$

$a_4 = 3$ for a normal curve. For a flat-topped curve, $a_4 < 3.0$. For a peaked curve, $a_4 > 3.0$.

Open End

This refers to a special case where the limits of the first or last class in a frequency distribution are indefinite or undefined, such as "greater than 65..." Such distributions are treated in other chapters in this book.

Multimodal

A multimodal distribution is one with two or more peaks (or modes).

Other Types of Nonnormality

These types are *hyperbolic*, *parabolic*, *exponential*, and *logarithmic*, and pertain to data that are distributed according to mathematical relationships (known or unknown) which differ from the normal equation (4). Examples of these are growth curves.

Although most actual distributions are not normal in nature, we will for the greater part consider normal distributions in this chapter for the purpose of simplification. Quite often, unless a distribution is tremendously skewed, assumptions of normality give acceptable results.

1.6. IMPORTANT CONSIDERATIONS

The concepts we have discussed are basic to the understanding and appreciation of the statistical approach. However, these must be examined together with several other fundamental considerations before we may proceed to statistical designs applicable to the coatings industry.

Population—Universe

First among these is a consideration of the population or universe of values from which one or several sets of samples are drawn for examination. As previously stated, since we cannot actually measure the entire population of values, say the acid value of every gram of fatty acid in a tank car shipment, we must substitute readings on a series of samples for the prediction of the acid value of the shipment.

The acid value predicted will approach the *true value* for the shipment to an extent governed by the type and scope of the sampling procedures and by the actual uniformity of the shipment.

Probability

The concept of probability is continually used in statistics and its meaning and importance should be firmly established. It may be defined as follows: If an event may happen in A ways, and fail to happen in B ways, and all of these ways are mutually exclusive (that is, uncorrelated or independent of each other) and equally likely to occur, the probability of the event happening is

$$P = \frac{A}{A + B}$$

or A chances of success in $A + B$ total trials.

A simple example for clarification is as follows: The probability of choosing an off-specification drum of fatty acid in 1-drum samples from a shipment of 60 drums containing 3 off-standard drums located at random in the group is 3 in 60 or 1 chance in 20.

Probability may also be considered as relative frequency of occurrence in the long run. Thus,

$$p = \lim \left(\frac{f}{f_{tot}} \right) f_{tot} \rightarrow \infty$$

or

$$p = \lim_{n \rightarrow \infty} \left(\frac{f}{N} \right) \qquad (1.23)$$

This is quite useful when considering areas under a frequency distribution curve, as shown in Figure 1.4. Here, it is shown that 99.73% of all the values are included in the range $\overline{X} \pm 3s$. Therefore, in a set of values that is distributed normally, the probability of any individual value being in the range of $\overline{X} \pm 3s$ is 99.73 chances in 100. Similarly for the $\overline{X} \pm 2s$ and the $\overline{X} \pm 1s$ ranges, $p = 95.45$ and 68.27%, respectively.

Most statistical tables are similarly based in that they show the relationship of the probability of occurrence of a particular variable or concept (as measured by areas under curves), with distances along the abscissas of these curves (usually in units of s).

Significance Levels

In our further study of significance of differences we shall consider the significance of the difference between a sample value and an assumed population value, and the significance of the difference between two sample values. The procedure for testing the significance of the difference may be summarized in three steps (2):

1. Set up the hypothesis (i.e., a *null hypothesis*) that the true difference is zero (i.e., the samples have been drawn from the same population).
2. On the basis of this hypothesis, determine the probability that such a difference as the one observed might occur because of sampling variations.

3. Draw a conclusion concerning the reasonableness of the hypothesis.

If such an observed difference could hardly have occurred by chance, we have cast much doubt upon the hypothesis of step 1. We therefore abandon the hypothesis and conclude that the observed difference is significant. However, if such an observed difference could very often occur because of chance, we have cast very little doubt upon the hypothesis. We therefore continue to regard the hypothesis as tenable and conclude that the difference is probably not significant.

The question will arise as to what level of probability we shall choose to decide for or against our null hypothesis. Some authorities prefer the criterion of 5 chances out of 100, the 5% or 0.05 level of significance. Others insist on a 1% level. Perhaps most satisfactory of all is to ascertain that an observed sample mean might occur because of chance, and then to decide whether or not the probability is small enough for the particular problem at hand. Critical probability levels should be lowest for tests of events or properties whose failure could be tragic or extremely costly.

Degrees of Freedom (df)

We are almost always dealing with samples, both small and large, and not with infinite parent populations. Our estimates of population means and dispersions will always be subject to error, dependent in part on sample size.

Also, the mere calculation of a sample average $\overline{X}$ and its further use in statistical analysis restricts the data. When the arithmetic mean is computed, one *degree of freedom* is lost, since the value of any one of the items is defined by knowledge of the value of the mean and of the remaining items. In other words, if for a series of items the sole requirement set up is that $\Sigma X = 0$ (i.e., the mean has been determined), the values of all of the other items save one may be arbitrarily set down; all but one are "free to vary". The value of the other item is determined by the above requirement. In more general language, the number of degrees of freedom is the number of deviations (items of n) minus the number of constants determined from the sample and used to fix the points from which those deviations are measured. The use of $n - 1$ in the above expression is particularly important when n is small. When n is large, it matters little whether we divide by n or $n - 1$ (5).

The rule that the "number of degrees of freedom allotted to error is one less than the number of measurements" does not hold generally. In fact, it applies only when the fixed "true" value is the only systematic effect in the data. Usually, in experimental work several true values are built into the experiment. Each will require a degree of freedom and only the remaining degrees of freedom can be used to display the effects of error alone (6).

Therefore, when the expression degrees of freedom is used, it must be recognized that the full meaning is "the number of degrees of freedom of a set of measurements available for estimating error."

It will be seen that consideration of degrees of freedom is quite important in such tests as analysis of variance and linear regression, since useful results can only be computed when the proper restrictions are placed on the data.

1.7. RELIABILITY, SIGNIFICANCE—*t* TESTS

The comparison of means of sample sets is an everyday occurrence. If statistics did no more than show how to systematically compare these means, it would make an important contribution. It goes far beyond, however, and provides the techniques that demonstrate how means and variations of all types of data can be compared quickly and effectively with clearly stated *confidence limits*.

Standard Deviation of the Universe

Assume the existence of an infinite number of paints for which gloss readings have been obtained from time minus infinity to the present. This distribution of data will have an average X' and a standard deviation σ. Since σ can never actually be measured, we attempt to estimate it by sampling from the parent universe. For a sample of n, our estimate of σ is given by

$$\hat{\sigma} = \sqrt{\frac{\Sigma(X_i - \bar{X})^2}{n - 1}}$$

$$= \sqrt{\frac{\Sigma(X - \bar{X})^2}{df}} = s\sqrt{\frac{n}{n - 1}} \tag{1.24}$$

As $n \to \infty$, as the sample size gets greater, $\hat{\sigma}$ (and s) $\to \sigma$—the true dispersion of the universe.

Standard Deviation of the Mean

As we proceed to take sets of n samples each from the parent universe, we notice that the standard deviations of the means so obtained are considerably less than the individual standard deviations. This relationship is

$$s_{\bar{X}} = \frac{s}{\sqrt{n}} \quad \text{or} \quad \frac{\hat{\sigma}}{\sqrt{n}} \tag{1.25}$$

If the distribution is not exactly normal, the distribution of means of random samples tends to normality as the size of each sample is increased.

Standard Deviation of the Standard Deviation

For small samples, deviation of sets of samples from a universe is

$$s_s = \frac{s}{\sqrt{2n}} \tag{1.26}$$

This relationship is useful for checking reproducibility of precision from one test to another.

Difference of Means—Large Samples

Tests were made of a sample of 30 bags of white lead chosen at random from a series of shipments totalling 1000 bags. The quality characteristic examined was tinting strength, which for the previous shipments had averaged 176. The 30-bag sample had an average tinting strength of 180, with a standard deviation of 8.0 units. Was this sample representative of former shipments or was there a significant difference between them?

Here H_0, the null hypothesis, is that there is no significant difference between $\overline{X}' = 176$ for the population and $\overline{X} = 180$ and $s = 8.0$ for the sample. Therefore, we hypothesize that the true difference between $\overline{X}'$ and $\overline{X}$ is zero (they are equal) and we determine what the probability is that their difference might be as much as 4.0 units of tinting strength due to change alone.

An estimate of the standard deviation of the universe is first obtained:

$$\hat{\sigma} = s\sqrt{\frac{n}{n-1}} = 8.0\sqrt{\frac{30}{29}} = 8.13$$

Then we estimate the standard deviation of the mean of 30 samples from this universe, thus

$$s_{\overline{X}} = \frac{\hat{\sigma}}{\sqrt{n}} = \frac{8.13}{\sqrt{30}} = 1.484$$

We then proceed to express the observed difference between means in terms of the standard deviation of the mean by making the following ratio:

$$\frac{\overline{X}_1 - \overline{X}_2}{s_{\overline{X}}} = \frac{x}{s_{\overline{X}}} = \frac{4.0}{1.484} = 2.70$$

This assumes that the sampling distribution of

$$\frac{\overline{X'} - \overline{X}}{s_{\overline{X}}}$$

is normal, which is essentially true when n is large, but not when n is small. Appendix Table A.1 is the tabulation of areas under the normal curve as a function of

$$\frac{X - \overline{X}}{s} \quad \text{or} \quad \frac{x}{s}$$

For $x/s = 2.70$, the crosshatched area in Table A.1 is equal to 0.4965. (The total area under the curve $= 1.0$, which is a summation of all probabilities or frequencies involved. See Ref. 7.)

This means that 49.65 out of 100 samples would fall between $\overline{X'} - \overline{X} = 0$ and $\overline{X'} - \overline{X} = +4.0$ and consequently 0.5000–0.4965 (since 0.5000 is one-half the area under the normal curve) or 0.35 sample in 100 would occur beyond $+4.0$.

If the means were identical for the $s_{\overline{X}}$ calculated, we would get a difference as large as $+4.0$ units only 0.35% of the time. Since this is considerably below both of the usual acceptance levels (5 and 1%), we have shown the means to be statistically dissimilar and have thus rejected the null hypothesis.

Difference of Means—Small Samples

The solution of this problem for small samples is a great landmark in the development of statistical methods. It was accomplished in 1908 by the English chemist, W. S. Gosset, who wrote under the name of Student (6).

Student devised a statistic, t, and a resultant table called the t table (Table A.2). This gives, for various degrees of freedom, the probability of exceeding the listed limiting values of t if sample sets tested come from the same population and show differences which are due to chance causes (7).

$$t = \frac{\overline{X_1} - \overline{X'}}{s_{\overline{X}}} \tag{1.27}$$

Several examples will illustrate the use of the t tables:

Example 1.1. An established method for running acid values has given a "population" mean of 0.230. A new method has been developed and checked on 16 samples, with $\overline{X} = 0.250$ and $s = 0.080$. Are the two methods significantly different?

Again, the null hypothesis is that there is no significant difference in the methods.

$$\overline{X}' = 0.230 \qquad\qquad n = 16$$

$$\hat{\sigma} = 0.080\sqrt{\frac{16}{15}} \qquad = 0.0826$$

$$s_{\overline{X}} = \frac{0.0826}{\sqrt{16}} \qquad = 0.0206$$

$$t = \frac{\overline{X} - \overline{X}'}{s_{\overline{X}}} \qquad = \frac{0.250 - 0.230}{0.0206}$$

$$t = \frac{0.020}{0.0206} \qquad = 0.97$$

Referring to Table A.2, for df $= 16 - 1 = 15$, the probability of $t = 0.866$ being exceeded due to chance alone is $p = 0.2$. For $t = 1.341$, $p = 0.1$. Therefore, by interpolation, the probability that the difference in means is due to chance causes alone, and not any bias or assignable causes, is about 0.17, 17%, or 17 chances in 100. This figure is well above the 5% limit and so we accept our null hypothesis.

Example 1.2. Test method A for rosin acid determination, which uses an indicator for the endpoint, gives the following results on eight standard samples as compared to test method B, which uses potentiometric titration on six of the same standard samples:

$$\overline{X}_A = 4.83 \qquad \overline{X}_B = 4.06$$

$$s_A = 0.95 \qquad s_B = 0.86$$

$$n_A = 8 \qquad n_B = 6$$

Do these two test methods give statistically equivalent results? The null hypothesis says that $E(\overline{X}_A - \overline{X}_B) = 0$, ($E =$ expected value), that is, there is no significant difference in the methods.

Here, we must determine the standard deviation of the difference of the means; therefore equation (1.27) for this case is rewritten as

$$t = \frac{\overline{X}_A - \overline{X}_B - 0}{s_{\overline{X}_A - \overline{X}_B}} \qquad\qquad (1.28)$$

$$\hat{\sigma}_A = 0.95\sqrt{\frac{8}{7}} = 1.015 \qquad s_{\overline{X}_A} = \frac{1.015}{\sqrt{8}} = 0.359$$

$$\hat{\sigma}_B = 0.86\sqrt{\frac{6}{5}} = 0.941 \qquad s_{\overline{X}_B} = \frac{0.941}{\sqrt{6}} = 0.385$$

The standard error of the difference of two means is given by

$$s_{\bar{X}_A - \bar{X}_B} = \sqrt{(s_{\bar{X}_A})^2 + (s_{\bar{X}_B})^2} \tag{1.29}$$

$$= \sqrt{(0.359)^2 + (0.385)^2} = \sqrt{0.129 + 0.148} = \sqrt{0.277} = 0.527$$

Then

$$t = \frac{4.83 - 4.06}{0.527} = \frac{0.77}{0.527} = 1.46$$

In this case, degrees of freedom $= (n_A - 1) + (n_B - 1)$ or df $= (8 - 1) + (6 - 1) = 12$.

For $t = 1.46$ and df $= 12$, Table A.2, by interpolation, shows a probability equal to 0.09 or 9 times in 100 that this difference in means could have been caused by chance variations. The null hypothesis is again upheld.

In Example 1.2, for utmost precision in a case where n_A is not equal to n_B and both n's are small, the following relationship (Eq. 1.30) should more properly be used to weight the two standard deviations by their respective degrees of freedom to determine $s_{\bar{X}_A - \bar{X}_B}$.

$$s_{\bar{X}_A - \bar{X}_B} = \sqrt{\frac{(n_A + n_B)\left[\Sigma(X_A - \bar{X}_A)^2 + \Sigma(X_B - \bar{X}_B)^2\right]}{n_A n_B\left[(n_A - 1) + (n_B - 1)\right]}} \tag{1.30}$$

However, for the sake of simplicity, equation (1.29) will be used, although we recognize that we are not on quite as solid ground for extremely small values of n.

The t test is extremely valuable and is used again and again in statistical analysis. As a matter of interest, it is readily apparent that the t function for df $= \infty$ approaches the same value as the normal curve.

1.8. CONFIDENCE LIMITS

In Example 1.2, consider a series of samples from the same population. We may say that we can state the means of each of these samples with a known precision. However, how close do the means that we calculate actually approach the true mean of the universe or even agree with each other?

We can estimate this agreement by examination of the form of the t test.

$$t = \frac{\bar{X}' - \bar{X}}{\sigma_{\bar{x}}} = \frac{d}{0.0206}$$

For $\overline{X} = 0.250$ and df $= 15$, let $\overline{X}' - \overline{X} = d$, which is a difference that will yield a value of $t = 2.131$ corresponding to $p = 0.025$. (Since Table A.2 shows single tail areas of t, p should be doubled or $p = 0.05$ for the entire distribution area.) Thus

$$2.131 = \frac{d}{0.0206} \quad \text{or} \quad d = 0.0439$$

This relationship can now be restated as follows: The probability of a mean from a large series of such tests falling in the range of $\overline{X} \pm d$ or 0.250 ± 0.044 (or 0.206–0.294) is 95 in 100 or 95% ($1.00 - 0.05$). Thus, a prediction has been made regarding the *confidence limits* within which the mean would lie if based on a very large number of experiments.

The *95% confidence limits* as calculated are commonly used as an expression of the degree of confidence which the statistician or experimenter has in the conclusions. Other confidence limits for other percentages are easily calculated in a similar manner.

It is interesting to note that Figure 1.4 shows confidence limits for normal curve areas corresponding to $\overline{X} \pm 1$, 2, and 3σ.

1.9. RELIABILITY OF MEASURES OF DISPERSION

Sample Standard Deviation

The reliability of a sample s (or difference between two sample sigmas) may be tested in a manner analogous to that used for means, but using equation (1.26) for s_s. For large samples ($n > 30$), a t test based on s_s will be used and areas under the normal curve or from the Student t distribution may be used, depending on the number of degrees of freedom.

However, for small values of n (under 30), the use of s_s is not valid since the distribution of s_s is highly skewed for $n < 30$.

Chi-Square

For small samples, we may use a statistic called *chi-square* or χ^2. Tables of chi-square values provide an answer to the question: What is the probability of obtaining a fit, due only to chance, which is as poor as or worse than that observed? The distribution of sample values of s or actually the variance s^2 may be put in the form

$$\chi^2 = \frac{ns^2}{(\sigma)^2} \tag{1.31}$$

where the distribution of the population is assumed to be normal. The distribution of the chi-square function has been determined in the usual form of a probability–degrees of freedom relationship and is shown as Table A.3.

Using this relationship [equation (1.31)] and Table A.3, we may determine confidence limits for samples of sigmas from a population with a known σ or, conversely, we may determine limits within which σ may confidently be expected to fall if we know s (as is more usually the case).

The chi-square test is probably the most accepted test for determining the goodness of fit between the observed and the expected (theoretical) distribution of data in sampling from a population. However, there are better tests, and once familiar with the chi-square test, the reader who decides to dig deeper should examine the others as well.

z—Variance Ratio

The square of the standard deviation s^2, the variance (or mean square), is a most valuable concept and will be used a great deal in all that follows.

For testing the significance of the difference between two standard deviations (or variances), when n_1 and n_2 are small and not equal to each other, R. A. Fisher, the father of modern statistical methods, suggested a transformation, thus,

$$z = \ln\frac{s_1}{s_2} = \frac{1}{2}\ln\frac{s_1^2}{s_2^2} = 1.15729\log\frac{s_1^2}{s_2^2} \qquad (1.32)$$

where $s_1 > s_2$ (to make z positive).

Tables of z exist; the significance of the difference of two sigmas would be tested by first setting up the null hypothesis that they are from the same population. Then we would determine if the probability of the z function so calculated is comparable to what would exist if the two sigmas differed due to chance variations alone.

However, the z test has been largely supplanted by a more modern test, called the F test. (Note: Both z and F tests actually require the use of the population for complete accuracy. However, σ is rarely available and sample sigmas may be used, providing n is not too small.)

F—Variance Ratio

The F variance ratio is

$$F = \frac{s_1^2}{s_2^2} \qquad (1.33)$$

It is used in a manner similar to the z ratio and its functions are shown in Table A.4. The F ratio is illustrated by the following example: In the white lead shipment previously referred to, tinting strength determinations from the 30-bag sample had a value of $s = 8.0$. An additional 20-bag sample is taken from another car of the same shipment. This sample has $s = 5.5$. Do these samples have the same variability?

The variances are set up so that the subscript 1 is associated with the larger value. This has the effect of making F a positive number.

$$s_1 = 8.0 \qquad\qquad s_2 = 5.5$$
$$s_1^2 = 64.00 \qquad\qquad s_2^2 = 30.25$$
$$n_1 = 30 \qquad\qquad n_2 = 20$$
$$df_1 = 29 \qquad\qquad df_2 = 19$$

$$F = \frac{s_1^2}{s_2^2} = \frac{64.00}{30.25} = 2.12$$

Referring to Table A.4,

$$\text{For } df_1 = 29 \text{ and } df_2 = 19$$
$$\text{For } p = 0.05 \qquad F = 2.08$$
$$p = 0.01 \qquad F = 2.86$$

The value of F calculated is very close to the 5% limit, indicating that even if the variances were equivalent, 1 time in 20 they could differ as shown by chance alone. If this amount of risk is too great—if perhaps it caused too high a percentage to violate a tinting strength specification—then sampling can be continued to determine, by further F testing, a more accurate estimate of the reliability of the dispersion of the test results.

This completes our discussion of several of the basic statistical tools which we can use to help us increase our experimental efficiency, precision, and accuracy. Subsequent chapters will introduce us to further applications of these various techniques and to a number of new techniques.

Concluding this section, one word of warning must be given (8). In all good experimental work, it is essential to ask the right questions before the right answer can be obtained. In statistical methods, it is essential to know the right questions before the experiment is designed. For it is on the basis of those questions that the work will be planned, and on this depends not only the answers obtained but the types of answers. A poor experiment will not become a good one merely because statistical methods are used; and additionally these methods still require the same, and perhaps more, rigorous attention to detail on the part of the experimenter. They should not be used automatically either, for there is little point in making laborious computations where the results are obvious. Statistical methods are not a panacea for all ills, but a powerful instrument in the hands of those who wish to use them in the cause of progress.

From this discussion of basic concepts, the organization of the remainder of the book's topics follow the following sequence: Chapter 2 introduces the principles of probability followed by a discussion of probability or frequency distributions, or the categories into which data fall, in Chapter 3. To a degree, Chapter 3 is an expansion of Chapter 1. Chapter 4 specifically addresses the critical importance of the size of the data sample and its treatment. The viewpoint here is not from the aspect of what size sample need be taken for specific tests, but from the more important viewpoint of what influence the size of the sample selected has on test results, population parameters and sample statistics, the frequency distribution, and similar aspects.

Chapter 5 is linked to Chapter 2 and considers in depth an extremely useful topic for the chemist and engineer, but one infrequently discussed in some texts on statistics. This is the subject of probability plotting and the applications of graphical statistics. Chapter 6 is an expansion of topics touched on in Chapters 1 and 2 and discusses the subject of hypothesis testing in practical terms. This chapter is followed by the related subjects of statistical intervals (Chapter 7), acceptance sampling (Chapter 8), and comparisons of populations (Chapter 9).

Chapter 10 introduces the principles of regression analysis involving regression lines, types of variables, the linear regression model and its analysis, confidence limits, and correlation. Chapter 11 expands on correlation, discussed in Chapter 10, as applied to nonparametric testing, an extremely useful uncomplicated statistical technique for preliminary investigations, shortcut methods, small sample sizes, and subjective data. The four chapters following are concerned specifically with experimental design, with Chapter 12 introducing the subject, Chapter 13 treating experimental design and analysis in depth, and the following two chapters (14 and 15) expanding on optimization as introduced in Chapter 13.

Chapter 16 concerns a topic of ever-increasing importance to the industrial chemist and engineer, that of statistical quality control. The four final chapters are general. They deal with topics with which the industrial chemist or engineer must frequently be concerned or which are commonly used in treating data (such as Chapter 17, Sequential Methods, and Chapter 18, Error), developing equations representative of physical laws (Chapter 19), and developing graphical methods for the solution and visualization of both rational and empirical equations by employing the principles of nomography (Chapter 20). Nomographs find widespread use in both the world of the chemist and that of the engineer, but aside from their presentation (as charts), the principles underlying their classification and construction are to be found only in specialized texts. Hopefully this introduction in Chapter 20 will elucidate this important although somewhat tangential facet of the chemical and engineering field.

REFERENCES

1. Youden, W. J., Statistical design, *Ind. Eng. Chem.*, April 1954.
2. Croxton, F. E., and Crowden, D. J., *Applied General Statistics*, Prentice-Hall, New York, 1939.
3. King, J. R., *Probability Charts for Decision Making*, Team, Inc., P.O. Box 25, Tamworth, NH.
4. Schrumpf, W. J., Carter, R. M., and Hader, R. J., "A Rapid Method of Evaluating Check Resistance of Furniture Lacquer Films," ACS Meeting, Dallas, TX, April 1956.
5. Freeman, H. A., *Industrial Statistics*, Wiley, New York, 1942.
6. Youden, W. J., *Statistical Methods for Chemists*, Wiley, New York, 1951.
7. Fisher, R. A., and Yates, F., *Statistical Tables*, Hafner Press, New York, 1948.
8. Touchin, H. R., Some applications of statistical methods to exposure trials, part 1, *J.O.C.C.A.*, December 1953.

2

PROBABILITY

ANTHONY A. SALVIA

Behrend College
Pennsylvania State University
Erie, Pennsylvania

2.1. INTRODUCTION

In this chapter, we take up the study of probability. The word *probability* itself has a number of common interpretations in ordinary usage; we shall soon present a rather precise definition of it, and references to probability throughout this book shall be understood to apply to the definition we shall give.

It is best to sound a caution at the outset. The material we are about to cover is, for the uninitiated, perhaps somewhat difficult to assimilate. The difficulty seems to stem from two principle sources: (1) unfamiliarity with the notation and mathematical language employed, and (2) the usefulness of the material is not immediately apparent.

With respect to the former difficulty, we urge the reader to pay close attention to the details presented. The subject matter flows in a logical and orderly procession, but on first reading, so many new concepts are presented that difficulty in perceiving and understanding that flow may be encountered.

It may be helpful if we outline in advance the various points to be made in the chapter. The first section is devoted to developing a suitable *definition* of probability; it includes the notions, fundamental to the theory, of *sample space* and *event*. The second section discusses some of the laws of probability which are of frequent use; the chapter closes with a brief discussion of the law of large numbers, which is the basis of most of the layman's understanding (and *mis* understanding) of probability.

The second difficulty mentioned above, about the usefulness of the material, is not so easy to address. At this point, we simply call to the reader's attention the fact that the subject of probability stands, with respect to statistical methods, in a position analogous to that of geometry with respect to surveying or drafting. Without a fundamental grasp of probability theory, progress in the study of statistics is extremely slow and tedious. Mastery of the material in this chapter will aid in far better understanding of the subject matters to follow.

At the risk of delaying too long before plunging into the basics, it will be worthwhile to explore the above analogy just a bit further. The subject matter of (Euclidean) geometry consists of concepts known as *points, lines,* and so on, which, by assumption, have certain properties (*axioms*). Through the application of ordinary logic, other properties are predictable with certainty (*theorems*). The usefulness of geometry derives from the fact that, when objects in the real world which resemble geometric points and lines are observed, they behave in approximately the same way that the geometry would predict. In short, the mathematics gives one an abstract *model* which fits one's perception of the "real" world.

Likewise, probability theory provides models of situations of uncertainty which one may examine in the light of real-world experience; we then choose, in a sense that will be clear later, one or more of these models which "best" fit the real world.

2.2. INITIAL CONCEPTS

The first attempts* to develop a systematic theory of probability evolved from studies of various games of chance. In retrospect, such games are rather simple to analyze (at least conceptually), primarily because of a sort of *symmetry* that usually exists. Let us suppose, for example, that we take an ordinary deck of playing cards, shuffle the deck several times, and then *randomly* choose a card. If we agree to call this sequence of activities a *random experiment*, it is clear that the result or outcome of that experiment will be one, and only one, of the 52 potential outcomes. Furthermore, in the absence of any other information about the experiment, it seems reasonable to assume that each of these 52 outcomes has the same chance of occurring as its fellows. If we use the word *probability* instead of *chance*, we have just agreed that all outcomes have the same probability. This agreement is occasionally dignified by being called the *principle of indifference*.

---

*Many standard introductory texts in probability theory describe the historical development of the subject, for example, Feller, *An Introduction to Probability Theory and Its Applications*, Vol. 1, Wiley, New York, 1968, or Cramer, *The Elements of Probability Theory*, Wiley, New York, 1955.

Now, label the outcomes as $E_1, E_2, \ldots, E_{52}$, in any manner whatsoever. For example, we may set up the correspondence

E_1 through E_{13} $\leftrightarrow$ 2, 3, ..., K, A of Spades
E_{14} through E_{26} $\leftrightarrow$ 2, 3, ..., K, A of Hearts
E_{27} through E_{39} $\leftrightarrow$ 2, 3, ..., K, A of Diamonds
E_{40} through E_{52} $\leftrightarrow$ 2, 3, ..., K, A of Clubs

This labeling permits a second description of the experimental result, which is, of course, equivalent to the first. Thus we could equally well say, "The experiment resulted in the draw of the 5 of Hearts," or "The experiment resulted in the outcome of event E_{17}." The statement that all outcomes have the same probability may be expressed as

$$P\{E_1\} = P\{E_2\} = \cdots = P\{E_{52}\}, \qquad (2.1)$$

where the notation $P\{E_j\}$ is to be read as "The probability of the event E_j." Note carefully that we have not as yet *defined* probability; we have really done nothing more than adopt a sort of shorthand way of describing the experiment under consideration. The equality expressed in equation (2.1) is an *assumption* that our common sense agrees with, owing to the symmetry of this particular experiment. It shall be most convenient in the sequel to let

$$P\{E_j\} = \tfrac{1}{52} \quad \text{for} \quad j = 1, 2, \ldots, 52$$

that is, we assign a numerical value of $\frac{1}{52}$ to each of the equal probabilities in equation (2.1). Generally, in an experiment of this sort with N outcomes, $E_1, E_2, \ldots, E_N$, we would let $P\{E_j\} = 1/N$.

The outcomes in the card-drawing experiment are normally called *simple events*. "Simple" here is used in the sense of indivisible: each performance of the experiment leads to one and only one such event. We may, of course, combine simple events in various ways. One of these is through the logical operation of *union*;* the (compound) event $E_i \cup E_j$ occurs if E_i or E_j or both occur. (Here, of course, E_i and E_j could not occur simultaneously unless $i = j$.) We find then, for example, that the probability of a spade is

$$P\{E_1 \cup \cdots \cup E_{13}\}$$

the probability of an Ace is $P\{E_{13} \cup E_{26} \cup E_{39} \cup E_{52}\}$, and so on. It

---

*A very brief summary of set algebra is provided in the Appendix to this chapter.

seems reasonable to define for all simple events E_i, E_j, $j \neq i$

$$P\{E_i \cup E_j\} = P\{E_i\} + P\{E_j\} \tag{2.2}$$

and to extend this recursively by means of

$$P\{E_i \cup E_j \cup E_k\} = P\{E_i\} + P\{E_j\} + P\{E_k\}$$

continuing in the obvious way. In this manner, we are led to the conclusions that the probability of a Spade is $\frac{13}{52}$, or $\frac{1}{4}$, and the probability of an Ace is $\frac{4}{52}$, or $\frac{1}{13}$.

In some earlier literature in the field of probability the simple events were called *cases*; the probability of some event occurring was then calculated as the ratio of the number of cases favorable to the event to the total number of cases. Such a formulation leads to the same numerical results as we obtained above for the card-drawing experiment. There are, however, difficulties with this formulation which we illustrate in the examples below.

Example 2.1 Counting Cases.   Two coins are tossed. What is the probability that one falls Heads, the other Tails? One may count three cases—both Heads, both Tails, or one of each. Thus, we might say the probability of each event is $\frac{1}{3}$. But one may equally well count four cases:

First Coin	Second Coin
H	H
H	T
T	H
T	T

and conclude that the probability of "one head, one tail" is $\frac{1}{2}$. Finally, one might say only two cases exist—the coins match or do not—and so the probability is $\frac{1}{2}$. We have obtained three results (two of which happen to coincide) by using three different plausible enumerations of cases.

Example 2.2 Tossing Tacks.   Consider an experiment in which a thumb tack is dropped onto a smooth horizontal surface. Suppose we pose the question "What is the probability that the tack comes to rest point up?" There are new difficulties to face in this experiment. To begin with, the number of potential cases may be considered infinite, since the tack's final position can vary continuously in three-dimensional space. Even if we simplify by a reduction to two cases, "point up" and "point not up," we may hesitate to assign a probability of $\frac{1}{2}$ to the event.

The reason for this hesitation stems from our intuitive grasp of the numerical value of the probability of an event as a measure of how often the event occurs. In card drawing, the value $\frac{1}{4}$ assigned the the event "Spade" seems reasonable; we somehow *expect* about one experiment in four to produce a Spade.* This intuitive base is lacking in the tack-tossing experiment. If probability is to be assigned according to our reasonable expectations, we would want to know much more about the geometry of the tack, the surface onto which it was dropped, and other pertinent experimental conditions, before assigning a value to the probability.

Considerations such as those illustrated in the examples above lead us to abandon the rather näive definition of probability as a ratio of favorable cases to total cases, and to seek an improved definition. The second example in particular might suggest that we proceed empirically toward a definition.

Let us suppose, then, that we repeat a specific experiment N times, and observe that the event of interest occurs n of these times. We may define the empirical probability of the event as

$$P_N = \frac{n}{N} \tag{2.3}$$

Note that $0 \leq P_N \leq 1$, and equation (2.3) would satisfy equation (2.2).
We may then define the probability of the event as

$$P = \lim_{N \to \infty} P_N \tag{2.4}$$

In the historical development of probability theory, equation (2.4) was used for a long period of time as the definition of probability. There are, however, two very serious objections to this definition:

1. Are we certain the limit exists?
2. If so, how is the limit determined?

The example below illustrates these objections.

Example 2.3 Prime Numbers. Consider an experiment in which we choose "at random" a positive integer and determine whether it is prime. What is the probability that integer is prime? The table below lists, for

---

*The basis of this expectation is an intuitive notion about something called the *law of large numbers*. Shortly we shall discuss this law.

various values of N, the number of primes n in the range $1 \le n \le N$

n	N	n/N
4	10	0.400
8	20	0.400
10	30	0.333
12	40	0.300
15	50	0.300
25	100	0.250
168	1,000	0.168
1229	10,000	0.1229

Clearly, if the ratio n/N is approaching some fixed constant P, N shall need to be considerably larger than 10,000 for us to have any idea about how large P might be. Furthermore, it is not at all clear what the phrase "at random" means here. For example, we have examined in the table above the "first" N integers; we can not determine the "last" N. There is no assurance that, say, the number of primes between 1 and 10,000 is anywhere near the number between 10,000,001 and 10,010,000.

The essential difficulty is that there is no reasonable way to sample randomly from the set of *all* positive integers; in practice we would always sample from the first N (for some fixed value of N).

Having come this far, it is clear that, in order to be able to develop a theory of some usefulness, we need first a method for defining probability which (a) is precise, and (b) is not subject to the difficulties illustrated above. We now proceed to that task.

Let S be a set which has as its elements descriptions of all* the possible outcomes of a given experiment. For illustration, we might have

Experiment	Elements of S
Tossing a coin	H, T
Drawing a card	A of Spades, K of Spades, ...
Measuring the percent by weight of material A in a product	A number X in the interval $0 \le X \le 100$

We shall call S the *sample space* of the experiment. Suppose A, B, C, and

---

*Actually, some outcomes may, for simplicity's sake, be omitted in the model one chooses. In the first experiment listed, it *is* possible that the coin might fall on its edge; nothing of importance is lost by omitting that contingency from the set S, however.

so on, are *subsets* of S. Probability is now defined as a function of these subsets, with $P(\ )$, the probability function associating a number with a subset, having the properties:

i. $P(A) \geq 0$ for any subset A.

ii. $P(A \cup B \cup C...) = P(A) + P(B) + P(C) + \cdots$, provided no two of the subsets have any common elements.

iii. $P(S) = 1$.

The first two properties can be expressed verbally by stating that the probability function is non-negative and additive; property iii normalizes the function to have a maximum value of 1.

In books on probability or mathematical statistics, one proceeds from these properties to prove a large number of theorems about probability. Here, however, we simply pause to point out that the preceding "definitions" of probability become, so to speak, special cases of our general definition, which turns out to be quite useful. Furthermore, the numbers 0 and 1 correspond to our intuitive notions of "impossibility" and "certainty";* varying degrees of likelihood then fall in the interval $[0, 1]$.

Let us consider this definition of probability with respect to the first sample space listed above. That space has four subsets,

$$A = \{H\}$$
$$B = \{T\}$$
$$C = \{H, T\} = S$$
$$D = \{\ \}$$

The last of these, which contains no elements, will usually be called the *empty set*, and we shall use the symbol $\varnothing$ to denote it. Thus the four subsets are A, B, S, and $\varnothing$. If we specify the function P as having the values

$$P(\varnothing) = 0$$
$$P(A) = \tfrac{1}{2}$$
$$P(B) = \tfrac{1}{2}$$
$$P(S) = 1$$

then P satisfies properties i–iii and is, therefore, a "legitimate" probability model. Note carefully that it is not the *only* possible model we could adopt;

---

* This is technically not exact, but for our purposes it will do.

for example,

$$P(\varnothing) = 0$$
$$P(A) = \tfrac{1}{4}$$
$$P(B) = \tfrac{3}{4}$$
$$P(S) = 1$$

is also valid as a model. Now, the first model says, among other things, that the probability of a Head is $\tfrac{1}{2}$, while the second model claims that this probability is $\tfrac{1}{4}$. Our definition of probability gives no indication of which model is "correct"; this type of inquiry is at the heart of *statistics*, not probability theory. So far as probability theory is concerned, *both* models (and indeed, infinitely many others) are valid. The choice of a "best" model is a problem of statistical inference.

In the second sample space listed above (for a card-drawing experiment), we see that S has 52 elements, and thus has 2^{52} distinct subsets. The task of assigning a value to the function P for each of these is overwhelming; fortunately, we do not need to do so. It is true that if we define 52 values $P_1, P_2, \ldots, P_{52}$ which are non-negative for the simple events (elements), and add up to 1, then the probabilities for the remaining $2^{52} - 52$ compound events are uniquely determined. (The reader will recall that P_j was previously defined as $P_j = \tfrac{1}{52}$.)

Finally, let us consider several examples based on the third sample space listed above, namely, the space of numbers X, $0 \le X \le 100$, where X represents the percentage of material A in a certain product. Clearly, $P(0 \le X \le 100) = 1$, since the set $0 \le X \le 100$ coincides with S, the entire sample space. Suppose, now, that for any set B_x (see Figure 2.1) of the form $0 \le X \le x \le 100$, the probability is given by the formula

$$P(B_x) = \frac{x}{100} \tag{2.5}$$

so that, for example,

$$P(0 \le X \le 20) = 0.2$$
$$P(0 \le X \le 90) = 0.9$$

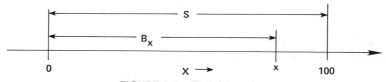

FIGURE 2.1.   Definition of B_x.

and so on. The formula (2.5) in effect implies a certain mathematical model which describes how the probability is spread (*distributed* might be a better word) over the sample space.

A priori, there is no particular reason why formula (2.5) should be valid. One might equally well postulate

$$P(B_x) = \frac{x^2}{10,000} \tag{2.6}$$

in which case

$$P(0 \le X \le 20) = 0.04$$
$$P(0 \le X \le 90) = 0.81$$

and so forth. Although it may not be readily apparent, axioms i–iii hold for both of these models.*

2.3. A PRECISE DEFINITION OF PROBABILITY

Among the rules that are consequences of our definition of probability, the most important for our purposes are

1. For any set A and its complement A',

$$P(A) + P(A') = 1 \tag{2.7}$$

2. For any two sets A, B

$$P(A \cup B) = P(A) + P(B) - P(AB) \tag{2.8}$$

Item 2 is sometimes called the *addition* law for probability; the notation AB refers to the *intersection* of the sets A and B, that is, the points common to both sets.

Example 2.4. A card is drawn from an ordinary deck. What is the probability it is a Jack or a Diamond?
Defining the events

A—Card is a Jack.
B—Card is a Diamond.

we have $P(A) = \frac{1}{13}$, $P(B) = \frac{1}{4}$, $P(AB) = \frac{1}{52}$, and so

$$P(A \cup B) = \frac{1}{13} + \frac{1}{4} - \frac{1}{52} = \frac{4 + 13 - 1}{52} = \frac{4}{13}$$

---

*The axioms do not hold if we consider all possible types of subset B_x. Essentially, what is required is that the subsets be of a certain type, and the collection of all of these form what is called a *Borel field*. In this book, only subsets belonging to Borel fields are used.

Example 2.5. Using the probability model (2.5), suppose the product is acceptable if X lies in the range $25 \leq X \leq 60$. What is the probability that this occurs?

Let us calculate the probability that the product is *not* acceptable. Define a as the event "X exceeds 60" and B as the event "X is less than 25." The probability of an unacceptable product is then

$$P(A \cup B) = P(A) + P(B)$$
$$= P(B'_{60}) + P(B_{25})$$
$$= (1 - P(B_{60})) + B(B_{25})$$
$$= 1 - \tfrac{60}{100} + \tfrac{25}{100} = \tfrac{65}{100}$$

The probability of acceptable product is, accordingly, $\tfrac{35}{100}$.

In Example 2.5 it was easier to calculate the probability of the complement of the desired event, and then use equation (2.7) to determine the desired probability; in other problems a direct calculation may be simpler.

Notice in Example 2.5 that the formula for $P(A \cup B)$ did not contain the term $P(AB)$; this followed from the fact that the sets A and B, as defined therein, had no points in common. Their intersection, then, was the empty set $\varnothing$, for which $P(\varnothing) = 0$.

As one might expect after hearing reference to an addition law, there is also a multiplication law for probability, which we will now proceed to develop. Our development begins with the immediately preceding example. Suppose that the product we are dealing with is manufactured one batch per day. In the above example we determined that the probability of an acceptable product (day's batch) was 0.35. Let us now ask what the probability is for two consecutive days' acceptable product. Let us assume that every day production starts afresh, without any dependence on the quality of the preceding day's production. Consider the possible two-day sequences, as displayed below:

Today

	.35 GOOD	.65 BAD
.35 GOOD		
.65 BAD		

Tomorrow

It seems reasonable that the sequence "Good, Good" will occur about 12%

of the time $(.35 \times .35 = .1225)$ and that this number, therefore, is a reasonable answer to the question as posed.

Proceeding in similar fashion, we may fill in the entire table as

		Today	
		.35 GOOD	.65 BAD
Tomorrow	.35 GOOD	.1225	.2272
	.65 BAD	.2275	.4225

Note that we had here events of the form

$$B = A_1 A_2$$

where A_i referred to a characteristic of the product on the ith day, and we found

$$P(A_1 A_2) = P(A_1) P(A_2). \tag{2.9}$$

You may raise a rather severe objection at this point; surely the results of any particular day's production will vary in some dependent way upon the previous day's; that is, the occurrence, say, of A_2 (good product on day 2) has a probability that should depend upon what happened on day 1. If in fact A_1 (good product, day 1) occurred, we might denote the probability of A_2 by

$$P(A_2 | A_1)$$

$[P(U|V)$ is read "the probability of U, given V"] whereas if A_1 did not occur, we would have

$$P(A_2 | A_1')$$

and we might not ordinarily expect any of the three numbers

$$P(A_2), P(A_2 | A_1), P(A_2 | A_1')$$

to be equal. In the event that A_1 and A_2 are independent of each other, all three are equal, and equation (2.9) is valid. Otherwise, we write

$$P(A_1 A_2) = P(A_1) P(A_2 | A_1) = P(A_2) P(A_1 | A_2) \tag{2.10}$$

and refer to equation (2.10) as the multiplication law for probability.

The multiplication law is frequently used in instances where it is otherwise difficult to calculate the probability of compound events, such as $A_1 A_2$. An example of this use follows.

Example 2.6. A deck of cards is well shuffled and a card is randomly selected. A second card is then selected without replacing the first. What is the probability that both cards are Diamonds?

Define events A_j as "jth card is a Diamond," for $j = 1$, 2. What is required is $P\{A_1 A_2\}$. By an easy calculation,

$$P\{A_1 A_2\} = P\{A_1\} P\{A_2 | A_1\} = \tfrac{13}{52} \cdot \tfrac{12}{51} = \tfrac{3}{17}$$

Notice that $P\{A_2 | A_1\}$ has a value of $\tfrac{12}{51}$, since the sample space for the second draw contains 51 elements (and we implicitly assume each has the same probability, $\tfrac{1}{51}$).

What is the probability that the *second* card is a Diamond? It may be thought at first that there is no way to ascertain the requested probability, since no information is provided about the first card. However, this is not the case, for, with the events as defined immediately above, we have

$$A_2 = A_2 S = A_2 \cap (A_1 \cup A_1')$$
$$= A_2 A_1 \cup A_2 A_1',$$

and so

$$P\{A_2\} = P\{A_2 A_1\} + P\{A_2 A_1'\}$$
$$= P\{A_1\} P\{A_2 | A_1\} + P\{A_1'\} P\{A_2 | A_1'\}$$
$$= \tfrac{13}{52} \cdot \tfrac{12}{51} + \tfrac{39}{52} \cdot \tfrac{13}{51}$$
$$= \tfrac{1}{4}$$

The rather surprising result in the second part of Example 2.6 is that the probability of obtaining a Diamond on the second draw, in the absence of information about the first draw, is the same as the probability of a Diamond on the first draw, that is,

$$P\{A_2\} = P\{A_1\}$$

Likewise we would find

$$P\{A_{52}\} = P\{A_{51}\} = \cdots = P\{A_1\}.$$

Example 2.6, although framed in the setting of a simple card-drawing experiment, has rather far-reaching implications. What it says is this: If we do *not* have data (no knowledge of the first card) we do *not* adjust our original sample space. The probability of some specified event occurring is independent of the experimental procedure. On the other hand, when we *do* have data, the sample space is changed after each performance of the experiment.

We are ready now to define the very important notion of *statistical independence*. The events A and B are *statistically independent* (we will often shorten this to *independent*) if

$$P\{AB\} = P\{A\}P\{B\}$$ (2.11)

From equation (2.11) we may deduce

$$P\{A|B\} = P\{A\} \quad \text{and} \quad P\{B|A\} = P\{B\}$$

that is, the conditional probability of either event, given the other, is the same as its unconditional probability. Knowledge about A does not change our mind about B, and vice-versa. It would seem that such independence must be relatively rare, but, in fact, most of this book will consider situations in which independence does hold.

The multiplication law also finds frequent use in the following form. Equating the last two members of equation (2.10), we may solve for $P\{A_2|A_1\}$:

$$P\{A_2|A_1\} = P\{A_2\} \cdot \frac{P\{A_1|A_2\}}{P\{A_1\}}$$ (2.12)

This relationship is usually called Bayes' theorem. Its use is illustrated in Example 2.7.

Example 2.7. A plant has three production lines producing white paint, all operating to the same specifications. The third line is used only during periods of heavy demand, while the first two are used daily. Each of the first two account for 45% of production.

Occasionally, the blending apparatus malfunctions, producing unacceptable product. Maintenance records show that the down time for such malfunctions is 2% for line 1, 2.5% for line 2, and 4% for line 3.

A recently produced gallon of paint is opened, and it is found to be improperly blended. What is the probability that it was produced in line 3? Define the events

$$A_i = \{\text{paint produced by line } i\}, \quad i = 1, 2, 3$$

$$B = \{\text{improper blending}\}$$

The given information is as follows:

$$P\{A_1\} = 0.45, \qquad P\{A_2\} = 0.45, \qquad P\{A_3\} = 0.10$$
$$P\{B|A_1\} = 0.02, \qquad P\{B|A_2\} = 0.025, \qquad P\{B|A_3\} = 0.04$$

The quantity we are seeking to determine is $P\{A_3|B\}$. Now

$$P\{A_3|B\}P\{B\} = P\{B|A_3\}P\{A_3\}$$

The right-hand side is already known:

$$P\{B|A_3\}P\{A_3\} = 0.04 \times 0.10 = 0.004$$

The quantity $P\{B\}$ may be determined by noting that

$$B = BA_1 \cup BA_2 \cup BA_3$$

since *some* line produced the paint, and so

$$
\begin{aligned}
P\{B\} &= P\{BA_1\} + P\{BA_2\} + P\{BA_3\} \\
&= P\{B|A_1\}P\{A_1\} + P\{B|A_2\} + P\{B|A_3\} + P\{A_3\} \\
&= 0.02 \times 0.45 + 0.025 \times 0.45 + 0.04 \times 0.10 \\
&= 0.0090 + 0.01125 + .004 \\
&= 0.02425
\end{aligned}
$$

Inserting this into the equation above and solving, we have

$$P\{A_3|B\} = \frac{0.004}{0.02425} = 0.165$$

Note that $P\{A_3|B\} > P\{A_3\}$; that is, our assessment of the likelihood that line 3 produced the paint is changed by the observation of the improper blending. It seems natural to have the probability increasing, since line 3 has the poorest performance record. Note also that we can write

$$P\{A_3|B\} = \frac{P\{B|A_3\}}{P\{B|A_1\}P\{A_1\} + P\{B|A_2\} + P\{B|A_3\}P\{A_3\}} \cdot P\{A_3\}$$

The multiplier to $P\{A_3\}$ indicates how we revalue the probability of the event A_3, given the information B.

2.4. THE LAW OF LARGE NUMBERS

Now that we have a rather precise definition of probability it is of interest to see how this definition accords with our intuitive notion of probability. Intuitively, we interpret statements such as "The probability of a salary increase next year is 90%" to mean that it is quite likely that an increase will be received, whereas a low probability would make us believe that an increase is most unlikely. Thus, the numerical value of the probability of an event would seem somehow to relate to the frequency with which we might expect that event to occur. This intuitive grasp of the notion of probability is made more precise by a consideration of what has come to be called the *law of large numbers*. We shall now describe that law. Suppose that a certain random experiment is fixed. The result of the experiment is measured, and it is possible to ascertain whether a given event A has occurred. Let us suppose further that the probability of occurrence of the event A is some unknown number p, which of course lies in the range 0 to 1. If the experiment is repeated a large number of times (in later chapters we shall become more precise about the meaning of the word "large"), we would expect that in about $100p\%$ of all repetitions of the experiment, the given event would occur. Suppose that in N repetitions, n occurrences take place. The law of large numbers states essentially that the difference between the fraction n/N and the unknown number p must essentially* approach 0 as N approaches infinity.

In the next chapter, we shall see the workings of this law in connection with the so-called *binomial distribution*.

Example 2.8. A coating is being tested for abrasion resistance. A large number of test panels are prepared and subjected to abrasion. It is determined that only 3% fail performance specifications. We feel that the true probability of failure must therefore be "near" 3%. Of course, how near depends somewhat on what is meant by large. In practice, one would want, typically, rather strong assurance that the failure probability does not exceed some targeted value as an upper limit. Ways of obtaining such assurance will become clear in later chapters. For example, we will learn that, if 100 panels produce 3 failures in a properly controlled experiment, it is almost certain that the process itself is producing no more than about 8% defective product.

BIBLIOGRAPHY

Special References

Hahn, G. J., and Shapiro, S. S., *Statistical Models in Engineering*, Chapter 2, Wiley, New York, 1967.

Langley, R., *Practical Statistics*, Chapter 2, Dover, New York, 1971.

Longley-Cook, L. H., *Statistical Problems*, Chapter 6, Barnes & Noble, New York, 1971.

---

*The reasons for this rather curious statement of the law will only be clear after a reading of the next chapter.

Mack, C., *Essentials of Statistics for Scientists and Technologists*, Chapter 3, Plenum Press, New York, 1975.

Meyer, S. L., *Data Analysis for Scientists and Engineers*, Part III, Wiley, New York, 1975.

Myers, B. L., and Enrick, N. L., *Statistical Functions*, Chapter 2, Kent State University Press. Kent, OH, 1970.

Singh, J. *Great Ideas of Modern Mathematics: Their Nature and Use*, Chapter 9, Dover, New York, 1959.

Young, H. D., *Statistical Treatment of Experimental Data*, Chapter 2, McGraw-Hill, New York, 1962.

General References

Chung, K. L., *A Course in Probability Theory*, Harcourt Brace Jovanovich, New York, 1968.

Drake, A. W., *Fundamentals of Applied Probability Theory*, McGraw-Hill, New York, 1967.

Eisen, M., *Introduction to Mathematical Probability Theory*, Prentice-Hall, Englewood Cliffs, NJ, 1969.

Feller, W., *An Introduction to Probability Theory and Its Applications*, Vol. I, 3rd ed., Wiley, New York, 1968; Vol. II, 2nd ed., 1971.

Harris, B., *Theory of Probability*, Addison-Wesley, Reading, MA, 1966.

Kyburg, H. E., Jr., *Probability Theory*, Prentice-Hall, Englewood Cliffs, NJ, 1969.

Lamperti, J., *Probability: A Survey of the Mathematical Theory*, W. A. Benjamin, New York, 1966.

Rényi, A., *Probability Theory*, North-Holland, Amsterdam, 1971.

Revesz, P., *The Laws of Large Numbers*, Academic Press, New York, 1968.

APPENDIX: SET THEORY

Introduction

Set theory, since its origination in the late eighteenth century, has developed to a degree of importance accomplished by few mathematical theories. Devised to give mathematicians a clearer look at the internal logic of their subject, it now permeates most of their advanced papers. Aside from this, it has been applied in almost all fields of mathematics plus finding practical applications in the fields of probability, game theory, physics, and even chemistry. Essentially without qualification, the concept of a set can be stated to form the foundation of probability and statistics, and of mathematics in general.

A set is simply a collection of well-defined entities all of which have something in common, such as the set of all human beings. Obviously there are subsets, for example, of all males and even further subdivisions such as all male Americans, adults, and so on, plus members of noninclusive sets such as females.

Fundamental Concepts

A set can be thought of as a collection of objects, called elements or members of a set. For example, all the vowels in the English alphabet can be considered as a set. In general, a set is denoted by capitals such as A, B, C and an element by a lower case letter such as a, b, c.

We say that the element a *belongs to a set* C, in symbols $a \in C$. If a does not belong to C we write $a \notin C$. If both a and b belong to C, we write $a, b \in C$. For a set to be *well defined*, a common necessary assumption, we must be able to determine whether a particular object does or does not belong to the set. In principle it serves for the needs of mathematics to restrict the members of sets to sets only, because very importantly, numbers, functions, points, and so on may also be conceived of as sets.

A set can be defined by actually listing its elements (such as the vowel set a, e, i, o, u) or if this is not possible, by describing some property held by all members and by no nonmembers. These definitions are not mutually exclusive, as elements may be defined by listing *or* by property, for example, $\{ x|x$ is a vowel$\}$ is the set of all elements x such that x is a vowel. However, for the set $\{ x|x$ is a triangle in a plane$\}$ the listing method obviously cannot be used. A further example of a set would be the possible numbers that can come up on a single die if we toss a pair of ordinary dice. Here the points are elements of the set $\{1, 2, 3, 4, 5, 6\}$. Another example would be $F = \{ x|x \leq 5\}$ where F is the set of all x's less than or equal to 5. The principle of extensionality states that sets which have the same members are considered equal, for example, if a set S'' belongs to S' and S' to S, where S' is a subset of S, and so on, then S'' belongs to S. This can be written: if $S'' \subset S'$ and $S' \subset S$, then $S'' \subset S$. An equivalent way of stating the same thing would be if S "contains" S' and S' contains S'', then S contains S''. In symbols, if $S \supset S'$, and $S' \supset S''$, then $S \supset S''$. It is then presupposed that with respect to every object x, it shall be definite whether $x \in s$ or not.

Other important concepts are subset, equivalence and order.

Subsets

If every element of a set A also belongs to a set B we call A a *subset* of B, in symbols, $A \subset B$ (A is contained in B) or $B \supset A$ (B contains A). Therefore, $A \subset A$ (A belongs to A) for every set A.

If $A \subset B$ and $B \subset A$ we call A and B equal and note $A = B$. In such a case A and B have exactly the same elements. If A is not equal to B we note $A \neq B$.

For example, if $A = \{1, 2, 3, \ldots, 10\}$ and $B = \{6, 10\}$, then $B \subset A$ or is a subset of A, since 6 and 10 are in both set B and in A the larger set.

If B has at least one member which does not belong to A, A is referred to as a *proper subset*. In notation, if $A \subset B$ but $A \neq B$, we call A a *proper subset* of *B*. For example, $\{a, i, u\}$ is a proper subset of $\{a, e, i, o, u\}$. Also, in tossing a die the possible outcomes where the die comes up *even* are elements of the set $\{2, 4, 6\}$ which is a proper subset of all possible outcomes $\{1, 2, 3, 4, 5, 6\}$ of the die. Usually a subset of *A* is defined by a property *P* which is meaningful for the members of *A*, namely as the set of those $x \in A$ which satisfy *P*. For instance, for the set *I* of all positive integers and the property "prime" there is obtained the set of all prime numbers, which is a subset of *I*. If no member of *s* has the property, there must be admitted a set which contains no member, referred to as the empty set or null set, $\varnothing$.

The Universal Set or Null Set

The *universal set U*, also symbolized by *W* to distinquish it from $\cup$ (union), contains all the members considered. The difference, $U - s$, is called the complement of *s*. For many purposes, we restrict our discussion to subsets of the universal set *W*, also called the universal space. The elements of a space are often called points of the space.

It is useful to also consider a set having no elements. This is referred to as the *empty set* or *null set* and is noted as $\varnothing$. It is a subset of any set. For example, the set of all real numbers such that $x^2 = -1$, written $\{x | x^2 = -1\}$, is the null set since there are no real numbers whose squares are equal to -1. Another example is the set of all natural numbers greater than 4 and less than 5.

Equivalence and Order

If a one-to-one correspondence exists between the members of set *s* and set *t*, *s* is said to be equivalent to *t*, symbolically $s \sim t$. Also, $t \sim s$ and $s \sim t$, $t \sim u$ together imply $s \sim u$. A set *D* which is equivalent to the set $I = \{1, 2, \ldots, k, \ldots\}$ of all positive integers is called denumerable. Denoting by d_k the member of *D* which corresponds to the integer *k*, *D* can be written in the form $D = \{d_1, d_2, \ldots, d_k, \ldots\}$. The set of all positive integers *I* as such is a *plain set*, but it becomes an *ordered set* when a succession or order of its members is introduced. An ordered set is called *well ordered* if every ordered and nonempty subset has a first member. For example $(1, 2, 3, \ldots)$ is well ordered, whereas $(\ldots, 3, 2, 1)$ is not since the subset of all odd integers, as well as the set itself, has no first member.

Finite and Infinite Sets

Sets may be finite or infinite. When finite, they are written as follows: $\{0, 1, 2, 3\}$, $\{2, 4, 6, 8\}$, that is they end in a definite number. Where the number of elements cannot be tabulated, we have an infinite set, for example, $\{1, 2, 3, \ldots\}$. Another example would be the set of all the points on a line segment.

Set Operations

Two major and one minor operations between sets are fundamental. These operations are in many respects analogous to arithmetic number operations such as addition, subtraction, and multiplication. The operations are commutative (by definition) and associative.

Union

Consider the set of all elements (or points) which belong to either set A or set B or both set A and set B. This common set is called the union of A and B and is written $A \cup B$. The union of two or more sets $s_1, s_2, \ldots$ is the set of those members which are contained in at least one set s_n, and is designated as $s_1 \cup s_2$. For example, with set $A = \{1, 2, 3, 4, 5, 6\}$ and set $B = \{-2, -1, 0, 3, 4, 5\}$, $A \cup B = \{-2, -1, 0, 1, 2, 3, 4, 5, 6\}$.

Intersection

Corresponding to union is the operation of *intersection* which yields the set $s_1 \cap s_2$ of the members contained in each of the sets s_m. Two sets s_1 and s_2 such that $s \cap s_2 = \varnothing$, that is, which have no elements in common, are called *disjoint sets*. The intersection of two or more sets yields those elements common to those two or more sets. For example, take the two sets of natural numbers, set $A = \{1, 2, 3, 4, 5, 6\}$ and set $B = \{2, 4, 6, 8\}$. The intersection of sets A and B is $A \cap B = \{2, 4, 6\}$. As a further example consider the sets $A = \{x | x < 5\}$ and $B = \{x | x > 2\}$. The result $A \cap B = \{x | 2 < x < 5\}$.

Difference

Of somewhat lesser importance is difference $B - A$, which is essential only in the case $A \subset B$. Here $B - A$ is the set of the members of B which are not contained in A.

Complement

If $B \subset A$ then $A - B$ is referred to as the *complement* of B relative to A and is symbolized as B'_A (or occasionally as B^*_A). If $A = W$, the universal set, we refer to $Q - B$ as the complement of B, noted as B'(or B^*). The complement of $A \cup B$ is denoted by $(A \cup B)'$.

For example, if $A = \{1, 2, 3\}$, $B = \{3, 4, 5\}$, and $C = \{1, 2, 3, 4, 5, 6, 7\}$, then $A \cap B = \{3\}$. The complement of A is A' (every number in the universal set not included in A). For A, B, and C these numbers are $4, 5, 6, 7$. Therefore, the intersection of A with B is $A \cap B = \{4, 5\}$. Similarly $B' \cap A = \{1, 2\}$.

To illustrate these operations, visualize two partially overlapping circles contained within a rectangle (a so-called Venn diagram developed by John Venn in 1880). Here each enclosed circular area is a set, A and B, the overlapping area is C, and the area outside the circles is D. The union of set A with set B, $A \cup B$, is represented by the total area bounded by the two circles. It would be areas $A + B - C$. The area C is the intersection of the two sets and is $A \cap B$, or the area common to the two sets. The complement of $A \cup B$ is the set D. Finally, the union of A, B, and D, written $D \cup (A \cup B)$ would be W, the set represented by the entire rectangle.

The Algebra of Sets

$A \cup B = B \cup A$

$A \cup (B \cup C) = (A \cup B) \cup C = A \cup B \cup C$

$A \cap B = B \cap A$

$A \cup (B \cap C) = (A \cap B) \cap C = A \cap B \cap C$

$A \cap (B \cup C) = (A \cap B) \cup (A \cap C)$

$A - B = A \cap B'$

If $A \subset B$, then $A'i \supset B'$ or $B' \subset A'$

$A \cup \emptyset = A, \qquad A \cap \emptyset = \emptyset$

$A \cup W = W, \qquad A \cap W = A$

$(A \cup B)' = A' \cap B'$

$(A \cap B)' = A' \cup B'$

$A = (A \cap B) \cup (A \cap B')$

$A \cup (B \cap C) = (A \cup B) \cap (A \cup C)$

Cartesian Product

If s and t are sets without common members, the *Cartesian product* $s \times t$ is defined as the set of all pairs $\{\sigma, \tau\}$ with $\sigma \in s$ and $\tau \in t$. For example, if $s = \{1, 2, 3\}$ and $t = \{4, 5\}$, then $s \times t = \{\{1, 4\}, \{1, 5\}, \ldots, \{3, 5\}\}$, or a set with $3 \times 2 = 6$ members.

Sample Spaces

A set W which consists of all possible outcomes of a random experiment is called a *sample space* and each outcome is called a sample point. Sample spaces can be pictured graphically where outcomes can be designated numerically. For example, if we toss a coin twice and use 0 to represent tails and 1 to represent heads, the sample space can be portrayed as in

Cartesian coordinates for further throws as, for example, $(0, 1)$, $(1, 1)$, $(0, 0)$, $(1, 0)$, and so forth.

If a sample space has a finite number of points it is called *finite sample space*. If it has as many points as there are natural numbers $1, 2, 3, \ldots$, it is called a *countably infinite sample space*. If it has as many points as there are in some intervals on the X axis, such as $0 \leq X \leq 1$, it is called a *noncountably infinite sample space*. A sample space which is finite or countably infinite is referred to as a *discrete sample space*, while one which is noncountably infinite is called a *nondiscrete* or *continuous sample space*.

A function (in the statistical sense) can be regarded as a set. For instance, a single-valued function $y = f(X)$ can be regarded as a set of ordered pairs (x, y) in which different pairs contain different values of x but not necessarily different values of y.

BIBLIOGRAPHY

Spiegel, M. R., *Probability and Statistics*, Chapter 1, Schaum's Outline Series, McGraw-Hill, New York, 1975.

Singh, J., *Great Ideas of Modern Mathematics: Their Nature and Use*, Dover, New York, 1959.

3

$\overline{}$

DISTRIBUTION FUNCTIONS

ANTHONY SALVIA

Behrend College
Pennsylvania State University
Erie, Pennsylvania

H. EARL HILL

Lord Corporation
Erie, Pennsylvania

3.1. THEORETICAL PRINCIPLES

In the development of probability in the previous chapter, you may have noticed that three distinct stages were involved:

1. As background, a "random experiment" is performed. The results of this experiment are not predictable with certainty.
2. Every conceivable result is labeled in some orderly fashion; the aggregate of all of these unique, identifiable outcomes forms the sample space for the experiment.
3. A function P( ) is defined over the sample space.

We see that the experiment, in effect, determines a sample space; the probability function P is arbitrary within the constraints i–iii as listed in

Chapter 2. Consider, now, a series of random experiments such as these:

Experiment	Labeling of Outcomes
Tossing a coin	H, T
Birth of a child	M, F
Status of an order to be delivered	On time, late
Quality of a manufactured product	Acceptable, unacceptable
Observing a door	Open, closed
Jury verdict	Guilty, not guilty
Employee's attendance	Present, absent

Clearly, there is a commonality to each of these; the experiment produces only two possible results. However, each generates its own sample space, and all of the sample spaces are different.

There is at this point an opportunity for a tremendous simplification; suppose we decide always to use a "code" for experiments of this type. In this code, one of the two possible results will be labeled 0 (zero) and other result labeled 1. We define, accordingly, a variable X that has only two possible values; either $X = 0$ or $X = 1$. Provided we remember how the assignment of X values is made, there is then no difference in reporting experimental results between the statements "The coin fell Heads" and "$X = 1$." We can then say

$$P\{X = 1\} = P\{H\}$$
$$P\{X = 0\} = P\{T\}$$

A variable, such as X, whose value depends on the result of a random experiment, is called a *random variable*. Effectively, what we have done is to provide a common sample space for all experiments of the type shown above, namely, the two integers 0 and 1. We shall, in the sequel, refer to such experiments as Bernoulli experiments (for historical reasons); the discrete random variable X is called a Bernoulli variable.

At stated above, the probability content of the sample space is concentrated on the two integers 0 and 1. It will be convenient to imagine the probability to be assigned to the entire set of real numbers. This can be done quite simply by the assignment

$$P\{X = 1\} = p \, (0 \leq p \leq 1)$$
$$P\{X = 0\} = 1 - p \qquad\qquad (3.1)$$
$$P\{X = x\} = 0, \quad \text{if} \quad x \neq 0 \quad \text{or} \quad 1$$

In equation (3.1) we have defined a function, say f, over all the real numbers, that is,

$$f(0) = 1 - p, \qquad f(1) = p, \qquad f(x) = 0, \qquad x \neq 0, 1 \qquad (3.2)$$

This function f satisfies the two conditions

i. $f(x) \geq 0$
ii. $\sum_x f(x) = 1$

Such functions, designated by a lower case f, are called *probability density functions*, or simply *density functions*. Thus, equation (3.2) is the Bernoulli density function. It can be written somewhat more succinctly as

$$f(x) = \begin{cases} p^x(1-p)^{1-x}, & x = 0, 1 \\ 0, & \text{otherwise} \end{cases} \qquad (3.3)$$

The graph of this function as seen in Figure 3.1 is rather stark. Generally, to improve clarity, we erect rectangles over the points of nonzero probability, as shown in Figure 3.2.

A graph such as Figure 3.2 is called a histogram; you will note that it is constructed in such a way that the total area of all the rectangles is one square unit. Remember, though, that this is simply a visual aid; the probability is concentrated at two points, 0 and 1, and is not (in this instance) spread over the range -0.5 to $+1.5$.

FIGURE 3.1. Bernoulli density function.

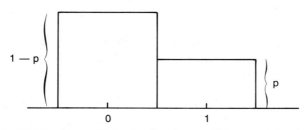

FIGURE 3.2. Bernoulli density function, auxiliary rectangles shown.

It is sometimes convenient to think of these concentrated probabilities as point masses located along the x axis. A natural inquiry, then, is to determine the center of gravity (center of mass) of the system. Using a familiar formula from mechanics, we find the center of mass μ to be

$$\mu = \frac{\Sigma x f(x)}{\Sigma f(x)} = \frac{0 \cdot (1 - p) + 1 \cdot p}{(1 - p) + p} = \frac{p}{1} = p$$

In statistics, μ is termed the *arithmetic mean*, or simply the *mean*, of the random variable X. Since $\Sigma f(x) = 1$, the calculation of the denominator is superfluous, and we therefore define

$$\mu = \sum_x x f(x) \tag{3.4}$$

It is also of frequent value to measure the dispersion of probability around this mean value; by far the most common measure employed is the *variance*, denoted by σ^2, and defined as

$$\sigma^2 = \sum_x (x - \mu)^2 f(x) \tag{3.5}$$

For the previously mentioned Bernoulli random variable,

$$\sigma^2 = \sum_x (x - \mu)^2 f(x) = (0 - p)^2 \cdot (1 - p) + (1 - p)^2 \cdot p$$

$$= p^2(1 - p) + p(1 - p)^2$$
$$= p(1 - p)(p + 1 - p)$$
$$= p(1 - p)$$

The positive square root of the variance is called the *standard deviation*; for the Bernoulli variable,

$$\sigma = \sqrt{p(1 - p)}$$

We wish presently to generalize slightly the types of computation involved in determining μ and σ^2; before doing so, an example or two may serve to clarify the ideas presented so far.

Example 3.1.   Paint is packaged 4 gal to a carton. From experience, it has been discovered that an occasional defective lid will be used, with resultant spillage. Suppose X is the number of defective lids found per carton.

Historical data show the following:

Value of X	Number of Cartons
0	9,500
1	400
2	100
3 or 4	0
	10,000

We adopt as a probability density function

$$f(x) = \begin{cases} 0.95, & x = 0 \\ 0.04, & x = 1 \\ 0.01, & x = 2 \\ 0, & \text{all other } x \end{cases}$$

The average number μ of defective lids per carton is then

$$\mu = \sum_x xf(x) = 0 \text{X}.95 + 1 \text{X}.04 + 2 \text{X}.01 = 0.06$$

and the variance is

$$\sigma^2 = \sum (X - \mu)^2 f(x) = (0 - 0.06)^2 \text{X}.95 + (1 - 0.06)^2 \text{X}.04$$
$$+ (2 - 0.06)^2 \text{X}.01$$
$$= 0.0762$$

Example 3.2. In a large manufacturing operation, it has been suggested that X, the number of absentees per day, has a probability density function of the form

$$f(x) = \frac{e^{-\lambda}\lambda^x}{x!}, \qquad x = 0, 1, 2, \ldots \tag{3.6}$$

Here, $x! = 1 \cdot 2 \cdot 3 \cdot x$, with the convention that $0! = 1$; e is the base of natural logarithms (having an approximate value 2.71828) and λ is some unknown positive constant. By methods that need not concern us at present, it is determined that

$$\mu = \lambda, \qquad \sigma^2 = \lambda$$

Example 3.2 raises some intriguing questions. To begin, notice that the range of possible values for X is infinite. (The notation $x = 0, 1, 2, \ldots$ is taken to mean that x is a nonnegative integer.) Clearly, if X represents the number of absent employees, X can never be greater than the total number of persons employed, which of course is finite. Remember, however, that equation (3.6) is only a *model* of the real-world situation. As such, it cannot be expected to be exact. All we can really expect is that the model will allow more or less accurate *predictions* about the matter at hand, namely, employee absenteeism. For very large values of x, $f(x)$ is, for all practical purposes, indistinguishable from zero, anyway.

A second, and more basic question concerns the rationale for the particular model chosen. We shall in due time motivate this choice; at this point it suffices to state that experience has shown that a model of type (3.6) is often a good "fit" to a phenomenon like employee absenteeism.

Still another question, granting for the moment that we accept equation (3.6) as a reasonable model, concerns the "unknown positive constant," λ. Such constants will occur in nearly every probability model we shall have occasion to consider; they are termed *parameters*. A good portion of the subject matter of statistics deals with the estimation of parameters; in the instance in question one would undoubtedly obtain data from past attendance records and use this data to compute a numerical value that would be used for λ. (As it turns out, a good procedure to follow here is to take for λ the average number of absentees per day.)

As we mentioned prior to the above examples, it is useful to generalize somewhat the type of operations involved in the computation of μ and σ^2. To this end, suppose we have the density function $f(x)$ of some random variable X, and any* other function $g(x)$. The sum

$$\sum_x g(x)f(x)$$

may not always exist (typically, it is an infinite series); when it does, we call it the *mathematical expectation* or *expected value* of $g(x)$, and denote it as

$$E[g(X)] = \sum_x g(x)f(x) \tag{3.7}$$

Using this E notation, we readily observe that

$$\mu = E[X]$$
$$\sigma^2 = E\left[(X - \mu)^2\right]$$

We state without proof several properties of the operator E:

(P1) $E[C] = C$ for any constant C
(P2) $E[cg(X)] = cE[g(X)]$
(P3)* $E[g_1(X) + g_2(X)] = E[g_1(X)] + E[g_2(X)]$

---

*The function g is completely arbitrary, except that the sum $\sum g(x)f(x)$, if it is an infinite series, must be absolutely convergent, that is, $\sum |g(x)| f(x)$ must be finite.
*By mathematical induction we may extend (P3) to

$$E\left[\sum_{i=1}^{n} g_i(X)\right] = \sum_{i=1}^{n} E[g_i(X)]$$

The expectations $E[X^K]$ and $E[(X - \mu)^K]$ are called the Kth *moment* and the Kth central moment, respectively. Thus, the mean μ is the first moment and the variance σ^2 is the second central moment. The third and fourth central moments are denoted

$$\mu_3 = E\left[(X - \mu)^3\right]$$
$$\mu_4 = E\left[(X - \mu)^4\right]$$

In practice it is convenient to deal with so-called β-values:

$$\beta_1 = \frac{\mu_3^2}{(\sigma^2)^3}, \qquad \beta_2 = \frac{\mu_4}{\sigma^4}$$

or the coefficients of skewness and excess

$$\gamma_1 = \frac{\mu_3}{\sigma^3}, \qquad \gamma_2 = \beta_2 - 3$$

For a random variable with a density function that is symmetric about its mean μ, $\gamma_1 = 0$. Departures from the value $\gamma_1 = 0$ are indicative of asymmetry or skewness. Values $\gamma_1 < 0$ indicate that the density function has a relatively long tail to the left of μ; $\gamma_1 > 0$ indicates a relatively long tail to the right.

The use of the constant 3 in the definition of γ_2 facilitates comparisons with the *normal* distribution, which will be defined shortly. Essentially, γ_2 measures "flatness'; of the distribution, positive values indicating distributions flatter than the normal and negative values indicating distributions more peaked than the normal.

Generally, we shall have little need for any higher order moments or central moments.

We now make use of properties (P1)–(P3) to develop a formula for σ^2 which is often simpler to use than the one just discussed.

$$
\begin{aligned}
\sigma^2 &= E\left[(X - \mu)^2\right] \\
&= E\left[X^2 - 2\mu X + \mu^2\right] \\
&= E\left[X^2\right] - 2\mu E\left[X\right] + \mu^2 \\
&= E\left[X^2\right] - \mu^2 \qquad\qquad\qquad (3.8)
\end{aligned}
$$

The use of equation (3.8) avoids the need for subtracting μ from each X in computing the sum (3.5). Similar computations may be made for higher moments as well; for example,

$$\mu_3 = E\left[(X - \mu)^3\right] = E\left[X^3 - 3X^2 + 3X\mu^2 - \mu^3\right]$$
$$= E\left[X^3\right] - 3\mu E\left[X^2\right] + 2\mu^3$$

Earlier, we mentioned that the probability concentrations along the x axis can be considered as point masses, the mass at the number x being equal to $f(x)$. It shall frequently be useful to consider not only the mass (or probability) at a single point, but also the accumulated probability to the left of that point. That is, for any number x, we wish to consider the probability content of the real number axis up to and including x.

Clearly, that content is a function of x; it is called the *distribution function* of the random variable X, and is generally denoted by a capital letter that corresponds to the lower case letter used for the density function, for example,

$$F(x) \quad \text{and} \quad f(x)$$

These are related by

$$F(x) = \sum_{t \le x} f(t) \tag{3.9}$$

For example, the Bernoulli random variable has a distribution function

$$F(x) = \begin{cases} 0, & x < 0 \\ 1 - p, & 0 \le x < 1 \\ 1, & 1 \le x \end{cases}$$

Note that the graph of this function is a step function that (in this case) consists of three horizontal segments and two jumps, one of size $1 - p$ (at $x = 0$), the second of size p (at $x = 1$).

In the example of paint packaging (Example 3.1), we would have

$$F(x) = \begin{cases} 0 & x < 0 \\ 0.95 & 0 \le x < 1 \\ 0.99 & 1 \le x < 2 \\ 1.00 & 2 \le x \end{cases}$$

Let us now turn our attention to another series of random experiments:

1. Measuring the height of a "randomly selected" adult male.
2. Measuring reaction time.
3. Measuring weight percent of a specified ingredient in a compound.

All of these experiments result in a number X which can vary along a continuous scale of values; we refer to X here as a *continuous* random variable. To avoid confusion, the random variables we discussed earlier are by contrast generally termed discrete.

Much of the preceding discussion about discrete random variables is also applicable to continuous ones. The requirements i $f(x) \geq 0$ and ii $\Sigma f(x) = 1$ become

i. $f(x) \geq 0$
ii. $\int_{-\infty}^{\infty}(x)\, dx = 1$

and expectation is defined as

$$E[g(X)] = \int_{-\infty}^{\infty} g(x)f(x)\, dx$$

The distribution function F is related to the density function f by means of

$$F(x) = \int_{-\infty}^{x} f(t)\, dt$$

and it is easy to show that the amount of probability contained within an interval $a \leq x \leq b$ is $F(b) - F(a)$.*

We now consider some examples of *continuous* random variables.

Example 3.3. Based upon an extensive data analysis, it has been determined that the thickness of a particular coating is not uniform, but varies due to the coating process. The thickness X at any particular point on the surface, measured in mils, is a random variable with density function

$$f(x) = 0.002xe^{-0.001x^2}, \qquad x \geq 0$$

Notice that the definition of $f(x)$ here is not complete; there is no indication of its values for $x < 0$. Generally speaking, when a density function is given by a formula such as the one above, we assume $f(x) = 0$ for the values of x not specified.

Let us determine the average thickness:

$$\mu = E[X] = \int_{-\infty}^{\infty} xf(x)\, dx = \int_{0}^{\infty} 0.002x^2 e^{-0.001x^2}\, dx$$

---

*Strictly speaking, $\Pr\{a \leq X \leq b\} = F(b) - F(a) + \Pr\{X = a\}$. For continuous random variables, $\Pr\{X = a\} = 0$ for any a, however.

This integral may be evaluated by changing variables:

$$u = 0.001x^2$$
$$du = (0.002x)\, dx$$
$$x = (1000u)^{1/2}$$

Then

$$\int_0^\infty 0.002x^2 e^{-0.001x^2}\, dx = \int_0^\infty (1000u)^{1/2} e^{-u}\, du$$
$$= \sqrt{1000} \int_0^\infty u^{1/2} e^{-u}\, du$$

Recalling that the gamma function, $\Gamma(t)$, is defined as

$$\Gamma(t) = \int_0^\infty u^{t-1} e^{-u}\, du$$

we see that

$$\mu = \sqrt{1000}\, \Gamma(3/2) = 31.62 \times 0.88623 \simeq 28.0$$

Suppose that specifications call for a minimum thickness of 20 mils. We can readily determine the probability of meeting this requirement.

$$\Pr\{ X \geq 20 \} = \int_{20}^\infty 0.002x e^{-0.001x^2}\, dx$$

As before, we let

$$u = 0.001x^2$$
$$du = 0.002x\, dx$$

Then

$$\Pr\{ X \geq 20 \} = \int_{0.4}^\infty e^{-u}\, du = e^{-0.4} = 0.67$$

approximately. This random variable is a special case of what is called a *Weibull* variable.

Example 3.4. A measuring device produces a reading that may differ from the true value being measured. In many cases it is reasonable to assume that the measurement error X is a random variable with density function

$$f(x) = \begin{cases} \dfrac{1}{\theta}, & -\dfrac{\theta}{2} \leq x \leq \dfrac{\theta}{2} \\ 0, & \text{otherwise} \end{cases}$$

If this is the case, the average error is

$$\mu = E[X] = \frac{1}{\theta} \int_{-\theta/2}^{\theta/2} x \, dx = 0$$

The variance is given by

$$\sigma^2 = E[(X - \mu)^2] = \frac{1}{\theta} \int_{-\theta/2}^{\theta/2} x^2 \, dx = \frac{\theta^2}{12}$$

Example 3.5. A wide variety of physical measurements have been found to possess the so-called "normal" distribution. The density function for such a variable is

$$f(x) = \frac{1}{\sqrt{2\pi}\,\sigma} \exp\left[-\frac{1}{2}\frac{(x - \mu)^2}{\sigma^2}\right], \qquad -\infty < x < \infty$$

In the sequel we shall have frequent occasion to "assume a normal distribution."

It is often the case that a random experiment has outcomes that are more appropriately described by two or more numbers, rather than just a single number; for example, we might measure the height, weight, and shoe size of a randomly chosen individual. Again, we may measure color and gloss of some product. Such experiments lead us quite naturally to consider density functions of several random variables. For definiteness we shall in the following discussion consider two continuous random variables; the discussion applies, however, to any number of continuous variables. If integrals are replaced by sums, the discussion also applies to discrete variables.*

The density function of a pair of random variables (X_1, X_2) is a function $f(x_1, x_2)$ which satisfies the conditions:

i. $f(x_1, x_2) \geq 0$
ii. $\int_{-\infty}^{\infty}\int_{-\infty}^{\infty} f(x_1, x_2) \, dx_1 \, dx_2 = 1$

To distinguish this f as referring to more than one variable, we often call it the *joint* density function. If $g(x_1, x_2)$ is properly behaved we define

$$E[g(X_1, X_2)] = \int_{-\infty}^{\infty} \int_{-\infty}^{\infty} g(x_1, x_2) f(x_1, x_2) \, dx_1 \, dx_2$$

*Situations in which both types of variables occur are of course theoretically possible; in practice such situations are rare.

For example, the mean of X_2 is obtained by selecting $g(x_1, x_2) = x_2$. Denoting the two means by μ_1 and μ_2, the quantity $E[X_1 - \mu_1)(X_2 - \mu_2)]$ is called the *covariance* of X_1 and X_2; we use the symbol σ_{12} for this:

$$\sigma_{12} = E[(X_1 - \mu_1)(X_2 - \mu_2)]$$

If σ_1^2 and σ_2^2 are the variances of the variables X_1, X_2, we define the *correlation coefficient* ρ_{12} as

$$\rho_{12} = \frac{\sigma_{12}}{\sigma_1 \sigma_2}$$

Usually, with only two variables involved, the subscripts on ρ_{12} are omitted.

The joint *distribution* function $F(x_1, x_2)$ may be obtained as

$$F(x_1, x_2) = \int_{-\infty}^{x_2} \int_{-\infty}^{x_1} f(t_1, t_2) \, dt_1 \, dt_2$$

Conversely, given F, one has

$$f(x_1, x_2) = \frac{\partial^2 F(x_1, x_2)}{\partial x_1 \, \partial x_2}$$

Example 3.6. Suppose the joint density function of X_1, X_2 is

$$f(x_1, x_2) = \begin{cases} x_1 + x_2 & \begin{array}{l} 0 \le x_1 \le 1 \\ 0 \le x_2 \le 1 \end{array} \\ 0 & \text{otherwise} \end{cases}$$

The mean of X_1 is

$$\mu_1 = E[X_1] = \int_0^1 \int_0^1 (x_1^2 + x_1 x_2) \, dx_1 \, dx_2 = \tfrac{7}{12}$$

By symmetry, $\mu_2 = \tfrac{7}{12}$ as well.

To determine the covariance, note that

$$\begin{aligned} \sigma_{12} &= E[(X_1 - \mu_1)(X_2 - \mu_2)] \\ &= E[X_1 X_2 - \mu_1 X_2 - \mu_2 X_1 + \mu_1 \mu_2] \\ &= E[X_1 X_2] - \mu_1 \mu_2 - \mu_2 \mu_1 + \mu_1 \mu_2 \\ &= E[X_1 X_2] - \mu_1 \mu_2 \end{aligned}$$

so that

$$\sigma_{12} = E[X_1 X_2] - \tfrac{7}{12} \cdot \tfrac{7}{12}$$

Further,

$$E[X_1 X_2] = \int_0^1 \int_0^1 (x_1^2 x_2 + x_1 x_2^2)\, dx_1\, dx_2 = \tfrac{1}{3}$$

so that

$$\sigma_{12} = \tfrac{1}{3} - \tfrac{49}{144} - \tfrac{1}{144}$$

Integration yields also

$$F(x_1, x_2) = \frac{x_1 x_2 (x_1 + x_2)}{2}, \qquad 0 \le x_1 \le 1, \qquad 0 \le x_2 \le 1$$

so that, for example,

$$\Pr\{ X_1 \le \tfrac{1}{2}, X_2 \le \tfrac{3}{4} \} = F(\tfrac{1}{2}, \tfrac{3}{4}) = \tfrac{15}{64}$$

On many occasions we are given a joint density, but primary interest is focused on only a subset of the total number of variables involved. Suppose, for instance, that in Example 3.6 we wished to determine $\Pr\{ X_1 \le \tfrac{1}{2}\}$ without considering the value of X_2. Clearly,

$$\Pr(X_1 \le \tfrac{1}{2}) = \Pr(X_1 \le \tfrac{1}{2}, X_2 \le 1) = F(\tfrac{1}{2}, 1)$$
$$= \tfrac{3}{8}$$

Likewise, $\Pr\{ X_1 \le x_1 \} = x_1(x_1 + 1)/2$ for any x_1 in the interval $0 \le x_1 \le 1$. This expression, then, must be the distribution function for X_1, and so

$$f(x_1) = \frac{d}{dx_1} \frac{x_1(x_1 + 1)}{2} = x_1 + \frac{1}{2}, \qquad 0 \le x_1 \le 1$$

The steps we followed in obtaining $f(x_1)$ were rather simple: in $F(x_1, x_2)$, we set x_2 at its "effective" upper limit, $x_2 = 1$. Differentiating the resulting expression yielded $f(x_1)$.

It is usually simpler in this type of problem to begin with $f(x_1, x_2)$ and integrate with respect to the "unwanted" variable; thus

$$\int_{-\infty}^{\infty} f(x_1, x_2)\, dx_2 = \int_0^1 (x_1 + x_2)\, dx_2 = x_1 + \tfrac{1}{2}$$

as before. To avoid confusion, the density function of X_1 so obtained is usually subscripted as f_1. That is, in general,

$$f_1(x_1) = \int_{-\infty}^{\infty} f(x_1, x_2) \, dx_2$$

Likewise, of course,

$$f_2(x_2) = \int_{-\infty}^{\infty} f(x_1, x_2) \, dx_1$$

It is often convenient to call f_1 and f_2 *marginal* density functions to provide further distinction from the *joint* density function f.

As a further illustration of this notation, if we begin with, say, $f(x_1, x_2, x_3, x_4)$, then $f_{23}(x_2, x_3)$ is the marginal joint density of X_2 and X_3, which is obtained by

$$f_{23}(x_2, x_3) = \int_{-\infty}^{\infty} \int_{-\infty}^{\infty} f(x_1, x_2, x_3, x_4) \, dx_1 \, dx_4$$

Very often, we are given or wish to assume the value of one (or more) of the variables, and seek to determine the density of the other(s) *using this information*. We omit here the details and derivation, and simply define the conditional density function of X_1, given $X_2 = x_2$ by means of

$$f(x_1|x_2) = \frac{f(x_1, x_2)}{f_2(x_2)}, \qquad f_2(x_2) \neq 0$$

In Example 3.6 we had

$$f(x_1, x_2) = x_1 + x_2$$
$$f_2(x_2) = x_2 + \tfrac{1}{2}$$

and so

$$f(x_1|x_2) = \frac{2(x_1 + x_2)}{2x_2 + 1}, \qquad 0 \le x_1 \le x_2$$

We may likewise define conditional *expectation* as

$$E[g(X_1)|x_2] = \int_{-\infty}^{\infty} g(x_1) f(x_1|x_2) \, dx_1$$

For example, the conditional mean of X_1 is given by

$$E[X_1|x_2] = \int_0^1 \frac{2x_1(x_1 + x_2)}{2x_2 + 1} \, dx_1 = \frac{2 + 3x_2}{3 + 6x_2}$$

This, of course, differs from the unconditional mean $E[X_1] = \frac{7}{12}$. Note that

$$E[X_1|\tfrac{1}{4}] = \frac{2 + \tfrac{3}{4}}{3 + \tfrac{6}{4}} = \tfrac{11}{18}, \quad \text{for example.}$$

The determination of $E[X_1|x_2]$ provides us with a function of x_2; in statistics this function is referred to as the *regression* function; it forms the basis for much of the material to be found in Chapters 10 and 14.

Let us consider now a further example, beginning with

$$f(x_1, x_2) = 4x_1 x_2, \qquad 0 \le x_1 \le 1, \qquad 0 \le x_2 \le 1$$

The marginal density of X_1 is

$$f_1(x_1) = \int_0^1 4x_1 x_2 \, dx_2 = 2x_1$$

and, similarly, $f_2(x_2) = 2x_2$. We then have

$$f(x_1|x_2) = \frac{4x_1 x_2}{2x_2} = 2x_1 \equiv f_1(x_1)$$

and

$$f(x_2|x_1) = \frac{4x_1 x_2}{2x_1} = 2x_2 \equiv f_2(x_2)$$

Clearly, the conditional densities carry precisely the same information as the marginal density (in which the "other" variable was, in effect, ignored). In this situation it would appear that information about either of the variables in no way influences the distribution of the other: X_1 and X_2 are *statistically independent*.

Note that we may write

$$f(x_1, x_2) = f_1(x_1)f_2(x_2)$$

The relationships above are of the same sort as

$$P\{A|B\} = P\{A\}$$
$$P\{B|A\} = P\{B\}$$
$$P\{AB\} = P\{A\} P\{B\}$$

which we used to define independence in the preceding chapter. The only differences are that random variables have replaced events, and density functions have replaced probability functions.

The Example 3.7 illustrates many of the concepts we have just discussed.

Example 3.7. The corrosion resistance X_1 and gloss X_2 of a plastic coating are random variables with the joint probability density function (pdf)

$$f(x_1, x_2) = \frac{1}{100} \exp\left[-\left(\frac{x_1}{20} + \frac{x_2}{5}\right)\right], \qquad x_1 \geq 0, \qquad x_2 \geq 0$$

The marginal pdf of X is

$$f_1(x_1) = \int_0^\infty f(x_1, x_2)\,dx_2 = \frac{1}{20} e^{-x_1/20}$$

Likewise, $f_2(x_2) = \frac{1}{5} e^{-x_1/50}$. Since $f(x_1, x_2) = f_1(x_1)f_2(x_2)$, the random variables X_1 and X_2 are independent. Let us determine $E[X_1^k]$:

$$E[X_1^k] = \frac{1}{20} \int_0^\infty x_1^k e^{-x_1/20}\,dx_1 \tag{3.10}$$

In equation (3.10), set $u = x_1/20$. Then $x_1 = 20u$, $dx_1 = 20\,du$, and

$$E[X_1^k] = \frac{1}{20} \int_0^\infty 20^k u^k e^{-u} \cdot 20\,du$$

$$= 20^k \int_0^\infty u^k e^{-u} = \Gamma(k+1) \cdot 20^k \tag{3.11}$$

If k is an integer, equation (3.11) becomes

$$E[X_1^k] = k! 20^k$$

Setting $k = 1, 2, 3$, we obtain

(a) The mean: $\mu_1 = 1! 20^1 = 20$.
(b) The variance: $E[X_1^2] = 2! 20^2 = 800$
   and so $\sigma^2 = 800 - (\mu_1)^2 = 400$.
(c) The third central moment:

$$\mu_3 = E[X_1^3] - 3\mu_1 E[X_1^2] + 2\mu^3$$

$$= 3! 20^3 - 3 \cdot 20 \cdot 800 + 2 \cdot 20^3$$

$$= 16,000$$

From (b) and (c), $\gamma_1 = \mu_3/\sigma^3 = 16,000/8000 = 2$. The positive value indicates a long right tail, which is evident from the graph of $f_1(x_1)$ (Figure 3.3).

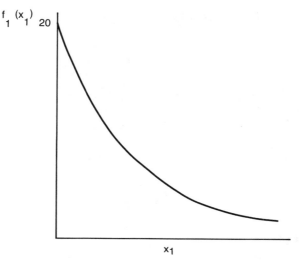

FIGURE 3.3. Marginal density function, $f_1(x_1)$.

With the basis of distribution theory now established, we proceed to an examination of a variety of random variables and their corresponding distributions which occur frequently in industrial applications.

For a discrete random variable the probability function $p(x_i)$ was defined as the probability associated with the value x_i. As discrete probability distributions can be plotted like histograms, so cumulative relative frequency distributions for continuous random variables lead to cumulative probability distributions. For, as the number of probable events approaches infinity, the probability distribution functions become increasingly "dense," until we can form a continuous probability distribution as the limit the discrete probability distribution approaches as the "divisioning" in the process approaches zero. Probability for a continuous random variable then may be regarded as the relative area under the curve defined by the probability density function (pdf), in other words, the first derivative of the cumulative distribution (cdf). Plots of the pdf and cdf for various values of their parameters and the discrete or continuous variables are given in Refs. 1, 2, and 5 for the numerous statistical distributions, along with comprehensive discussions of the statistical distributions.

The following sections present a summary of a number of the common distributions. Several of these are ones frequently encountered by the coatings chemist or engineer such as the familiar normal, binomial, or Poisson, while others are presented for completeness or as background information since some distributions are special cases of other distributions. A brief summary of several of the sampling distributions such as Student's t is also presented since these are sometimes not thought of in the ordinary sense as being part of distribution theory but only as values for testing purposes.

3.2. CONTINUOUS DISTRIBUTIONS

Normal Distribution

The normal (or Gaussian) distribution is one of the most well known and important probability functions. Unlike the binomial and the Poisson distributions, it is continuous and always symmetrical.

The normal distribution pdf is given by

$$f(x) = y = \frac{1}{\sigma\sqrt{2\pi}} e^{-(x-\mu)^2/2\sigma^2} \tag{3.12}$$

where μ is the mean and σ is the standard deviation. The normal cdf is

$$F(x) = \frac{1}{\sigma\sqrt{2\pi}} \int_{-\infty}^{x} e^{-(x-\mu)^2/2\sigma^2} \, dx \quad \text{for} \quad -\infty < x < \infty \tag{3.13}$$

The normal distribution has the following characteristics:

1. The area bounded by the distribution and the X axis is identically equal to 1.
2. The area bounded by the distribution and the X axis in the interval between $x = a$ and $x = b$ where $a < b$ is identically equal to the probability that X lies between a and b.

The properties of the normal distribution are the arithmetic mean μ, the variance σ^2, and the population standard deviation σ. The moment coefficient of skewness is 0, and the moment coefficient of kurtosis, a_4, is 3.

When expressed in terms of the standardized variable z,

$$z = \frac{x - \mu}{\sigma}$$

formula (3.12) becomes

$$y = \frac{1}{\sigma\sqrt{2\pi}} e^{-z^2/2} \tag{3.14}$$

This is the so-called "standard form" and is the equation when the mean is 0 and the variance is 1.

The normal distribution is completely defined by the two parameters, the mean and the standard deviation. The parameter μ is the location parameter of a normal distribution, while σ is a measure of its spread, scatter, or dispersion. Thus, a change in μ merely moves the curve right or left without changing its profile, while a change in σ widens or narrows the curve without changing the location of its center.

The normal distribution provides a model for numerous distributions experienced in all types of statistical work. Other distributions, such as the binomial, to be covered later, are closely related to the normal distribution. For example, if for a binomial random variable x, n and p are not zero, nor near zero, and n is large, the binomial distribution can be approximated by the normal distribution, since $(x - np)/npq$ is approximately normal with 0 mean and unit variance. The normal distribution as the limit of the binomial distribution is a reasonable approximation of the binomial distribution if both np and nq are greater than 5.

Log-Normal Distribution

The log-normal or logarithmic normal distribution has had a confusing existence because of the varied nomenclature used to refer to it in statistical literature (Ref. 2, Chapter 9). Early work on development of this distribution was sporadic and scattered. However, this situation has changed in recent years and the log-normal distribution is now well documented (3, 4, 5).

The log-normal cumulative distribution function is

$$F(x) = \Phi\left(\frac{\log x - \mu}{\sigma}\right) x > 0 \qquad (3.15)$$

where Φ is the standard normal distribution function, μ is the value of the logarithms of the random variable x, and σ is the standard deviation of the logarithms of x.

The probability density function is

$$y = \frac{1}{\sigma_g\sqrt{2\pi}} \exp\left[-\frac{(\ln x - \mu)^2}{2\sigma_g^2}\right] \quad \text{for} \quad -\infty < x < \infty \qquad (3.16)$$

where y is the frequency of the random variable x, μ is the mean of the logarithms of x, and σ_g is the logarithmic or geometric standard deviation.

Basically Equation (3.16) states that the logarithms of the observed variables are normally distributed. In other words, a skewed frequency plot of, say, particle diameter becomes similar in shape to a normal distribution when frequency is plotted against the logarithm of particle diameter. A probability plot of data such as from a particle size distribution can be used directly to obtain, from the 50% point, the median an estimate of the geometric mean x. This plot also serves to obtain an estimate of the geometric dispersion g which plays an equivalent role to σ for the normal distribution.

The log-normal distribution finds its main application in coatings in describing the particle size distributions of pigments and emulsions. It may also be used to determine the optical performance of a white pigment as a function of the particle size volume distribution. It also finds some use in chemical testing. Details of plotting, parameter estimation, and interpretation of plots of the log-normal distribution may be found in Ref. 2. Plots of the probability density function for the log-normal distribution vary from curves resembling the normal distribution to curves of varying degrees of right skew depending on the values of the mean μ and the standard deviation σ. As with the normal distribution, linear probability plots may be extrapolated to obtain estimates outside the bounds of the observed variable. Also confidence limits may be determined as well as an idea of the number of a given variable, for example, pigment particle size, below, above or between given sizes.

3.3. DISCRETE PROBABILITY DISTRIBUTIONS

Binomial Distribution

The binomial distribution is a natural outcome when events depend on a fixed probability of occurrence p, and when the number of trials is limited and independent. The binomial distribution provides a suitable model for many statistical distributions that occur in economics, nature, business, and so on. If the probability of an event occurring at any trial is p, and n trials are made, then the probability of exactly x successes in the n trials is

$$f(x) = \frac{n!}{x!(n-x)!} p^x q^{n-x} \qquad (3.17)$$

where

$$q = 1 - p$$

and

$$x! = x(x-1)(x-2)(x-3)\ldots 1$$

Here, p is the probability of success, that is, the probability that an event will happen in any single trial; q is the probability that it will *fail* to happen in any single trial, usually called the probability of failure and equal to $1 - p$; and $f(x)$ is the probability that the event will happen exactly x times in n trials. Here x is defined only as an integer as $x = 0, 1, 2, \ldots, n$. The "binomial" reflects the fact that as x takes on integer values from 0 through n, the corresponding probabilities are given by the terms in the binomial expansion

$$(p + q)^n = q^n + {}_nC_1 pq^{n-1} + {}_nC_2 p^2 q^{n-2} + \cdots + p^n \qquad (3.18)$$

The statistics of the binomial distribution are: mean $\mu = np$; variance $\sigma^2 = npq$; and standard deviation $\sigma = \sqrt{npq}$.

When $p = q$, this distribution is symmetrical and $a_3 = 0$. When n becomes large it approaches the normal distribution (provided neither p nor q are close to zero). On the other hand, if n is large and the probability p of occurrence of an event is close to zero, $q = (1 - p)$ is close to 1, and we refer to this as a "rare event." In these situations the binomial distribution can be closely approximated by the Poisson distribution by letting $\lambda = np$. Comparison of the mean, variance, skewness, and coefficient of excess when $\lambda = np$, $p = 0$, and $q \simeq 1$ shows that the binomial distribution properties (e.g., mean, variance) are approximately equal to those of the Poisson distribution.

The Bernoulli, multinomial, hypergeometric, and geometric distributions can be considered as generalizations or special cases of the binomial distribution. The *Bernoulli* distribution considers a single experiment with two possible outcomes, such as heads or tails with a coin toss. For the Bernoulli, $x_i = 1$ for the probability p of success, and $x_i = 0$ for $(1 - p)$. The mean is p, and the variance $p(1 - p)$.

The *multinomial* distribution considers cases in which there are several mutually exclusive classes. For example, of the possible outcomes of boy–girl, there are also subclasses of brown hair, black hair, and so on.

The *hypergeometric* distribution is concerned with sampling without replacement from a population with a specific composition, for example, the number of defective items in a sample drawn without replacement from a finite population. It considers the case of M objects of one kind, N objects of another kind, and K objects of kind M found in a drawing of N objects. The N objects are drawn from the population after each drawing. Meyer (Ref. 1, Chapter 23) discusses this distribution in detail. The hypergeometric distribution finds use in problems of a sampling nature, such as what percentage of voters will vote for a certain candidate. However, due to its inconvenience, alternatively approximations to it, such as the normal or Poisson, are used.

The *geometric* distribution is related to the Bernoulli in that it concerns the number of failures before the first success in a sequence of Bernoulli trials. This could be, for example, the number of tails before the first head in coin tossing.

Poisson Distribution

A probability distribution widely encountered in practical probability work is the Poisson distribution. It is particularly appropriate to statistical studies where an event can occur more than once, for example, the number of loose lids per hour in a large production batch of paint. The Poisson distribution is a good model for describing random phenomena where the probability of occurrence is small and constant. It finds frequent use in quality control and reliability testing. It is useful in probability problems where there is no distinct specification of sample size. Classically it serves as the model for cases such as the number of insurance claims per year, the number of flaws in similar pieces of material, and so on. A requirement for use of the Poisson distribution model is that the number of events occur independently and at a constant average rate.

This distribution is defined by the density function

$$f(x) = \frac{\lambda^x e^{-\lambda}}{x!} \tag{3.19}$$

where λ is a constant of the distribution, x is discrete values $0, 1, 2, 3$, and so on, and e is as before. When λ is small , the distribution is reversed J-shaped, but when λ is large, the curve resembles the normal distribution.

Some properties of the Poisson distribution are: mean $\mu = \lambda$; variance $\sigma^2 = \lambda$; and standard deviation $\sigma = \sqrt{\lambda}$.

The binomial distribution is approximated by the Poisson distribution where the number of trials n is large, and the probability of success p of a single trial is small. In other words, if in the binomial distribution, $\lambda = np$ and n approaches infinity, then to keep np finite ($= \lambda$), p or $(1 - p)$ must approach zero. Under these conditions, the properties of the binomial distribution listed earlier approach those of the Poisson distribution as a limit. In addition, the Poisson distribution is less skewed and approaches the normal distribution as the limit for large values (approaching infinity) of the mean. The Poisson can also be used to approximate the hypergeometric distribution.

Plots of the Poisson probability function are quite asymmetrical for small values of λ (e.g., 0.5), but become increasingly symmetrical (resembling the normal distribution) for larger values (e.g., $\lambda = 6.0$) of λ. An excellent discussion of the Poisson distribution may be found in Meyer (Ref. 1, Chapter 24).

3.4. SAMPLING DISTRIBUTIONS

We have established that a collection of means of samples taken from a population will be normally distributed, also, that these samples will have the same overall mean as the population they come from and a standard deviation, or standard error, equal to the population standard deviation divided by the square root of n. Other statistics can be calculated from samples. We now proceed to look at a few of these sampling statistics and the nature of several sampling distributions.

Student's *t* Distribution

Student's t distribution is given as

$$t = \frac{\bar{x} - \mu}{s/\sqrt{n}} \tag{3.20}$$

If we consider samples of size n selected from a normal distribution with mean μ, and if we compute t given the sample mean and sample standard deviation s, the sampling distribution for t can be obtained. This distribution is given by

$$Y = \frac{Y_0}{\left[1 + t^2/(n-1)\right]^{n/2}} = \frac{Y_0}{\left(1 + t^2/k\right)^{(k+1)/2}} \tag{3.21}$$

Here Y_0 is a constant depending on n and is such that the area under the t distribution is 1. The constant $k = (n-1)$ is the number of degrees of freedom. This distribution is called Student's t distribution. It is used for examining the means of samples when we do not know the population variance and must use the variance of the sample. Note that for large values of k or n ($n > 30$) the curves closely approximate the normal distribution:

$$Y = \frac{1}{\sqrt{2\pi}} e^{-t^2/2} \tag{3.22}$$

Details of the background of this critical distribution, its origin in 1908, and the personal aspects of the originator, W. S. Gosset, make very interesting reading and are to be found in Refs. 6, 7, 8, and 9.

We can define the 95 and 99% confidence intervals by using a table of t distributions. Specifically, if $-t_{0.975} + t_{0.975}$ are the values of t for which 2.5% of the area lies in each tail of the t distribution, then a 95% confidence level for t is

$$-t_{0.975} < \frac{\bar{X} - \mu}{s} \sqrt{n-1} < t_{0.975} \tag{3.23}$$

Clearly, then, μ is expected to lie in the interval

$$\bar{X} - t_{0.975}\left(\frac{s}{\sqrt{n-1}}\right) < \mu < \bar{X} + t_{0.975}\left(\frac{s}{\sqrt{n-1}}\right) \tag{3.24}$$

with 95% confidence.

In general, we can represent the confidence limits for population means by

$$\bar{X} \pm t_c \frac{s}{\sqrt{n-1}} \tag{3.25}$$

For this distribution, the chi-square random variable is defined by

$$\chi^2 = \frac{ns^2}{\sigma^2} = \frac{(X_1 - \bar{X})^2 + (X_2 - \bar{X})^2 + \cdots + (X_n - X^2)}{\sigma^2} \quad (3.26)$$

The chi-square distribution is defined by

$$f(x) = \frac{X^{(k/2)-1}e^{-x/2}}{[(k/2) - 1]!2^{k/2}} \qquad x > 0 \qquad (3.27)$$

Here $k = n - 1$ is the number of degrees of freedom. Plots of the chi-square pdf for $k \geq 4$ are right skewed, while the plot for $k = 2$ is a reverse J-shape. The mean is k and the variance, $2k$. The chi-square distribution is used for statistical tests on assumed normal distribution samples.

As was done with the normal, we can define 95% and 99% or other confidence limits and intervals for chi-square by using a table of the chi-square distribution. In this manner we can estimate, within specified confidence limits, the population standard deviation σ in terms of the sample standard deviation s. If $\chi^2_{0.025}$ and $\chi^2_{0.095}$ have the values of χ^2 for which 2.5% of the area lies in each tail of the distribution, the 95% confidence interval is

$$\chi^2_{0.025} < \frac{ns^2}{\sigma^2} < \chi^2_{0.975} \qquad (3.28)$$

We can see that σ is estimated to be in the interval

$$\frac{s\sqrt{n}}{\sqrt{\chi^2_{0.975}}} < \sigma < \frac{s\sqrt{n}}{\sqrt{\chi^2_{0.025}}} \qquad (3.29)$$

with 95% confidence.

The chi-square test is probably the most accepted test for determining whether there is significance between the observed and the expected value in sampling a population. It suffices to mention that there are better tests, and the reader desiring to dig deeper can consult the numerous references in standard tutorial statistical texts. An excellent discussion of the chi-square distribution may be found in Meyer (Ref. 1, Chapters 26 and 32).

The *F* Distribution

The F test, which was mentioned in Chapter 1, is, like Student's *t*, actually derived from a distribution function. The *F* test is a distribution used to decide if two samples have come from a population with the same variance. It is based on a ratio of variances as

$$F = \frac{s_1^2}{s_2^2} \tag{3.30}$$

It was named after R. A. Fisher, the eminent mathematician.

A random variable is said to have the F distribution when its probability density function is

$$f(F; n_1, n_2) = \frac{\Gamma[(n_1 + n_2)/2](n_1/n_2)^{n_1/2} F^{(n_1-2)/2}}{\Gamma(n_1/2)\Gamma(n_2/2)(1 + n_1 F/n_2)^{(n_1+n_2)/2}} \tag{3.31}$$

where n_1 and n_2 are the degrees of freedom, respectively, of two independent variables and Γ is the gamma function described later on. The F table gives percentile values of the chances of the observed sample variances as a ratio of s_1^2/s_2^2 being $\leq F$.

The mean of the F distribution is given by

$$\mu = \frac{n_2}{n_2 - 2} \qquad (n_2 > 2) \tag{3.32}$$

and the variance by

$$\sigma^2 = \frac{2n_2^2(n_1 + n_2 - 2)}{n_1(n_2 - 4)(n_2 - 2)^2} \qquad (k_2 > 4) \tag{3.33}$$

The F distribution is theoretically related to the chi-square distribution. For example, for the density function of F, Student's t when squared is the F ratio, with $n_1 = 1$ and n_2 the number of degrees of freedom associated with Student's t. The chi-square is equal to Fn_1, when $n_2 \rightarrow \infty$, with n_1 degrees of freedom. The common F table represents a table of the cumulative F distribution, tabulating against n_1, n_2 the values of $F(n_1, n_2)$. An example of this use has been given in Chapter 1.

3.5. MISCELLANEOUS DISTRIBUTIONS

The Exponential Distribution

The exponential distribution is a special case of the Weibull distribution (see Chapter 5) and the gamma distribution to be covered later. It is also referred to as the "negative exponential" distribution. In contrast to the

Poisson distribution, which considers the probability of x occurrences in a period of time if the occurrences occur at random, the exponential distribution considers the intervals of time between successive random happenings. In other words, the exponential distribution is the model for the time for a single outcome to occur if events occur independently at a constant average rate. It finds wide use in the statistical analysis of random failure of a component, as contrasted to wearout failure. The exponential distribution is given by the Weibull distribution, when the Weibull shape parameter β equals one.

The exponential probability density function is given by

$$f(x) = \lambda e^{-\lambda x} \quad \text{for} \quad x \geq 0, \qquad \lambda > 0 \tag{3.34}$$

The cumulative distribution function is

$$F(x) = 1 - e^{-\lambda x} \qquad x \geq 0 \tag{3.35}$$

The mean is given by $\mu = 1/\lambda$ and the standard deviation by $\sigma^2 = 1/\lambda^2$. Occasionally λ is taken as $1/\theta$ giving for the distribution function, for example, $F(x) = 1 - e^{-x/\theta}$.

Extreme Value Distributions

Extreme value distributions are divided into three classes: type I, the Gumbel or double exponential distributions; type II, the Cauchy or log-extreme value distribution; and type III, the Weibull distribution. Extreme value distributions are all skewed distributions. Type I can also be considered to include as special cases the normal, log-normal, exponential, chi-square, logistic, and range distributions.

Extreme value distributions occur widely in such cases as meteorological and geophysical phenomena, the strength properties of materials, the film strength of polymers, life testing, sales analysis, and industrial engineering, to name only a few. Additional details may be found in Refs. 2, 10, and 11.

The cumulative distribution function and the probability density function for the extreme value distribution are given by

$$\Phi(x) = \int_{-\infty}^{y} e^{-e^{-y}} \tag{3.36}$$

and

$$\phi(x) = \alpha e^{-y-e^{-y}} \tag{3.37}$$

with $-\infty < y < +\infty$ and $y = \alpha(x - \mu_0)$. Here y is the normalized unit variable, x is the value of the random variable, μ_0 is the mode, and $1/\alpha$ is the slope. As with other distributions, probability statements, for example, the area under a cumulative probability curve between the mode and $\pm n\sigma$, and confidence intervals, may be determined.

The logarithmic extreme value distribution (type II) displays a similar relationship to the extreme value distribution as does the log-normal to the normal. In other words, the logarithms of a log-extreme value distribution are extreme value distributed. Log-extreme value distributions are highly skewed. A common use for the log-extreme value distribution is in the statistics of particle size of powders. Due to this use, it is sometimes referred to as the Rosin–Rammler distribution. The cumulative distribution function and the density function are given by

$$\Pi(x) = \int_0^x \exp\left[-(\mu_0/x)^k\right] \qquad 0 \le x < \infty \tag{3.38}$$

and

$$\pi(x) = -k\left(\frac{\mu_0}{x}\right)^{k-1} \exp\left[-(\mu_0/x)^k\right] \tag{3.39}$$

Probability statements, confidence intervals, and so on, are applied as usual.

The Weibull distribution, discussed in more detail in Chapter 5, is also known as a type III extreme value distribution. It is widely applicable to such diverse applications as pigment particle size, the size distribution of aerosol particles, corrosion testing, sales analysis, film strength, and general industrial engineering. Its cumulative distribution function is

$$F(x) = 1 - \exp\left[-\frac{(x-\gamma)^\beta}{\alpha}\right] \tag{3.40}$$

and the probability density function is

$$f(x) = \frac{\beta}{\alpha}(x-\gamma)^{\beta-1}\exp\left[-\frac{(x-\gamma)^\beta}{\alpha}\right] \tag{3.41}$$

The scale parameter α may also be expressed as η^β. The mean and variance of the Weibull distribution are, respectively,

$$\mu = \eta\Gamma\left(\frac{1}{\beta} + 1\right) \tag{3.42}$$

and

$$\sigma^2 = \eta^2\left\{\Gamma\left(\frac{2}{\beta} + 1\right) - \left[\Gamma\left(\frac{1}{\beta} + 1\right)\right]^2\right\} \tag{3.43}$$

Gamma Distribution

The gamma distribution as described in Refs. 1, 5, and 11, is a generalized factorial function leading to a family of essentially extreme value functions for variables bounded at one side, that is, $0 \leq x < \infty$. For such cases, x is then a threshold value somewhat similar to the location parameter γ of the Weibull distribution. It considers problems such as the time required for a specific minimum number of events to occur. It is a basic distribution of statistics. Special cases of it are, for example, the exponential distribution and sampling distributions such as the chi-square and Student's t distribution. Some illustrative applications are the number of machines needing repair, the time between stock reorder points, and the time between consecutive maintenance operations.

The CDF, after Ref. 5, is

$$F(x; n, \lambda) = \frac{\lambda^n}{\Gamma(n)} \int_0^x t^{(n-1)} e^{-\lambda t} \, dt \qquad (3.44)$$

and the pdf is

$$f(x; n, \lambda) = \frac{\lambda^n}{\Gamma(n)} x^{(n-1)} e^{-\lambda x} \qquad (3.45)$$

where $\Gamma(n)$ is the familiar gamma function used, for example, in determining the Weibull standard deviation. The values of the other parameters are not simply described, and can best be studied by referring to Ref. 2, Chapter 17 for more complete details. As with other distributions, probability plots and confidence intervals can be derived.

Beta Distribution

The beta distribution is a useful model for variants whose values are limited to a finite interval. Like the gamma distribution, it is a basic distribution of statistics.

If, for the gamma function, we take α as n and λ as $1/\beta$, the beta density function (Ref. 5, 11) is given by

$$f(x) = \frac{x^{\alpha-1}(1-x)^{\beta-1}}{B(\alpha, \beta)} \qquad (3.46)$$

where α, $B > 0$ and $B(\alpha, \beta)$ is the beta function. The beta distribution can also be defined, in view of its relationship to the gamma function, as

$$f(x) = \frac{\Gamma(\alpha + \beta)}{\Gamma(\alpha)\Gamma(\beta)} x^{\alpha-1}(1-x)^{\beta-1} \qquad 0 < x < 1 \qquad (3.47)$$

The beta distribution has perhaps the widest variety of distributional shapes of all distributions. Due to this, it is used to represent a large number of physical variables. Several examples (Ref. 5, Chapter 3) of many of its uses are the daily proportion of defective units on a production line and the estimated time to complete a project phase in the well-known PERT scheduling.

REFERENCES

1. Meyer, S. L., *Data Analysis for Scientists and Engineers*, Wiley, New York, 1975.
2. King, J. R., *Probability Charts for Decision Making*, Industrial Press, New York, 1971.
3. Hald, A., *Statistical Theory with Engineering Applications*, Wiley, New York, 1952.
4. Aitchison, J., and Brown, J. A. C., *The Lognormal Distribution*, Cambridge University Press, New York, 1957.
5. Hahn, G. J., and Shapiro, S. S., *Statistical Models in Engineering*, Wiley, New York, 1962.
6. Mack, C., *Essentials of Statistics for Scientists and Technologists*, Plenum Press, New York, 1975.
7. Langley, R., *Practical Statistics—Simply Explained*, Dover, New York, 1971, p. 160.
8. Student, *Biometrika*, Vol. 6, 1908, pp. 1–25.
9. Student, *Biometrika*, Vol. 6, 1908, p. 1; *Collected Papers* (*2*) (E. S. Pearson and J. Wishart, Eds.), Biometrika Office, University College, London, 1947.
10. Gumbel, E. J., *Statistics of Extremes*, Columbia University Press, New York, 1958.
11. Myers, B. L., and Enrick, N. L., *Statistical Functions*, Kent State University Press, Kent, OH, 1970.
12. Young, H. D., *Statistical Treatment of Experimental Data*, McGraw-Hill, New York, 1962.

4

SAMPLE SIZE

GARY E. MEEK

College of Business Administration
University of Akron
Akron, Ohio

4.1. INTRODUCTION

A statement that is heard commonly whenever statistical analyses are reported is "The sample is not large enough." That statement is often misleading since the obvious answer is "Not large enough for what?." The sample size required for any statistical procedure is contingent upon what is to be inferred from the data and the confidence level to be associated with that inference. In some cases a sample size of 4 may suffice while in others 40,000 or more samples may be required to attain the desired accuracy and confidence level. In this chapter we shall develop formulas for determining the sample size required to estimate a parameter to within a specified deviation for a given probability level.

In addition, we will derive the formula for finding the sample size required for a test to have a given power (see Chapter 7) against a specified alternative. All of our formulas will assume that either the standard deviation is known or an upper bound can be determined for it. In many situations, the assumption of a known σ_x will not be true. For one-sample cases involving unknown standard deviations and two-sample tests, we present tables that can be used in particular situations.

4.2. TERMINOLOGY

Before we commence our discussion of sample size there are some general terms that require defining. The first of these is the concept of a population of items. By population we mean:

Definition 4.1. A population is the set of all items of interest in a given situation.

For example, a population might be the set of all electron tubes produced by a specific supplier or it might be the set of all possible viscosity readings that could be associated with a given type of paint produced by a paint manufacturer.

In general, it will be impossible to inspect every member of a population for a specific attribute or to obtain a measurement on every member. Also, the distributional characteristics of the population such as the population mean, μ, and standard deviation, σ, are usually unknown and must be estimated. Due to the impossibility of inspecting every item in the population these estimates must be based on the information contained in a subset of the population. That is, on a sample of items:

Definition 4.2. A sample of items is any subset of the population of interest.

There are various methods available for selecting a sample from a population. In order to make inferences (probability statements based on the sample results) statistics usually requires what is known as a simple random sample. For infinite populations this means that every item in the population has the same chance of being selected each time. If the population is finite, then a sample is said to be a simple random sample if all possible samples of the same size are equally likely to occur. Generally, a simple random sample will be representative of the population as a whole, but not always. Simple random sampling always admits the possibility of obtaining a set of values that come from one end of the population, that is, from a set of extremes. But, with simple random samples we are able to calculate the probability of this happening.

The probability that a random sample comes from a set of extreme values in the population is called the significance level and is denoted by α. The probability that it is representative of the population is called the confidence level and is denoted by $1 - \alpha$. For a thorough discussion of these terms see Chapters 6 and 7. Our development of formulas for determining sample sizes assumes that $1 - \alpha$ (or α) has been specified in advance and that a maximum distance (error) from the value being estimated is given.

4.3. THE DETERMINATION OF *n* FOR ESTIMATING μ (OR SAMPLE SIZE)

Before developing the equation for calculating *n* in this situation we make some general comments about the effect of varying the sample size. Recall that the standard error of the sample mean, namely $\sigma_{\bar{x}}$, is given by $\sigma_x / \sqrt{n}$ if the population is infinite and by $(\sigma_x / \sqrt{n})\sqrt{(N - n/N - 1)}$ if the population is finite. This is nothing more than the standard deviation of the sample mean and as such represents the maximum distance one expects to be from μ approximately 68% of the time. Note that for either an infinite or a finite population the only variable in the formula is the sample size, n. In either case increasing the value of n decreases $\sigma_{\bar{x}}$. Thus, the larger the sample size the closer we expect our estimate to be to the true value. Curves representing the distributions of $\bar{X}$'s for various sample sizes are given in Figure 4.1. Note that as n increases the corresponding curves become more and more peaked with less and less area in the tails. In the limiting case, if it were possible to take an infinite sample, we would have nearly perfect information and a value for $\bar{X}$ that would be identical to μ for all practical purposes.

To determine our sample size, recall that the central limit theorem states that the distribution of the sample mean, our estimate of μ, can be approximated with a normal distribution even though the original population may be non-normal. The approximation improves as the sample size increases. We know also that for normal distributions $(1 - \alpha) \times 100\%$ of the values for the variable of interest will be within $z_{\alpha/2}$ standard devia-

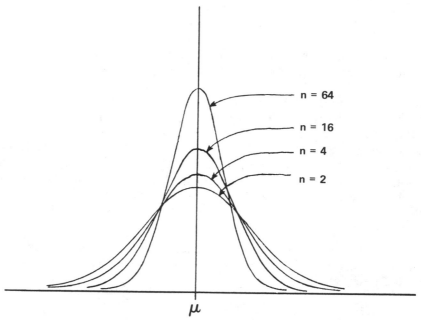

FIGURE 4.1. Illustrations of the effect of sample size on the distribution of the sample mean.

tions of the true mean. The number $z_{\alpha/2}$ is the value from the standard normal table such that $P(-z_{\alpha/2} \le Z \le +z_{\alpha/2}) = 1 - \alpha$. Therefore, we expect that $\overline{X}$ will be within a distance,

$$\pm z_{\alpha/2} \sigma_{\overline{x}}$$

of μ_x for $(1-\alpha) \times 100\%$ of our samples. Note that, correspondingly, we also expect it to be farther away from the above value $\alpha \times 100\%$ of the time.

Utilizing this fact we can specify a maximum error that we would like to attain with a probability of $1 - \alpha$ and then solve for the sample size required to attain it. That is, we specify that

$$\text{error} = z_{\alpha/2} \sigma_{\overline{x}} = z_{\alpha/2} \left(\sigma_x / \sqrt{n} \right) \tag{4.1}$$

where it is assumed that the population is quite large with respect to the sample size (generally at least 20 times as large). Ignoring the finite population correction factor $N - n/N - 1$ completely will simply result in a sample size that may actually be larger than necessary. In equation (4.1), the error is specified, $z_{\alpha/2}$ is found from the normal table, and σ_x is a known value. Thus, we have one equation in one unknown. Solving for the required sample size, n, gives

$$n = \left(\frac{z_{\alpha/2} \sigma_x}{\text{error}} \right)^2 \tag{4.2}$$

If σ_x is unknown then it may be replaced by a prior estimate of it or, to ensure the minimum sample size, an upper bound for σ.

Example 4.1. Brand X Paint Co. has been having problems with coagulation of one of its paints. Numerous complaints have been received from customers that the paint in question solidifies sometime in the first 6 to 8 weeks after being made. Brand X would like to determine the actual time to coagulation to within 1 week with a probability of .98. By comparing complaints and purchase dates the company is certain that the standard deviation of the time to coagulation is no more than 4 weeks. How many batches should be tested to meet Brand X's requirements for the estimate of the time?

Solution. We have that $1 - \alpha = .98$ which implies that the corresponding number of standard deviations is $z_{.01} = 2.326$. To estimate the time to within 1 week implies that

$$\text{error} = 1.00$$

The standard deviation is unknown but a bound of 4 weeks is given.

Therefore, using equation (4.2), we have

$$n = \left[\frac{2.326(4)}{1} \right]^2 = (9.304)^2 = 86.564$$

or 87 batches.

Note that even if the value in example (4.1) had been 86.26 we would have rounded upward to 87. Standard procedure in the determination of sample sizes is to round upward since we generally are specifying the probability to be at least $1 - \alpha$. If we were to round downward then the actual probability would be slightly less than $1 - \alpha$.

4.4. AN EASY ESTIMATE FOR σ

The formula that is generally used to estimate the population standard deviation from a simple random sample is

$$s = \sqrt{\frac{\sum\limits_{i=1}^{n} (x_i - \bar{x})^2}{n - 1}}$$

Solutions of this equation can be quite time consuming. If time is of the essence a fast approximation of this value can be obtained by considering the sample range; that is, the difference between the largest and smallest values in the sample. We define this to be

$$R = \text{largest observation} - \text{smallest observation} \qquad (4.3)$$

For a normal distribution we know that approximately 95% of all values lie within two standard deviations of the mean, 99.74% are within three standard deviations of μ, and so on. Therefore, if it can be assumed that the population of interest can be reasonably approximated by a normal curve then σ can be estimated by R/c_n, where c_n is a constant related to the sample size. Values for the divisor, c_n, are given in Table 4.1.

TABLE 4.1. Values of c_n for Using R/c_n as an Estimate of σ (Adapted from Ref. 1.)

Sample Size (n)	Divisor for $R(c_n)$
1–10	$\sqrt{n}$
11–32	4
33–100	5
101–500	6
≥ 501	6.5

4.5. SAMPLE SIZE REQUIRED TO ESTIMATE A PROPORTION

In many situations it will be of interest to attempt to estimate the proportion of items in the population that have a particular attribute (such as being defective). If we desire our estimate to be within a specified distance of the true proportion, p, with a high probability, $1 - \alpha$, a large sample will generally be required. Thus, we restrict our considerations to large sample properties of binomial distributions; namely the normal approximation.

True Proportion Completely Unknown

As in Section 4.3 we will use error to denote the maximum difference of our estimate from p with a confidence level of $1 - \alpha$. We note that to use equation (4.2) (our estimate for p is given by X/n = no. of successes/no. of observations and is simply a form of a sample mean) requires a value for σ_x. For a binomial distribution the standard deviation of a single observation is given by

$$\sigma_x = \sqrt{p(1 - p)}$$

but p is the unknown value that we are trying to estimate.

The fact that a value for p is required in advance poses no major difficulty since, if we can determine the maximum possible value for σ_x, then the resulting sample size will be the maximum required to estimate p with the given precision. The actual error resulting from this sample size will generally be less than the specified value. To determine the maximum

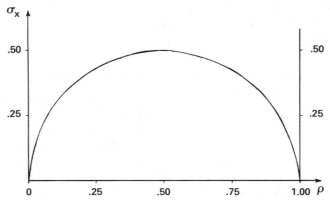

FIGURE 4.2. Graph of $\sigma_x = p(1 - p)$ for a binomial random variable, X.

possible value of σ_x in this case we could use the calculus but instead we merely refer to the graph of σ_x versus p as given in Figure 4.2. Note that the curve appears to peak (i.e., reach its maximum value) at $p = .5$. It can be proven that this is actually the case. Thus we have that for any binomial distribution the standard deviation of a single observation is always less than or equal to $\sqrt{.5(.5)} = .5$. That is,

$$\sigma_x = \sqrt{p(1 - p)} \le \sqrt{.5(.5)} = .5$$

Therefore the sample size required to estimate p within a given error with a probability of at least $1 - \alpha$ is

$$n = \left(\frac{z_{\alpha/2}}{\text{error}}\right)^2 p(1 - p) \le \left(\frac{z_{\alpha/2}}{\text{error}}\right)^2 (.5)(.5)$$

or

$$n \le \frac{1}{4}\left(\frac{z_{\alpha/2}}{\text{error}}\right)^2 \tag{4.4}$$

That is, if we take a sample of the maximum size as indicated by equation (4.4) then the probability of being within the prescribed distance of the true value of p is at least $1 - \alpha$. Table 4.2 gives maximum sample sizes required for confidence levels of .90, .95, and .99 with specified errors or distances from p, in either direction, of .01 to .10 in increments of .01.

TABLE 4.2.   Maximum Sample Sizes Required to Estimate P Within $\pm$ Error with Probability $\ge 1 - \alpha$

	Confidence Level $= 1 - \alpha$				
Error Bound	.80	.90	.95	.98	.99
.01	4,109	6,766	9,604	13,526	16,590
.02	1,028	1,692	2,401	3,382	4,148
.03	457	752	1,068	1,503	1,844
.04	257	423	601	846	1,037
.05	165	271	385	542	664
.06	115	188	267	376	461
.07	84	139	196	277	339
.08	65	106	151	212	260
.09	51	84	119	167	205
.10	42	68	97	136	166

Example 4.2. In a recent production run of one of their paints Brand X Paint Company found, after the run was completed, that the wrong labels were placed in the labeling machine for part of the run. Thus some gallons were labeled as white when they should have been labeled brown. Brand X does not know how many of the 100,000 gallons are mislabeled and would like to estimate the true proportion to within .06 with a probability of .95. If they have no idea as to the actual proportion, what is the maximum sample size required to satisfy the conditions?

Solution. The stated conditions are that the error is to be less than or equal to .06 and the confidence level is to be at least .95. Since there is no indication of the possible proportion of mislabeled cans we must use the maximum possible value of σ_x, that is, .5. For $1 - \alpha = .95$, $z_{\alpha/2} = 1.96$ and we have, using equation (4.4), that

$$n \le \frac{1}{4}\left(\frac{1.96}{.06}\right)^2 = 266.78$$

or

$$n \le 267 \text{ cans}$$

Note that using $\sigma = .5$ in Example 4.2 assumes the maximum possible sample size. If p is known to be bounded away from .5 then we could use a smaller sample and still satisfy our requirements.

A Bound on p is Available

By a bound on p we mean a value (other than .5) such that we are certain that p is either less than the value or greater than the value. For example, in trying to estimate the proportion of defective items produced by a process we may be 100% confident that the proportion of inoculations that are successful using a particular serum is not less than .75. In cases where a bound can be specified the sample size (and hence the sampling cost) will be reduced by using the corresponding bound in calculating σ_x rather than by using .5. Letting p_B denote the bound then the maximum standard deviation becomes

$$\max \sigma_x = \sqrt{p_B(1 - p_B)} \tag{4.5}$$

Substituting σ_x from equation (4.5) for $\frac{1}{4}$ in equation (4.4) gives a required sample size of

$$n = p_B(1 - p_B)\left(\frac{z_{\alpha/2}}{\text{error}}\right)^2 \tag{4.6}$$

To illustrate the effect that the bound has on n we revisit Example 4.2

Example 4.3. Suppose that Brand X Paint Company is certain that no more than 25,000 of the 100,000 cans are mislabeled. They still desire their estimate to be within .06 of the actual value with a probability of at least .95. What sample size is required in light of the additional information?

Solution. The only change occurring in this situation is in the value used for σ_x. If no more than 25,000 gal are mislabeled then $p \le .25$ or $p_B = .25$. Thus, the maximum standard deviation becomes

$$\max \sigma_x = \sqrt{.25(.75)}$$

Correspondingly,

$$n = .25(.75)\left(\frac{1.96}{.06}\right)^2$$
$$= 200.08 \quad \text{or} \quad 201$$

The information that $p \le .25$ allows us to reduce our sample size from 267 to 201 cans or a 24.7% reduction while maintaining the same precision and reliability.

By examining the formulas for determining n and/or inspecting Table 4.2 you can see that the sample size is influenced by three basic factors: (1) the value used for σ; (2) the confidence level chosen; (3) the specified maximum error. The sample increases with increases in either σ or the confidence level (or both) and decreases with an increase in the allowable error. This is true regardless of the parameter being estimated.

4.6. SAMPLE SIZE REQUIRED FOR TESTING μ WITH BOTH α, β SPECIFIED

The terminology of this section is that of Chapter 6. In our discussion the alternative hypothesis, H_1, will correspond to the conclusion we would prefer to reach and the null hypothesis, H_0, will normally be the opposite. H_0 is always assumed to be true until there is sufficient evidence to the contrary. The measure of what constitutes sufficient evidence is the significance level, α, with the evidence being sufficient to reject H_0 if its probability of occurrence under H_0 is less than α.

Known Variance (σ^2)

In all hypothesis testing situations there are two ways in which one can make an error. The first is to reject H_0 when it is actually true and is called a type I error. The probability of a type I error is controlled through the specification of α. The other way of making a mistake is by failing to reject H_0 when it is false. This is called committing a type II error and its probability of occurrence is denoted by β. It is more difficult to control the probability of a type II error because it is a function of the alternative values under consideration. As the alternative value changes so does the value of β. If a specific value of the alternative is of interest then a maximum value for β can be stated and the sample size necessary to ensure that neither α nor β is exceeded can be determined. Before developing this formula we consider the following example to illustrate the interrelationships between α, β, n, and μ_1 for a specified μ_0.

Example 4.4. Suppose that we desire to test the hypothesis, $H_0: \mu \leq 25$ against the alternative $H_1: \mu > 25$, and are interested in the specific values $26, 27, 28, 29$ under the alternative, H_1. (a) Assume that α is specified at .05, σ is known to be 4, and the sample size to be selected is 16, then determine β for each of the alternative values. (b) Assume $\alpha = .05$, $\sigma = 4$, $\mu_1 = 27$, and determine β for sample sizes of 4, 16, 32, and 64. (c) Assume $\sigma = 4$, $\mu_1 = 27$, $n = 16$, and determine β for α levels of .20, .10, .05, and .01.

Solution. (a) We have that $\sigma_{\bar{x}} = 4/\sqrt{16} = 1.0$ and that the critical or deciding value for $\overline{X}$ is $\mu_0 + z_\alpha \sigma_{\bar{x}}$, that is,

$$\mu_0 = z_\alpha \sigma_{\bar{x}} = 25 + 1.645(1) = 26.645$$

Thus,

$$\beta_{\mu 1} = P(\text{not rejecting } H_0 | H_1 \text{ is true})$$
$$= P(X < 26.645 | \mu_1)$$
$$= P(Z < (26.645 - \mu_1)/\sigma_{\bar{x}})$$

For each of the alternative values $\sigma_{\bar{x}} = 1.0$. At $\mu_1 = 26$ we have

$$\beta_{26} = P\left[Z < \frac{(26.645 - 26)}{1}\right] = P(Z < +.645)$$
$$= .7406 \text{ (from the normal table)}$$

That is, there is a 74% chance that we will fail to reject $H_0: \mu \leq 25$ when μ is actually 26. Correspondingly, for $\mu_1 = 27$ we have

$$\beta_{27} = P\left[Z < \frac{(26.645 - 27)}{1}\right]$$
$$= P(Z < -.355) = .3613$$

For the other two alternatives we have

$$\beta_{28} = P(Z < -1.355) = .0877$$

and

$$\beta_{29} = P(Z < -2.355) = .0092$$

As you can see, with everything else held constant β decreases as the specific alternative gets farther and farther from μ_0. Thus the chance of making a type II error decreases as the distance between μ_1 and μ_0 increases. This is illustrated in Figure 4.3.

(b) In this situation μ_1 is held constant and n is permitted to vary. The calculation of β is similar to part (a) with $\mu_1 = 27$ but $\sigma_{\bar{x}}$ changes with the sample size. The values of $\sigma_{\bar{x}}$ for $n = 4, 16, 32,$ and 64 are $2, 1, .707,$ and $.5$, respectively. Thus, for $n = 4$ we have the critical value for $\bar{X}$ as

$$\mu_0 + \frac{z_\alpha \sigma_x}{\sqrt{4}} = 25 + 1.645(2) = 28.29$$

Hence, for $n = 4$,

$$\beta_{27} = P(\bar{X} < 28.29 | \mu = 27)$$
$$= P\left(Z < \frac{(28.29 - 27)}{2}\right)$$
$$= P(Z < .645) = .7406$$

For $n = 16$ we had $\beta_{27} = .3613$ in part (a). In the remaining two cases we have $n = 32$ implies $\mu_0 + z_\alpha \sigma_x / \sqrt{32} = 26.163$ with $\beta = .1182$, and $n = 64$ implies $\mu_0 + z_\alpha \sigma_x / \sqrt{64} = 25.8225$ with $\beta_{27} = .0092$. These results are illustrated in Figure 4.4. We see that for a fixed μ_1 the chance of a type II error decreases as n increases with α and σ_x held constant.

(c) In this case $\sigma_{\bar{x}} = 1$, $\mu_1 = 27$, $\mu_0 = 25$, and the critical value of $\bar{X}$ varies as α varies. For α's of $.2, .1, .05,$ and $.01$ the respective critical values are:

$$\alpha = .20 \qquad \mu_0 + z_{.20}\sigma_{\bar{x}} = 25.842$$
$$\alpha = .10 \qquad \mu_0 + z_{.10}\sigma_{\bar{x}} = 26.282$$
$$\alpha = .05 \qquad \mu_0 + z_{.05}\sigma_{\bar{x}} = 26.645$$
$$\alpha = .01 \qquad \mu_0 + z_{.01}\sigma_{\bar{x}} = 27.326$$

Thus, the corresponding value for β_{27} is for $\alpha = .2$,

$$\beta_{27} = P\left[Z < \frac{(25.842 - 27)}{1}\right]$$
$$= P(Z < -1.158) = .1235$$

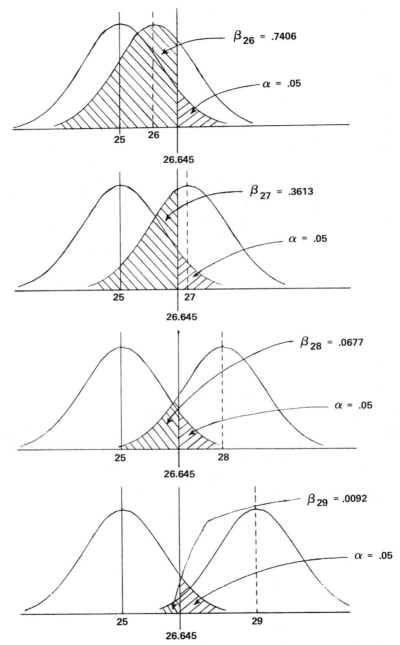

FIGURE 4.3. Illustration of the effect on β as the distance between μ_1 and μ_0 increases.

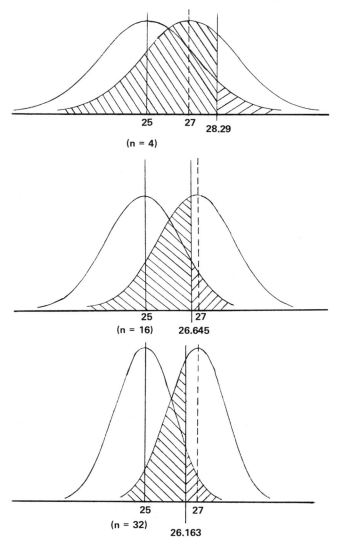

FIGURE 4.4.   Illustration of the effect on β as the sample size increases. (*continued on page 100*)

For $\alpha = .1$,

$$\beta_{27} = P(Z < .718) = .2365$$

For $\alpha = .05$, we had

$$\beta_{27} = .3613$$

and for $\alpha = .01$,

$$\beta_{27} = P(Z < +.326) = .6279$$

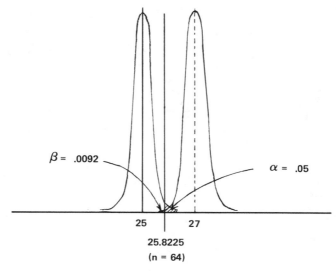

$\beta = .0092$

$\alpha = .05$

25   27

25.8225

(n = 64)

FIGURE 4.4. (*continued*).

These are illustrated in Figure 4.5 where it is easily seen that β decreases as α increases and vice versa when all else is held constant.

To determine the necessary sample size when maximum values are specified for both α and β we introduce the following notation. Let $\bar{X}_\alpha$ be the critical value for the sample mean under H_0 and $\bar{X}_\beta$ be the critical value for the sample mean when the specific alternative, μ_1, is true. For the hypothesis set,

$$H_0 : \mu \leq \mu_0 \qquad H_1 : \mu > \mu_0$$

$\bar{X}_\alpha$ and $\bar{X}_\beta$ are given by

$$\bar{X}_\alpha = \mu_0 + z_\alpha \left(\frac{\sigma_x}{\sqrt{n}} \right) \tag{4.7}$$

and

$$\bar{X}_\beta = \mu_1 - z_\beta \left(\frac{\sigma_x}{\sqrt{n}} \right) \tag{4.8}$$

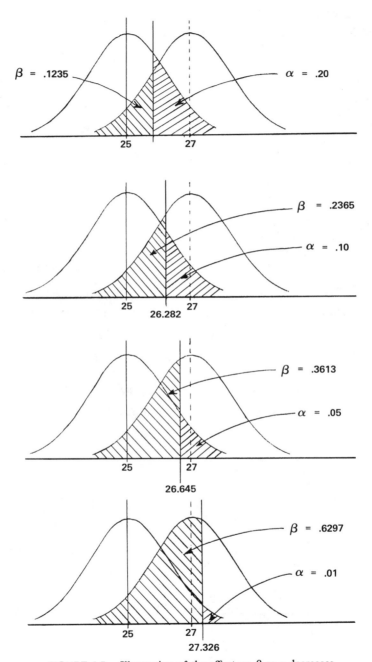

FIGURE 4.5. Illustration of the effect on β as α decreases.

For specified α, β, μ_1, and σ these two values must be identical; that is, $\overline{X}_\alpha$ must equal $\overline{X}_\beta$ (see Figure 4.6). Equating equations (4.7) and (4.8) gives

$$\mu_0 + z_\alpha\left(\frac{\sigma_x}{\sqrt{n}}\right) = \mu_1 - z_\beta\left(\frac{\sigma_x}{\sqrt{n}}\right)$$

or

$$\left(\frac{\sigma_x}{\sqrt{n}}\right)(z_\alpha + z_\beta) = \mu_1 - \mu_0$$

and

$$n\left[\frac{(z_\alpha + z_\beta)\sigma}{(\mu_1 - \mu_0)}\right]^2 \qquad (4.9)$$

Equation (4.9) holds for either of the possible one-sided alternative hypotheses; that is, for either $H_1: \mu > \mu_0$ or $H_1: \mu < \mu_0$. The values to be used for z_α and z_β will always be the positive values.

If the alternative hypothesis is two-sided of the form, $H_1: \mu \neq \mu_0$; that is, it is of interest to detect a shift in either direction with equal probability, then for a specific μ_1 the sample size required for a given α and β is

$$n = \left[\frac{(z_{\alpha/2} + z_\beta)\sigma}{(\mu_1 - \mu_0)}\right]^2 \qquad (4.10)$$

Note that the only difference between equations (4.9) and (4.10) is the use of $z_{\alpha/2}$ rather than z_α.

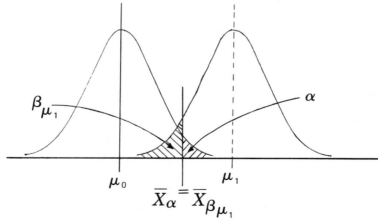

FIGURE 4.6. Illustration of $\overline{X}_\alpha$ and $\overline{X}_\beta$ for testing $H_0: \mu \leq \mu_0$ against $H_1: \mu > \mu_0$ with μ_1 specified.

Example 4.5. The drying time for Brand X's standard oil base interior paint averages 12 hours and has a standard deviation of 2 hours. The laboratory has been experimenting with an additive to decrease the average drying time. Management believes the use of the additive is justified if using the additive will decrease the drying time by at least 1.5 hours. If α is set at .01 and it is desired to detect the desired decrease with a probability of .95 or higher, how large a sample will be required? (Assume the standard deviation of drying time is unchanged by the additive.)

Solution. We have $\mu_0 = 12$, $\mu_1 = 12 - 1.5 = 10.5$, and $\sigma = 2$. For $\sigma = .01$ and $\beta = .05$, z_α and z_β are 2.326 and 1.645, respectively. Thus, the required sample size is

$$
n = \left[\frac{(2.326 + 1.645)2}{(10.5 - 12)} \right]^2
$$

$$
= (5.295)^2 = 28.03 \quad \text{or} \quad 29
$$

Note that in Example 4.5 we again rounded upward for our final sample size to ensure that the true values for α and β are no larger than specified. To see the effect of the rounding we calculate the values of α and β. For the values indicated for μ_0 and μ_1 the critical value for $\overline{X}$ is

$$
\mu_0 - z_\alpha \left(\frac{\sigma}{\sqrt{n}} \right) = 12 - 2.326 \left(\frac{2}{\sqrt{29}} \right)
$$

$$
= 11.1361 \text{ hours}
$$

Thus,

$$
\beta_{10.5} = P\left(\overline{X} \geq 11.1361 | \mu_1 = 10.5, \sigma_{\overline{x}} = .3714 \right)
$$

$$
= P[Z \geq (11.1361 - 10.5)/.3714]
$$

$$
= P(Z \geq 1.713) = .0433
$$

and

$$
\alpha = P\left(\overline{X} < 11.1361 | \mu_0 = 12, \sigma_{\overline{x}} = .3714 \right)
$$

$$
= P[Z < (11.1361 - 12.00)/.3714]
$$

$$
= P(Z < 2.326) = .01
$$

σ unknown.

In many practical situations the process standard deviation will be unknown. If this is the case there are two alternatives available. The first possibility is to estimate σ from a pilot sample and to use the formulas of the last section (4.6) realizing that the sample size obtained will be slightly

TABLE 4.3. Number of Observations for t Test of Mean

The entries in this table show the numbers of observations needed in a t-test of the significance of a mean in order to control the probabilities of errors of the first and second kinds at α and β respectively.

Level of t-test

Value of D = δ/σ	0.01 (Single α=0.005 / Double α=0.01) β=0.01	β=0.05	β=0.1	β=0.2	β=0.5	0.02 (α=0.01 / α=0.02) β=0.01	β=0.05	β=0.1	β=0.2	β=0.5	0.05 (α=0.025 / α=0.05) β=0.01	β=0.05	β=0.1	β=0.2	β=0.5	0.1 (α=0.05 / α=0.1) β=0.01	β=0.05	β=0.1	β=0.2	β=0.5
0.05																				
0.10																				
0.15																				122
0.20										139					99					70
0.25					110				115	90				128	64			139	101	45
0.30				134	78			109	85	63			119	90	45		122	97	71	32
0.35			125	99	58		101	85	66	47		109	88	67	34		90	72	52	24
0.40		115	97	77	45		81	68	53	37	117	84	68	51	26	101	70	55	40	19
0.45		92	77	62	37	110	66	55	43	30	93	67	54	41	21	80	55	44	33	15
0.50	100	72	63	51	30	90	66	55	43	25	76	54	44	34	18	65	45	36	27	13
0.55	83	63	53	42	26	75	55	46	36	21	63	45	37	28	15	54	38	30	22	11
0.60	71	53	45	36	22	63	47	39	31	18	53	38	32	24	13	46	32	26	19	9
0.65	61	46	39	31	20	55	41	34	27	16	46	33	27	21	11	39	28	22	17	8
0.70	53	40	34	28	17	47	35	30	24	14	40	29	24	19	9	34	24	19	15	8
0.75	47	36	30	25	16	42	31	27	21	13	35	26	21	16	9	30	21	17	13	7
0.80	41	32	27	22	14	37	28	24	19	12	31	22	19	15	8	27	19	15	12	6
0.85	37	29	24	20	13	33	25	21	17	11	28	21	17	13	7	24	17	14	11	6
0.90	34	26	22	18	12	29	23	19	16	10	25	19	16	13	7	21	15	13	10	5
0.95	31	24	20	17	11	27	21	18	14	9	23	17	14	11	6	19	14	11	9	5
1.00	28	22	19	16	10	25	19	16	13	9	21	16	13	10	6	18	13	11	8	5

TABLE 4.3. (continued)

Level of *t*-test

Level of *t*-test	0.01					0.02					0.05					0.1				
Single-sided test, α =	0.005					0.01					0.025					0.05				
Double-sided test, α =	0.01					0.02					0.05					0.1				
β =	0.01	0.05	0.1	0.2	0.5	0.01	0.05	0.1	0.2	0.5	0.01	0.05	0.1	0.2	0.5	0.01	0.05	0.1	0.2	0.5
$D = \dfrac{\delta}{\sigma}$																				
1.1	24	19	16	14	9	21	16	14	12	8	18	13	11	9	6	15	11	9	7	
1.2	21	16	14	12	8	18	14	12	10	7	15	12	10	8	5	13	10	8	6	
1.3	18	15	13	11	8	16	13	11	9	6	14	10	9	7		11	8	7	6	
1.4	16	13	12	10	7	14	11	10	9	6	12	9	8	7		10	8	7	5	
1.5	15	12	11	9	7	13	10	9	8	6	11	8	7	6		9	7	6		
1.6	13	11	10	8	6	12	10	9	7	5	10	8	7	6		8	6	6		
1.7	12	10	9	8	6	11	9	8	7		9	7	6	5		8	6	5		
1.8	12	10	9	8	6	10	8	7	7		8	7	6			7	6			
1.9	11	9	8	7	6	10	8	7	6		8	6	6			7	5			
2.0	10	8	8	7	5	9	7	7	6		7	6	5			6				
2.1	10	8	7	7		8	7	6	6		7	6				6				
2.2	9	8	7	6		8	7	6	5		7	6				6				
2.3	9	7	7	6		8	6	6			6	5				5				
2.4	8	7	7	6		7	6	6			6									
2.5	8	7	6	6		7	6	6			6									
3.0	7	6	6	5		6	5	5			5									
3.5	6	5	5			5														
4.0	6																			

smaller than required. The second option is to express the absolute difference between μ_0 and μ_1 in terms of the number of sample standard deviations. That is, let

$$D = |\mu_1 - \mu_0|/s \qquad (4.11)$$

where s is an estimate of the population standard deviation. Table 4.3 gives the sample sizes required for various values of D, α, and β.

For example, assume the hypotheses of interest are

$$H_0: \mu = \mu_0 \qquad H_1: \mu \neq \mu_0$$

and it is desired to detect an absolute difference of .4 (i.e., $|\mu_1 - \mu_0| = .4$). If the preliminary sample yields an s of .5 then the sample size required when $\alpha = .05$ and $\beta = .01$ is found (by entering Table 4.3 with $D = .4/.5 = .8$, $\alpha = .05$, and $\beta = .01$) to be 31. If we assume $\sigma = .5$ and use equation (4.10) we obtain a sample size of 28.7 or 29 observations. The increase in the sample size occurring with the use of s is due to the fact that s is only an estimate of σ and does not contain perfect information about the process variability.

If it is impossible to estimate σ via a preliminary sample then we may stipulate the standardized distance between μ_0 and μ_1 directly. That is, we simply specify the value for D, the number of standard deviations between μ_0 and μ_1, and use that value with the appropriate α and β in Table 4.3.

4.7. SAMPLE SIZE REQUIRED FOR TESTING THE DIFFERENCE BETWEEN TWO MEANS

To test the difference between the means of two normal distributions we either must know the population variances or else must assume that the variances are the same. Suppose that the first population is distributed normally with mean μ_1 and variance σ_1^2 while the second is normal with mean μ_2 and variance σ_2^2. Assume that σ_1^2 and σ_2^2 are both known values and we wish to test the hypothesis set

$$H_0: \mu_1 \leq \mu_2 \qquad H_1: \mu_1 > \mu_2$$

or, equivalently,

$$H_0: \mu_1 - \mu_2 = 0 \qquad H_1: \mu_1 - \mu_2 > 0 \qquad (4.12)$$

For either hypothesis set, the critical value for the average difference,

$\overline{D} = \overline{X}_1 - \overline{X}_2$, at a significance level of α is given by

$$\overline{D}_\alpha = 0 + z_\alpha \sqrt{\frac{\sigma_1^2}{n} + \frac{\sigma_2^2}{n}}$$

$$= z_\alpha \sqrt{\frac{\sigma_1^2 + \sigma_2^2}{n}} \qquad (4.13)$$

where we are assuming equal sample sizes. Correspondingly, for a specified distance between μ_1 and μ_2 (say D) under H_1 the value of $\overline{D}$ corresponding to β would be

$$\overline{D}_\beta = D - z_\beta = \sqrt{\frac{\sigma_1^2}{n} + \frac{\sigma_2^2}{n}}$$

$$= D - z_\beta \sqrt{\frac{\sigma_1^2 + \sigma_2^2}{n}} \qquad (4.14)$$

As before, the points corresponding to equations (4.13) and (4.14) must coincide. Therefore, setting the equations equal to each other and solving for n gives

$$n = \left(\frac{z_\alpha + z_\beta}{D}\right)^2 (\sigma_1^2 + \sigma_2^2) \qquad (4.15)$$

If $\sigma_1^2 = \sigma_2^2$, equation (4.15) reduces to

$$n = 2\left(\frac{z_\alpha + z_\beta}{D}\right)^2 \sigma^2 = 2(z_\alpha + z_\beta)^2 \left(\frac{\sigma}{D}\right)^2 \qquad (4.16)$$

For two-sided alternatives use $z_{\alpha/2}$ in place of z_α in Equation (4.16).

Example 4.6. Brand X Company is considering a second additive to decrease the drying time of its oil base paint and would like to test for differences in drying times between the two additives. It is desired to detect a difference of 2 hours with a probability of at least .95 when α is .05. Brand X Company believes that the standard deviation will be unchanged by either additive. What sample size should be used for each?

Solution. Since the standard deviation is assumed to be unchanged, $\sigma_1^2 = \sigma_2^2 = (2)^2 = 4$ (from Example 4.5). The power of the test is to be at least .95 and $\alpha = .05$ for

$$H_0: \mu_1 - \mu_2 = 0 \qquad H_1: \mu_1 - \mu_2 \neq 0$$

TABLE 4.4. Number of Observations for t Test of Difference Between Two Means

The entries in this table show the number of observations needed in a t-test of the significance of the difference between two means in order to control the probabilities of the errors of the first and second kinds at α and β respectively.

Level of t-test

Value of D = δ/σ	Single-sided α = 0.005 / Double-sided α = 0.01					Single-sided α = 0.01 / Double-sided α = 0.02					Single-sided α = 0.025 / Double-sided α = 0.05					Single-sided α = 0.05 / Double-sided α = 0.1				
β =	0.01	0.05	0.1	0.2	0.5	0.01	0.05	0.1	0.2	0.5	0.01	0.05	0.1	0.2	0.5	0.01	0.05	0.1	0.2	0.5
0.05																				
0.10																				
0.15																				
0.20																				137
0.25															124					88
0.30										123					87					61
0.35					110					90					64				102	45
0.40					85					70				100	50			108	78	35
0.45				118	68				101	55			105	79	39		108	86	62	28
0.50				96	55			106	82	45		106	86	64	32		88	70	51	23
0.55			101	79	46		106	88	68	38		87	71	53	27	112	73	58	42	19
0.60		101	85	67	39		90	74	58	32	104	74	60	45	23	89	61	49	36	16
0.65		87	73	57	34	104	77	64	49	27	88	63	51	39	20	76	52	42	30	14
0.70	100	75	63	50	29	90	66	55	43	24	76	55	44	34	17	66	45	36	26	12
0.75	88	66	55	44	26	79	58	48	38	21	67	48	39	29	15	57	40	32	23	11
0.80	77	58	49	39	23	70	51	43	33	19	59	42	34	26	14	50	35	28	21	10
0.85	69	51	43	35	21	62	46	38	30	17	52	37	31	23	12	45	31	25	18	9
0.90	62	46	39	31	19	55	41	34	27	15	47	34	27	21	11	40	28	22	16	8
0.95	55	42	35	28	17	50	37	31	24	14	42	30	25	19	10	36	25	20	15	7
1.00	50	38	32	26	15	45	33	28	22	13	38	27	23	17	9	33	23	18	14	7

TABLE 4.4. (*continued*)

The entries in this table show the number of observations needed in a *t*-test of the significance of the difference between two means in order to control the probabilities of the errors of the first and second kinds at α and β respectively.

	Level of *t*-test																				
	0.01					0.02					0.05					0.1					
Single-sided test	$\alpha=0.005$					$\alpha=0.01$					$\alpha=0.025$					$\alpha=0.05$					
Double-sided test	$\alpha=0.01$					$\alpha=0.02$					$\alpha=0.05$					$\alpha=0.1$					
$\beta=$	0.01	0.05	0.1	0.2	0.5	0.01	0.05	0.1	0.2	0.5	0.01	0.05	0.1	0.2	0.5	0.01	0.05	0.1	0.2	0.5
Value of $D=\dfrac{\delta}{\sigma}$																				
1.1	42	32	27	22	13	38	28	23	19	11	32	23	19	14	8	27	19	15	12	6
1.2	36	27	23	18	11	32	24	20	16	9	27	20	16	12	7	23	16	13	10	5
1.3	31	23	20	16	10	28	21	17	14	8	23	17	14	11	6	20	14	11	9	5
1.4	27	20	17	14	9	24	18	15	12	8	20	15	12	10	6	17	12	10	8	4
1.5	24	18	15	13	8	21	16	14	11	7	18	13	11	9	5	15	11	9	7	4
1.6	21	16	14	11	7	19	14	12	10	6	16	12	10	8	5	14	10	8	6	4
1.7	19	15	13	10	7	17	13	11	9	6	14	11	9	7	4	12	9	7	6	3
1.8	17	13	11	10	6	15	12	10	8	5	13	10	8	6	4	11	8	7	5	
1.9	16	12	11	9	6	14	11	9	8	5	12	9	7	6	4	10	7	6	5	
2.0	14	11	10	8	6	13	10	9	7	5	11	8	7	6	4	9	7	6	4	
2.1	13	10	9	8	5	12	9	8	7	5	10	8	6	5	3	8	6	5	4	
2.2	12	10	8	7	5	11	9	7	6	4	9	7	6	5		8	6	5	4	
2.3	11	9	8	7	5	10	8	7	6	4	9	7	6	5		7	5	5	4	
2.4	11	9	8	6	5	10	8	7	6	4	8	6	5	4		7	5	4	4	
2.5	10	8	7	6	4	9	7	6	5	4	8	6	5	4		6	5	4	3	
3.0	8	6	6	5	4	7	6	5	4	3	6	5	4	4		5	4	3		
3.5	6	5	5	4	3	6	5	4	4		5	4	3	3		4	3			
4.0	6	5	4	4		5	4	4	3		4	4	3			4				

109

where $D = |\mu_1 - \mu_2| = 2$ hours. Thus, $z_{\alpha/2} = 1.960$ and $z_\beta = 1.645$. Substituting these values into equation (4.16) gives

$$n = 2(1.96 + 1.645)^2(\tfrac{2}{2})^2$$
$$= 25.992 \quad \text{or} \quad 26$$

That is, 26 test samples should be prepared using each additive for a total of 52 observations.

In most practical situations the population variances will not be known in advance. If it can be assumed that they are equal (but unknown), then equation (4.16) can be used to obtain an approximate sample size by specifying D in terms of the number of standard deviations apart. That is, by letting $D = k$ where k is a specific constant, then equation (4.16) becomes

$$n = 2(z_\alpha + z_\beta)^2 \left(\frac{\sigma}{k\sigma}\right)^2$$
$$= 2\left(\frac{z_\alpha + z_\beta}{k}\right)^2 \tag{4.17}$$

Use of equation (4.17) when σ is unknown will yield a sample size somewhat smaller than that actually required for the specified β since the sample variance would be used in the actual test.

Table 4.4 gives approximate sample sizes adjusted to compensate for the standard deviation's being unknown. It is based on the utilization of D expressed as a number of standard deviations. For example, if $\alpha = .05$, $\beta = .05$, and $D = 1.5$ for a two-sided alternative then the table yields a common sample size of 13. That is, $n_1 = n_2 = 13$. The table values are determined under the assumption of common variances and the distance is measured in numbers of standard deviations (in our example, 1.5σ).

4.8. SAMPLE SIZES REQUIRED FOR TESTING VARIANCES

We limit our discussion of sample sizes for testing variances to the use of Tables 4.5 and 4.6. The reason for this limitation is that the formulas involve the chi-square and F distributions, respectively, and are much more complicated than those involving the Z and t distributions.

Testing a Single Variance

For testing a single variance, that is, hypotheses of the basic form,

$$H_0: \sigma^2 \leq \sigma_0^2 \qquad H_1: \sigma^2 > \sigma_0^2$$

TABLE 4.5. Number of Observations Required for the Comparison of a Population Variance with a Standard Value Using the Chi-Square Test

The entries in this table show the value of the ratio R of the population variance σ_1^2 to a standard variance σ_0^2 which is undetected with frequency β in a χ^2 test at significance level a of an estimate s_1^2 of σ_1^2 based on ϕ degrees of freedom.

ϕ	$a = 0.01$				$a = 0.05$			
	$\beta = 0.01$	$\beta = 0.05$	$\beta = 0.1$	$\beta = 0.5$	$\beta = 0.01$	$\beta = 0.05$	$\beta = 0.1$	$\beta = 0.5$
1	42,240	1,687	420.2	14.58	24,450	977.0	243.3	8.444
2	458.2	89.78	43.71	6.644	298.1	58.40	28.43	4.322
3	98.79	32.24	19.41	4.795	68.05	22.21	13.37	3.303
4	44.69	18.68	12.48	3.955	31.93	13.35	8.920	2.826
5	27.22	13.17	9.369	3.467	19.97	9.665	6.875	2.544
6	19.28	10.28	7.628	3.144	14.44	7.699	5.713	2.354
7	14.91	8.524	6.521	2.911	11.35	6.491	4.965	2.217
8	12.20	7.352	5.757	2.736	9.418	5.675	4.444	2.112
9	10.38	6.516	5.198	2.597	8.103	5.088	4.059	2.028
10	9.072	5.890	4.770	2.484	7.156	4.646	3.763	1.960
12	7.343	5.017	4.159	2.312	5.889	4.023	3.335	1.854
15	5.847	4.211	3.578	2.132	4.780	3.442	2.925	1.743
20	4.548	3.462	3.019	1.943	3.802	2.895	2.524	1.624
24	3.959	3.104	2.745	1.842	3.354	2.630	2.326	1.560
30	3.403	2.752	2.471	1.735	2.927	2.367	2.125	1.492
40	2.874	2.403	2.192	1.619	2.516	2.103	1.919	1.418
60	2.358	2.046	1.902	1.490	2.110	1.831	1.702	1.333
120	1.829	1.661	1.580	1.332	1.686	1.532	1.457	1.228
∞	1.000	1.000	1.000	1.000	1.000	1.000	1.000	1.000

EXAMPLES

Testing for an increase in variance. Let $a = 0.05$, $\beta = 0.01$, and R = 4. Entering the table with these values it is found that the value 4 occurs between the rows corresponding to $\eta = 15$ and $\eta = 20$. Using rough interpolation it is indicated that the estimate of variance should be based on nineteen degrees of freedom.

Testing for a decrease in variance. Let $a = 0.05$, $\beta = 0.01$, and R = 0.33. The table is entered with $a' = \beta = 0.01$, $\beta' = a = 0.05$, and R' = 1/R = 3. It is found that the value 3 occurs between the rows corresponding to $\eta = 30$. Using rough interpolation it is indicated that the estimate of variance should be based on 26 degrees of freedom.

TABLE 4.6. Number of Observations Required for the Comparison of Two Population Variances Using the F Test

The entries in this table show the value of the ratio R of two population variances σ_2^2/σ_1^2 which remains undetected with frequency β in a variance ratio test at significance level α of the ratio s_2^2/s_1^2 of estimates of the two variances, both being based on ϕ degrees of freedom.

ϕ	$\alpha = 0.01$				$\alpha = 0.05$				$\alpha = 0.5$			
	$\beta = 0.01$	$\beta = 0.05$	$\beta = 0.1$	$\beta = 0.5$	$\beta = 0.01$	$\beta = 0.05$	$\beta = 0.1$	$\beta = 0.5$	$\beta = 0.01$	$\beta = 0.05$	$\beta = 0.1$	$\beta = 0.5$
1	16,420,000	654,200	161,500	4052	654,200	26,070	6,436	161.5	4,052	161.5	39.85	1.000
2	9,801	1,881	891.0	99.00	1,881	361.0	171.0	19.00	99.00	19.00	9.000	1.000
3	867.7	273.3	158.8	29.46	273.3	86.06	50.01	9.277	29.46	9.277	5.391	1.000
4	255.3	102.1	65.62	15.98	102.1	40.81	26.24	6.388	15.98	6.388	4.108	1.000
5	120.3	55.39	37.87	10.97	55.39	25.51	17.44	5.050	10.97	5.050	3.453	1.000
6	71.67	36.27	25.86	8.466	36.27	18.35	13.09	4.284	8.466	4.284	3.056	1.000
7	48.90	26.48	19.47	6.993	26.48	14.34	10.55	3.787	6.993	3.787	2.786	1.000
8	36.35	20.73	15.61	6.029	20.73	11.82	8.902	3.438	6.029	3.438	2.589	1.000
9	28.63	17.01	13.06	5.351	17.01	10.11	7.757	3.179	5.351	3.179	2.440	1.000
10	23.51	14.44	11.26	4.849	14.44	8.870	6.917	2.978	4.849	2.978	2.323	1.000
12	17.27	11.16	8.923	4.155	11.16	7.218	5.769	2.687	4.155	2.687	2.147	1.000
15	12.41	8.466	6.946	3.522	8.466	5.777	4.740	2.404	3.522	2.404	1.972	1.000
20	8.630	6.240	5.270	2.938	6.240	4.512	3.810	2.124	2.938	2.124	1.794	1.000
24	7.071	5.275	4.526	2.659	5.275	3.935	3.376	1.984	2.659	1.984	1.702	1.000
30	5.693	4.392	3.833	2.386	4.392	3.389	2.957	1.841	2.386	1.841	1.606	1.000
40	4.470	3.579	3.183	2.114	3.579	2.866	2.549	1.693	2.114	1.693	1.506	1.000
60	3.372	2.817	2.562	1.836	2.817	2.354	2.141	1.534	1.836	1.534	1.396	1.000
120	2.350	2.072	1.939	1.533	2.072	1.828	1.710	1.352	1.533	1.352	1.265	1.000
∞	1.000	1.000	1.000	1.000	1.000	1.000	1.000	1.000	1.000	1.000	1.000	1.000

we transform the hypothesis set into the equivalent form

$$H_0: \frac{\sigma^2}{\sigma_0^2} \leq 1 \qquad H_1: \frac{\sigma^2}{\sigma_0^2} > 1$$

Then, for a specified ratio under the alternative, say $R = \sigma_1^2/\sigma_0^2$, and desired levels of α (.01 or .05) and β (.01, .05, .1 or .5), we can refer to Table 4.5. This table is based on R, α, and β with R being the entries in the body of the table. The degrees of freedom, $n - 1$, are then read from the left-hand column in the table and identified by ϕ.

Example 4.7.

$$H_0: \sigma^2 \leq \sigma_0^2 \qquad H_1: \sigma^2 > \sigma_0^2$$

and it is of specific interest to detect alternative values such that $\sigma_1^2 \geq 2\sigma_0^2$. If α and β are specified to be .05 and .1, respectively, what size sample should be selected?

Solution. We have that $R = \sigma_1^2/\sigma_0^2 = 2$ with $\alpha = .05$ and $\beta = .10$. Referring to Table 4.5 we find values for R of 2.129 and 1.919 under the appropriate combination for α and β. These correspond to ϕ of 30 and 40, respectively. Interpolating gives ϕ of approximately 37 and a corresponding sample size of 38.

Comparing Two Population Variances

The basic hypotheses for comparing two population variances are

$$H_0: \sigma_1^2 = \sigma_2^2 \qquad H_1: \sigma_1^2 \neq \sigma_2^2$$

As in the one-sample situation we convert the hypotheses to the form

$$H_0: \sigma_2^2/\sigma_1^2 = 1 \qquad H_1: \sigma_2^2/\sigma_1^2 \neq 1$$

and express the specific alternative of interest in terms of the ratio $R = \sigma_2^2/\sigma_1^2$. This ratio assumes that σ_2^2 is the larger of the two population variances.

Table 4.6 is indexed by values for α (.01, .05, and .5), β (.01, .05, .1, and .5) with the ratio, R, given in the body of the table. The left-hand margin then yields values for the degrees of freedom (ϕ) associated with each sample, that is, values for $n - 1$. Use of this table requires that samples of the same size will be selected from each population.

Example 4.8. It is desired to test

$$H_0: \sigma_1^2 = \sigma_2^2 \qquad H_1: \sigma_1^2 \neq \sigma_2^2$$

at $\alpha = .05$ with the probability of failing to detect a ratio as large or larger than 2.5 being no more than .10. What sample sizes are required to do so?

Solution. We have $R = 2.5$ with $\alpha = .05$ and $\beta = .10$. Referring to Table 4.6 we find, for the specified α and β, that 2.5 is between 2.549 and 2.141. These values correspond to $\phi = 40$ and $\phi = 60$, respectively. Using linear interpolation we obtain

$$\phi = n - 1 = 42$$

Thus, samples of size $n = 43$ should be selected from each population of interest.

4.9. SOME FURTHER CONSIDERATIONS

In this chapter we have attempted to answer the basic question "How large a sample should be taken?" As you saw, the answer to this question depends on the information available and the precision required of the procedure. For interval estimates it requires that a minimum error be specified, a value obtained for the standard deviation of the variable of interest, and a confidence or reliability level be selected. If the resulting sample size is economically infeasible then one must sacrifice either precision (use a larger error) or reliability (use a smaller confidence level). When the standard deviation is known, higher confidence levels also require larger sample sizes. If σ is unknown then it must be estimated in some way with the use of the estimate requiring a larger sample size due to the resultant loss of information.

In testing situations the only way one can control both α and β for a specified difference under the alternative is by determining the sample size. The smaller the difference between H_0 and H_1 the larger the required sample size for fixed α and β. If all else is fixed, small α's yield large values for β and vice versa. Note that β, the probability of a type II error, is always taken as one sided when determining n. The reason for this is that, even for a two-sided alternative, any logical value for β will be smaller than .5 and thus the entire acceptance region will fall in one tail of the alternative distribution.

REFERENCES

1. Burbridge, H. K., A "trick" for the trade, *Quality/The Magazine of Product Assurance*, June 1976, pp. 29–30.

2. Davies, O. L., *Design and Analysis of Industrial Experiments*, 2nd ed., Hafner Publishing Company, New York, 1963, pp. 12–42.

3. Guenther, W. C., *Concepts of Statistical Inference*, McGraw-Hill, New York, 1965.

4. Rickmers, A. D., and Todd, N. H., *Statistics, An Introduction*, McGraw-Hill, New York, 1967.

GRAPHICAL STATISTICS

JAMES R. KING

TEAM
Tamworth, New Hampshire

Basically, pairs of points (x, y) are plotted using x as the abscissa and y as the ordinate. When we plot the equation $y = 2 + 3x$ the related (x, y) pairs yield a graph of y versus x, as in Figure 5.1. More generally, $y = a + bx$, where x is an independent variable; y is a variable dependent on x; a is the intercept of the graph line on the y axis; and b is the slope of the graph line.

Analytically, the (x, y) pairs describe a straight line in a plane. The intercept, a, locates the start of the line and is thus sometimes called the location parameter. The slope, b, defines the direction of the line away from its located origin. Also, for any value of x, there is a unique value of y which occurs. When these (x, y) values are plotted, the result is called a linear graph. Due to the rules of algebra, the relationships are strict and any particular expression of $y = a + bx$ yields a unique line, totally described, when a and b are given.

If we now modify our designations of x and y somewhat, we can define a probability graph which is basic to graphical statistics. The first modification is to change x from an ordinary number to a probability number. In particular, we will change x to a set of values called the cumulative probability values. The second modification is to change y from a simple dependent variable to an ordered random variable, which has a special kind of dependence on x.

5.1. GRAPHICAL DATA ANALYSIS

The significance of the previous introductory statements may not be directly obvious. However, a simple example will illustrate some of the

115

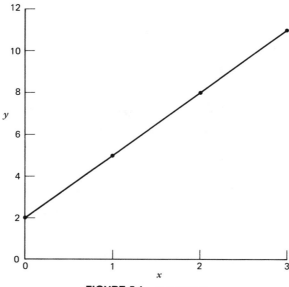

FIGURE 5.1. *y* versus *x*.

statistically useful differences from ordinary algebraic concepts. As an example, suppose that we are interested in weight control of pints of a packaged paint product. In order to assess the effectiveness of weight control, we take a sample of 19 packages from a factory filling machine, determine the weight of the contents by difference, and obtain the results in ounces shown in Table 5.1.

The 19 numerical values obtained from the weighings are a set of observed data. These data are random values of a sample of packages selected from a potentially large number of presumably identically filled items. These values, as recorded after each weighing, are not too intelligible. However, we can get a better idea of what they represent by rearranging the as-observed values into some different formats.

First, we can arrange the values of Table 5.1 into numerical order, from smallest to largest, as in Table 5.2. Here, we can see directly a smallest value and a largest value. Half-way down in this tabulation, we see a kind

TABLE 5.1.   Weight of Paint Product in Ounces

16.48	16.28	16.32	16.42
16.38	16.52	16.34	16.72
16.62	16.48	16.47	16.17
16.91	16.39	15.97	16.03
16.15	16.26	16.43	

TABLE 5.2.   Weight of Paint Product

	y	n	x_i	x_j
Low	15.97	1	.05	5%
	16.03	2	.10	10
	16.15	3	.15	15
	16.17	4	.20	20
	16.26	5	.25	25
	16.28	6	.30	30
	16.32	7	.35	35
	16.34	8	.40	40
	16.38	9	.45	45
Middle	16.39	10	.50	50
	16.42	11	.55	55
	16.43	12	.60	60
	16.47	13	.65	65
	16.48	14	.70	70
	16.48	15	.75	75
	16.52	16	.80	80
	16.62	17	.85	85
	16.72	18	.90	90
High	16.91	19	.95	95

of representative, or typical, value. By simple rearrangement of the original as-observed data, we have derived a sense of low, middle, and high values of the sample data.

Secondly, we can tabulate the data as in Table 5.3, which is obtained by grouping the measured values. Now, we have a frequency tally, or histogram, which illustrates the distribution of the data points quite dramatically.

Detailed methods of preparing data by grouping are given in King.[1] Additional useful methods of presenting data are given in Refs. 2 and 3.

Neither Table 5.2 nor Table 5.3 resembles the (x, y) graph that we started out with. In fact, we appear to have only a set of values in y, the

TABLE 5.3.   Grouped Values — Paint Weights

Cell Interval	Frequency
15.80–15.99	I
16.00–16.19	II
16.20–16.39	IIIIII
16.40–16.59	IIIII
16.60–16.79	III
16.80–16.99	I

ordered random variable. Where are the x values? The x values can be established readily because they are a direct function of n, the sample size. The cumulative probability value, x, corresponding to a particular ordered value of y for a specific sample size, n, is given by

$$x_i = \frac{y_i}{n+1} \tag{5.1}$$

The corresponding cumulative percentage, or percentile, is given by

$$x_j = \frac{y_j \times 100}{n+1} \tag{5.2}$$

where

x_i = cumulative probability of occurrence of the ith ordered variable
y_i = value of ith item of ordered data
x_j = cumulative percentage of occurrence of the jth ordered variable
y_j = value of jth item of ordered data
n = sample size, or number of observations

The values of x_i and x_j for the ordered weight data are given in Table 5.2. Figure 5.2 is a plot of this data as an (x, y) graph on a suitable probability plotting paper. In this format, the data appear as a linear graph in (x, y) coordinates representing weight versus percentage of occurrence. A straight line has been fitted to these data by dividing the plotted points into an upper half and a lower half. This line is the median regression line. Detailed requirements for fitting regression lines to probability plots are given in Ref. 1.

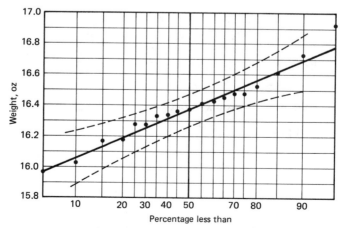

FIGURE 5.2.   Probability plot of packaged weights.

5.2. DATA INTERPRETATION USING A PROBABILITY PLOT

The straight line that can be fitted to a probability plot has many useful and interesting properties. Some particular ones are:

1. By locating the intersection of the 50% vertical of the cumulative percentage scale and the graph line and projecting this point to the y axis, we obtain a direct estimate of the mean value of the sample. The mean obtained from the graph is 16.39 compared to 16.386 obtained by conventional calculation (2, 4, 5).

2. The difference between the y values at the intersection of the 50 and 16% verticals yields an estimate of the standard deviation of the sample data. On the graph, the standard deviation estimate is 0.25 compared to 0.225 by calculation (2, 5). The value of the standard deviation is a direct function of the slope of the fitted straight line. Visually, therefore, on similarly scaled plots, small slopes indicate small variation while large slopes indicate large variation.

3. In the case of packaged commercial products, the weight marked on the package implies that the contents do not amount to less than the marked weight. In this example, we have weighed the contents of nominally 1-lb containers. How good is the container weight control? If one now enters the y scale at 16 oz and projects horizontally to the graph line, and then projects down to the percentage scale, we estimate that about 7% of the packages appear to contain less than 1 lb.

4. Now, the other side of the coin is that excessive overfill may occur in the attempt to avoid underweight contents. Excessive overfill increases costs and decreases profit. Figure 5.2 shows that the average fill is 16.39 oz, or 2.44% overfill.

5. If subsequent changes in fill techniques were to be evaluated, similar probability plots could be made which would permit quick direct comparison to prior data.

These examples are a demonstration of the capability for fast incisive analysis directly from probability plots. Consider that for the kinds of questions posed here, a probability plotting paper is equivalent to an analog computer with the probability scale comparable to a stored program. A set of ordered data plotted against corresponding probability values is an input. The graph line is a visual output which allows direct estimation of the statistical parameters of the population represented by a sample of data as well as answering a wide variety of pertinent questions about alternative interpretations of the set of data. In the particular example presented, we found 7% underfill and 2.44% average overfill. The

population of packaged weights is estimated to fall in the range of
16.39 ± 0.75 oz.

Many other details of applications of probability graphs are illustrated
in Ref.1. Examples of categorical problems in quality control and reliability which are amenable to probability graphing are given in Refs. 2 and 5.

5.3. MORE ABOUT DISTRIBUTIONS

The previous example was based on the normal distribution. However,
other distributions that occur frequently such as the binomial, log-normal,
extreme value, and Weibull, in their proper applications, have special
meanings which can be exploited in data analysis and statistical problem
solving. Typical curves for these distributions are shown in Figure 5.3.

The great flexibility of probability graphs is that a simple algebra can be
utilized to construct a probability graph for any distribution provided that
one has the specific probability plotting paper for a particular distribution.
A probability plotting paper is a specially dimensioned graph paper which
provides one axis giving probability or percentage coordinates for a particular distribution and a second axis for displaying the scale of the ordered y
values. The underlying assumptions and detailed mathematics for most
distributions can be found in Refs. 4 and 6. The particular advantage of
using preprinted probability plotting papers is that the need for extensive
tables is eliminated.

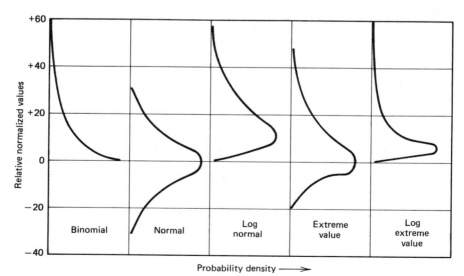

FIGURE 5.3. Comparative appearance of different types of statistical distributions.

5.4. COMPARATIVE ANALYSIS USING PROBABILITY GRAPHS

Figure 5.4 shows sample data for tensile strength of 20-min cure of SBR vulcanizates adapted from Ref. 7 plotted on two different probability plotting papers. The left plot is on normal probability paper while the right one is on extreme value probability paper. Now, suppose we were required to meet a minimum requirement of 18 kg/cm^2 even though none of the observed values were less than 21 kg/cm^2. By extrapolating the fitted straight line, we can obtain estimates of the probability of a tensile strength less than 18 kg/cm^2. The normal plot extrapolation estimates that about 0.3% of the specimens represented by this sample would be less than 18 kg/cm^2 or that typically 3/1000 would fail.

On the other hand, the extreme value plot estimates that only 0.001% would be less than 18 kg/cm^2 or about 1/100,000 would fail. This is a 300/1 ratio in the level of estimated failures. How can we interpret this? First, both plots appear to give a reasonable fit to the data. For a long time, the ability to fit a reasonable straight line to data on probability plotting paper was considered a pragmatically sufficient test of fit of that distribution represented by a particular probability plotting paper. Since the large majority of people in statistics will start out with normal distribution assumptions, normal probability paper is most frequently used. In the present example, most people would have obtained the result illustrated in the left plot and let it go at that.

As it turns out, data will frequently give a reasonably straight line plot on several probability papers so that we need criteria for choosing a probability paper. Table 5.4 gives some useful criteria for choosing a probability paper. These include known mathematical models for the underlying process being investigated, categorical and qualitative process descriptions, areas of application, and kinds of estimation problems involved. A review of these factors will greatly improve distribution choices.

Descriptions of some useful test statistics and directions for their application are contained in Ref. 6. This book covers a large number of types of distributions in relation to their areas of application in engineering and chemistry. The mathematical development shows different families of distributions in terms of uses and limitations.

In the present example, there are sound theoretical grounds for choosing the extreme value distribution over the normal (6, 8, 9). In addition, the fitted line for the extreme value plot falls directly on values at the mode, at the median and at the mean while the line on the normal plot clearly misses the mean, mode, and median at 50%. Thus, we accept the more optimistic estimate that only 1/100,000 specimens will fail the minimum tensile requirement. Economics of sampling requirements may be such that the working answer is that one does not care whether the estimate is 3/1000 or 1/100,000. However, for batch processes with irreversible reactions, inspection and tests lead to acceptance and rejection decisions.

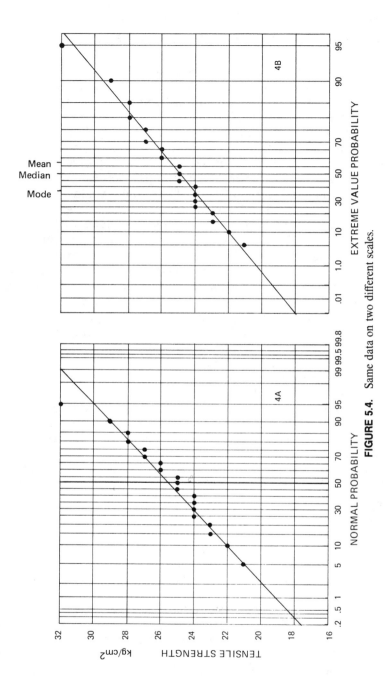

FIGURE 5.4. Same data on two different scales.

TABLE 5.4. Mathematical Models, Process Descriptions, and Statistical Distributions

Mathematical Operation	Mathematical Model	Process Description	Example	Resultant Statistical Distribution
Counting	$p = \dfrac{c}{n}$	Enumeration or classification	Inspection Sorting	Binomial
Addition	$f(y) = \sum\limits_{i}^{n} (x_i)$ or	Linear additive	Addition or subtraction of materials; i.e., cutting, weighing, etc., also mechanical assembly.	Normal
Multiplication	$f(y) = \sum\limits_{i}^{n} (x_i)$	Rate-dependent	Simple chemical processes; i.e., etching, corrosion, gaseous diffusion.	Log-normal
		Proportional response	Simple biological processes; i.e., growth rate. Simple economic processes; i.e., distribution of income.	
Simple exponentiation or	$f(y) = ax_0 + bx_1 + cx_2^2$	Algebraic polynomial	Complex processes involving the combined effects of a number of independent causes each with a different operational form; i.e., breaking strengths, meteorological and geophysical phenomena, electronic and chemical measurements, financial data.	Extreme value
addition of transcendental terms	$f(y) = e^{x_0} + e^{x_1} + e^{x_2}$	Solutions of linear differential equations with constant coefficients		

(continued)

123

TABLE 5.4. (*continued*)

Mathematical Operation	Mathematical Model	Process Description	Example	Resultant Statistical Distribution
Counting of time duration to an event	$f(x; n, \lambda) = \dfrac{\lambda^n}{\Lambda(n)} x^{(n-1)} e^{-\lambda x}$	Waiting time	Time required for an event(s) to occur or to obtain some service.	Gamma
Addition of squared normalized vectors	$f(y) = \sum_i^n \left(\dfrac{x_i}{\sigma_i}\right)^2$	Vector sums	Resultant value in a system of n-fold vector spaces from physics, space-time, and probability applications.	Chi-square
Multiplication of transcendental terms	$f(y) = e^{(x_0 \cdot x_1 \cdot x_3 \ldots)}$ $f(y) = e^{(x_1/x_2)(x_3/x_4)}$	Solutions of general differential equations Particle sizing	Complex exponential processes involving the interdependent effects of independent causes; i.e., breakage of particulate materials, solid state diffusion, chemical kinetics.	Log-extreme value
Sums, products, and powers of exponents of transcendental terms	$f(y) = e^{\left(\frac{\omega - x}{\omega - \mu}\right)^{\beta}}$	Solutions of differential equations with boundary conditions "Upper-limit" distributions	Processes involving limits and maxima-minima; i.e., life/failure distributions, bounded particle size distributions, and general potential, gradient, and field problems.	Weibull

Consequently, being able to assume a double-exponential extreme value distribution allows a prediction that the failure rate under such a distribution assumption is about 300 times lower than that predicted by the normal model. Against a stipulated AQL (average quality level), it would be possible to use much smaller samples with obvious cost savings, or an extremely high level of assurance that the process/product is satisfactory. A third alternative would be to choose an optimum sample size and assurance based on the OC (operating characteristic) curves for the sampling plan to be employed (2, 5).

5.5. CONFIDENCE INTERVALS

In Section 5.1 and Figure 5.2 a straight line was fitted to the "fill" data plotted on probability plotting paper. The question then would be how good is the yield estimate made from the "best-fit" line? However, to answer this question, we must first answer: How good is the fit of the "best-fit" line? This topic has received only scattered attention in the literature: by Gumbel for the extreme value distribution (10); by Kao for the Weibull distribution (11); by King for the log-normal distribution (1); and by Mann et al. for other distributions (12). Team Easy Analysis Methods (13) gave some methods for the Weibull distribution. However, here we will use the approach used in Ref. 9 and Wallis (10) and developed in Ref. 1 to obtain results equivalent to those presented in TEAM (13).

The Use of Confidence Intervals

We can assess the "best-fit" line by applying a confidence interval around the line. Consider that in conventional statistics, we symbolize a confidence interval for an estimate of the mean as $\overline{X} \pm ks$, where $\overline{X}$ is the mean, or average, and s is the sample standard deviation. The probability factor k is selected in such a way that the interval, $\pm ks$, will contain a designated percentage of present or future values. Values for k are obtained from tables of the cumulative normal distribution. In a geometrical sense, ks is linear and one-dimensional along a straight line.

However, we are interested in the confidence interval for an estimated regression line on probability plotting paper. The rationale is similar in that we wish to know about

$$\overline{Y}_p \pm k_1 s_{yp} \tag{5.3}$$

where

$\overline{Y}_p$ = the mean value of the line at some designated percentage point
k_1 = a probability factor based on Student's t distribution which is similar to but not identical to k
s_{yp} = the standard deviation of Y_p

The interval, $\pm k_1 s_{yp}$, will contain a designated percentage of values for $\overline{Y}_p$ directly corresponding to the conventional confidence interval. However, when these intervals are taken about the whole "best-fit" line, they result in defining a "region" about the line on the probability plot which is nonlinear and is two-dimensional. This is shown by means of broken lines above and below the "best-fit" line. The area between these broken lines is now the 95% confidence interval for the "best-fit" line as in Figure 5.2.

The confidence interval can also be used to evaluate sample data. For example, with a 95% confidence interval, we expect that at least 19 out of 20 values will fall inside the interval. If all the data values fall well inside the limits we conclude that they are quite consistent. Conversely, if several points fall outside the interval, we should become suspicious that: (1) the data represent a sample that is "bad" in some unknown way; (2) the data may be plotted on a probability paper that does not correctly represent the underlying statistical distribution or (3) the plotting or associated calculations are incorrect, as sometimes happens. One should also question each of the earlier estimates that were made. The cause or causes of the poor data fit should be determined to the extent possible since the causes may represent a serious problem in the process, the sampling procedures employed, measurements taken, or data handling which must be corrected before any valid and meaningful parameter estimates may be derived. Our experience indicates that the problem of "bad" data occurs so frequently that it is valuable to have a simple means of validating data. One significant advantage of probability plotting is ready identification of "bad" data.

Calculating Confidence Intervals

To obtain confidence intervals for a fitted line on probability plotting paper, we must solve the relationship from equation (5.3). Since we do not know s_{yp}, we use

$$k_1 s_{yp} = k_1 s_{y \cdot x} \sqrt{\frac{1}{n} + \frac{k^2}{n-1}} \qquad (5.4)$$

with k_1 and s_{yp} as in equation (5.3).

$s_{y \cdot x}$ = the sample standard deviation estimated from the "best-fit" line
  n = the sample size
  k^2 = the square of the normal probability factor

Now, there is no need to solve this messy expression endlessly. In an approach similar to Ref. 9 and TEAM (13), we preselect two confidence levels to work with, 80 and 95%. Then, we also preselect one set of

TABLE 5.5.   Confidence Factors for the Normal and Logarithmic Normal Distributions

Sample Size *n*	80% Confidence Factors Preselected Percentiles					95% Confidence Factors Preselected Percentiles				
	50	30–70	10–90	5–95	1–99	50	30–70	10–90	5–95	1–99
5	0.732	0.849	1.280	1.534	2.041	1.422	1.648	2.485	2.977	3.962
6	0.626	0.722	1.079	1.290	1.713	1.135	1.309	1.957	2.339	3.106
7	0.559	0.643	0.955	1.140	1.513	0.971	1.116	1.659	1.980	2.627
8	0.509	0.584	0.864	1.030	1.364	0.866	0.993	1.470	1.752	2.322
10	0.443	0.506	0.744	0.886	1.172	0.730	0.834	1.228	1.462	1.934
12	0.395	0.451	0.661	0.786	1.039	0.644	0.734	1.076	1.280	1.691
14	0.362	0.412	0.603	0.716	0.946	0.583	0.663	0.970	1.153	1.522
16	0.336	0.382	0.558	0.663	0.875	0.535	0.608	0.888	1.055	1.392
18	0.315	0.358	0.522	0.620	0.817	0.500	0.568	0.827	0.982	1.296
20	0.297	0.338	0.491	0.583	0.770	0.470	0.533	0.776	0.921	1.215
25	0.264	0.299	0.434	0.516	0.680	0.414	0.469	0.682	0.809	1.066
30	0.239	0.271	0.393	0.466	0.614	0.372	0.422	0.611	0.725	0.965
40	0.206	0.233	0.337	0.400	0.527	0.319	0.361	0.523	0.620	0.817
50	0.184	0.208	0.301	0.356	0.469	0.284	0.321	0.465	0.551	0.726
100	0.129	0.146	0.210	0.249	0.328	0.199	0.224	0.324	0.384	0.505

Intermediate values may be obtained by linear interpolation

percentage points to be used to locate the confidence limits about the fitted line, say at 1 and 99%, 5 and 95%, 10 and 90%, 30 and 70%, and at 50%. Since k_1 and the expression under the radical are sample size dependent, we can solve equation (5.4) for a range of sample sizes and tabulate the results as in Table 5.5.

5.6.   EXTREME VALUES, RETURN PERIODS, AND ENVIRONMENTAL CATASTROPHES

Introduction and Background

A simple algebra of probability graphs was developed in Sections 5.1 through 5.3. Applications to normal distributions were shown and a comparison of identical data on normal and extreme value probability papers was given without any development of the extreme value distribution.

The main distinguishing feature of extreme value distributions is that they are skewed, either to the right or left, as compared to the symmetry of the normal distribution, Figure 5.5. The name for this distribution came from applications by Gumbel (8) in the 1930s to extremes of human life and extreme intervals between radioactive emissions. However, interest in similar problems began with Laplace and Fourier in the 1820s. Little more

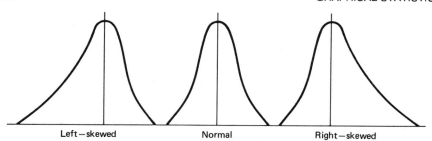

Left—skewed   Normal   Right—skewed

FIGURE 5.5. Symmetrical (normal) and extreme value curves.

was done until 1900 when von Bortkiewicz and Galton revived interest which blossomed in the 1920s in work by Charlier, Dodd, Fisher, Frechet, Griffith, von Mises, Pierce, and Tippett. Early practical applications were in analysis of annual floods, the largest daily rainfall, or the largest snowfall. Extended theory and applications can be found in Ref. 14. In 1957, Botts (15) showed that the double exponential form occurred in its own right by applying it to loss claims for 1 year under a mutual crop loss insurance plan. Haviland (16) extended applications to reliability of long-life engineering design. Hahn and Shapiro (6) illustrate other engineering applications. In 1968, Yabuta and Kase showed applications to vulcanized elastomers (7). Applications to a wide range of physical and industrial problems can be found in King (1).

Description

Extreme value distributions are also known as double exponential and Gumbel type I distributions. The cumulative distribution function is

$$\Phi(x) = \int_{-\infty}^{y} e^{-e^{-y}} \, dy$$

when

$$-\infty < y < +\infty$$

and

$$y = \alpha(x - \mu_0)$$

where

y = the normalized unit variable, usually called the *reduced variate*
x = the value of the random variable
μ_0 = the *mode*, the measure of central tendency for the extreme value distribution

α = when expressed as $1/\alpha$, the measure of variability, or dispersion, also called the *Gumbel slope*

In Section 5.1, (x, y) pairs for probability plotting were defined in terms of x for the probability value and y for the random variable. However, Gumbel's notation for the extreme value distribution, which has been widely adopted, uses y for the probability value and x for the random variable.

Description of Extreme Value Probability Paper

Extreme value probability paper has a probability scale for cumulative probability and a linear scale for the random variable. In addition, there are two auxiliary scales. One is for the reduced variate, y, and the other is for the return period which will be discussed separately later. Data are prepared for plotting by arranging the data in order of increasing values for right skewed distributions and by decreasing values for left skewed data. The probability plotting percentiles are obtained by dividing the number of the rank order position, n_x, for each data entry, by $(n + 1)$, where n is the sample size. This is demonstrated in Table 5.6.

TABLE 5.6. Tensile Strengths of Polyether Urethane Samples

Rank Order, n_x	Cumulative Probability, $n_x/(n + 1)$	Tensile Strength, psi	
		Standard	Experimental
1	.05	4970	5385
2	.10	5175	5670
3	.15	5200	5675
4	.20	5200	5800
5	.25	5250	5950
6	.30	5280	6025
7	.35	5290	6190
8	.40	5350	6220
9	.45	5400	6500
10	.50	5485	6500
11	.55	5490	6500
12	.60	5575	6500
13	.65	5630	6515
14	.70	5790	6550
15	.75	5800	6640
16	.80	5825	6695
17	.85	6000	6725
18	.90	6025	6820
19	.95	6250	7040

Example of an Extreme Value Analysis

An important class of situations resulting in extreme value distributed data arises from the combined effects of multiple variables where the relationships between the several variables can be expressed as a polynomial. Let us consider a common case such as that of a third-order polynomial:

$$y = a + bx + cx^2 + dx^3 \qquad (5.5)$$

As an example, consider the case of a coating development program. A customer wanted a polyether urethane coating with a minimum of 5000 psi tensile strength. The manufacturer's standard coating test results indicated that about 6.5% of the specimens tested would be less than 5000 psi as shown by line I in Figure 5.6. A new coating formulation was developed which gave the results shown by line II in Figure 5.6 which indicated that less than 0.01% of the specimens would have a tensile strength less than 5000 psi. However, this satisfactory result was the outcome of many attempts, most of which were unsuccessful.

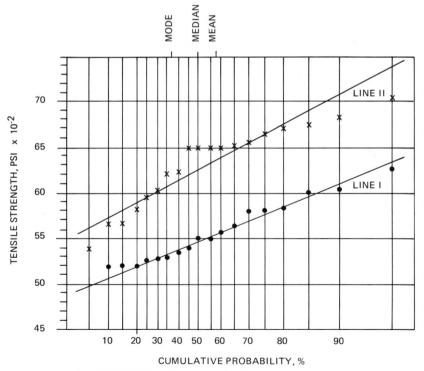

FIGURE 5.6.   Extreme value plots of tensile data.

In an attempt to understand the mechanisms that caused such variability in experimental results, a model was developed based on the third-order polynomial:

$$\text{tensile strength} = a + bt + cA + dV$$

where

a = a threshold parameter
b, c, and d = constants of proportionality
t = a function of coating thickness
A = a function of surface area
V = a function of material volume

The value of a is related to inherent properties of materials and also to the grade or quality of materials used. The value of t is related to the variation in coating thickness which is process dependent. A can vary with geometry and process variables that affect shrinkage and V is sensitive to permissible variation in tolerances. This approach to problem solving with statistics has come to be called statistical engineering.

Some Further Considerations

Besides having either right or left skew, extreme value data may consist of both positive and negative values. The possibility of both positive and negative values is often valuable in deciding whether or not the extreme value distribution assumption is to be preferred over the assumption of a logarithmic normal distribution which it often resembles (17). One important characteristic of the log-normal distribution is that it can have only positive increasing values.

Deciding between right- and left-hand skew is easily done by plotting simple histograms of the raw data using six to eight equally spaced cells. Detailed information for preparing histograms is contained in Refs. 1, 3, and 5. Extreme value histograms typically show a thin end and a thick end as in Figure 5.7. The thin end points in the direction of skew. When a left skewed distribution is observed, data arrangement is reversed from that for a right skewed distribution; that is, the data are arranged in order from largest to smallest. When plotted on extreme value probability paper, the plot will slope down from left to right, as in Figure 5.8. The value of the slope, $1/\alpha$, will be negative and the sign of any values required from the reduced variate scale, y, will be reversed in sign. If a left skewed data set is not reversed for plotting it will look like the dotted line superimposed on Figure 5.8, which is simply the left skewed data plotted incorrectly.

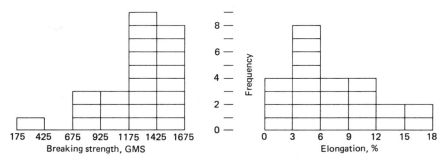

FIGURE 5.7. Extreme value histograms.

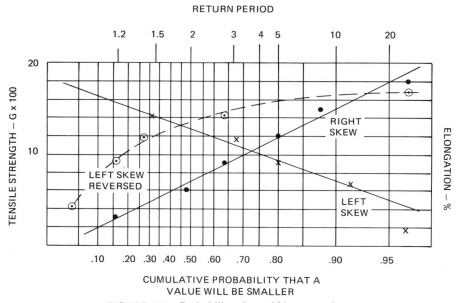

FIGURE 5.8. Probability plots of histogram data.

Extreme Values and Differential Equations

Many physical processes are mathematically described by differential equations. In the particular case of differential equations with constant coefficients the description of a process may be similar to

$$y = e^{x_1} + e^{x_2} + e^{x_3} \tag{5.6}$$

Such equations will also lead to extreme value distributions and problem solving may then be exploited in the same manner as equation (5.5). Application of this form to life testing and failure analysis is given in Ref. 1.

Description of the Return Period

The return period scale for extreme value probability papers is normally located at the top of the grid. The values on this scale are the reciprocals of $(1 - p)$, where p is the probability that a future observed value will be smaller. The return period represents the average time between recurrences of a value greater than the represented by the return period. That is, it represents a 50% chance of observing another value larger than the observed one.

For example, in Figure 5.8, project from 14 on the right-hand scale for percent elongation to the distribution line and then project vertically upward. This indicates a return period of 8, meaning that in 8 additional observations, there is a 50–50 chance that a value of 14 or more will occur. Alternatively, if a number of samples of 8 were tested, then values of 14 or greater would be expected to occur in about half of the samples.

The Return Period and Environmental Catastrophes

The return period has been widely used to analyze data for the size of the annual flooding of streams and watersheds. It is also applicable to maxima and minima of meterological data and, more recently, it has been found valuable for analyzing pollution data for air and water.

However, when the extreme value distribution is applied to distributions of extremes, it is particularly important that the interpretation of the return period be carefully made. Professor Brian M. Reich, then with the

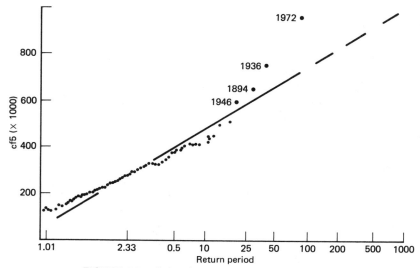

FIGURE 5.9. Susquehanna river floods at Harrisburg.

Department of Civil Engineering, Pennsylvania State University, became intensely concerned about the wide devastation caused by flooding of the Susquehanna River at Harrisburg in 1972 due to Hurricane Agnes. From historical data, the 1972 flood exceeded the 100-year expected flood by 30%, a large excess. Furthermore, there had been another 100-year flood in 1936, a span of only 36 years, Figure 5.9. Superficially, it seems that two 100-year floods should not occur so close together (see Ref. 18).

However, one must consider an alternative interpretation. That is, the probability of a 100-year flood in any single year is 1%, or a probability of .01. Over a span of years, the risk of occurrence increases as some function of the sample size, or, in this case, as the number of consecutive years of additional observation are added. The resultant combined probabilities have been presented by Professor Reich (18), based on earlier work of Beard and Markowitz and are included here as Table 5.7.

Observe that the probability of one or more floods greater than the 100-year flood in 36 years is about 30%. Even a 1000-year flood may occur within a 50-year period with nearly 5% probability of occurrence, about 1 chance in 20. Therefore, it is important to be aware of the cumulative effect of relatively small initial probabilities of occurrence in any one time period or in any one sample.

The current concern for ecology has led to greatly increased funds for studies of air and water pollution. Much of this data will be extreme value distributed. In order to set standards, to issue warnings, and to assess penalties, such data must be carefully evaluated and interpreted in a rational context or the results will be unrealistic standards and inadvertently punitive enforcement.

Preparation of Grouped Data

In the previous examples, data from small samples were presented and all the data from each sample were used to prepare the probability plots. However, much data occurs in large samples and is often presented as grouped data. Grouped data occur in two forms. The first form is where the variable of interest is discrete; that is, where it can take on only integer values, as in Table 5.8. The second form is where the variable is continuous, that is, it can take on noninteger values, that is, fractional or decimal values, as in Table 5.9.

First, we will demonstrate that these data have the skewed behavior that is characteristic of extreme value distributions. The cell frequencies are plotted in Figure 5.10 as histograms and both show a heavy end to the left and a thin tail to the right indicating right skew. Then, the data are plotted on extreme value probability paper, Figure 5.11. The plots are obtained by drawing a horizontal line representing the cumulative probability range for each cell against the number of stops in Figure 5.11*a* and against the

TABLE 5.7. Chances of Recurrence Based on Return Periods

	RETURN PERIOD, years →	2	5	10	20	50	100	200	500	1,000	2,000	5,000	10,000
% CHANCE OF GETTING ONE OR MORE SUCH OR BIGGER FLOODS IN THIS MANY YEARS	ANY ONE YEAR	50	20	10	5	2	1	.5	.2	.1	.05	.02	.01
	TEN YEARS		80	65	40	18	9.6	5	2	1	.5	.2	.1
	TWENTY-FIVE YEARS		99	94	71	40	22	12	5	2.5	1.2	0.5	.25
	FIFTY YEARS			99.9	90.5	61	39	22	9.5	4.8	2.3	1.0	0.5
	ONE HUNDRED YEARS					86	64	40	18	10	5	2	1

TABLE 5.8.   Grouped Data — Discrete Values

Number of Stops	Frequency Cell	Frequency Cum.	Cumulative Probability From	Cumulative Probability To
1	6	1–6	.011	.065
2	29	7–35	.076	.380
3	30	36–65	.391	.706
4	15	66–80	.717	.870
5	8	81–88	.880	.956
6	2	89–90	.967	.978
7	1	91		.989

TABLE 5.9.   Grouped Data — Continuous Values

Downtime Hours	Frequency Cell	Frequency Cumulative	Cumulative Probability From	Cumulative Probability To
0.00–0.99	1	1	.004	
1.00–.99	19	2–20	.008	.080
2.00–2.99	85	21–105	.084	.420
3.00–3.99	56	106–161	.424	.644
4.00–4.99	58	162–219	.648	.876
5.00–5.99	17	220–236	.880	.944
6.00–6.99	8	237–244	.948	.976
7.00–7.99	4	245–248	.980	.992
8.00–8.99	1	249		.996

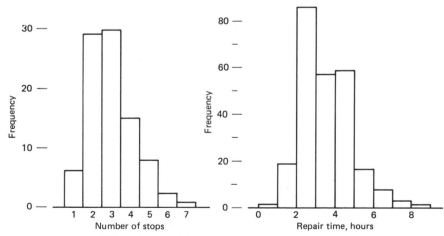

FIGURE 5.10.   Histogram of stops and repair time.

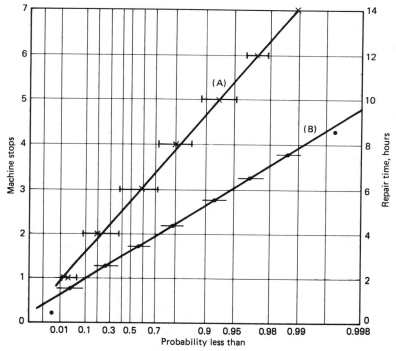

FIGURE 5.11. Probability plots of stops and repair time.

midpoint of each cell in Figure 5.11*b*. The middle of each horizontal probability range is established by measuring with a ruler and making a dot. The median regression line is then established by drawing a line through the dots so that as many dots as possible fall on the line and so that the remaining dots are evenly divided above and below the line. This then transforms the data into a straight line probability function which can be used for various kinds of statistical estimation. More detailed require-ments for data preparation of this type can be found in Ref. 1.

An Industrial Engineering Example

Figure 5.11 showed the results of a joint study of the number of machine stops per shift and the distribution of downtimes caused by machine stops, which are adapted from Berretoni (19). In this example, down time includes time to determine a failure after some level of malfunction, waiting time for the repairman, and time to repair. The plots for stops and down time are each extreme value distributed.

Such plots can be used to evaluate staffing requirements. For example, suppose that it is desirable to have enough repairmen available to cover at least 90% of possible repair requirements. This is a joint probability problem since both the number of stops and the downtime enter in. The requirement is determined as

$$P_s = 1 - (1 - p_j)^{1/2} \tag{5.7}$$

where

p_j = the joint probability required
P_s = the single probability required for each individual distribution

solving the example $P_s = 1 - (1 - p_j)^{1/2} = 0.68$.

That is, in order to cover for a joint probability of 90%, we must be able to handle the 68% levels of stops and time because the probability of simultaneous occurrences of stops and time equal to or greater than the 68% level of each factor is 90%.

Rounding off, we will estimate the values for stops and time at the 70% point. For stops, this value is 3.5 and for downtime, it is 4 hours. Thus, the 90% capability required is 3.5 × 4, or 14 manhours. Allowing for breaks and personal time, a work shift is approximately $7\frac{1}{3}$ hours. A 14 manhour requirement equals 1.9 men required, or 2, rounding off.

However, the median work load, when $p_j = 50\%$, occurs at $P_s = 0.29$, resulting in 2.25 × 2.6 = 5.85 manhours so that, half of the time, both men are idle more than half a shift. Worse yet, at the lower end, at the 10% level, the work load is only 1.3 manhours so that both men are hardly working at all 10% of the time.

The preceding suggests that staffing for repair work can be apparently inefficient when the staffing is dependent on some peak load criterion. Often, such inefficiency can be justified by cost-benefit studies when such studies show that the cumulative costs of downtime would exceed the excess costs of the repair department. However, the idle time shown above can be estimated along with the necessary times to provide repair coverage and the idle time can be usefully applied to provide preventive and corrective maintenance.

The potential of a properly conceived corrective maintenance program is considerable. A simple analysis of causes of stops and associated times to repair will show the proper mix of corrective actions required to reduce both the number of stops and the longest times to repair. For example, suppose that an effective maintenance program reduces the values at the 70% level by 20% for both stops and times to repair, then the 90% staffing level only requires about 9 hours. At this level, it would be feasible to have only one repairman and take care of the occasional overruns with casual overtime resulting in a substantial labor saving and increased operating efficiency. The important key to such potential savings, however, is the capability to analyze such a problem on a probabilistic basis which provides sound estimates of the risks and rewards. Techniques similar to this are used as the basic methods in operations research and analysis.

An Example from Sales Analysis

Berretoni (19) presented an interesting example of sales life for ethical drugs from a graduate student project. The underlying question was: "How long do new ethical drugs survive in the marketplace?" The approach used was to start with a drug catalog, circa 1950, and to determine which new products were introduced in that year. The sales life was considered to end when the drug was no longer listed in the catalog. Table 5.10 gives the results.

Figure 5.12 displays these results as a logarithmic extreme value plot. Originally, Berretoni used these results as an example of an application of the Weibull distribution. The cumulative distribution form of the Weibull is also given by the logarithmic extreme value as indicated by Nelson (20).

Analysis of Figure 5.12 indicates that about 20% of the products introduced about 1950 were gone from the catalog within 5 years and that about 40% were gone in 10 years. What is not known from this data is the breakdown of the pharmacological classifications of the drugs with early market failure as opposed to those with greater survival. If this were known, it would be possible to develop an associated plot of failure risk by classification equivalent to Figure 5.11 to estimate the risk–reward probabilities of specific categories of drug development in order to determine those categories with higher probabilities of payoff. Such categorization

TABLE 5.10. Sales Life for Ethical Drugs

Catalog Life, years	Frequency		Cumulative Probability	
	Cell	Cumulative	From	To
2	7	1–7	.009	.061
3	6	8–13	.070	.114
4	9	14–22	.123	.193
5	4	23–26	.202	.228
6	6	27–32	.237	.281
7	3	33–35	.289	.307
8	5	36–40	.316	.351
9	3	41–43	.360	.377
10	0			
11	1	44		.386
	Total sample	114		

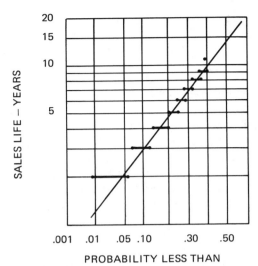

FIGURE 5.12.   Probability plot of drug sales life.

would be extremely pertinent since we now know that certain classes of chemotherapeutic drugs also lead to survival mutations of the bacterial and viral strains which are initially controlled.

Another Example of Logarithmic Extreme Value Behavior

One highly specialized, but useful, class of applications of the logarithmic extreme value distribution is in the field of particle size determination. This was probably first done in the analysis of the particle size distribution of coals where it is known as the Rosin–Rammler distribution (21, 22). The distribution parameters are used in studies of combustion efficiency of variously sized coals.

A second area of development was in the field of sprays and aerosols where it is known as the upper-limit distribution (23). It is interesting to note that we keep running into equivalent problems in different technological areas and that, because of different technological biases, the mathematical and statistical developments acquire different nomenclature and titles.

A third area of application is in powdered metals and refractory oxides where sintered objects are made. The particle size distribution is important in controlling the final sintered characteristics. Figure 5.13 shows the particle size distribution for a particular lot of powdered metal. The slope of the plot, downward to the right, indicates that this distribution has left skew; that is, the long tail of this distribution stretches out to the left, or in the direction of the smallest particle sizes.

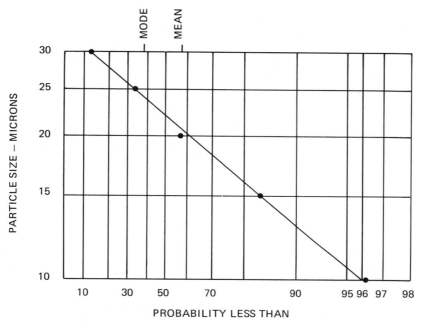

FIGURE 5.13. Plot of metal powder particle size.

The Weibull Distribution

We will close our consideration of extreme value distributions with a discussion of the Weibull distribution.

The Weibull distribution was named after Wallodi Weibull, a Swedish engineer, due to his publication of early practical applications in strength of materials, rupture of solids, and the distribution of the particle size of fly ash, in a Swedish engineering journal during 1939. His work became better known after 1951 when it was published in the *Journal of Applied Mechanics* (24). Because of what has followed since, it is interesting to note that the title of the 1951 paper was: "A Statistical Distribution of *Wide* Applicability," (emphasis added).

In the middle 1950s, there were more published applications: Leiblein and Zelen, of NBS, on fatigue life of ball bearings (25); J. H. K. Kao, at Cornell University, on vacuum tube life testing and failure analysis (11); and Johnson, at General Motors Research Labs, on wear, fatigue, and optimized life test plans (26). In 1963, Berretoni, of Case Western Reserve University, published applications for corrosion resistance, time lags for customers to return defective goods, number of machine stops per shift, time to leakage failure of dry cell batteries, rate of market failure of ethical drugs, and of the reliability of step motors and solid tantalum capacitors (19). About 1969, Nelson, of the General Electric Research and Development Center, showed applications to warranty problems and assessment of the effects of planned engineering changes in product life (27, 28).

In recent years, there has been a large amount of intensive study given to the Weibull distribution, resulting in a variety of mathematical derivations of the Weibull distribution which have, in turn, led to improved methods of parameter estimation. These developments have been summarized by Mann et al. (12) and by Thoman et al. (29). Most of this recent work has required third-generation computers to facilitate the large calculation efforts involved. However, some practical results have emerged which can be used to simplify and to improve graphical methods of Weibull parameter estimation.

Mathematics of the Weibull Distribution

The Weibull distribution is also known as a type III extreme value distribution. The Weibull equation has three parameters which make it a very general distribution capable of representing a wide variety of data. It has become particularly popular for representing failure data because of the many different shapes this distribution may assume. The simplest form of the cumulative distribution function, following Kao (11), is

$$F(x) = 1 - \exp\left[-\frac{(x - \gamma)^{\beta}}{\alpha} \right] \qquad (5.8)$$

where

x = time, number of cycles, or stress units
α = the scale parameter
β = the shape parameter
γ = the location parameter

The location parameter, γ, has the effect of moving the distribution along the time base as shown in Figure 5.14. When failures may be expected to begin as soon as an item is placed under stress or in use, then $\gamma = 0$. On the other hand, many items have or, at least, are expected to have, some period of failure-free operation and/or use, where x is then some positive time after zero. This failure-free period is often called the guarantee or warranty period for such products as have been designed and manufactured with the intent of providing some satisfactory lifetime under a broad spectrum of use conditions.

The shape parameter, β, determines how the shape of the Weibull failure function varies with time. When $\beta > 1$, we know that the failure rate is increasing with time; for $\beta = 1$, the failure rate is uniform or

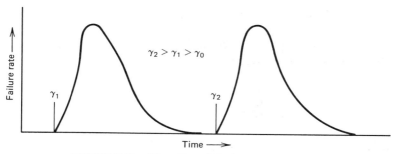

FIGURE 5.14. The effect of the location parameter.

constant over time, a condition also known as the exponential failure distribution; and, for $\beta < 1$, the failure rate is decreasing with time. These relationships are shown in Figure 5.15.

The scale parameter, α, also affects the shape of the Weibull distribution. For $\beta > 1$, increasing values of α cause the failure rate curve to flatten out, while for $\beta < 1$ increases in α also flatten out the failure rate curve for a fixed β as well as giving a relative failure rate which is always lower as shown in Figure 5.16.

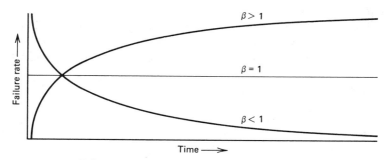

FIGURE 5.15. Effects of the shape parameter.

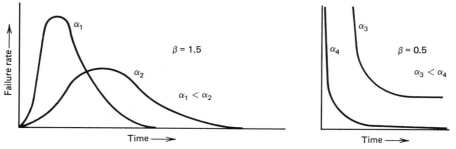

FIGURE 5.16. Effects of the scale parameter.

Description of Weibull Probability Paper

Weibull probability paper is derived from a rearrangement of equation (5.8). By transposing and inverting, we obtain

$$\frac{1}{1 - F(x)} = \exp\left[\frac{(x - \gamma)^{\beta}}{\alpha}\right] \tag{5.9}$$

Now, taking logarithms twice on each side, we have

$$\ln\ln\frac{1}{1 - F(x)} = \beta\ln(x - \gamma) - \ln\alpha \tag{5.10}$$

which has the linear form $y = a + bx$, as in Figure 5.1.

The values of $y = \ln\ln 1/[1 - F(x)]$ yield the right-hand vertical scale on Weibull paper, Figure 5.17. The actual values are multiplied by 100 to convert to percentages. The values of $\ln(x - \gamma)$ give the top horizontal scale. However, to facilitate plotting of real data, direct entry scales are provided. The left-hand vertical scale for $F(x)$ is scaled in cumulative percent. The bottom horizontal scale allows direct plotting of the observed times to failure or the value of other stress units.

It will be noted that the direct reading scales are reversed from the orientation which is generaly used for other probability papers. This format has been generally used since the first presentation by Kao (11) and has been widely followed in the literature of Weibull applications. In spite of the change in orientation of Weibull paper compared to more conventional probability papers, the same methods of preparing and plotting data which we have been covering can be used.

Weibull probability papers are available (30) with cumulative percentage scales covering 0.0001%, 0.01% and 1.0% to 99.9% combined with time/stress scales of 1, 2, 3, 5, and 7 logarithmic cycles.

Other features of Weibull paper are the principal ordinate, running vertically from 0.0 on the top scale to 1 on the bottom scale; and the principal abscissa from 0.0 on the right-hand scale to 63.2% on the left scale. There are also two circled cross marks toward the upper left quadrant of the paper in Figure 5.17. These features are used to estimate the several Weibull parameters.

Estimation of Weibull Parameters

In Figure 5.17, we show an arbitrary line to demonstrate the method of estimating Weibull distribution parameters graphically. In many cases, as previously discussed, γ is known to be 0 or may practically be assumed to

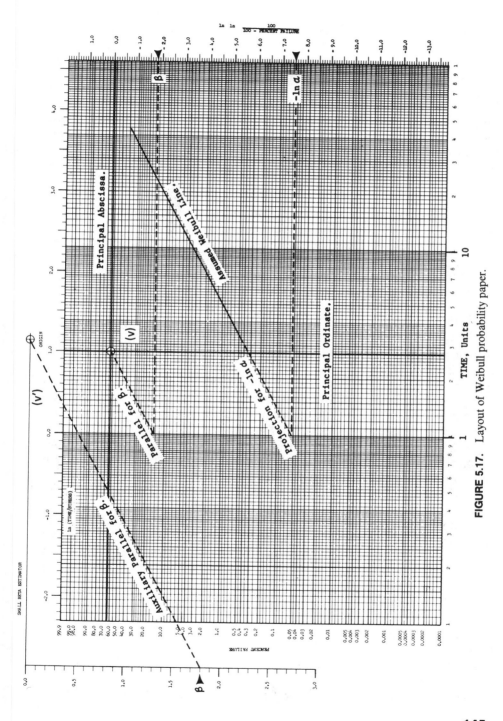

FIGURE 5.17. Layout of Weibull probability paper.

145

be 1. When the time to first failure is a definite time greater than 0, the simplest estimate of γ, or $\hat{\gamma}$, is the actual observed time at the occurrence of the first failure

$$\hat{\gamma} \cong t_1$$

where t_1 is the time to first failure.

The value for β is estimated from the slope of the Weibull line. An auxiliary line is drawn parallel to the Weibull line passing through the circled cross mark, Figure 5.17, (V) and down to the left until it intersects the principal ordinate. The point of intersection is then projected to the right-hand scale and the value of the intercept there is the estimate, $\hat{\beta}$, for that Weibull line. To improve estimates of small values of β, there is a second circled cross mark (V') above the main grid which may be used. In this case the auxiliary parallel line is drawn down only to the left vertical scale where the value for $\hat{\beta}$ can be read directly, as shown in Figure 5.17.

The value for α is estimated by drawing the Weibull line down to the left, if necessary, to intersect the principal ordinate. This intersection is also projected to the right-hand scale and the value of the intercept there is $-\ln \hat{\alpha}$. The value for $\hat{\alpha}$ is obtained from

$$-\ln \hat{\alpha} = -X$$
$$\hat{\alpha} = \text{anti ln } x$$

where x is the absolute value of the intercept, and anti $\ln(\ln^{-1})$ is of the natural logarithm.

Other parameters of the Weibull distribution can be estimated from the Weibull plot and Figure 5.18. The characteristic life, $\hat{\eta}$, is obtained from the intersection of the Weibull line and the principal abscissa, which passes through the 63.2% point on the left-hand scale and the 0.0 point on the right-hand scale. The mean of the distribution, $\hat{\mu}$, is estimated by first referring to Figure 5.18. Using the value of $\hat{\beta}$ obtained earlier, enter the bottom scale of Figure 5.18 and determine the percentage point at which the mean occurs for that value of $\hat{\beta}$. Then, enter the Weibull plot from the percentage point determined in Figure 5.18, project across to the Weibull line, and then down to the bottom scale to obtain the value of $\hat{\mu}$. The standard deviation, $\hat{\sigma}$, is also obtained by using Figure 5.18. Using $\hat{\beta}$, find the estimating factor on the upper scale. Multiply this factor by $\hat{\eta}$ to obtain $\hat{\sigma}$. Note that for small values of β, the value of the factor increases rapidly which causes very large and practically meaningless values for σ.

Fitting a Weibull Line

Various methods have been suggested for obtaining a "best-fit" straight line to data plotted on Weibull paper. In earlier publications, plotting positions based on $y_i/(n + 1)$ were recommended. However, for plotting

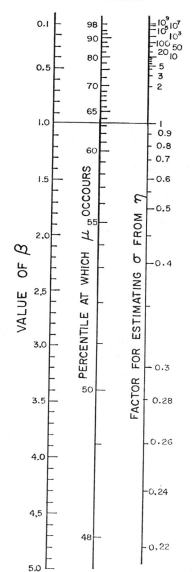

FIGURE 5.18. Nomograph of the relationship of β, μ, and σ.

data that is expected to be Weibull distributed, there is an improved method based on studies by Mann et al. (12). Plotting positions that yield "best-fit" Weibull lines that agree closely with calculated least-squares lines are obtained by using

$$x_i = \frac{y_i - 0.5}{n}$$

TABLE 5.11. 80% Confidence Intervals for the Weibull Distribution. Lower Limit (L) and Upper Limit (U) Expressed as Percentages at Preselected Percentile Points for Several Sample Sizes

Sample Size n	Preselected Percentiles																	
	1		5		10		30		50		70		90		95		99	
	L	U	L	U	L	U	L	U	L	U	L	U	L	U	L	U	L	U
10			1.0	21	3.0	28	15	50	31	69	50	85	72	97.0	79	99.0		
20			1.5	15	4.2	22	19	44	36	64	56	81	78	95.8	85	98.5		
30			1.8	13	4.9	19	21	42	39	61	58	79	81	95.1	87	96.2		
40			2.2	12	5.7	19	22	40	40	60	60	78	81	94.3	88	97.8		
50	0.23	5.0	2.5	11	6.3	18	23	39	41	59	61	77	82	93.7	89	97.5	95.0	99.77
100	0.32	3.0	3.1	10	7.4	17	25	35	45	55	65	75	83	92.6	90	96.9	97.0	99.68

where

x_i = cumulative probability of occurrence of the ith ordered variable
y_i = rank value of ith item of ordered data
n = sample size, or, number of observations

Further details may be found in TEAM (31).

Comparing Results for Two Weibull Tests

It is frequently of interest to compare results of two or more Weibull tests. Such comparisons may be between an unknown sample and a standard for process or product control. Or, one may evaluate effects of a change in a process or product; or, one may compare competitive products. An approach suggested by Johnson (32) is useful. Analytical comparison of Weibull data requires allowing for variations in β, in sample size differences and confidence intervals. Graphical comparison, using Table 5.11, is easier and more flexible. Table 5.11 is similar to the commonly used 95% confidence interval tables for obtaining confidence limits about the best-fit Weibull line except that it gives 80% confidence bounds.

Table 5.12 lists data for two 12-piece samples of polyether urethane from Hill (33). These samples were tested under substantially identical conditions to determine the existence of any difference between them.

The data for each sample are plotted, the Weibull line estimated, 80% confidence bounds for each line are added, and the bounds are drawn in as shown in Figure 5.19. Up to about 930 psi, the confidence regions overlap

TABLE 5.12.   Comparison of Two Polyether Urethane Samples

Rank Number	$100(n - 0.50)/n$ Plotting Position	Tensile Strength, psi			
		Sample A Data $- \hat{\gamma} = 5300$		Sample B Data $- \hat{\gamma} = 4800$	
1	04.17	5775	475	5385	585
2	12.50	5830	530	5800	1000
3	20.83	5875	575	6020	1220
4	29.17	6000	700	6100	1300
5	37.50	6025	725	6190	1390
6	45.83	6100	800	6220	1420
7	54.17	6225	925	6310	1510
8	62.50	6280	980	6420	1620
9	70.83	5340	1040	6515	1715
10	79.17	6440	1140	6550	1750
11	87.50	6450	1150	6625	1825
12	95.83	7200	1900	6640	1840

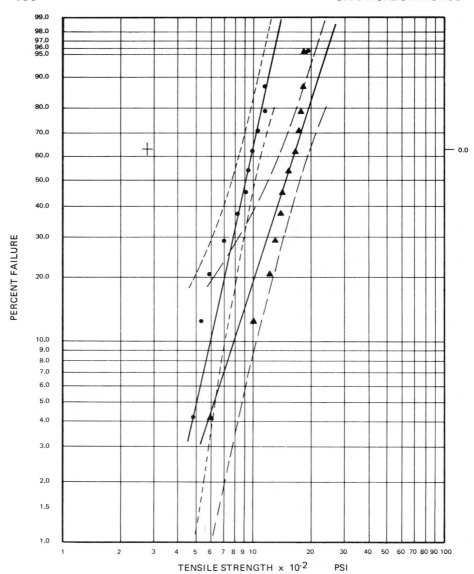

FIGURE 5.19. Comparison of two polymers for tensile strength at break using 80% confidence bounds.

and there is no practical difference in results but, after 930 psi, sample B looks increasingly better. Once the two confidence regions have diverged, there is 99% confidence that the results are significantly different and confidence increases as the divergence increases. If a more critical test is desired, 95% confidence intervals can be used (Table 5.11) and then, after a point of divergence, there is at least 99.95% confidence that a significant difference exists.

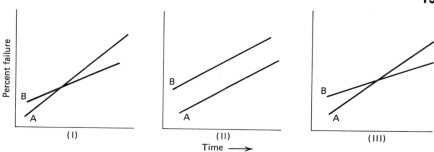

FIGURE 5.20. Typical patterns in Weibull comparisons.

There are three typical patterns that occur in comparisons of Weibull plots as shown in Figure 5.20. At *I*, the pattern is similar to Figure 5.19 with overlap in early time making it difficult to differentiate between the two samples. However, with time, there is increasing divergence of results so that A shows up better with more time. At II, the samples are reacting at two different performance levels with A everywhere better than B. However, confidence regions could overlap in early time because of the large relative uncertainty at small failure percentages. AT III, sample A may appear better in early time, but then, results converge and become indistinguishable for some time and eventually sample B looks better than A.

REFERENCES

1. King, J. R., *Probability Charts for Decision Making*, Revised Editions, TEAM, New Hampshire, 1981.

2. Enrick, N. L., *Quality Control and Reliability*, 7th ed., Industrial Press, New York, 1977.

3. Enrick, N. L., *Effective Graphic Communication*, Auerbach, Princeton, NJ, 1972.

4. Myers, B., and Enrick, N. L., *Statistical Functions*, Kent State University Press, Kent, OH, 1972.

5. Grant, E. L., and Leavenworth, R. S., *Statistical Quality Control*, 5th ed., McGraw-Hill, New York, 1980.

6. Hahn, G. J., and Shapiro, S. S., *Statistical Models in Engineering*, Wiley, New York, 1967.

7. Yabuta, S., and Kase, S., Distribution of tensile data of vulcanized rubbers, *Journal of Polymer Science*, Vol. 6, 1968, pp. 639–651.

8. King, J. R., Data analysis in physics of failure studies, *Trans. 19th Ann. Tech. Conf.*, ASQC, 1965.

9. Gumbel, E. J., *Statistical Theory of Extreme Values*, NBS, AMS #33, 1954.

10. Wallis, W. A., and Roberts, H. V., *Statistics, A New Approach*, Free Press of Glencoe, 1956.

11. Kao, J. H. K., A summary of some new techniques on failure analysis, *Proc. of 6th Nat. Symp. on R & QC in Electronics*, 1960.

12. Mann, N. R., Schaffer, R. E., and Singpurwalla, N. D., *Methods for Statistical Analysis of Reliability and Life Data*, Wiley, New York, 1974.

13. TEAM, *TEAM Easy Analysis Methods*, Vol. 3, No. 2, 1976.

14. Gumbel, E. J., *Statistics of Extremes*, Columbia University Press, New York, 1958.

15. Botts, R. R., Extreme value methods simplified, *Agricultural Economic Research*, Vol. 9, No. 3, July 1957.

16. Haviland, R. P., *Engineering Reliability and Long Life Design*, Van Nostrand, Princeton, N.J., 1964.

17. TEAM, *TEAM Easy Analysis Methods*, Vol. 1, No 1, Technical and Engineering Aids for Management, Tamworth, NH, 03886, 1974.

18. Reich, B. M., How frequently will floods occur?, *Water Resources Bulletin*, Vol. 9, No. 1, February 1973.

19. Berretoni, J. N., Practical Applications of the Weibull Distribution, *Industrial Quality Control*, Vol. 21, No. 2, August 1964.

20. Nelson, W., and Thompson, V. C., Weibull probability papers, *Journal of Quality Technology*, Vol. 1, No. 2, 1971.

21. Rosin, P., and Rammler, E., The laws governing the fineness of powdered coal, *Journal of the Fuel Institute* Vol. 7, 1933, p. 29.

22. Lapple, C. E., Particle size analysis and analyzers, *Chemical Engineering*, May 20, 1968.

23. Mugeles, R. A., and Evans, H. D., Droplet size distribution in sprays, *Industrial and Engineering Chemistry*, Vol. 43, No. 6, June 1951.

24. Weibull, W., *J. Appl. Mech.*, Vol. 18, 1951, p. 293.

25. Leiblein, J., and Zelen, M., Statistical investigation of the fatigue life of deep groove ball bearings, Research Paper 2719, *Journal of Research*, *NBS*, Vol. 57, 1956.

26. Johnson, L. G., *The Statistical Treatment of Fatigue Experiments*, General Motors Research Laboratories Report 202, 1959, Elsevier, New York, 1964.

27. Nelson, W. B., Hazard plotting methods for analysis of data with different failure modes, *Journal of Quality Technology*, July 1970.

28. Nelson, W. B., Hazard plotting for incomplete failure data, *Journal of Quality Technology*, Vol. 1, No. 1, 1969.

29. Thoman, D. R., Bain, L. J., and Antle, C. E., Inferences on the parameters of the Weibull distribution, *Technometrics*, Vol. 11, No. 3, August 1969.

30. Catalogue and Price List, TEAM, Box 25H, Tamworth, NH 03886.

31. TEAM, *TEAM Easy Analysis Methods*, Vol. 3, No. 1, 1976.

32. Johnson, L. G., *The Statistical Treatment of Fatigue Data*, Elsevier, New York, 1964.

33. Hill, H. E., Application of the Weibull distribution function in the coatings industry, part II, *Journal of Paint Technology*, Vol. 47, No. 606, July 1975.

6

HYPOTHESIS TESTING

JOHN V. KNOPP

1025 Reynolds Road
Johnson City, New York

> *There are lies, damn lies, and statistics.*
>
> MARK TWAIN

Possibly Mark Twain wrote the above quotation after seeking advice from an eminent statistician friend on whether or not to raise a bet with a given poker hand. The statistician, after taking into account the number of cards dealt, the cards showing, and the cards held by Twain, calculated that there was only a 1 in 20 chance that Mark Twain's opponent could have a better hand than Twain held. The statistician assured Mark Twain that there was only a 5% risk that he could lose. Mark Twain proceeded to bet and lose. After writing the above quotation, Mark Twain probably lost a statistician friend as well.

The above incident probably never happened; however, the story does serve to illustrate two points that are important in hypothesis testing. First, there are always risks involved in hypothesis testing. Hypothesis testing allows us to quantify the risk. Unfortunately, hypothesis testing can do nothing to eliminate risk. We will return to the discussion of risk later in the chapter.

The second point that the above story illustrates is that assumptions are used in hypothesis testing. In the above story, the statistician assumed among other things that poker was a game of chance. If you have ever played poker, you will recognize the fallacy of this assumption.

All tests presented in this chapter will be based on the assumption that:

1. The data are normally distributed.
2. A random sample has been obtained.

Neither assumption should be taken lightly. Not all data are normally

153

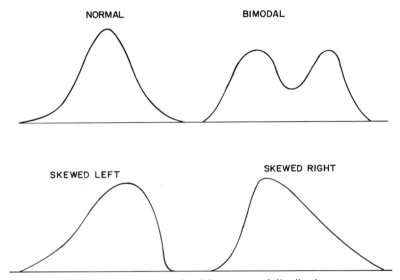

FIGURE 6.1.   Example of four types of distributions.

distributed. When a sufficiently large sample is taken, it is desirable to plot a histogram to detect non-normal distributions. A sample of a hundred data points or more is needed to obtain a clear picture of the actual distribution; however, small samples can be used to detect gross non-normal distribution.

In Figure 6.1, histograms of common distributions are shown. Two non-normal distributions that often occur are bimodal and skewed distributions. Bimodal distributions are the result of the sample containing two populations. When a bimodal distribution occurs, the factor causing the two populations should be eliminated or sampling should be from one of the populations. Skewed distributions are often the result of the measure used for the sample. For example, if the mass of a spherical material is normally distributed, the diameter will not be normally distributed but will be skewed toward the smaller diameters.

A technique that can be used to ensure that a near-normal sample is present is to use the distribution of means. The central limit theorem states that if a population has a finite variance and finite mean, as the sample size increases, the distribution for the sample mean approaches a normal distribution with mean μ and variance σ^2/n.

6.1.   STATEMENT OF THE HYPOTHESIS

In hypothesis testing, we are not concerned with the risk in winning at poker but are interested in determining whether or not one process or procedure is superior to or equivalent to another process or procedure in either mean value or variability. For example, in the photographic industry, we measure coating weight in terms of grams of silver per square meter of support. Any process or procedure which can reduce the amount of silver required and still provide the needed photographic properties is highly desirable. Hypothesis testing can be used to determine if the new process or procedure has:

1. Reduced the silver coating weight.
2. Provided the same photographic properties.
3. Provided a product with the same or less variability.

In establishing a hypothesis test, the first requirement is a statement of the null hypothesis. If the null hypothesis is rejected, then an alternative hypothesis must be available. The convention for stating the null hypothesis and alternative hypothesis is:

1. Null hypothesis. The test statistics are equal; that is, there is no difference between the old process and the new process.
2. Alternative hypothesis. The test statistics are not equal; that is, there is a difference between the old and new process.

If, for example, the mean values of the new and old process were being compared, the null hypothesis would be that the two means are equal or

$$\mu_1 = \mu_2$$

and the alternative hypothesis would be that the two means are not equal or

$$\mu_1 \neq \mu_2$$

Hypotheses are stated in the above manner since rejection of the null hypothesis requires acceptance of the alternative hypothesis. Often very little data are required to reject the hypothesis that two things are equal and accept inequality. On the other hand, an infinite amount of data may be required to reject the hypothesis that two means are unequal, that is, that two means are equal.

By not rejecting the null hypothesis, we are not stating equality. We are only stating that there is insufficient data to establish inequality.

There are cases where we are only interested in establishing whether or not one mean is greater than another mean. In this case, the null hypothesis would be stated as

$$\mu_1 = \mu_2$$

and the alternate hypothesis would be stated as

$$\mu_1 > \mu_2$$

6.2. TYPES OF ERRORS AND RISKS

There are two types of errors associated with any hypothesis test:

1. Type I error is that of rejecting the null hypothesis when it should have been accepted. Associated with the type I error is an alpha (α) risk. The α risk is the probability of being wrong when the null hypothesis is rejected.
2. Type II error is that of accepting the null hypothesis when it should have been rejected. Associated with the type II error is a beta (β) risk. The β risk is the probability of being wrong when the null hypothesis is accepted.

Unfortunately, the α and β risks are not independent. The α and β risks and sample size are related. For a given sample size, one risk can only be decreased by increasing the other. Conversely, for a given α and β risk, the sample size is fixed.

The interdependence of the α risk, β risk, and sample size can clearly be shown using normal curves. In Figure 6.2, a normal distribution is shown with a 1.645 cutoff. Ninety-five percent of the data from the population should fall in the unshaded area and 5% in the shaded area. The shaded area is referred to as the critical area.

By chance only 5% of the data points could fall in the critical area. The 1.645 can represent a cutoff point for a hypothesis test. For example, if you had a process that you knew was normally distributed with variance σ^2 and mean of μ and you suspected the level had shifted to a high value, then you could define a hypothesis test using one data point:

Null hypothesis	New level = μ
Alternate hypothesis	New level > μ
α risk	5%

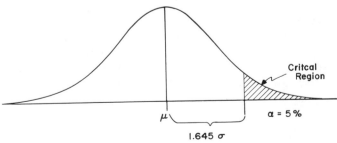

FIGURE 6.2.   One-tailed 5% critical region for a normal distribution.

If the data point fell in the critical region, then the null hypothesis would be rejected and the alternate hypothesis accepted.

Normally, you would not use one piece of data to arrive at a decision, but would use the mean of a sample. As the sample size increases, the distance of the critical region from the old mean, μ in this example, would decrease inversely proportional to the square root of the sample size. This reduction in the critical region's distance from the old mean is depicted in Figure 6.3 for different sample sizes.

As the sample size increases, the difference between the critical region and the old mean becomes unimportant. Once you have established the unimportant difference, you are in a position to define the β risk. In Figure 6.4, two normal curves with variances σ^2 are shown. One normal curve is centered about the old mean of μ and the second normal curve is centered

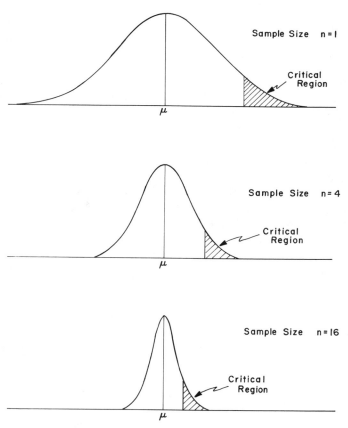

FIGURE 6.3. Reduction of critical distance from the mean as the sample size increases.

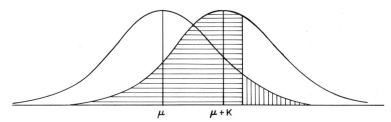

FIGURE 6.4. α and β critical regions for a sample of size 1 and a .8 σ/K ratio. The α region of .05 is shown as vertical lines. The β region of .67 is shown as horizontal lines.

about $\mu + K$ where K is the minimum value for a level shift to be important. For a single data point, the critical region for the α risk of 5% in Figure 6.4 is the same as the critical region in Figure 6.2. If the single data point fell in the area shaded with vertical lines, the null hypothesis would be rejected with an α risk of 5%.

The β risk for the hypothesis would be the area shaded with horizontal lines. For the example shown in Figure 6.4, the β risk would be 67%. If the new level were at $\mu + K$, there would be a 67% chance of the data point falling in the critical region for the β risk.

For the situation shown in Figure 6.4, a single data point would not provide a strong hypothesis test since the β risk would be very poor. The way to improve (reduce) the β risk without changing K or α risk is to increase the sample size. In Figure 6.5, the α and β risks for three sample sizes are shown.

Before conducting an experiment, the null hypothesis and alternative hypothesis must be stated. Further, the α and β risks must be selected. The α and β risks determine the sample size to be used. In the previous example, if the α risk were selected as 5% and the β risk were selected as 10%, then the sample size would be

$$1.645\frac{\sigma}{\sqrt{n}} + 1.282\frac{\sigma}{\sqrt{n}} = K$$

or

$$n = \frac{(2.927)^2\sigma^2}{K^2} = 5.48 \simeq 6$$

where

$$\frac{\sigma}{K} = .8$$

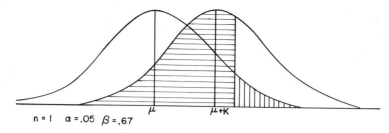

n = 1   α = .05   β = .67

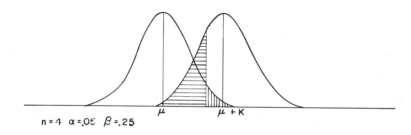

n = 4   α = .05   β = .25

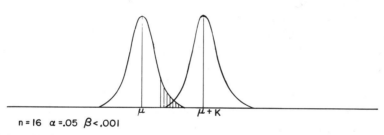

n = 16   α = .05   β < .001

FIGURE 6.5. Reduction of β risk as a function of sample size when the α risk is held constant.

The examples used so far have been for single-sided tests. For a double-sided test, the null hypothesis and alternative hypothesis would be

| Null hypothesis | New level $= \mu$ |
| Alternative hypothesis | New level $\neq \mu$ |

For the double-sided test, there are two critical regions for the α risk as shown in Figure 6.6.

For the previous example, the variances were known. In actual cases, the variance is not known and the t distribution must be used. However, before discussing the t distribution, it is appropriate to provide some insights into practical coating problems.

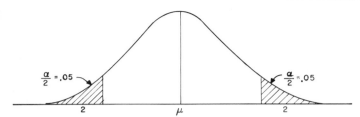

FIGURE 6.6.   Critical regions for double-sided test ($\alpha = .10$).

When a process is stopped and then restarted, a level shift can occur. Further, if a continuous process is fed by a batch operation, then level shifts will occur in the output of the process due to batch variation. Thus, the output of a film coating process is not truly random but will have both short- and long-term variations. Any experiment conducted on a coating process must reflect both long- and short-term variations.

For example, suppose you were interested in determining if a process change in coating affected your output. You might run one batch with process A and a second batch with process B. Now, no matter how many data points you collected, you would still have only one data point for a hypothesis test if batch-to-batch or startup variations are equal to or larger than the within batch variation. Considering the example shown in Figure 6.7 where five pairs of batches were used, the within batch distribution is shown as normal curves around the two levels.

Suppose that six rolls are coated from each batch and the mean value for each kettle is used to determine if there is a difference between the two processes. Analysis of the first two batches, Figure 6.7a, might indicate that A is greater than B. However, analysis of the two batches might indicate that B is greater than A, Figure 6.7b. Analysis of additional batches might indicate no difference between the processes.

For the example shown in Figure 6.7, there is no difference between the two processes. All curves were plotted using random numbers. The within-batch and batch-to-batch variations are equal and the means of both processes are equal. The different variances within batches and levels are the effect of chance.

By ignoring the batch-to-batch variation, misleading conclusions could be reached. The distribution of the means of the two processes in the five pairs of kettles is shown in Figure 6.7f. Even if the difference shown in Figure 6.7f was real, it is doubtful that the difference between process A and process B would be important since the batch-to-batch variation is so large.

An analysis of the two populations shown in Figure 6.7a might indicate a difference between A and B. However, the difference would be due to batch-to-batch variation and not to the two different processes.

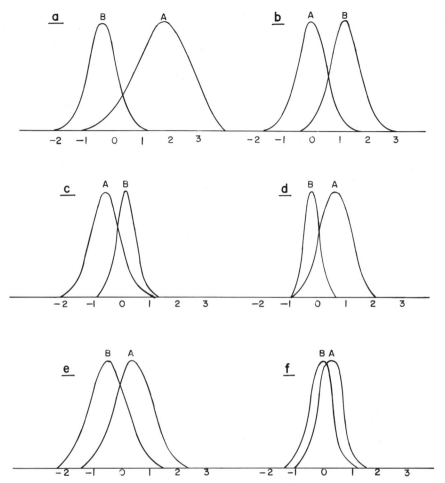

FIGURE 6.7. Possible distributions of sample within pairs of kettles when the kettle-to-kettle variation is as large as the within-kettle variation.

6.3. *t* DISTRIBUTION AND HYPOTHESIS TEST ON THE MEAN WITH UNKNOWN VARIANCES

When the variance of a process is known, the normal distribution can be used for hypothesis tests on the mean. For hypothesis tests on the mean when the variance is unknown, the *t* distribution is used. Note, the *t* distribution still assumes that the data are normally distributed.

The *t* distribution and normal distribution are both symmetrical about the mean. Further, for large sample sizes, the *t* distribution and normal distribution are essentially indistinguishable. Although the tables for the *t* and normal distributions are presented in a different format, both distributions are used by selecting critical regions under the distributions. In Figure 6.8, four *t* and normal distributions for different sample sizes are shown with their respective 0.10 critical regions.

Note that for small sample sizes, the spread of the *t* distribution is considerably larger than the spread of the normal distribution. This difference in spread occurs because with the normal distribution we know the variance. However, with the *t* distribution we do not know the variance and a calculated variance must be used. The inaccuracy in the calculated variance results in less accuracy in the value of the mean. As the sample

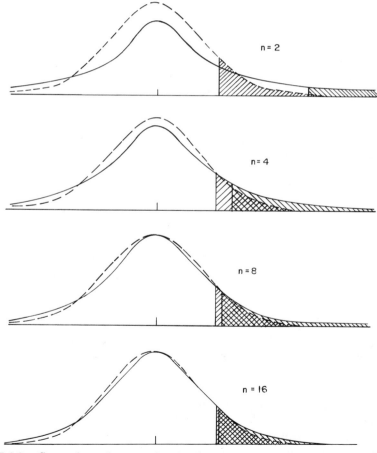

FIGURE 6.8.   Comparison of a normal and *t* distribution and their .10 critical regions for different sample sizes. As the sample size increases, the *t* distribution approaches a normal distribution.

size increases, the calculated variance better approximates the actual variance and the t distribution approaches the normal distribution.

In Appendix A.3, the cumulative t or "Student's" distribution is given for difference in α risks and degrees of freedom. Values of ν, degrees of freedom, are found in the left-hand column. The desired $1 - \alpha$ values are found in the upper row for a double-sided test and the lower row for a single-sided test.

There are three situations where a hypothesis test on the mean is used:

1. Test of sample mean against a known standard.
2. Test of two means.
3. Test of paired data.

Each of the above tests can be either single or double sided. In a single-sided test, you are only interested in determining if a mean is larger (smaller) than another mean or standard. For example, suppose you are interested in developing a harder coating and you have a formulation that you think will provide a harder coating. You are only interested in changing the formulation if it will provide a harder coating; therefore, you would conduct a single-sided test. For this example, the null and alternative hypotheses would be

Null hypothesis	New coating = standard coating
Alternative hypothesis	New coating > standard coating

In a double-sided test on the mean, you are interested in determining if a mean differs from another mean or standard. For example, suppose a questionable lot of new material is being tested; you would want to know if the raw material change will shift the process mean. For this case, you would be interested in both a positive and negative shift—a double-sided test. The null and alternative hypotheses would be

Null hypothesis	New mean = old mean
Alternative hypothesis	New mean $\neq$ old mean

6.4. COMPARISION OF A SAMPLE MEAN AGAINST A KNOWN STANDARD

There are many cases when a sample mean must be compared to a known standard or aim point. One such case would be the introduction of a new process. You want the mean response or responses of the new process to "equal" a standard or aim point. Equality in this case is acceptance of the null hypothesis and the overriding concern is that there is not a significant difference between the sample mean and the aim point or standard. The significant difference, estimated variance, α risk, and β risk are used to select the sample size for the test.

The calculations used in a hypothesis test are simple and straightforward. For convenience, the equations for a comparision of a sample mean against a known standard are presented in Table 6.1. An example of a hypothesis test of a sample mean against a known standard is presented below:

1. For a new product, a coating thickness of 6.8 to 7.2 μm is desired. Historically, the standard deviation for coating thickness has been 0.10 μm. Management is only willing to accept a 5% risk of rejecting a good product but is willing to accept a 10% risk of accepting a bad product.

2. Null hypothesis: $\mu = 7.0$
   Alternative hypothesis: $\mu \neq 7.0$
   α risk $= .05$
   β risk $= .10$

3. The needed sample can be obtained from Table 4.3 (Chapter 4) where

$$\frac{\delta}{s} = \frac{.2}{.10} = 2$$

   or sample size of 5.

TABLE 6.1. Comparison of Mean with Known Standard

	Double-Sided	Single-Sided
Null hypothesis	$\mu = \mu_0$	$\mu = \mu_0$
Alternative hypothesis	$\mu \neq \mu_0$	$\mu < \mu_0$ or $\mu > \mu_0$
Test statistic	$t = \dfrac{\overline{X} - \mu}{s_{\overline{X}}}$	
Definition of terms	n is the sample size	
	ν is the degrees of freedom, $\nu = n - 1$	
	$\overline{X}$ is the sample mean	
	μ_0 is the standard	
	$s_{\overline{X}}$ is the standard error of the sample	
	s is the standard deviation of the sample	
	$\overline{X} = \sum\limits_{i=1}^{n} \dfrac{X_i}{n}$	
	$s_{\overline{X}} = \dfrac{s}{\sqrt{n}} = \left[\sum\limits_{i=1}^{n} \dfrac{(\overline{X} - X_i)^2}{n(n-1)} \right]^{1/2}$	

If the test statistic, t, is greater than the appropriate one- or two-tailed value in the t table, Appendix A.3, for ν df and for $1 - \alpha$, the null hypothesis is rejected.

4. The following sample was obtained: 6.98, 7.19, 7.30, 7.15, and 7.18.
5. Referring to Table 4.3, the calculations are

$$\overline{X} = \frac{6.98 + 7.19 + 7.30 + 7.15 + 7.18}{5} = 7.160$$

$$s_{\overline{X}}^2 = .00329 \qquad s_{\overline{X}} = .057$$

$$t_\nu = \frac{7.160 - 7.00}{.057} = 2.807$$

6. The tabulated t for a two-tailed test with 4 degrees of freedom (df) and an α risk of .05 is 2.7764 so the null hypothesis is rejected. That is, we accept the alternative hypothesis that the new process does not equal the aim point.

I know we made the correct decision. Random numbers from a normal distribution with a mean of 7.21 and a variance $(.1)^2$ were used in the example. Note, however, that we were rather close to not making the right decision. The calculated and tabulated t values are nearly equal. We specified a risk of .10 and therefore accepted a one-in-ten chance of accepting the null hypothesis when we should have rejected it.

Suppose after reporting the results, you are informed by management that the thickness limits of 6.8 to 7.2 are too tight. (The original limits may have resulted in too many batches being rejected by quality control as being out of specification.) The new limits are 6.6 to 7.4. What do you do now? You are left with several options:

1. You can quit in disgust. Not a very practical solution.
2. You can calculate a new sample size and rerun the experiment. Again not a very practical solution but preferred over (1).
3. You can reanalyze the data.

One way to reanalyze the data is to place a confidence interval about the best estimate of the process mean. For the above example, this estimate is 7.16 μm. Confidence intervals will be covered in another chapter. However, they are of considerable utility in conjunction with hypothesis testing.

The confidence interval about the mean is found by centering a t distribution about the mean and calculating the distance to the critical region—see Figure 6.9. For the above example, the two-tailed 95% confidence interval would be

$$7.16 - t_{.025} \times s_{\overline{X}} < \mu < 7.16 + t_{.025} \times s_{\overline{X}}$$

or

$$7.001 < \mu < 7.318$$

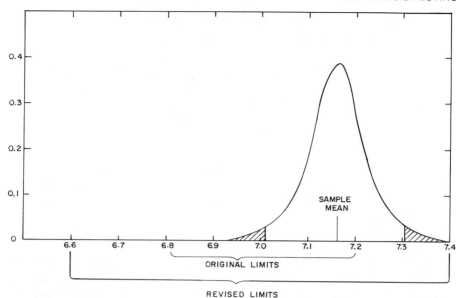

FIGURE 6.9. 95% confidence interval compared to original limits and revised limits used in the example given in "Comparison of Sample Mean Against a Known Standard."

There is only a 5% risk that the true mean μ falls outside the 95% confidence interval. Since the confidence interval falls within the new limits of 6.6 to 7.4, there is less than a 5% risk that the process mean is actually outside the limits.

Note also that the aim point of 7.0 μm is outside the confidence interval for the mean. This we should have expected since we rejected the null hypothesis. The confidence interval provided two pieces of information at a 95% confidence level:

1. The process mean is not equal to the aim point of 7.0 μm. This we knew from the original null hypothesis.
2. The process mean is within the new limits of 6.6 to 7.4 μm. This we did not discover in the original hypothesis test since our sample size was based on limits of 6.8 to 7.2 μm.

6.5. COMPARISON OF TWO MEANS

A common problem in the industrial environment is to determine if a process change or raw material change has shifted the process level. For such a problem, the means of two samples are compared. The null hypothesis in this case would be that the two means are equal:

$$\mu_1 = \mu_2$$

and the alternative hypothesis would be that the two means are not equal:

$$\mu_1 \neq \mu_2$$

or that one mean is larger (smaller) than the other:

$$\mu_1 > \mu_2$$

Again the major concern is that an important difference or shift is present. The significant difference, estimated variances, α risk, and β risk determine the needed sample size.

Again for convenience, the equations for a comparison between sample means are presented in Table 6.2. An example of a hypothesis test for two means is presented below:

1.   On a *pilot* coater, a new raw material A is being investigated as a possible replacement for raw material B. Since raw material A is cheaper

TABLE 6.2.   Comparison of Two Means

	Double-Sided	Single-Sided
Null hypothesis	$\mu^1 = \mu^2$	$\mu^1 = \mu^2$
Alternative hypothesis	$\mu^1 \neq \mu^2$	$\mu^1 > \mu^2$ or $\mu^2 > \mu^1$
Test statistic	$t = \dfrac{(\overline{X}_i - \overline{X}_2)}{s_{\overline{X}_1 - \overline{X}_2}}$	
Definition of terms	n_1 is the sample size for $\overline{X}_1$	
	n_2 is the sample size for $\overline{X}_2$	
	$\overline{X}_1$ is the sample mean for population 1	
	$\overline{X}_2$ is the sample mean for population 2	
	$s_{\overline{X}_1 - \overline{X}_2}$ is the estimate of the pooled standard error	
	s_p is the pooled estimate of standard deviation	
	s_1^2 is the variance of sample 1	
	s_2^2 is the variance of sample 2	
	ν_1 df for sample 1	
	ν_2 df for sample 2	

$$s_{\overline{X}_1 - \overline{X}_2} = s_p \sqrt{\frac{1}{n_1} + \frac{1}{n_2}} = \sqrt{\frac{\nu_1 s_1^2 + \nu_2 s_2^2}{\nu_1 + \nu_2}} \sqrt{\frac{1}{n_1} + \frac{1}{n_2}}$$

If the test statistic, t, is greater than the appropriate one- or two-tailed value in the t table, Appendix A.3, for $\nu_1 + \nu_2$ df and for $1 - \alpha$, the null hypothesis is rejected.

than B, it is hoped that the two raw materials provide equal output responses. One such output response is the density of the coated material.

The standard deviation of the process is known to be .10 density units and an important difference in product quality is .25. Management is willing to accept a 5% risk of rejecting the new raw material when it is good and a 5% risk of accepting the raw material when it is bad.

2. Null hypothesis: $\mu_A = \mu_B$
    Alternative hypothesis: $\mu_A \neq \mu_B$
    α risk = .05
    β risk = .05
    Significant difference = 0.25
    Estimated standard deviation = 0.10

3. The needed sample size can be obtained from Table 4.4 (Chapter 4) where

$$\frac{\delta}{s} = \frac{.25}{.10} = 2.5$$

or a sample size of 6.

4. The following data were obtained:

Raw Material A	Raw Material B
3.01	2.86
3.02	3.02
2.96	3.08
2.94	2.92
3.09	2.95
2.97	2.90

5. Referring to Table 6.2, the calculations are

$$\overline{X}_A = 2.998$$
$$\overline{X}_B = 2.955$$
$$s_{\overline{X}_1 - \overline{X}_2} = .028$$
$$t = 1.536$$

The tabulated t for a two-tailed α risk of 5% and 10 df is 2.228—see Appendix A.3.

6. Since the calculated test statistic is less than the tabulated t value for an α risk of 5% and 10 df, the null hypothesis is accepted.

As in the case of comparing a mean to a known standard, confidence intervals can be used to compliment a hypothesis test. The 95% confidence interval for the above example is

$$\left(\overline{X}_A - \overline{X}_B\right) \pm 2.2281 \times s_{\overline{X}_1 - \overline{X}_2} \quad \text{or} \quad -.019 < \mu_A - \mu_B < .086$$

The above confidence interval implies that there is only a 1-in-20 chance that the difference between the level due to raw material A and raw material B is less than $-.019$ or greater than $.086$. Note that zero is within the 95% confidence interval. If zero were not in this confidence interval, the null hypothesis would have been rejected!

6.6. COMPARISON OF PAIRED DATA

There are times when a comparison of two means is desired; however, due to uncontrollable variations, it is not possible to obtain a large enough sample size. For example, suppose you wish to compare two drying conditions and are only able to obtain two drying conditions per batch of material. Since batch-to-batch variations would mask the drying effect, many runs would be required.

One way around the above problem is to use paired data. Each batch would have one pair of data and the difference for each pair can be used in a hypothesis. The null hypothesis in this case would be that the mean difference between pairs is zero:

$$\bar{d} = 0$$

and the alternative hypothesis would be that the mean difference is not equal to zero,

$$\bar{d} \neq 0$$

or that the difference is less than or greater than zero.

As in the previous two sections, one major concern is that an important difference is present. The significant difference, estimated variance, α risk, and β risk determine the needed sample size. For convenience, the equations for testing paired data are given in Table 6.3.

An example of a hypothesis test for paired data is presented below:

1. On a *production* coater, a new drying condition A is being tested to determine if it is superior in hardness to an old drying condition B. It is known that batch-to-batch variations are large relative to the within-batch variations. Therefore the difference between drying conditions A and B is being tested for significance.

The within-batch standard deviation is 10 units, and a significant difference of 16 units is considered important. Management is willing to accept a 5% risk of rejecting the new drying condition when it is better and a 10% risk of accepting the new drying condition when it is not better. Since only a harder product is considered important, a single-sided test is used.

2. Null hypothesis: $\bar{d} = 0$
    Alternative hypothesis: $\bar{d} > 0$
    α risk $= .05$ (one-sided)

β risk = .10
Significant difference = 16
Estimated standard deviation = $10 \times \sqrt{2}$ = 14.14

3.   The needed sample size can be obtained from Table 4.3 (Chapter 4) where

$$\frac{\delta}{s} = \frac{16}{14.14} = 1.13$$

or a sample size of 9.

4.   The following data were obtained:

Batch	Drying A	Drying B	Difference A–B
1	75	76	−1
2	136	106	30
3	64	44	20
4	85	57	28
5	107	76	31
6	135	118	17
7	75	67	8
8	106	92	14
9	105	103	2

TABLE 6.3.   Comparison of Paired Data

	Double-Sided	Single-Sided
Null hypothesis	$\bar{d} = 0$	$\bar{d} = 0$
Alternative hypothesis	$\bar{d} = 0$	$\bar{d} > 0$ or $\bar{d} < 0$
Test statistics	$t = \dfrac{\bar{d}}{s_{\bar{d}}}$	
Definition of terms	n is the number of pairs ν is the degrees of freedom, $n - 1$ $\bar{d}$ is the difference between the pairs $s_{\bar{d}}$ is the standard error of the difference between the pairs s_d is the standard deviation of the difference between the pairs $s_{\bar{d}} = \dfrac{s_d}{\sqrt{n}}$	

If the test statistic, t, is greater than the appropriate one- or two-tailed value in the t table, Appendix A.3, for ν df and for $1 - \alpha$, the null hypothesis is rejected.

5.   Referring to Table 6.3, the calculations are

$$\bar{d} = 16.6$$
$$s\bar{d} = 3.97$$
$$t = 4.18$$

The tabulated t for a single-sided α risk of 5% and 8 df is 1.860—see Appendix A.3.

6.   Since the calculated test statistic is greater than the tabulated t value for an α risk of 5% and 8 df, the null hypothesis is rejected. That is, we conclude that drying A gives a harder product.

The result obtained from the above example would have been different if the difference between two means had been tested:

Null hypothesis $\qquad\qquad \mu_A = \mu_B$
Alternative hypothesis $\qquad \mu_A > \mu_B$

The calculation for testing the difference between the two means is

$$\bar{X}_A = 98.7$$
$$\bar{X}_B = 82.1$$
$$s_A = 25.9$$
$$s_B = 74.4$$
$$t = 1.397$$

The tabulated t for a single-sided t tail at an α risk of 5% and 16 df is 1.7459. In other words, the null hypothesis would have been accepted. By using the difference in paired data rather than the difference in means we minimized the effects of batch-to-batch variation.

6.7.   HYPOTHESIS TEST ON THE VARIANCE

Sometimes a difference in level is not as important as a difference in variability. When variability is a concern, a hypothesis test on the variance can be performed. A hypothesis test on the variance differs from a hypothesis test on the mean in several ways.

1.   Rather than a difference between variances, a ratio of variances is used:

$$\frac{s_1^2}{s_2^2}$$

2. The distribution associated with variances is not symmetric.
3. Normally only a single-sided test is used.

The distribution of means is normally distributed. The distribution of variances is a chi-squared, χ^2 distribution. The χ^2 distribution differs from the normal distribution in three ways:

1. There are no negative numbers since a variance is always positive.
2. The distribution is skewed.
3. There are as many χ^2 distributions as sample sizes.

Although the χ^2 distribution has a different shape than the normal distribution, critical regions are still used in a hypothesis test.

The chi-squared distribution is related to the t distribution squared:

$$\chi_\nu^2 = \sum_1^n t^2 = \sum_1^n \left(\frac{X - \mu}{\sigma} \right)^2$$

For unknown mean, the χ^2 distribution is equal to

$$\chi_\nu^2 = \Sigma \left(\frac{X - \overline{X}}{\sigma} \right)^2 = \frac{\nu s^2}{\sigma^2}$$

Note that the χ^2 distribution is the ratio of two variances: a known variance σ and an unknown variance s^2.

In Appendix A.4 the chi-squared distribution is given for different degrees of freedom and α risks. Values of ν df are found in the left-hand column. The desired risk values are found in the upper row. For a double-sided test, the $\alpha/2$ and $1 - \alpha/2$ columns are used. For a single-sided test, the $1 - \alpha$ column is used. A χ^2 distribution is shown in Figure 6.10 for 4 df. In Figure 6.10, the double-sided critical region is shaded for $\alpha = .10$.

As stated previously, normally a single-sided test is used for a hypothesis test on the variance and in this section only single-sided tests will be considered. An example of a single-sided test on the variance is presented below:

1. A new coating process is being investigated. It is important to ensure that the new coating process does not produce a more variable coating weight. The present coating weight is known to have a standard deviation of 0.1 μm. A process that is four times as variable is considered unacceptable. Your department is willing to accept a 5% risk of rejecting a good process and a 10% risk of accepting a bad product with a variance four times as great as the present process.

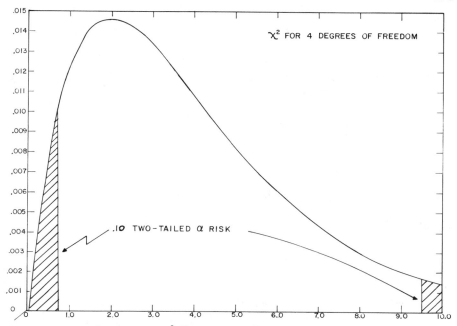

FIGURE 6.10. χ^2 distribution with two-tailed α risk of .10.

2. Null hypothesis: $\sigma^2 = .01$
   Alternative hypothesis: $\sigma^2 > .01$
   α risk $= .05$
   β risk $= .10$

3. The needed sample size can be determined by using Table 4.5 (Chapter 4). In the top rows of Table 4.5, various values of α and β risks are given. The degrees of freedom are listed in the left column. For a given α and β risk, the ratio of the calculated-to-known variances is given. For this example, a ratio of 4 or less is needed. Under an α risk of 5% and a β risk of 10%, the necessary degrees of freedom for a ratio of 4 or less is 10. The sample size must be one more or 11.

4. The following sample was obtained:
   $- .107, .083, -.110, .069, -.101, -.146, .077, .006, .049, -.063$
   and .024

5. The calculations are

$$\overline{X} = -.0199$$

$$s_{\overline{X}}^2 = (.0868)^2$$

$$\chi_{10}^2 = \frac{\nu s_{\overline{X}}^2}{\sigma^2} = \frac{10(.0868)^2}{(.1)^2} = 7.534$$

6. The tabulated χ^2 value for a 5% risk and 10 df is 18.307. The null hypothesis is accepted, that is, the new variance is assumed equal to the old variance.

7. Before leaving this example, it is instructive to calculate a confidence interval about the new standard deviation. The 90% confidence interval for the new variance is:

$$\frac{\chi^2_{95,10}(.0868)^2}{10} < \sigma^2_{new} < \frac{\chi^2_{.05,10}(.0868)^2}{10}$$

or

$$(.0545)^2 < \sigma^2_{new} < (.1174)^2$$

The 90% confidence interval for the standard deviation would be

$$.0545 < \sigma_{new} < .1174$$

The χ^2 distribution is used when an unknown variance is compared to a known variance or standard. For cases where two unknown variances are compared, an F distribution is used. Like the χ^2 distribution, the F-distribution is not symmetric. The shape of the F distribution is a function of the degrees of freedom for both samples (see Figure 6.11). As one of the sample sizes approaches infinity, the F-distribution approaches a χ^2 distribution.

In Appendix A.5, the F distribution is given for different degrees of freedom and α risks. Two tables are presented in Appendix A.5: A table for $\alpha = .05(95\%)$ and $\alpha = .01(99\%)$. For each table, the degrees of freedom associated with the variance in the numerator are given in the first row, and the degrees of freedom associated with the variance in the denominator are given in the left column. The values corresponding with the proper α risk and degrees of freedom indicate the critical ratio of the two variances.

An example of a single-sided test in the ratio of two variances is given below:

1. A new method is being proposed for preparing coating solutions. Although the new method is cheaper and more efficient than the old method, there is fear that the new method will result in a more variable minimum density in the coated product. No exact measurement is available on the variance of the old method; however, variability has not been a problem. Further, it is suspected that a threefold increase in variance would not present problems. You are willing to accept a 10% risk of rejecting the new method as being more variable when it is not, but only willing to accept a 5% risk of accepting the new method when it is three times as variable as the old method.

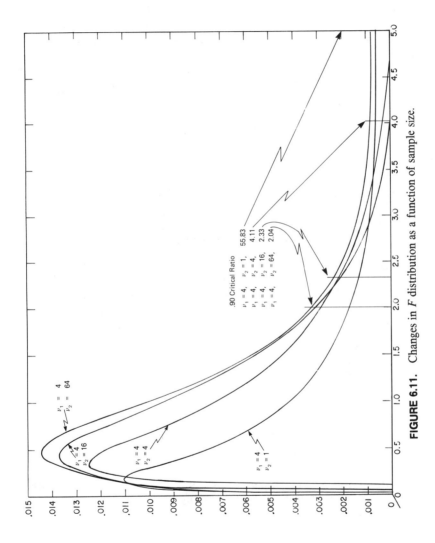

FIGURE 6.11. Changes in F distribution as a function of sample size.

.90 Critical Ratio

$\nu_1 = 4$,	$\nu_2 = 1$,		55.83
$\nu_1 = 4$,	$\nu_2 = 4$,		4.11
$\nu_1 = 4$,	$\nu_2 = 16$,		2.33
$\nu_1 = 4$,	$\nu_2 = 64$,		2.04

$\nu_1 = 4$
$\nu_2 = 64$

$\nu_1 = 4$
$\nu_2 = 16$

$\nu_1 = 4$
$\nu_2 = 4$

$\nu_1 = 4$
$\nu_2 = 1$

2.   Null hypothesis: $\sigma^2_{new} = \sigma^2_{old}$
     Alternative hypothesis: $\sigma^2_{new} > \sigma^2_{old}$
     α risk $= .10$
     β risk $= .05$

3.   The needed sample can be determined by using Table 4.6 (Chapter 4). In the top two rows of Table 4.6, various values of α and β risks are given. In the left-hand column, the degrees of freedom (identified by ϕ) are given. Note, it is assumed that $\nu_1 = \nu_2 = \nu$. For a given α and β risk, the ratio of the two variances is given. For this example, a ratio of three or less is needed. There is no column for an α risk of .10 and a β risk of .05; however, the column with an α risk of .05 and a β risk of .10 can be used. The needed sample size would be 31.

4.   The following two samples were obtained:
Old method

.21,   .18,   .25,   .16,   .19,   .25,   .10,   .24,   .19,   .20,   .23,
.20,   .18,   .23,   .21,   .16,   .15,   .12,   .21,   .22,   .24,   .20,
.27,   .17,   .16,   .17,   .16,   .17,   .16,   .20,   .22,   .14,   .17.

New method

.20,   .17,   .19,   .19,   .18,   .20,   .24,   .19,   .22,   .22,   .17,
.18,   .15,   .09,   .19,   .10,   .15,   .15,   .09,   .17,   .21,   .21,
.22,   .26,   .26,   .28,   .24,   .31,   .19,   .20,   .11.

5.   The calculations are

$$\bar{X}_{old} = .193 \qquad\qquad \bar{X}_{new} = .191$$

$$s^2_{old} = (.040)^2 \qquad\qquad s^2_{new} = (.052)^2$$

$$F = \frac{(.052)^2}{(.040)^2} = 1.69$$

6.   The tabulated F value for an α risk of .10 and two samples with 30 df is 1.6065. Since the tabulated F is less than the calculated F, the null hypothesis is rejected.

Before making any firm conclusion about a hypothesis test, the data should be analyzed to ensure that all the assumptions are met. In the above example, it was assumed that the data were normally distributed. The data are normally distributed since the data were generated from a normal random number generator. However, the new method has two variances present: a short-term variation and a long-term variation. A control chart plot of the data shows both variations (see Figure 6.12).

Before rejecting the null hypothesis, I would want to know what caused the level shift, and more important, could the level shift, long-term variation be eliminated. Suppose, for example, that the level shift was due to a

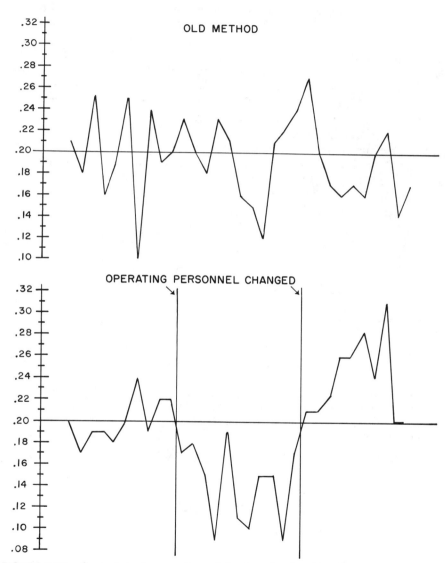

FIGURE 6.12. Control chart comparison of the data for the old and new method used in the example given in "Hypothesis Test on the Variance."

change in operating personnel. Perhaps the level shift could be reduced or eliminated by better defined operating procedures.

6.8. CONCLUSION

Get your facts first, and then you can distort them as much as you please.
                                                                    MARK TWAIN

I thought it would be appropriate to end this chapter as I started it with a quote by Mark Twain. As I cautioned in the beginning of this chapter, there are risks involved in any hypothesis test. Further, there are many assumptions behind a hypothesis test. One wrong assumption can result in an incorrect conclusion for a hypothesis test.

In starting any hypothesis test, the first operation is to decide on the question or questions to be answered. Second, the hypotheses are stated, and then the risks and sample size are determined. Then the experiment is run and the data analyzed. Do not decide on the question(s) after the data have been obtained. Too often a bushel basket full of data is collected, and then questions are asked.

Information can be obtained from a bushel basket full of data; however, a hypothesis test cannot be conducted in this fashion. The data can be analyzed for trends, direction for future experimentation, and so on, but the data cannot be analyzed as though it were a hypothesis test.

REFERENCES

1. Mood, A. F., *Introduction to the Theory of Statistics*, McGraw-Hill, New York, 1950.
2. Rickmers, A. D., and Todd, H. N., *Statistics, an Introduction*, McGraw-Hill, New York, 1967.
3. Natrella, M. G., *Experimental Statistics*, National Bureau of Standard Handbook 91, Government Printing Office, Washington, D.C., 1963.
4. Freund, R. A., Graphic process control, *Industrial Quality Control*, Vol. 28, No. 7, January 1962, pp. 1–8.
5. Freund, R. A., Acceptance control charts, *Industrial Quality Control*, Vol. 14, No. 4, October 1957, pp. 13–23.

7

STATISTICAL INTERVALS

GERALD J. HAHN

General Electric Company
Corporate Research & Development
Schenectady, New York

Chemists and engineers have come to appreciate that few things in life are known exactly. The most they can do is to use the results of a random sample from a given population to obtain an estimate of some quantity of interest and construct an interval to contain the unknown quantity with a specified probability. This chapter describes three types of statistical intervals which arise in many practical applications and indicates when each should be used. The three intervals are:

1. A confidence interval to contain the mean of the sampled population (or to contain some other population value, such as the standard deviation).
2. A tolerance interval to contain a specified proportion of the sampled population.
3. A prediction interval to contain the values of all of a specified number of future observations from the sampled population.

Many nonstatistical users of statistics are well acquainted with confidence intervals for the population mean and for the population's standard deviation. Some are also aware of tolerance intervals, but most nonstatisticians know very little about prediction intervals, despite their practical importance. A frequent mistake is to calculate a confidence interval to contain the population mean when the actual problem calls for a tolerance interval or a prediction interval. At other times, a tolerance interval is used when a prediction interval is needed. Such confusion is understandable, since most texts on statistics discuss confidence intervals in detail, make limited reference to tolerance intervals, and almost never cover prediction intervals (except in regression analysis applications). This is unfortunate because tolerance intervals and prediction intervals are needed almost as frequently in industrial applications as confidence intervals and the procedure for constructing such intervals, given the required tabulations, is no more difficult than that for constructing confidence intervals.

In this chapter, procedures for constructing the various types of intervals are described and their use illustrated. The types of situations in which each interval applies is emphasized in the discussion.

It will be assumed that one is dealing with a random sample of independent observations from a normally distributed population (normal population, for short); and the sample mean, or arithmetic average, $(\bar{x})$, and the sample standard deviation (s) are calculated from the values of n given observations $x_1, \ldots, x_n$, by the well-known expressions

$$\bar{x} = \sum_{i=1}^{n} \frac{x_i}{n}$$

and

$$s = \left[\sum_{i=1}^{n} \frac{(x_i - \bar{x})^2}{n-1} \right]^{1/2}$$

From this information, it is desired to make probability statements concerning the population from which the sample was selected or the results that one might expect in some future random sample from the same population, or both. In constructing prediction intervals, it is further assumed that the future sample is selected randomly from the given normal population and independently of the first sample.

7.1. CONFIDENCE INTERVALS

Confidence Interval to Contain the Population Mean

The mean μ is the most commonly employed value to describe a population. It is frequently used to characterize product performance or as a standard to compare competing processes. This second use is especially appropriate when it can be assumed that each of the processes has the same statistical variability (as measured by its standard deviation); in that case, differences between normal populations can be described completely by the differences between their means.

The sample mean $\bar{x}$ is an estimate of the unknown population mean μ, but, because of sampling fluctuations, differs from μ. However, it is possible to construct a statistical interval, known as a confidence interval, to contain μ. A confidence interval provides a range which one can claim, with a specified degree of confidence, contains an unknown value. More specifically, if one constructs 95% confidence intervals to contain the value of μ in many applications, one will, in the long run, correctly include the true value in 95 cases out of 100. However, in 5 cases out of 100, the calculated confidence interval will fail to contain μ. The degree of confidence, 95% in the preceding statement, is known as the associated confidence level. In addition to the 95% confidence level, 90% and 99% confidence levels are also often used (see below).

A 95% confidence interval to contain the mean μ for a normal population is calculated as

$$\bar{x} \pm c_M(n)s$$

where $c_M(n)$ is given in the second column of Table 7.1 as a function of n, the sample size.

Example 7.1. The following readings of product performance have been obtained from a random sample of five observations from a normal population: 51.4, 49.5, 48.7, 49.3, and 51.6. The sample mean and sample standard deviation are calculated from these values as

$$\bar{x} = (51.4 + \cdots + 51.6)/5 = 50.10$$

and

$$s = \left\{\left[(51.4 - 50.10)^2 + \cdots + (51.6 - 50.10)^2\right]/(5 - 1)\right\}^{1/2} = 1.31$$

For this example, $c_m(5) = 1.24$ from Table 7.1 and the 95% confidence interval for μ is

$$50.10 \pm (1.24)(1.31)$$

Consequently, one can be 95% confident that the interval 48.48 to 51.72 contains the unknown value of μ.

TABLE 7.1. Factors for Calculating Two-Sided 95% Statistical Intervals for a Normal Distribution

Number of given observations	Factors for confidence interval to contain the population mean μ	Factors for tolerance interval to contain at least 90% 95% and 99% of the population			Factors for prediction interval to contain the values of all of 1, 2, 5, 10, and 20 future observations				
n	$c_M(n)$	$c_{T,90}(n)$	$c_{T,95}(n)$	$c_{T,99}(n)$	$c_{P,1}(n)$	$c_{P,2}(n)$	$c_{P,5}(n)$	$c_{P,10}(n)$	$c_{P,20}(n)$
4	1.59	5.37	6.37	8.30	3.56	4.41	5.56	6.41	7.21
5	1.24	4.28	5.08	6.63	3.04	3.70	4.58	5.23	5.85
6	1.05	3.71	4.41	5.78	2.78	3.33	4.08	4.63	5.16
7	0.92	3.37	4.01	5.25	2.62	3.11	3.77	4.26	4.74
8	0.84	3.14	3.73	4.89	2.51	2.97	3.57	4.02	4.46
9	0.77	2.97	3.53	4.63	2.43	2.86	3.43	3.85	4.26
10	0.72	2.84	3.38	4.43	2.37	2.79	3.32	3.72	4.10
11	0.67	2.74	3.26	4.28	2.33	2.72	3.24	3.62	3.98
12	0.64	2.66	3.16	4.15	2.29	2.68	3.17	3.53	3.89
15	0.55	2.48	2.95	3.88	2.22	2.57	3.03	3.36	3.69
20	0.47	2.31	2.75	3.62	2.14	2.48	2.90	3.21	3.50
25	0.41	2.21	2.63	3.46	2.10	2.43	2.83	3.12	3.40
30	0.37	2.14	2.55	3.35	2.08	2.39	2.78	3.06	3.33
40	0.32	2.05	2.45	3.21	2.05	2.35	2.73	2.99	3.25
60	0.26	1.96	2.33	3.07	2.02	2.31	2.67	2.93	3.17
∞	0	1.64	1.96	2.58	1.96	2.24	2.57	2.80	2.02

A two-sided 95 percent interval is $\bar{y} \pm c(n)s$, where $c(n)$ is the appropriate tabulated value and $\bar{y}$ and s are the mean and the standard deviation of the given sample of size n.

Use of Other Confidence Levels

The second column of Table 7.1 provides factors for constructing a 95% confidence interval to contain μ. Standard texts in statistics, such as Ref. 1, or a statistical handbook, such as Ref. 2, provide procedures for obtaining similar confidence intervals based upon confidence levels other than 95%.

The selection of the appropriate confidence level depends on the specific application and the importance of any decision that might be made based on the calculated confidence interval. For example, at the outset of a research project, one might be willing to take a reasonable chance of drawing incorrect conclusions and, thus, use a relatively low confidence level, such as 90% or even 80%, since one's initial conclusions will be verified by subsequent analyses. On the other hand, if one is about to report the final results of the project, and perhaps recommend building an expensive new plant or releasing a new product for wide general use, one would generally wish to have a high degree of confidence. Large samples tend to lead to small confidence intervals and small samples result in large confidence intervals. Sometimes one might want to compromise by using relatively high confidence levels with large samples and lower confidence levels with small samples. This implies that the more data one has, the surer one would like to be of one's conclusions.

One-Sided Confidence Bounds

A two-sided confidence interval is calculated when both a lower and an upper bound are required. In some situations, however, only a lower confidence bound *or* only an upper confidence bound is required. For example, a manufacturer who needs to establish a warranty on the average weight of a packaged product or on the mean time to failure of a system usually requires only a lower confidence bound. The concept of confidence level remains the same, irrespective of whether one is dealing with a two-sided confidence interval or a one-sided confidence bound, although details of the calculation differ slightly.

In particular, the lower endpoint of a 95% confidence interval provides a one-sided lower 97.5% confidence bound and the upper endpoint of a 95% confidence interval provides an upper 97.5% confidence bound. Thus, in Example 7.1, one can be 97.5% confident that the true value of μ exceeds 48.48; similarly, one can be 97.5% confident that the true value of μ is less than 51.72.

One-sided confidence bounds to contain μ for confidence levels other than 97.5% can be obtained using procedures described in standard tests in statistics, such as Ref. 1, or in a statistical handbook, such as Ref. 2.

Confidence Intervals on Other Quantities

The preceding discussion has dealt with confidence intervals to contain the mean μ of a normal population. The same concepts apply for constructing confidence intervals on other population parameters; for example,

1. A confidence interval to include the standard deviation σ of a normal population (the standard deviation being a measure of the spread of variability of the population).
2. A confidence interval to contain a population percentage, such as the percentage of defective units in a fabricated lot.

Specific procedures for constructing such confidence intervals are provided in standard texts and handbooks on statistics, such as Refs. 1 and 2.

7.2. TOLERANCE INTERVALS

Tolerance Interval to Contain a Specific Proportion of the Population

Many applications require an interval to contain a specific proportion p of the population, instead of, or in addition to, a confidence interval to contain the population mean μ. Such an interval is required, for example, by a manufacturer of a mass production item who needs to establish limits to contain at least a specified (large) proportion of his product with a high degree of confidence. For example, assume readings have been obtained on some performance characteristic on a random sample of 25 transistors from a particular production process. Based on the resulting data, a tolerance interval provides limits to contain, with a known degree of confidence, the readings of at least a specified proportion of the population of units from the process.

For a normal population, if μ and σ are known, it is also known that 90% of the population values are located in the interval

$$\mu \pm 1.64\sigma$$

(and 95% of the population values are in the interval $\mu \pm 1.966$). However, if only sample estimates, $\bar{x}$ and s, of the population values μ and σ are available, added uncertainty is introduced and the best that one can do is construct an interval which one can claim with a specified degree of confidence, say 95%, contains at least a specified percentage, such as 90%, 95%, or 99% of the population. Such an interval is called a 95% tolerance interval and can be calculated for a normal population by using the factors $c_{T,90}(n)$, $c_{T,95}(n)$, and $c_{T,99}(n)$, shown in columns 3, 4, and 5 of Table 7.1 for population percentages of 90%, 95%, and 99%, respectively. For example, it can be stated with 95% confidence that the interval

$$\bar{x} \pm c_{T,90}(n)s$$

contains at least 90% of a normal population.

The fact that two percentage values are associated with a tolerance interval is sometimes confusing to the user. One of these refers to the percentage of the population that the interval is to contain. The second specifies the confidence level associated with the calculated interval. When μ and σ are known exactly (which is highly unlikely), an interval to contain a specified percentage of the population may still be of interest, but, in this case, there is no longer any uncertainty associated with the population percentage contained in the interval.

Example 7.2. A 95% tolerance interval to contain 90% of the sampled normal population may be calculated for the data in Example 7.1 as

$$50.10 \pm (4.28)(1.31)$$

or

$$44.49 \quad \text{to} \quad 55.71, \quad \text{using } c_{T,90}(5) = 4.28$$

Thus, one may be 95% confident that at least 90% of the sampled population is contained in the interval 44.49 to 55.71.

Similarly,

1. A 95% tolerance interval to contain 95% of the population is

$$50.10 \pm (5.08)(1.31)$$

   or

$$56.75 \quad \text{to} \quad 43.45, \quad \text{using } c_{T,95}(5) = 5.08$$

2. A 95% tolerance interval to contain 99% of the population is

$$50.10 \pm (6.63)(1.31)$$

   or

$$41.41 \quad \text{to} \quad 58.79, \quad \text{using } c_{T,99}(5) = 6.63$$

Use of Other Confidence Levels in Constructing Tolerance Intervals and One-Sided Tolerance Bounds

Although a 95% tolerance interval is, perhaps, more frequently used than any other, one might sometimes wish to construct tolerance intervals involving other confidence levels. Tabulations of factors to obtain 90 and 99% tolerance intervals for a normal population are given in Refs. 1 and 2. Also, frequently, one requires a lower tolerance bound only, that is, a lower limit on the population value or an upper tolerance bound only, rather than a two-sided tolerance interval. Such one-sided tolerances bounds cannot be constructed directly from the tabulations for two-sided tolerance intervals. Instead, specially constructed factors for obtaining one-sided tolerance bounds for a normal population need be used. Such factors are tabulated in Ref. 3.

7.3. PREDICTION INTERVALS

Prediction Intervals to Contain All of *k* Future Observations

Sometimes, instead of requiring an interval that one can expect to include a specified proportion of the population (tolerance interval), one desires an interval that will contain the next observation, or each of a small number of future observations from the given population, with a high probability. This requires the third type of interval, a prediction interval.

Such an interval may be thought of as an astronaut's interval. A typical astronaut who has been assigned to undertake a specific number of flights, and who needs to decide how much fuel to take along each time, is not interested in what will happen on the average in the population of all space flights, of which his happens to be a random sample (confidence interval on the population mean), or even what will happen on at least 90 or 99% of such flights (tolerance interval). His main concern is what might happen in the one or small number of flights in which he will be personally involved. Similarly, a systems manufacturer who needs to warranty the performance of each unit in an order of three units, based upon his past experience on units of the same type, would wish a prediction interval to contain the values of a specified performance parameter for all three units with a high probability. Thus, in contrast to tolerance limits which are frequently required by a manufacturer of a mass production item, prediction intervals are of most interest to a manufacturer of a relatively small number of units. Such intervals are also desired by the typical customer who purchases one or a small number of units of a given product and is concerned with predicting the performance of the particular units he has purchased (in contrast to the long-run performance of the process from which the sample has been selected).

Prediction intervals to contain all of k future observations from a normal population based on the results of n past observations can be calculated with the help of the factors in the last five columns of Table 7.1. These provide values of the factor $c_{P,k}(n)$ such that all of k future observations from the same normal population will be located in the interval

$$x \pm c_{P,k}(n)s$$

with a probability of .95 for $k = 1, 2, 5, 10$, and 20.

Example 7.3. Consider again the random sample of five observations from the normal population of Example 7.1. Now assume that it is desired to obtain a 95% prediction interval to contain the values of both of two further randomly selected observations from the same population. For this case $k = 2$ and $n = 5$; from Table 7.1 the factor $c_{P,2}(5) = 3.70$.

Thus, two future units from the sampled population will be located in the interval

$$50.10 \pm (3.70)(1.31)$$

or

$$45.25 \quad \text{to} \quad 54.95, \quad \text{with a probability of } 0.95.$$

Similarly, a 95% prediction interval to contain:

1. A single future observation with 95% probability is

$$50.10 \pm (3.04)(1.31)$$

   or

$$46.12 \quad \text{to} \quad 54.08, \quad \text{using } c_{P,1}(5) = 3.04$$

2. All of 10 future observations with a 95% probability is

$$50.10 \pm (5.23)(1.31)$$

   or

$$43.25 \quad \text{to} \quad 56.95, \quad \text{using } c_{P,10}(5) = 5.23$$

From the preceding results, and also from inspection of the factors in Table 7.1, it is noted that, as one might expect, the shortest prediction interval is the one to contain a single future observation; the size of the interval increases as the number of future observations to be contained increases.

Use of Other Probability Levels in Constructing Prediction Intervals and One-Sided Prediction Bounds

One might sometimes wish to construct prediction intervals involving probability levels other than 95%. Tabulations of factors for 99% prediction intervals for a normal population are given in Ref. 3.

Also, frequently one requires a lower prediction bound only, or an upper prediction bound only, to contain all of k future observations. One-sided prediction bounds cannot be constructed directly from the tabulations for two-sided prediction intervals except for the special case of $k = 1$ (i.e., single future observation). Instead, specially constructed factors for obtaining one-sided prediction bounds need be used. Such factors are tabulated in Ref. 3.

Some Further Prediction Intervals

Prediction intervals to contain the values of all of k future observations are only one kind of prediction interval. In some applications, other kinds of prediction intervals are required. Some examples are:

1. A prediction interval to contain the *mean* of k future observations with a specified probability (see Ref. 3).
2. A prediction interval to contain the standard deviation of k future observation with a specified probability (see Ref. 3).
3. A prediction interval to contain at least k' out of k future observations with a specified probability, where $k' < k$ (see Ref. 4).

7.4. COMPARISON OF RESULTS FROM EXAMPLES

The relative lengths of the interval in Example 7.1 and the first interval in Examples 7.2 and 7.3 are compared in Figure 7.1. It is seen that for a sample of 5, the 95% confidence interval to contain the population mean is appreciably smaller than both the tolerance interval to contain at least 90% of the population and the 95% prediction interval to include both of two future observations. Also, the 95% tolerance interval to include at least 90% of the population is somewhat larger than the 95% prediction interval to contain both of two future observations.

Inspection of the tabulations of the factors in Table 7.1 indicates that a 95% confidence interval to contain the population mean is always smaller than the other two intervals, but that the relative sizes of the tolerance and prediction intervals depend on the proportion of the population to be covered by the tolerance interval and the number of future observations to

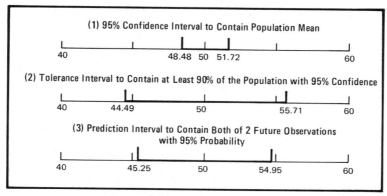

FIGURE 7.1. Comparison of lengths of statistical intervals for examples.

be contained in the prediction interval. Also, unlike the other two intervals, the length of a confidence interval approaches zero as the sample size increases; in fact, the interval converges to the point μ.

7.5. WHICH INTERVAL DO I USE?

As illustrated, statisticians have developed a variety of intervals for helping answer relevant questions from given data. However, the engineer or chemist must decide what the relevant questions are in a given application. Once the questions to be answered have been clearly stated, it should be relatively easy to decide on the correct intervals. Thus, the decision of which interval(s) to use must be made by the analyst based upon his understanding of the problem.

For example, say one is concerned with the number of miles per gallon of gasoline obtained by a particular type of automobile under specified conditions. If all that is needed is an estimate of average gasoline consumption, perhaps to compare with that for another type of car, then a confidence interval to include the population mean might be most appropriate. On the other hand, if the manufacturer must pay a penalty for each automobile with gasoline consumption below a value to be specified by him, then a lower tolerance bound would be desired. Finally, the purchaser of one of the automobiles, would be most interested in a prediction interval to contain the gasoline consumption of a single future automobile.

Table 7.2 provides a categorization which might be helpful in selecting the proper interval. The intervals are classified by area of inference and characteristic of interest. Thus, one must decide whether the main concern is in drawing inferences about the population from which the sample was selected or in predicting the results of a future sample from the same

TABLE 7.2.   Categorization of Some Statistical Intervals

Characteristic of Interest	Area of Inference	
	Description of Population	Prediction of Results of Future Sample
Location (measured by the mean)	Confidence interval to contain the population mean	Prediction interval to contain the mean of k future observations
Spread (measured by the standard deviation)	Confidence interval to contain the population standard deviation	Prediction interval to contain the standard deviation of k future observations
Enclosure region	Tolerance interval to contain a population proportion	Prediction interval to contain all of k future observations

population. Also, one must decide whether the major interest is in characterizing the location or spread of the population or future sample or in developing some form of enclosure region.

7.6.   IMPORTANCE OF UNDERLYING ASSUMPTIONS

The preceding intervals apply only for the population from which the given sample was randomly selected. For example, if the available data were taken from past production, the resulting inferences apply to future production only to the extent to which past units and future units come from the same population. It is also assumed that the given data is from independent samples. This would generally require, for example, that each observation is on a different sampled unit as opposed to repeat observations on the same sampled unit.

As previously indicated, the specific procedures given in this chapter are based on the assumption of normality for the sampled population. The assumption is generally not critical in the construction of a confidence interval to contain the population mean (due to the central limit theorem of statistics). However, it is important in many other situations, such as in obtaining a confidence interval to include the population standard deviation and in constructing a tolerance interval to include a specified proportion of the population or a prediction interval to include all members of a future sample. When the assumption of normality cannot be met, one might wish to turn to so-called distribution-free methods. These methods are generally not as informative as ones based upon the assumption of an underlying distribution model, such as the normal distribution, for the sampled population. However, they do have the advantage of not requiring such an assumption. Specific distribution-free methods, such as the construction of distribution-free tolerance intervals, are discussed in Refs. 1 and 2.

7.7. DIGGING DEEPER

As indicated, standard books on elementary engineering statistics, such as Ref. 1, and handbooks such as Ref. 2, provide detailed discussions of confidence intervals, and some, including Refs. 1 and 2, also consider tolerance intervals. A survey of prediction intervals is provided in Ref. 5. In addition, Ref. 3 provides a comprehensive comparison of statistical intervals for a normal population and a discussion of methods for constructing the various intervals. More detailed tabulations than these given here and further references to more extensive tabulations are also provided.

REFERENCES

1. Dixon, W. J., and Massey, F. J., Jr., *Introduction to Statistical Analysis*, 3rd ed., McGraw-Hill, New York, 1969.
2. Natrella, M. G., *Experimental Statistics*, National Bureau of Standards Handbook 91, U.S. Government Printing Office, 1966.
3. Hahn, G. J., Statistical intervals for a normal population, *Journal of Quality Technology*, Vol. 2, No. 3, July 1970, pp. 115–125; Vol. 2, No. 4, October 1970, pp. 195–206.
4. Hall, I. J., and Prairie, R. R., One-sided prediction intervals to contain at least *m* of *k* future observations, *Technometrics*, Vol. 15, No. 4, November 1973, pp. 897–914.
5. Hahn, G. J., and Nelson, W. B., A survey of prediction intervals and their applications, *Journal of Quality Technology*, Vol. 5, No. 4, October 1973, pp. 178–188.

8

ACCEPTANCE SAMPLING

FRANK CULLEN

Loyola College
Baltimore, Maryland

8.1. BACKGROUND

In many manufacturing and clerical processes a major problem is judging the quality of incoming material. Judgment may have to be made: (a) at the "front door," (b) at inspection points within the flow of the process, (c) at the "end of the line," or (d) at the warehouse or in the showroom.

Some companies take pride in that their product is reviewed 100% at all inspection points even though it has been demonstrated that the application of correct sampling techniques to the review process often results in better and more economic inspection.

There are psychological and physical reasons why 100% inspection seldom produces 100% results. Here are some of the reasons:

(a) Routine operations tend to bring on a certain amount of fatigue—even with "breaks" a human does not work at uniform efficiency all day.

(b) When employees know that someone at the end of the production line is going to review all the product, when a defect is noted there may be little concern under the assumption that all defectives will be removed by the final inspection.

(c) Workers performing 100% review tend to become bored and lose interest in what they are doing.

(d) Workers tend to lose the basic concept of "Do it right the first time" if they know all their work will be reviewed but all mistakes will not be noted.

(e)   Slack worker attitudes tend to develop when there is much rework; consequently overtime costs may skyrocket.

(f)   100% inspection, no matter how good, is costly and time consuming.

Now since the development of the thumb, which enabled him to make things more easily, man has always performed some kind of judgment on his finished work. Imagine the cave man designing a club, roughing it out, finishing it off, and then quality testing it against a tree for balance, weight, and the effect it would have on an animal's (or neighbor's) skull!

A comprehensive outline of the development of the techniques of acceptance sampling is found in Ref. 2, but a brief history of progress in the use of this tool of statistical quality control is given here.

Before the advent of World War II only a very few companies performed scientific inspection as applied to either "in process" or incoming inspection. Statistical techniques that applied to acceptance sampling had been described in the scientific literature by Bartky, Coggins, Deming, Dodge, Jordan, Molina, Neyman, Pearson, Romig, Shewhart, Thorndyke and others with many papers appearing in the *Bell System Technical Journal.*

However, the impetus to put these sampling techniques into wide use was provided by World War II. Until this time the inspector, using the tool of 100% review, was the "King of the Roost" in many plants and many inspectors lobbied strongly within their companies against the use of sampling.

But the military, faced with a dangerous nationwide situation in the conduct of the war, needed immediate economic and consistent production of materials and, at the same time, uniform quality production not dependent on the whims of individuals in the many companies they represented.

To reach this goal they encouraged the use of known sampling techniques and helped to underwrite the development of even better methods. Some of the sampling plans developed during the course of the war were considered important enough to warrant secret classification and were not released for general use until after peace was declared.

To further implement the quality control program the sampling plans developed by the military were published in the form of Standard Inspection Procedures. Training programs were instituted throughout the country to provide understanding of the application of these plans and to encourage manufacturers to use them. These "short-course" training programs were quite successful and after the war both the trainers and the trainees took their knowledge into industry and government, but most importantly, onto the campus. Courses in statistical quality control and related subjects, such as industrial statistics, were given in many colleges and good quality control texts, some of which have become standard works, were published.

The American Society for Quality Control, which was organized in 1946–1947, with much help from "graduates" of the short courses, is now one of the larger (36,000) professional engineering groups in the United States. There are similar groups overseas and there is also an International Society for Quality Control.

As an example of what can result from adoption of statistical quality control techniques, examine the progress of Japanese industry following World War II. Broad use of statistical quality control greatly aided this nation to a very rapid economic recovery and forward movement into world leadership in the quality of manufactured product. American quality experts (particularly Juran and Deming) played a significant role in the Japanese quality control movement. It seems strange to observe overseas industry readily practicing and profiting from the U.S. quality gospel, while U.S. industry often avoids quality as a plague.

Pressure from the U.S. government, particularly the military establishment and the space agencies, in mandating that quality control programs be a part of each contract placed emphasis on the use of scientific quality control techniques and systems and resulted in the adoption of quality programs by companies who, without persuasion, may not have initiated quality control programs.

Adoption of acceptance sampling systems has removed much individual judgment from the inspectors' baliwick as to what to inspect, how many to inspect, when to inspect, and what to do about rejected lots. Actually the modern inspector (or any technician who reviews product quality) is a professional employee and not just another part of the inspection process. Largely through the use of quality control techniques the inspector is now able to make better and more consistent judgments from industry to industry, plant to plant, and person to person.

8.2. ACCEPTANCE SAMPLING PHILOSOPHY

Acceptance sampling is a method of classifying lots of material, submitted for inspection or review, as either "good" or "bad." This important tool of quality control is used to assure the customer, based on a sample, that the material he is receiving will not have a deleterious effect on his product or process. Conversely, it may be used by the supplier to decide if an overall batch of items is acceptable to ship. While acceptance plans may not have a direct effect on improving quality, any producer who has much of his product rejected because it does not conform to review standards will probably try to improve his product.

A little philosophy applied to the sampling concept tells us that if all product submitted for inspection was good (bad) a sample of one would suffice to make a proper judgment as to its acceptability. Since there are few such processes extant we usually need to examine a sample larger than one to make a proper judgment as to product quality. How many items we really need to examine in order to make a probably correct judgment is one of the questions this chapter will attempt to answer.

Fortunately in most production or clerical processes more good material is produced than bad. However, the proper design of a sampling plan assures that no matter what the true quality of the product submitted for inspection the basic quality after review, in the long run, will be no worse than some quality value specified by the customer.

If the sampling plan is designed and used correctly then the better the quality of the material submitted for review the better chance that material has of being accepted; conversely the poorer the quality of the material submitted the greater is the chance that the material will be rejected.

Further if the design of the sampling plan specifies that all rejected lots be inspected 100%, that the bad items identified be removed and replaced with good items, then the overall effect will be to keep input at a constant rate of acceptable product which will not exceed some specified upper limit of percent defective.

The provision for 100% review of rejected lots is called detailing, screening, or rectifying inspection. In the sample designs considered in this chapter, it is generally assumed that this provision is in force.

Without a provision for screening, most of the time the acceptance plan merely sorts out the good lots from the bad lots. Including the concept of rectifying inspection can make the acceptance sampling plan a part of the total quality process.

Four basic types of acceptance plans are discussed in this chapter:

(a)   Attribute sampling plans.
(b)   Chain sampling plans.
(c)   Continuous sampling plans.
(d)   Variables sampling plans.

Acceptance sampling plans are detailed in table and graphic form. Graphically the actual operational plan shows as a long, sweeping reverse "S" curve, skewed to the right with the abscissa indicating the true percent defective (p') in the material being submitted for inspection and the ordinate representing the probability of accepting (Pa) a lot with (p') quality if such a lot is submitted for inspection.

The overall long-term effect of the sampling plan is graphed as a "mound-shaped" curve with the abscissa representing (p') and the ordinate giving the average outgoing quality (AOQ). The maximum point of the "mound-shaped" curve represents the poorest quality in percent defective that the plan will allow and this is named the average outgoing quality limit (AOQL).

Construction of these curves is carried out in detail in succeeding sections.

8.3.   SOME CONCEPTS NECESSARY TO DESIGN AN ACCEPTANCE SAMPLING PLAN FOR ATTRIBUTES

Attribute (discrete) plans are based on results of an inspection which classifies each item being inspected as "good" or "bad" even though a measurement may be made to determine its acceptability.

For illustrative purposes suppose a freight car containing 1000 drums of a pigment is received from a regular supplier. The accounting department will not okay receipt of shipment and release to production until it has assurance that the drums contain the proper amount (pounds) of pigment; so if short weight is encountered some adjustment must be made (if overweight the supplier should also be notified).

One thousand drums of material forms a "large" inspection lot. (A large lot may be defined as one in which the lot size N has a ratio of 10:1 or better when compared to sample size n).

Probably this lot was formed by the producer arbitrarily and conveniently. Limitations to lot size may have been, for example, the availability of types of freight cars. Assume that all this material has been produced from a continuous process in statistical control and that the material in the car was all produced, and the drums all filled, at about the same time. These assumptions are quite important and the more than can be verified as to the closeness to truth of the assumptions the more confidence that will result in the results produced after application of the plan.

Sampling plans are classified as single, double, or sequential (multiple) plans. Each type will be described and a comparison given as to the advantages and disadvantages. All three types of plan give about the same protection and the choice of a plan is generally a matter of administrative convenience.

Design of a Single Sampling Plan

The first question to be asked of the accounting department by the person responsible for the sample design is what tolerance is to be allowed? What constitutes an acceptable drum of material as regards weight? How far may the weight of each drum depart from some standard? In quality control parlance the question might be put how is "exactly alike" to be defined? This concept should be a part of the specification and clearly defined in any contract between a producer and a consumer.

In any inspection process an assumption must be made that the inspector will do the job properly and find all the materials inspected which do not meet specifications. Unless there is electronic equipment that will perform the inspection perfectly, measurement errors may occur even if every item is inspected and the resultant information recorded. In planning for any review it is well to note that the better the quality of the material being inspected the greater the chance that bad product will not be detected. After all, most non mechanical inspections are monotonous and employees performing inspection may be lulled by a continuous flow of good product to not expect to see bad product. Reviewers often form an image of perfection and see the image, not the item being inspected. Thus when there is a small amount of bad material present in a rather large lot and a defective does appear it may very well be missed. Consequently as we go to sample inspection we must be prepared to accept a certain amount of bad product if it is present.

This tolerance value, usually expressed in terms of percent defective, is called the acceptable quality level (AQL). It should be looked upon as an inspection standard and represents the lowest level of quality that the consumer will accept as the process average (PA) of the producer. This value should be known to all parties and should be specified in all contracts that deal with the quality of the product.

Now when sampling is used there is no guarantee that the items selected in the sample will represent the true lot quality; thus the plan may not accept all lots that meet the standard. So a value must be set that will be consistent throughout the use of the plan to specify the chance that a good lot will be rejected. This value is called the producer's risk and is often set at 5%. It is symbolized as α. (It is also called an error of the first kind, or type I error.)

It is also possible, due to the vagaries of sampling, that a bad lot may be identified as a good lot. So a value must be set for the probability of this occurence. This value is called the consumers risk and is very often set at 10%. It is symbolized as β. (It is also defined as an error of the second kind, or type II error.)

The discriminating power of the sampling plan is described by the operating characteristic (OC) curve which details what happens in the sampling process if, or as, the true quality of the product being inspected changes.

OC curves are characterized as type A and type B. The type A OC curve gives the chance of making a decision to accept or reject a lot on a "one at a time" basis, whereas type B gives the chance of a decision based on review of a long series of lots that come from a continuous process, in statistical control, operating in a random manner and centered about a specified process average. In essence, the type B curve is applied where the customer is receiving material on a regular basis from regular suppliers about whose product characteristics something is known while type A curves are put to work when odd lots are being purchased or where it is thought that the producer does not have his process in statistical control.

On a theoretical basis type B plans use the binomial distribution or a Poisson approximation as covered in earlier chapters. Type A plans are based on use of the hypergeometric distribution.

Type B curves assume sampling from an infinite population or populations that are so large that the withdrawal of sample items does not materially reduce the overall population; thus the probability of selection of each succeeding sample item is not changed significantly.

Type A curves assume sampling from populations where the probability of selection of each sample item is significantly changed.

As a matter of practice type B curves are substituted for type A when the lot size in relation to the sample size is 10:1 or more. While the difference in type A and type B curves is essentially theoretical the designer of the sampling plan should be aware of the difference in results if the wrong model is specified.

8.4. WHY "PERCENTAGE" IS SELDOM SATISFACTORY

It is often the practice of those who have not developed the habit of "statistical thinking" to take a fixed percentage of the lot to be inspected (most often 10%) as the size of the sample and to adopt an acceptance number of zero. This is thought to be a "common sense" approach to the problem, but examination from a "probability sense" point of view demonstrates that it is a poor method of deciding how many items to review.

TABLE 8.1. Fixed Percentage (10%) Sampling Demonstration. Body of Table Gives the Probability of Accepting (P_a) a Lot Submitted for Inspection for a Given True Process Average (p')

If the True Percent Defective Is p'	Lot A $N = 50$ $n = 5$ $c = 0$	Lot B $N = 100$ $n = 10$ $c = 0$	Lot C $N = 300$ $n = 30$ $c = 0$	Lot D $N = 1000$ $n = 100$ $c = 0$
	P_a, Probability of Acceptance[a]			
.01	.95	.90	.74	.37
.02	.90	.82	.55	.14
.03	.86	.74	.41	.05
.04	.82	.67	.30	.02
.05	.78	.61	.22	.007
.10	.61	.37	.05	.000
.15	.47	.22	.01	.000
.20	.37	.14	.002	.000

[a] P_a is calculated from Molina's tables (1).

For example, using the 10% sample criteria, if lot sizes were established as $N = 50$, 100, 300, and 1000 respectively, then the corresponding sample size to be used for each lot would be $n = 5$, 10, 30, and 100. For each sample assume an acceptance number (sometimes called "decision number") of $c = 0$.

Table 8.1 and Figure 8.1 demonstrate and compare the detailed operation of each plan. Since the smaller lot size has by far the greater chance of having poor quality lots accepted (61% of the time this plan will accept lots with 10% defective) a producer might very well try to keep lot sizes as small as possible.

A basic principle of sampling is that sample size is usually more important than the size of the population of interest (lot) from which the sample is drawn. To illustrate this concept assume lots of $N = 50$, 100, and 1000 and for each lot $n = 20$ and $c = 0$, thus allowing lot size to vary but holding sample size constant. The resulting OC curves (Table 8.2 and Figure 8.2) demonstrate that by keeping the sample size constant protection does not vary as much as where the sample size changes with the lot size. Note that the P_a for plan C was calculated using both Poisson and hypergeometric distributions and that the Poisson allows a little higher P_a for lots with over 3% defective. This comparison gives some idea of what

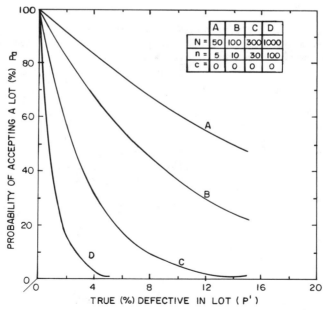

FIGURE 8.1. Fixed percentage (10%) sampling demonstration (reference Table 8.1).

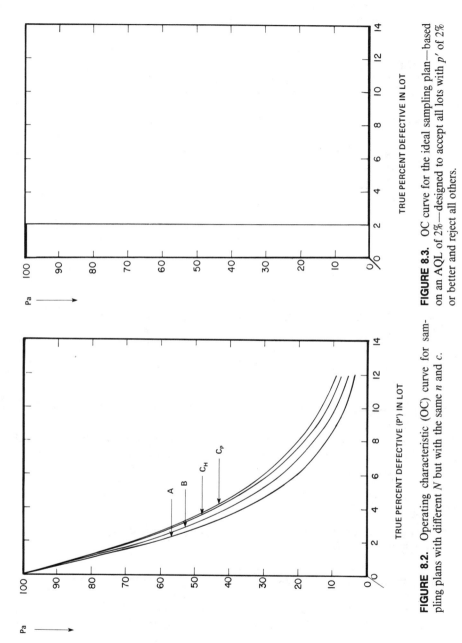

FIGURE 8.2. Operating characteristic (OC) curve for sampling plans with different N but with the same n and c.

FIGURE 8.3. OC curve for the ideal sampling plan—based on an AQL of 2%—designed to accept all lots with p' of 2% or better and reject all others.

201

TABLE 8.2. Comparison of Sampling Plans Where Lot Sizes (N) Are Different but Each Lot Uses the Same Sample Size (n) and Acceptance Number (c)

If the True Percent Defective Is p'		Lot			
		A^a	B^a	Cp^b	Ch^a
	Lot Size N:	50	100	1000	1000
	Sample Size n:	20	20	20	20
	Acceptance No. c:	0	0	0	0
		P_a			
.01		—	.80	.819	.816
.02		.60	.64	.670	.665
.03		—	.51	.549	.541
.04		.36	.40	.449	.438
.05		—	.32	.368	.355
.06		.21	.25	.301	.287
.07		—	.20	.247	.231
.08		.12	.16	.202	.186
.09		—	.12	.165	.149
.10		.07	.095	.135	.119
.11		—	.074	.111	.095
.12		.04	.057	.091	.076

[a] Hypergeometric.
[b] Poisson.

happens to the operation of a sampling plan when different mathematical models are used.

If it were possible to design an ideal sampling plan which would accept 100% of the time all lots which meet quality standards and reject all others, the OC curve would look like the curve presented in Figure 8.3.

8.5. PRELIMINARY STEPS IN ACCEPTANCE SAMPLING PLAN DESIGN

Before designing an acceptance sampling plan (or selecting one from one of the several available schemes,* i.e., Dodge–Romig, MIL-STD-105-D, etc.), it is wise to have some information on the quality policies of the producer and consumers and from the accounting, production, engineering,

*A sampling scheme is a complete set of sampling plans and instructions for their use.

and sales groups. If a laboratory is involved they should also be represented. Quality control should not be a "one man" show. Remember, too, that the statistical information that results from inspecting an individual lot does not really add much to the knowledge of the quality inherent in the process being inspected, but that information gathered over the long run tends toward the truth.

The following example involves adoption of a type B attributes plan and assumes sampling from a continuous process in statistical control with "large" lot size. It may not be common practice to design sampling plans exactly as shown here and in succeeding sections, but the basic principles of sample design are most easily illustrated by the following procedures.

STEP 1. Make a list of all the variables to be inspected and decide how they are to be inspected. The "how" may have a strong effect on the sample size and whether an attribute or a variables plan is needed.

STEP 2. Decide on the lot size to be used. For this type B plan assure that the lots will be homogeneous and will have been manufactured or produced about the same time under essentially the same conditions. For economy of inspection try to make the lots as large as possible.

STEP 3. Assume the product to be inspected comes from a continuous process in statistical control with screening of all rejected lots. The AOQ concept is introduced here, that is, all rejected lots must be screened and the defectives replaced with good items. Then by adding together the input from the adjusted lots (those rejected but corrected) and that of the lots accepted under the plan, the percentage of bad product still left in the lot can be calculated. The limiting value, AOQL, may then be calculated for the plan. See Table 8.6 and Figure 8.5.

STEP 4. Decide who will be responsible for screening rejected lots, where the screening will be done, and who will pay, or be charged, for this operation. (The latter concept may be particularly important for plans used internally.)

STEP 5. Decide on the type plan to use; that is, single, double, or multiple.

STEP 6. At this point decide whether to design an original plan or use an existing sample scheme. There is some advantage in do-it-yourself planning, particularly for better understanding, flexibility, and as an aid in training those who will use the plan. However, available sampling plans are comparatively easy to use and are quite adaptable to a wide variety of situations. Their standardization features are excellent and of course their use (MIL-STD-105, for example) may be mandated.

STEP 7. In any case the process average of the product being reviewed must be estimated (sometimes for several variables) and the OC curve analyzed.

STEP 8.   Decide how the sample is to be selected. Use only methods that will satisfy the requirement that samples be drawn in a random manner; that is, a sample that is drawn in such a way that every item in the universe of interest (lot) has a known probability of selection.

STEP 9.   Assure continuing analysis of the results of inspection following the plan. Assure one individual is given responsibility for this review and regularly submits an analysis to all concerned with the operation of the plan.

8.6.   DESIGNING THE SAMPLING PLAN (SINGLE SAMPLE)

To develop the plan for the inspection problem discussed in Section 8.3 the following steps may be taken:

STEP 1.   The variable to be inspected is weight (an oversimplified inspection variable but general enough in nature for illustrative purposes) with each sample item to be weighed to the nearest pound.

STEP 2. $N = 1000$ — (rather arbitrary lot size).

STEP 3.   Assume containers are filled directly from the output of a continuous process in statistical control with a specified process average and the controls are adjusted so that on the average no more than 1% of the containers will fill underweight. Rejected lots will be screened and corrected (in actuality of course, some dollar value might be adjusted) and the supplier will pay for screening.

STEP 4.   The consumer's quality control department will design the sampling plan and provide for monitoring the process.

STEP 5.   Samples will be selected as they are being unloaded and set aside for inspection.

The process described here is a very simple one but all the basic concepts are covered. To extend these concepts to other variables becomes only a matter of practicality.

The plan must satisfy the following statement: The consumer is willing to accept, 10% of the time, lots which have 5% of the containers underweight (the producer need not be told this but of course his own quality control department can calculate this risk—particularly if a standard plan is used.)

The producer is to aim at controlling the filling process so that the weight of material in no more than one container in a hundred will be less than tolerance and thus expects material of this quality to be accepted 95% of the time.

Symbolically the plan* reads as follows:

Producer's process average percent defective	$P_1 = .01$
Consumers lot tolerance percent defective	$P_2 = .05$
Producers risk (the chance of having a lot with only 1% bad quality rejected)	$\alpha = .05$
Consumers risk (the chance of having a lot with as much as 5% bad quality accepted)	$\beta = .10$
Lot size	$N = 1000$

Since sample size (n) and acceptance number (c) are integers it is unlikely we can compute an "exact" plan and the requirements must be "jiggled" a bit to get a good approximation. Burstein (2) has developed tables that are useful in designing single sampling plans. He gives six tables of values for the consumers risk—$\beta(0.5, 1\%, 2.5\%, 5\%, 10\%, 20\%)$ and then values of $\dot{c}$ and $\bar{m}$ for each combination of α and β.

These terms are defined as follows:

1. $\bar{m}$ is the upper confidence limit for the (based on c and $\gamma = 1 - \beta$) parameter m of a Poisson distribution.
2. $\dot{c}$ is the acceptance number.
3. $s = \bar{m}/m_1$ where m_1 is the lower confidence limit for the parameter m of Poisson distribution; based on $c + 1$ and $\gamma = 1 - \alpha$.

To estimate sample size n and acceptance number c we proceed as follows:

1. To estimate c calculate Burstein $\hat{s}$. The ratio between P_1 and P_2, s, is sometimes called the operating ratio. Each value of s applies to many different sampling plans which depend only on choice of P_1 and P_2. This ratio is particularly valuable since it is proven that for single sampling plans and a choice of c, upper and lower limits can be found that will always include s and if n is small, $5 \times (c + 1)$, the s values will be close to the upper limit. However, it is a good idea to "fiddle" with the tabled values for choice of c and n as demonstrated.

$$s = \frac{P_2}{P_1} \times \frac{2 - P_1}{2 - P_2}$$

$$= \frac{.05}{.01} \times \frac{2 - 0.01}{2 - 0.05} = 5.103$$

*Table 8.3, Ref. 3, p. 403 ($\beta = .10$, $\alpha = .05$) is used in this example.

2.   In Table 8.3, find an s as close as possible to $s = 5.103$. Using $\beta = 10\%$, $\alpha = 5\%$ we locate

$$
\begin{array}{lll}
c = 2 & \text{for} & s = 6.509 \\
c = 3 & \text{for} & s = 4.890
\end{array}
$$

3.   We can now solve for n in a number of ways.

(a)   Interpolate to get the closest value of c and solve for n as follows:

$$
\begin{array}{ll}
s = 6.509 & c = 2 \\
\hat{s} = 5.103 & c = c + x \\
s = 4.890 & c = 3 \\
\end{array}
$$
$$
x = 0.87 = 2.9 \cong 3
$$

Calculate

$$
n = \frac{\overline{m}}{P_2} - \frac{\overline{m} - c}{2}
$$

$\overline{m}$ for $c = 3 = 6.6808$

Then

$$
n = \frac{6.6808}{.05} - \frac{6.6808 - 3}{2}
$$
$$
= 133.6 - 1.8
$$
$$
= 131.8 \cong 132
$$
$$
n = 132 \qquad c = 3
$$

Using Molinas tables (1), or working out the first four terms of the Poisson with $p'_n = 1.32$, yields for $P_1 = .01$ a probability of acceptance of .955, $p'_n = 6.6$ yields for $P_2 = .05$ a probability of acceptance of .104. This nearly matches the requirements. However we may wish to know what will happen if we use an acceptance number of $c = 2$.

$$
\overline{m} \quad \text{for} \quad c = 2 = 5.3223
$$

and

$$
n = \frac{5.3223}{.05} - \frac{5.3223 - 2}{2}
$$
$$
106.4 - 1.7 = 104.7 \cong 105
$$

For the plan $n = 105$, $c = 2$: $p'_n = 1.05$ yields for $P_1 = .01$, $P_a = .91$, $p'_n = 5.25$ yields for $P_2 = .05$, $P_w = .105$.

TABLE 8.3. Values of *S* and *c* for Determining Sample Size for Acceptance Sampling; p_1 and $p_2 \leq .25$. From Burstein (6) *Tables for Attribute Sampling*

$\dot{c}$	$\overline{m}$	0.5%	1%	2.5%	5%	10%	20%
				S			
0	2.3026	459.359	229.090	90.947	44.890	21.854	10.319
1	3.8897	37.584	26.184	16.059	10.946	7.314	4.718
2	5.3223	15.753	12.206	8.603	6.509	4.829	3.467
3	6.6808	9.939	8.115	6.130	4.890	3.829	2.909
4	7.9936	7.416	6.249	4.924	4.057	3.286	2.587
5	9.2747	6.035	5.195	4.212	3.549	2.943	2.376
6	10.532	5.170	4.520	3.742	3.206	2.704	2.225
7	11.771	4.578	4.050	3.408	2.957	2.528	2.111
8	12.995	4.148	3.705	3.158	2.768	2.392	2.021
9	14.206	3.822	3.440	2.962	2.618	2.283	1.949
10	15.407	3.565	3.229	2.806	2.497	2.194	1.889
11	16.598	3.358	3.058	2.677	2.397	2.120	1.838
12	17.782	3.187	2.915	2.569	2.312	2.057	1.794
13	18.958	3.043	2.795	2.477	2.240	2.002	1.756
14	20.128	2.920	2.692	2.398	2.177	1.954	1.723
15	21.292	2.814	2.603	2.328	2.122	1.912	1.693
16	22.452	2.721	2.524	2.267	2.073	1.875	1.667
17	23.606	2.640	2.455	2.213	2.029	1.841	1.643
18	24.756	2.567	2.393	2.164	1.990	1.811	1.621
19	25.903	2.502	2.337	2.120	1.954	1.783	1.602
20	27.045	2.443	2.287	2.080	1.922	1.758	1.584
21	28.184	2.390	2.241	2.044	1.892	1.735	1.567
22	29.320	2.342	2.200	2.011	1.865	1.714	1.552
23	30.453	2.297	2.162	1.980	1.840	1.694	1.537
24	31.584	2.257	2.126	1.952	1.817	1.676	1.524
25	32.711	2.219	2.094	1.926	1.795	1.659	1.512
26	33.836	2.184	2.064	1.902	1.775	1.643	1.500
27	34.959	2.152	2.036	1.879	1.757	1.628	1.489
28	36.080	2.122	2.009	1.858	1.739	1.614	1.479
29	37.199	2.094	1.985	1.838	1.723	1.601	1.469
30	38.315	2.067	1.962	1.819	1.707	1.589	1.460
31	39.430	2.042	1.940	1.801	1.692	1.577	1.451
32	40.543	2.019	1.920	1.785	1.679	1.566	1.443
33	41.654	1.997	1.900	1.769	1.665	1.556	1.435
34	42.764	1.976	1.882	1.754	1.653	1.546	1.428
35	43.872	1.957	1.865	1.740	1.641	1.536	1.421
36	44.978	1.938	1.848	1.727	1.630	1.527	1.414
37	46.083	1.920	1.833	1.714	1.619	1.519	1.408
38	47.187	1.903	1.818	1.701	1.609	1.510	1.401
39	48.289	1.887	1.804	1.690	1.599	1.503	1.396

The table is headed by $\beta = 10\%$ and α.

TABLE 8.4. Four Possible Choices of Calculating *n* to Build a Sampling Plan with $c = 2$, $c = 3$, $\alpha = .05$, $\beta = .10$

	$c = 2$	$c = 3$
$\beta = .10$	$\dfrac{\overline{m}}{P_2} = \dfrac{5.3223}{.05}$ $= 106.4$ *n* for plan A $= 106$	$\dfrac{\overline{m}}{P_2} = \dfrac{6.6808}{.05}$ $= 133.6$ *n* for plan B $= 134$
$\alpha = .05$	$\dfrac{(\overline{m}/s)}{P_1} = \dfrac{(5.3223/6.509)}{.01}$ $= 81.7$ *n* for plan C $= 82$	$\dfrac{(\overline{m}/s)}{P_1} = \dfrac{(6.6808/4.890)}{.01}$ $= 136.6$ *n* for plan D $= 137$

In this case the consumer's risk is almost matched exactly but the producer will face a much tighter plan.

(b) Following a method suggested in Duncan (2) we can use both α and β in calculating sampling plans.

Taking combinations of c and P_1 and P_2 we get four possible plans. From these we can then select the one that best suits the purpose. S is the ratio between P_1 and P_2 for a plan which will have an acceptance number c. Since we are given $\overline{m}$, which in this example is equal to $P = .10$, we may calculate $P = .95$ by dividing $\overline{m}/s$.* Sample size calculations will then be

$$n = \frac{\overline{m}}{P_2} \quad \text{for} \quad \beta = .10$$

$$n = \frac{\overline{m}/s}{P_1} \quad \text{for} \quad \alpha = .05$$

The plans are demonstrated in Table 8.4.

Only the P_a for each plan at true percent defective values of .01 and .05 are needed to compare to the specifications. However, to examine the overall effect of the plans, Table 8.5 and Figure 8.4 give the complete OC curve for all four plans.

The bases for the OC curves for these four plans are given in Table 8.5. Included in this table are p'_n values for each plan so that the reader may follow the method of constructing OC curves using the Poisson distribution. Close approximations to these values may be obtained using Molina's table (1).

---

*For all practical purposes the $(\overline{m} - 2)/2$ factor may be and has been omitted.

TABLE 8.5. Data for Plotting the OC Curves for the Four Proposed Sampling Plans in Table 8.4

	Concentration on 10% Consumer Risk				Concentration on 5% Producer's Risk			
	$n = 106$ (A) $c = 2$		$n = 134$ (B) $c = 3$		$n = 82$ (C) $c = 2$		$n = 137$ (D) $c = 3$	
	np'	P_a	np'	P_a	np'	P_a	np'	P_a
.01	1.06	.908	1.34	.953	.82	.950	1.37	.950
.02	2.12	.644	2.68	.718	1.64	.773	2.74	.705
.03	3.15	.390	4.02	.430	2.46	.554	4.11	.412
.04	4.24	.205	5.36	.218	3.28	.363	5.48	.204
.05	5.30	.102	6.70	.099	4.10	.224	6.85	.090
.06	6.36	.048	8.04	.041	4.92	.132	8.22	.036
.07	7.42	.022	9.38	.016	5.74	.075	9.59	.014
.08	8.48	.009	10.72	.007	6.56	.041	10.96	.005
.09	9.54	.004	12.06	.002	7.38	.017	12.33	.002
.10	10.60	.002	13.40	.001	8.20	.012	13.70	.001

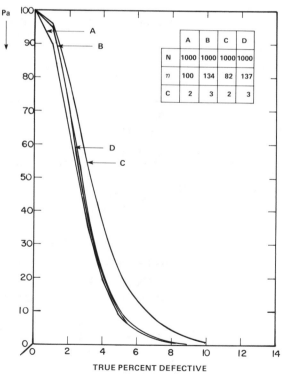

FIGURE 8.4. OC curves for four sampling plans in Table 8.5.

TABLE 8.6. Calculation of Average Outgoing Quality for the Sampling Plans in Table 8.5

True Percent Defective on Lot p'	AOQL = .0116 Plan A[a]		AOQL = .0126 Plan B[a]		AOQL = .0153 Plan C[a]		AOQL = .0122 Plan D[a]	
	P_a	AOQ[b]	P_a	AOQ[b]	P_a	AOQ[b]	P_a	AOQ[b]
.01	.90	.009	.96	.010	.95	.010	.95	.010
.02	.65	.013	.71	.014	.78	.0156	.71	.014
.03	.38	.011	.43	.011	.54	.0162	.41	.012
.04	.21	.008	.21	.008	.36	.014	.20	.008
.05	.10	.005	.10	.005	.22	.011	.09	.005
.06	.05	.003	.04	.002	.13	.008	.04	.002
.07	.02	.001	.016	.001	.08	.006	.014	.001
.08	.009	.001	.007	.001	.04	.003	.005	—
.09	.004	—	.002	—	.02	.002	.002	—
.10	.002	—	.001	—	.01	.001	.000	—

[a] $\text{AOQL} = y\left(\dfrac{1}{n} - \dfrac{1}{N}\right).$
[b] $\text{AOQ} = (P_a)(P').$

Another aid in selecting a sampling plan is to calculate the maximum percent of poor product that will, in the long run, remain in the lots submitted for inspection after review and screening. This maximum possible value of percent defective in the outgoing quality is known as the average outgoing quality limit (AOQL).

The average outgoing quality is first calculated by multiplying the true lot percent defective (p') by the probability of accepting (P_a) that percent defective in the plan under consideration. Using the P_a from the body of Table 8.6 which portrays data for the four plans being considered, the average outgoing quality (AOQ) is calculated and illustrated in Table 8.6 and Figure 8.5. Two examples are shown here:*

Plan A   True percent defective = .01
                Probability of acceptance = .90
                AQL = (.01)(.908) = .009
Plan B   True percent defective = .02
                Probability of acceptance = .644
                AQL = (.02)(.644) = .0129, etc.

*To be more precise the formula should read

$$\text{AQL} = \frac{(P_a)(P')(N - n)}{N} = \frac{(.02)(.65)(894)}{1000} = .0116$$

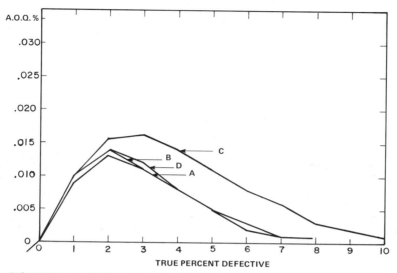

FIGURE 8.5. AOQL for each of the four proposed sampling plans (Table 8.6).

Dodge and Romig (4) provide a table (see Table 8.7) based on the Poisson distribution for use in calculation of the AOQL using the following formula

$$\text{AOQL} = y\left(\frac{1}{n} - \frac{1}{N}\right)$$

TABLE 8.7. Factors to Calculate AOQL Using Values of y for Given Values of c

Given c	y
0	.3679
1	.8400
2	1.371
3	1.942
4	2.544
5	3.168
6	3.812
7	4.472
8	5.146
9	5.831

Abridged, with permission, from Table 2-3, p. 39, *Sampling Inspection Tables*, Dodge–Romig (7).

Plan A = AOQL was estimated from Figure 8.5 as approximately .0129 and rounded to .013.

Using the Dodge–Romig formula we get

$$y = 1.371$$

when $c = 2$ with $n = 106$ and $N = 1000$. The AOQL = 1.371(.0094 − .001) = .0116. The ideal lot ratio N/n (A = 9.4, B = 7.5, C = 12.2, D = 7.3) to sample size should be 10:1 or better to warrant correctness of the assumption.

To obtain more information for a good decision on choosing a sampling plan, another approach is to investigate how much inspection results from each of the four plans under consideration. To obtain the average items inspected (AII) use this method:

Sample size (n) plus the product of the probability of rejecting a lot at a specified percent defective (P_a) and the lot size (N) minus the sample (n) or

$$AII = n + [(P_a)(N - n)]$$

For plan A at 1% defective,

$$AII = 106 + 1[(1 - .90)(1000 - 106)]$$
$$106 + [(.10)(894)] = 195.4 = 195$$

AII (shown in detail in Table 8.8) is included in the comparison of the four plans given in Table 8.9.

TABLE 8.8. Calculation of Average Number of Items Inspected per Lot for Each of the Four Proposed Plans

True Percent Defective in Lot p'	Plan			
	A	B	C	D
.01	195	169	128	180
.02	419	385	284	387
.03	660	628	504	646
.04	812	818	670	827
.05	911	913	798	922
.06	955	965	881	965
.07	982	986	927	987
.08	992	994	963	996
.09	996	998	972	998
.10	998	999	991	1000

TABLE 8.9. Comparison of Four Proposed Sampling Plans

	Plan			
	A	B	C	D
Lot size (N)	1000	1000	1000	1000
Sample size (n)	106	134	82	137
Acceptance number (c)	2	3	2	3
Probability of accepting P_1 quality (.01)	.90	.96	.95	.95
Probability of accepting P_2 quality (.05)	.10	.10	.22	.09
AOQL (from chart)	.013	.014	.016	.014
AOQL [from $y(1/n - 1/N)$]	.0116	.0126	.0153	.0122
Average number of items inspected per lot at $p' = .01$	195	169	128	180

We now have the information for a fairly complete summary and comparison of the four plans and may then use this data to select the plan best suited for the immediate purpose.

Analysis of Sampling Plan

Plan A is a little rough on the producer who would face the probability of having, on the average, 1 in each 10 of his good lots rejected, and paying the cost of the unnecessary screening of 10,000 items. This plan has the lowest AOQL and the highest AII.

Plan C is rough on the consumer who would face the probability of accepting nearly 1 lot with 5% defective of every 5 such lots submitted. This plan has the highest AOQL and the lowest AII.

Plan D favors the consumer in that he has only a 9% chance of accepting quality at the 5% level. The AOQL is .014 which is not as good as in plan A but matches plan B. The AII is fairly high in comparison to plan B but the producer must assume most of the inspection cost. The actual amount of inspection for the consumer is only about 2% more than in plan B.

Plan B satisfies both the consumer and producer with the producer having a little edge for P_1. The AOQL matches plan D; the AII is the second lowest of the plans.

The writer would be satisfied with plan B, particularly if the producers assured high quality. A "hard rock" consumer might try for plan A. He is protected for his P_2 level, and the plan gives the best AOQL. The AII is high but the consumer assumes only about 50% of the inspection cost and the inspection cost is about 20% less than the next highest sample size.

The reader is invited to form his own judgment, since this is only an exercise to illustrate that selecting a proper sampling plan has many ramifications.

The writer sees some advantage in "fiddling" with this type data.

1. It helps the planner to a better understanding of how acceptance sampling operates.
2. It helps the planner to explain the plan's operation to others.
3. Controlled flexibility in planning for acceptance of material will help to assure overall plant quality.

8.7. DESIGNING A DOUBLE SAMPLING PLAN

Again the simplest kind of sampling plan will be demonstrated. A double sampling plan operates as follows:

$$\text{Lot size} = N \qquad \begin{array}{l} \text{First sample size} = n_1, \text{acceptance number} = c_1 \\ \text{Second sample size} = n_2, \text{acceptance number} = c_2 \end{array}$$

A sample is taken with size equal to n_1. After inspection if the number of defective items is equal to or less than c_1 the lot is accepted immediately; if the number of defectives exceeds c_2 the lot is rejected immediately. If the number of defectives in the first sample is greater than c_1 but does not exceed c_2 a second sample of size n_2 is taken and inspected. If, after inspection, the total number of defectives for both samples is equal to or less than c_2 the lot is accepted; if the total number of defectives in both samples exceeds c_2, the lot is rejected.

Note that in practice it is customary to curtail inspection on the second sample when c_2 is exceeded to save inspection costs since a good estimate of the producer's quality is obtained from the completely inspected first sample for the long run. This makes calculation of the average amount of inspection (AAI) for double sampling a little different than in single sampling (demonstrated later).

The probability of accepting a lot after both samples are inspected is equal to the probability of accepting the lot on the first sample added to the probability of accepting the lot on the second sample. But in this more complicated plan, more information is needed. To simplify calculations it is good practice to have the second sample size proportionate to the first sample size. In this example the ratios 1:1 and 1:2 will be considered. Use of this constant ratio approach in designing sampling plans, if we use type B OC curves, allows use of the Poisson distribution. If P_1 and P_2 have a constant ratio, the sampling plans will have about the same OC curve as we change n_1.

Using acceptance criteria similar to those we used in the single sample demonstration, let $N = 1000$, the process average $(P_1) = .01$, consumer tolerance $(P_2) = .05$, producer's risk $(\alpha) = .05$, consumer's risk $(\beta) = .10$.

As in calculating the single sampling plan, use is made of several tables developed to simplify the design of sampling plans. Again these tables are based on the operating ratio P_2/P_1.

Tables 8.10 and 8.11 are demonstrated as follows: Using Table 8.10, given $N = 1000$,

$$P_1 = .01$$
$$P_2 = .05$$
$$\alpha = .05$$
$$\beta = .10$$
$$n_2 = 2n_1$$

Solve for n_1 and n_2, c_1 and c_2.

$$R = \frac{P_2}{P_1} = \frac{.05}{.01} = 5$$

TABLE 8.10. Values for Use in Designing a Sampling Plan—Given P_1 and P_2 with the Relationship of n_2 as $2n$, $\alpha = .05$, $\beta = .10$

Plan Number	R^a	Acceptance Numbers		Approximate Values of $p'n$, for		Approximate (ASN)/n for
		c_1	c_2	$P = .95$	.10	.95 Point[b]
1	11.90	0	1	.21	2.50	1.170
2	7.54	1	2	.52	3.92	1.081
3	6.79	0	2	.43	2.96	1.340
4	5.39	1	3	.76	4.11	1.169
5	4.65	2	4	1.16	5.39	1.105
6	4.25	1	4	1.04	4.42	1.274
7	3.88	2	5	1.43	5.55	1.170
8	3.63	3	6	1.87	6.78	1.117
9	3.38	2	6	1.72	5.82	1.248
10	3.21	3	7	2.15	6.91	1.173
11	3.09	4	8	2.62	8.10	1.124
12	2.85	4	9	2.90	8.26	1.167
13	2.60	5	11	3.68	9.56	1.166
14	2.44	5	12	4.00	9.77	1.215
15	2.32	5	13	4.35	10.08	1.271
16	2.22	5	14	4.70	10.45	1.331
17	2.12	5	16	5.39	11.41	1.452

Adapted from Duncan (2); p. 189.
[a] $R = p'/p'1$—operating ratio.
[b] ASN is without curtailment on second sample.

TABLE 8.11. Values for Use in Designing a Sampling Plan—Given P_1 and P_2 with the Relationship $n_1 = n_2$, $\alpha = .05$, $\beta = .10$

| Plan Number | R^a | Acceptance Numbers | | Approximate Values of p/n for | | Approximate (ASN)/n for 0.95 Point[b] |
		c_1	c_2	$P = .95$	.10	
1	14.50	0	1	.16	2.32	1.273
2	8.07	0	2	.30	2.42	1.511
3	6.48	1	3	.60	3.89	1.238
4	5.39	0	3	.49	2.64	1.771
5	5.09	1	4	.77	3.92	1.359
6	4.31	0	4	.68	2.93	1.985
7	4.19	1	5	.96	4.02	1.498
8	3.60	1	6	1.16	4.17	1.646
9	3.26	2	8	1.68	5.47	1.476
10	2.96	3	10	2.27	6.72	1.388
11	2.77	3	11	2.46	6.82	1.468
12	2.62	4	13	3.07	8.05	1.394
13	2.46	4	14	3.29	8.11	1.472
14	2.21	3	15	3.41	7.55	1.888
15	1.97	4	20	4.75	9.35	2.029
16	1.74	6	20	7.45	12.96	2.230

Adapted from A. J. Duncan, (2), p. 188.
[a] $R = p_2'/p_1'$—operating ratio.
[b] ASN is without curtailment on second sample.

Plan number 5 is close to this ratio, and gives $c_1 = 1$, $c_2 = 4$. Solving for n and holding the consumer point of view ($P_2 = .05$, $\beta = .10$) we get for plan 5

$$\frac{P^1 n_1 .10}{P_2} = \frac{3.92}{.05} = 78.4 = 78$$

Since our ratio is $n_1 : n_2 = 2N$, the plan reads as follows:

$$n_1 = 78 \qquad c_1 = 1$$
$$n_2 = 156 \qquad c_2 = 4$$

Average sample number (ASN) = (78)(1.359) = 106 (without curtailment). With curtailment ASN will be less and this is developed later.

Similarly if we wish to develop a plan with ratio $n_1 = n_2$, Table 8.11 is used with the following characteristics:

$$N = 1000$$
$$P_1 = .01$$
$$P_2 = .05$$
$$\alpha = .05$$
$$\beta = .10$$
$$n_1 = n_2$$

Again $R = .05/.01 = 5.0$.

This value falls between plan 4 and plan 5, and it is a toss up in choice. Assume we select plan 4. Then $c_1 = 1$, $c_2 = 3$. Again we hold for the consumer point of view and

$$\frac{P'n_{1/10}}{P_2} = \frac{4.11}{.05} = 82.2 = 82$$

The plan reads as follows:

$$n_1 = 82 \qquad c_1 = 1$$
$$n_2 = 82 \qquad c_2 = 3$$
$$\text{ASN} = (82)(1.169) = 96$$

The OC curve for a double sampling plan is developed in a manner similar to that used in single sampling, that is, using Molina's tables and noting that the probability of acceptance after both samples is the probability of acceptance on the first sample plus the probability of acceptance on the second sample. For the plan $n_1 = n_2 = 82$, $c_2 = 3$, set up the following column headings (see Table 8.12 and Figure 8.6).

1. True percent defective in the lot (p').
2. $P'n$.
3. $P_a(1, 1)$ on the first sample.
4. $P(2)$ on first sample times $P(0, 1)$ on the second.
5. $P(3)$ on first sample times $P(0)$ on the second.
6. P_a is equal to col. 3 + col. 4 + col. 5.

There is some interest in examining and calculating the average sample number (ASN) curve for double sampling with and without curtailed inspection, and comparing results to the straight line ASN which results

TABLE 8.12. Calculation of OC Curve for the Double Sampling Plan $n_1 = n_2 = 82$, $c_1 = 1$, $c_2 = 3$

(1) True Percent Defective in Lot	(2)	(3) $P(0,1)$ of First sample	(4) Plus $P(2)$ first × $P(0,1)$ second			(5) Plus $P(3)$ first × $P(0)$ second			(6)
P'	$P'n$		$P(2)$	$P(0,1)$	$P(2)P(0,1)$	$P(3)$	$P(0)$	Equals	P_as
.01	.82	.81	.14	.81	.11	.04	.45	.018	.94
.02	1.64	.53	.26	.53	.14	.14	.20	.028	.70
.03	2.46	.29	.26	.29	.08	.21	.08	.017	.39
.04	3.28	.16	.20	.16	.03	.22	.04	.008	.20
.05	4.10	.09	.14	.09	.01	.19	.02	.004	.10
.06	4.92	.04	.09	.04	.004	.15	.008	.001	.05
.07	5.74	.02	.06	.02	.001	.10	.004	—	.02
.08	6.56	.01	.03	.01	.0003	.07	.001	—	.01
.09	7.38	.005	.015	.005	—	.04	—	—	.01
.10	8.20	.002	.009	.002	—	.02	—	—	.004

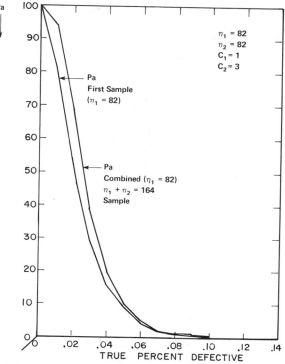

FIGURE 8.6. OC curve for the double sampling plan $n_1 = n_2 = 82$, $c_1 = 1$, $c_2 = 3$.

from using a single sampling plan to emphasize that one of the advantages of using double sample plans is that less inspection may be needed.

Using a single sample plan with a similar OC curve to the double sampling plans, all items (n) are inspected. For plan D, $n = 137$, $c = 3$.

The formula for calculating ASN using a double sampling plan with complete inspection of the second sampling is

$$\text{ASN} = n_1 + n_2(1 - P_1)$$

where P_1 is the chance that a decision will be made on the first sample (see Table 8.13).

Calculating the ASN for double sampling with curtailed inspection is a little more complicated since in addition to sample size and acceptance numbers we need to develop the following information for each value of the true fraction defectives (P_1) used in plotting the curve.

(a) The probability of getting exactly (x) defectives (P^E).
(b) The probability of getting (x) or more defectives (P^M).
(c) The probability of getting (x) or less defectives (P^L).

TABLE 8.13. Calculation of Average Sample Number Curve for the Double Sampling Plan $N = 1000$, $n_1 = 82$, $n_2 = 82$, $c_1 = 1$, $c_2 = 3$, with *Complete Inspection* of the second sample

True Percent Defectives in Lot, P^1	Probability of a Decision on the First Sample $P(0,1) + P(4) = (P_1)$			ASN $1 - P_1 + N_2(P_1)$	$\dfrac{N_1 + N_2(P_1)}{ASN}$
.01	.81	.009	.82	.18	15	97
.02	.53	.08	.61	.39	32	114
.03	.29	.24	.53	.47	39	121
.04	.16	.42	.58	.42	34	116
.05	.09	.59	.68	.32	26	108
.06	.04	.72	.76	.24	20	102
.07	.02	.82	.84	.16	13	95
.08	.01	.90	.91	.09	7	89
.09	.005	.94	.945	.055	5	87
.10	.002	.96	.962	.038	3	85

In Table 8.14 calculations are shown for computing ASN where $P^1 = .04$. Table 8.15 gives the ASN for various P^1 from .01–.10 and Figure 8.7 shows the ASN curves for single sampling and for double sampling with, and without, curtailed inspection. Curtailing inspection on the second sample will result in less inspection than in other plans. In practice it is wise to inspect all of the first sample to get the best estimate of the true lot quality.

Column headings for ASN calculations in Table 8.14 follow this formula:

$$ASN = n_1 + \sum_{c_s = c_1 + 1}^{c_2} P_{n_1}^E : c_s \left[n_2 P_{n_2}^L : c_2 - c_s + \frac{c_2 - c_s + 1}{p'} P_{n_2+1}^M : c_2 - c_s + 2 \right]$$

(n_1) = number of items in the first sample

(which is always completely inspected)

plus

$\left(\displaystyle\sum_{c_s = c_1 + 1}^{c_2} \right)$ for the summation of all possibilities of making a decision to accept

$\left(P_{n_1}^E : \text{ for } c_s \right)$ the probability of getting nondecision numbers on the first sample (exactly x defectives)

TABLE 8.14. Average Sample Number (ASN) for Double Sampling Plan $n_1 = n_2 = 82$, $c_1 = 1$, $c_2 = 3$ with Curtailed Inspection on the Second Sample[a]

(1) p^1	(2) $p^E n_1; c_s$	(3) $p^L n_2: c_2 - c_s$	(4) (3) 82	(5) $\dfrac{c_2 - c_s + 1}{p'}$	(6) $p^M n_2: c_2 - c_s + 2$	(7) (5)(6)	(8) (4 + 7)(2)	(9) Sum 82 + (8) ASN
.01	.15	.80	66	200	.05	10	76 11.4	
	.04	.44	36	100	.20	20	56 2.2 (13.6)	96
.02	.26	.51	42	100	.23	23	65 15.0	
	.14	.19	16	50	.49	24.5	41 5.7 (20.7)	103
.03	.26	.29	24	66.7	.46	30.7	55 14.3	
	.20	.08	6	33.3	.71	23.6	30 6.3 (20.6)	103
.04	.20	.16	13	50	.64	32	45 9	
	.22	.04	3	25	.84	21	24 5.3 (20.6)	103
.05	.14	.09	7	40	.78	31.2	38 5.3	
	.18	.016	1	20	.92	18.4	19 3.4 (14.3)	96

(continued)

TABLE 8.14. (*continued*)

(1) p^1	(2) $p^E n_1; c_s$	(3) $p^L n_2; c_2 - c_s$	(4) (3) 82	(5) $\dfrac{c_2 - c_s + 1}{p'}$	(6) $p^M n_2; c_2 - c_s + 2$	(7) (5)(6)	(4+7)	(8) (4+7)(2)	(9) Sum 82 + (8) ASN
.06	.09	.04	3	33.3	.87	29.0	32	2.9	
	.15	.008	1	16.7	.96	16.0	17	2.6	
								(5.5)	88
.07	.05	.02	2	28.5	.92	26.2	28	1.4	
	.10	.004	—	14.3	.98	14.0	14	1.4	
								(2.8)	85
.08	.03	.01	1	25	.96	24	25	.8	
	.07	.001	—	12.5	.99	12.4	12	.8	
								(1.6)	84
.09	.02	.005	—	22.2	.98	21.8	22	.4	
	.04	.001	—	11.1	.995	11.0	11	.4	
								(.8)	83
.10	.01	.002	—	20	.99	19.8	20	.2	
	.02	.000	—	10	.998	10.0	10	.2	
								(.4)	82

[a] For each p^1

(2) c_s	(3) $c_2 - c_s$	(5) $c_2 - c_s + 1$	(6) $c_2 - c_s + 2$
2	1	2	3
3	0	1	2

222

TABLE 8.15. Comparison of ASN for Double Sampling Plans with and Without Curtailed Inspection of Second Sample and a Single Sampling Plan with About Same OC Curve

	Average Sample Number (ASN)		
	Double Sampling Plan		
True Percent Defectives in Lot P'	With Curtailed Inspection of Second Sample (Table 14)	With Complete Inspection of Second Sample (Table 13)	Single Sampling Plan $n = 137, c = 3$ (Table 5)
.01	96	97	137
.02	106	114	137
.03	103	121	137
.04	96	116	137
.05	91	108	137
.06	88	102	137
.07	85	95	137
.08	84	89	137
.09	83	87	137
.10	82	85	137

multiplied by the sum of

$$\left(n_2 P_{n_2}^L : \quad \text{for} \quad c_2 - c_s \right) \qquad \text{the probability of getting } x \text{ or less defectives in the second sample}$$

plus

$$\left(\frac{c_2 - c_s + 1}{p'} \right) \qquad \begin{array}{l} = \text{the average number of pieces that will be inspect-} \\ \text{ed before a decision number is reached if the} \\ \text{given } p' \text{ is the true percent defective in the lot} \end{array}$$

multiplied by

$$\left(p_{n_2}^M : \quad \text{for} \quad c_2 - c_2 + 2 \right) \qquad \text{the probability of getting } x \text{ or more defectives in the second sample}$$

8.8. DESIGNING MULTIPLE SAMPLING PLANS

A multiple sampling plan operates in a manner similar to that of the double sampling plan except that more steps are used. At any step, if the number of defectives is equal to or less than the acceptance number the lot is accepted. At any step if the number of defectives is equal to or exceeds the rejection number the lot is rejected. If neither of these conditions is reached another sample is taken. Usually the plans are truncated to force a decision after the seventh or eighth sample. As in double sampling the first

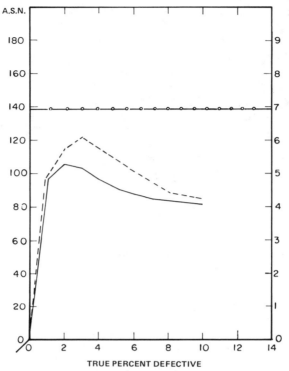

FIGURE 8.7. Comparison of ASN curve for single sampling (0) ($n = 137$, $c = 3$) and double sampling ($n_1 - n_2 = 82$, $c_1 = 1$, $c_2 = 3$) with complete (---) and curtailed (——) inspection.

sample is completely reviewed for the record but inspection is curtailed in subsequent samples as soon as the decision criteria is reached.

The procedure for designing a multiple sample plan is similar to designing a single and double plan if we specify P_1, P_2, α, β. Duncan (2) suggests use of tables developed by the Army Chemical Corps which make use of the operating ratio P_2/P_1. Specify

$$P_1 = .01$$
$$P_2 = .05$$
$$\alpha = .05$$
$$\beta = .10$$
$$R = \frac{P_2}{P_1} = \frac{.05}{.01} = 5$$

In Table 8.16 for $R = 5$ we are very close to plan No. 8. If we select the .095 value we divide $.31/.01 = 31$. If we select the 0.10 value we divide

TABLE 8.16. Values for use in Designing a Multiple Sample Plan

Number of Plan	R^a		Acceptance and Rejection Numbers									Approximate Values of $p'n_1$ for $P_a = .95$	.10	(ASN)/n_1 at Approximately $P_a = .95$
1	18.46	Ac	*	*	0	0	1	2	3			.048	.89	3.243
		Re	2	2	2	2	3	4	4					
2	12.15	Ac	*	*	*	0	0	1	2			.065	.79	4.373
		Re	2	2	2	3	3	4	5					
3	9.95	Ac	*	*	0	0	1	2	4			.10	1.00	3.461
		Re	2	2	2	3	3	4	5					
4	8.91	Ac	*	*	0	0	0	0	0	2		.088	.78	3.876
		Re	2	2	2	2	2	3	3	3				
5	8.06	Ac	*	*	0	0	0	0	0	1	2	.093	.75	4.077
		Re	2	2	2	3	3	3	3	3	3			
6	7.04	Ac	*	0	0	1	1	1	2	3		.18	1.27	2.828
		Re	2	3	3	3	4	4	4	4				
7	6.20	Ac	*	0	1	1	2	3	4			.24	1.48	2.515
		Re	2	3	3	3	4	5	5					
8	4.95	Ac	*	0	1	2	4	4	5			.31	1.55	2.606
		Re	2	3	4	5	6	6	6					
9	4.61	Ac	*	0	0	1	2	3	4	6		.31	1.43	3.268
		Re	3	3	4	4	5	6	7	7				
10	4.29	Ac	*	1	2	3	4	6	7			.68	2.93	1.727
		Re	4	5	7	9	10	11	11					
11	4.02	Ac	*	1	2	3	4	6	7			.47	1.89	2.380
		Re	3	4	5	6	6	8	8					
12	3.75	Ac	*	1	1	2	3	5	7			.56	2.11	2.839
		Re	3	4	5	6	6	8	8					
13	3.56	Ac	*	1	1	3	4	5	7	9		.59	2.10	2.872
		Re	3	5	6	7	8	9	10	10				
14	3.23	Ac	0	2	3	4	6	8	11			.96	3.10	2.218
		Re	4	5	8	9	10	12	12					
15	3.03	Ac	0	3	6	8	10	12	14			1.20	3.64	1.891
		Re	4	7	9	11	12	14	15					
16	2.69	Ac	1	3	6	9	11	14	17			1.56	4.20	1.839
		Re	5	7	10	13	15	18	18					
17	2.54	Ac	1	3	6	9	13	16	18			1.60	4.06	1.911
		Re	5	8	11	13	16	18	19					
18	2.35	Ac	1	5	7	10	13	17	22			2.00	4.70	1.982
		Re	6	9	12	16	19	21	23					
19	2.16	Ac	1	5	9	13	18	22	25			2.40	5.19	2.138
		Re	7	10	13	18	22	25	26					
20	1.94	Ac	3	8	13	18	24	30	36			3.74	7.26	1.967
		Re	8	15	20	25	30	34	37					

Adapted from Duncan, p. 204, (2).

[a] $R = p'_2/p_1$ = operating ratio.

TABLE 8.17. Comparison of Advantages and Disadvantages of Using Single, Double, and Multiple Sampling Plans

Factor	Single Sampling	Double Sampling	Multiple Sampling
Protection for rejection of inspection lots and acceptance of bad inspection lots		About equal	
Mean items inspected per inspection lot	Greatest	In between	Smallest
Variation of inspected items from inspection lot to inspection lot	None	Some	Some
Cost of sampling when samples can be drawn at once	Most expensive	In between	Least expensive
Cost of sampling when all samples must be drawn at once	Least expensive	Most expensive	In between
Estimate of average inspection-lot quality	Most precise	In between	Least precise
Amount of training	Least	In between	Most
"Extra chance to pass" philosophy	Poor	Good	Very good

$1.55/.05 = 31$. For simplicity round to $n = 30$ and the multiple sample plan will be as follows:

Sample No.	Cumulative n	Acceptance Number	Rejection Number
1	30	No decision	2
2	60	0	3
3	90	1	4
4	120	2	5
5	150	4	6
6	180	4	6
7	210	5	6

$$\text{ASN} = 30(2.606) = 78.2 = 78$$

Calculation of the complete OC curve for a multiple sampling plan is a laborious process and will not be discussed here. Duncan (2) has an excellent description of the approach to getting this information.

What plan to use is often a matter of administrative convenience and cost. Table 8.17 gives a summary of the advantages and disadvantages of single, double, and sequential sampling plans.

8.9. CHAIN SAMPLING PLANS

When destructive tests must be used to evaluate product for acceptance the use of small samples becomes mandatory. For lots that are quite small the acceptance number is often established as zero, which creates a high probability of rejecting good lots and may be unfair to the producer. Dodge (5) suggests the following procedure:

From each lot submitted for inspection, select and test a sample of size n. If all the items in the sample pass inspection accept the lot. If the sample contains two or more bad items reject the lot. If the lot contains only one bad item accept the lot if no defectives were found in i samples prior.

The basic formula that describes this type plan utilizes the first and second terms of the binomial distribution in describing the probability of obtaining zero and one defect, respectively.

$$P_2 = P(0, N) + P(0, N)[P(0, N)]^i$$

TABLE 8.18.   Calculation of the Operating Characteristic Curve for Typical Chain Sampling Plans Where:
Plan A $N = 8, C = 0, i = 1$
Plan B $N = 8, C - 0, i = 3$

True Percent Defective in Lot	Plan A	Plan B
.01	.992	.981
.02	.969	.937
.03	.936	.878
.04	.894	.811
.05	.848	.744
.06	.800	.681
.07	.749	.619
.08	.696	.561
.09	.645	.509
.10	.595	.461
.15	.377	.280
.20	.224	.170
.25	.127	.100

The underlying assumptions in using this plan are:

(a) Lots should be formed from a continuous process currently being received.
(b) The quality should be relatively constant from lot to lot.
(c) A good working relationship should exist with the producer (a good plan for within-plant use of acceptance sampling).

Calculation of one point in a typical chain sampling plan is illustrated below and the OC curves for typical chain sampling plans are shown in Table 8.18 and Figure 8.8.

For plan B, $n = 8$, $c = 0$, $i = 3$, at $p^1 = .01$.

$$P_2 = \left[\binom{8}{0}(.01)^0(.99)^8 \right] + \left\{ \left[\binom{8}{1}(.01)^1(.99)^7 \right] \left[\binom{8}{0}(.01)^0(.99)^8 \right]^3 \right\}$$

$$= [.9227] + \{[.0746][.7857]\} = .9813$$

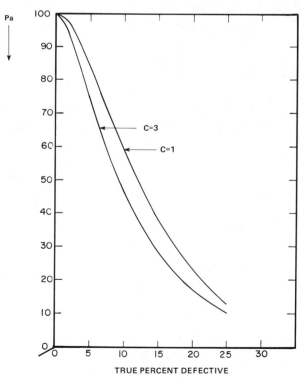

FIGURE 8.8. OC curves for a chain sampling plan with $n = 8$, $c = 0$, $i = 1$; $n = 8$, $c = 0$, $i = 3$.

8.10. SAMPLING PLANS OF THE DEPARTMENT OF DEFENSE — USING ABC-STD 105D

This sampling scheme was devised by an American–British–Canadian (ABC) Committee set up to establish an international standard. In the United States, the title is MIL-STD-105D.*

MIL-STD-105D is used most often when the supplier is under contract to the government and acceptance sampling is mandatory. However, the plans may be used by anyone who understands their use, in most situations where continuous protection against accepting bad product is desired.

Using the earlier criterion for sampling plan $N = 1000$, $P_1 = .01$, $P_2 = .05$, $\alpha = .05$, $\beta = .10$, we examine MIL-STD-105D for selection of a plan. We have the choice of using single, double, or multiple plans and each type will afford about the same protection. The choice will be based on the nature of the process being sampled and administrative considerations.

With single sampling (the other types follow the same method) we proceed as follows:

1. Select the desired AQL—in this case it will be equal to 1%.
2. Select the proper inspection level. If not already mandated, inspection levels are a function of product complexity and degree of discrimination necessary. There are three levels in MIL-STD-105. Ordinarily use level II at the start. Use level I for less complex and level III for more complex product.
3. The rules for switching inspection levels are as follows:
   (a) Switch from *normal* to *tightened* inspection when two out of five consecutive lots have been rejected on original inspection. *Continue tightened* inspection until five consecutive lots have passed original inspection.
   (b) Switch from *normal* to *reduced* inspection when *all* of the following conditions have been satisfied:
       (i) 10 original lots have been inspected under normal inspection, and none has been rejected.
       (ii) The total number of defectives in the 10 lots under consideration is equal to or less than the applicable number in Table 8.19 (VIII in MIL-STD-105D). If using double or multiple inspection, all samples inspected must be included.
       (iii) Production is at a steady rate.
       (iv) The consumer *wants* to institute reduced inspection.

See Duncan (2), pp. 236–245, for a good discussion of problems encountered during the course of the committee's deliberations.

TABLE 8.19.   Sample Size Code Letters

Lot or Batch Size			Special Inspection Levels				General Inspection Levels		
			S-1	S-2	S-3	S-4	I	II	III
2	to	8	A	A	A	A	A	A	B
9	to	15	A	A	A	A	A	B	C
16	to	25	A	A	B	B	B	C	D
26	to	50	A	B	B	C	C	D	E
51	to	90	B	B	C	C	C	E	F
91	to	150	B	B	C	D	D	F	G
151	to	280	B	C	D	E	E	G	H
281	to	500	B	C	D	E	F	H	J
501	to	1,200	C	C	E	F	G	J	K
1,201	to	3,200	C	D	E	G	H	K	L
3,201	to	10,000	C	D	F	G	J	L	M
10,001	to	35,000	C	D	F	H	K	M	N
35,001	to	150,000	D	E	G	J	L	N	P
150,001	to	500,000	D	E	G	J	M	P	Q
500,001	and	over	D	E	H	K	N	Q	R

    (c)   Switch from *reduced* to *normal* if any of the following occur:
        (i)   A lot or batch is rejected.
       (ii)   A lot is considered acceptable under special procedures. After this decision is made, normal inspection starts with the next lot inspected.
      (iii)   Production is irregular or delayed.
      (iv)   Any other conditions arise that warrant return to normal inspection.
    (d)   Discontinue inspection when 10 consecutive lots remain on tightened inspection until it is demonstrated that the quality of incoming lots is improved.

  4.   Select the lot size using criteria such as:
    (a)   Make lots as large as possible.
    (b)   Form them from natural parts of the process, that is, production lines, shifts, batches, and so on.
         NOTE.   Do not mix the product of different production lines, methods, shifts, batches, in forming inspection lots. Our lot size is $N = 1000$.

  5.   Using Table 8.19 select sample size code letter—in this case, since we are to start with normal inspection, we choose letter J, under inspection level II.

  6.   Since we are to use the single sample method, we use Table 8.20.

TABLE 8.20. Single Sampling Plans for Normal Inspection (Master Table)

Acceptable Quality Levels (normal inspection). Each entry shows "Ac Re".

Sample Size Code Letter	Sample Size	0.010	0.015	0.025	0.040	0.065	0.10	0.15	0.25	0.40	0.65	1.0	1.5	2.5	4.0	6.5	10	15	25	40	65	100	150	250	400	650	1000
A	2	↓	↓	↓	↓	↓	↓	↓	↓	↓	↓	↓	↓	↓	↓	↓	0 1	1 2	2 3	3 4	5 6	7 8	10 11	14 15	21 22	30 31	44 45
B	3	↓	↓	↓	↓	↓	↓	↓	↓	↓	↓	↓	↓	↓	↓	0 1	1 2	2 3	3 4	5 6	7 8	10 11	14 15	21 22	30 31	44 45	↑
C	5	↓	↓	↓	↓	↓	↓	↓	↓	↓	↓	↓	↓	↓	0 1	1 2	2 3	3 4	5 6	7 8	10 11	14 15	21 22	30 31	44 45	↑	↑
D	8	↓	↓	↓	↓	↓	↓	↓	↓	↓	↓	↓	↓	0 1	1 2	2 3	3 4	5 6	7 8	10 11	14 15	21 22	30 31	44 45	↑	↑	↑
E	13	↓	↓	↓	↓	↓	↓	↓	↓	↓	↓	↓	0 1	1 2	2 3	3 4	5 6	7 8	10 11	14 15	21 22	30 31	44 45	↑	↑	↑	↑
F	20	↓	↓	↓	↓	↓	↓	↓	↓	↓	↓	0 1	1 2	2 3	3 4	5 6	7 8	10 11	14 15	21 22	30 31	44 45	↑	↑	↑	↑	↑
G	32	↓	↓	↓	↓	↓	↓	↓	↓	↓	0 1	1 2	2 3	3 4	5 6	7 8	10 11	14 15	21 22	30 31	44 45	↑	↑	↑	↑	↑	↑
H	50	↓	↓	↓	↓	↓	↓	↓	↓	0 1	1 2	2 3	3 4	5 6	7 8	10 11	14 15	21 22	30 31	44 45	↑	↑	↑	↑	↑	↑	↑
J	80	↓	↓	↓	↓	↓	↓	↓	0 1	1 2	2 3	3 4	5 6	7 8	10 11	14 15	21 22	30 31	44 45	↑	↑	↑	↑	↑	↑	↑	↑
K	125	↓	↓	↓	↓	↓	↓	0 1	1 2	2 3	3 4	5 6	7 8	10 11	14 15	21 22	30 31	44 45	↑	↑	↑	↑	↑	↑	↑	↑	↑
L	200	↓	↓	↓	↓	↓	0 1	1 2	2 3	3 4	5 6	7 8	10 11	14 15	21 22	30 31	44 45	↑	↑	↑	↑	↑	↑	↑	↑	↑	↑
M	315	↓	↓	↓	↓	0 1	1 2	2 3	3 4	5 6	7 8	10 11	14 15	21 22	30 31	44 45	↑	↑	↑	↑	↑	↑	↑	↑	↑	↑	↑
N	500	↓	↓	↓	0 1	1 2	2 3	3 4	5 6	7 8	10 11	14 15	21 22	30 31	44 45	↑	↑	↑	↑	↑	↑	↑	↑	↑	↑	↑	↑
P	800	↓	↓	0 1	1 2	2 3	3 4	5 6	7 8	10 11	14 15	21 22	30 31	44 45	↑	↑	↑	↑	↑	↑	↑	↑	↑	↑	↑	↑	↑
Q	1250	↓	0 1	1 2	2 3	3 4	5 6	7 8	10 11	14 15	21 22	30 31	44 45	↑	↑	↑	↑	↑	↑	↑	↑	↑	↑	↑	↑	↑	↑
R	2000	0 1	1 2	2 3	3 4	5 6	7 8	10 11	14 15	21 22	30 31	44 45	↑	↑	↑	↑	↑	↑	↑	↑	↑	↑	↑	↑	↑	↑	↑

↓ = Use first sampling plan below arrow. If sample size equals, or exceeds, lot or batch size, do 100 percent inspection.

↑ = Use first sampling plan above arrow.

Ac = Acceptance number.

Re = Rejection number.

231

TABLE 8.21a. Tables for Sample Size Code Letter: J. Chart J — Operating Characteristic Curves for Single Sampling Plans

(Curves for double and multiple sampling are matched as closely as practicable)

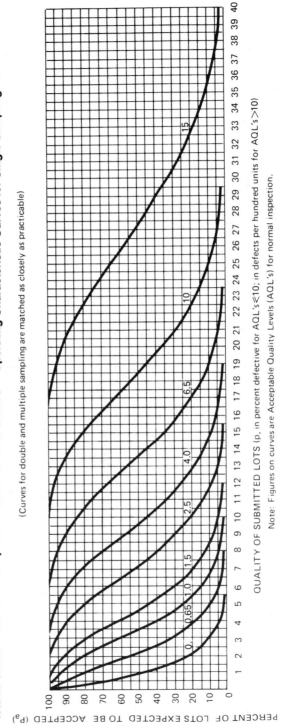

QUALITY OF SUBMITTED LOTS (p, in percent defective for AQL's ≤10; in defects per hundred units for AQL's >10)

Note: Figures on curves are Acceptable Quality Levels (AQL's) for normal inspection.

PERCENT OF LOTS EXPECTED TO BE ACCEPTED (P_a)

TABLE 8.21b. Tabulated Values for Operating Characteristic Curves for Single Sampling Plans

Acceptable Quality Levels (normal inspection)

P_a	p (in percent defective)										p (in defects per hundred units)											
	0.15	0.65	1.0	1.5	2.5	4.0	X	6.5	X	10	0.15	0.65	1.0	1.5	2.5	4.0	X	6.5	X	10	X	15
99.0	0.013	0.188	0.550	1.05	2.30	3.72	4.50	6.13	7.88	9.75	0.013	0.186	0.545	1.03	2.23	3.63	4.38	5.96	7.62	9.35	12.9	15.7
95.0	0.064	0.444	1.03	1.73	3.32	5.06	5.98	7.91	9.89	11.9	0.064	0.444	1.02	1.71	3.27	4.98	5.87	7.71	9.61	11.6	15.6	18.6
90.0	0.132	0.666	1.38	2.20	3.98	5.91	6.91	8.95	11.0	13.2	0.131	0.665	1.38	2.18	3.94	5.82	6.79	8.78	10.8	12.9	17.1	20.3
75.0	0.359	1.202	2.16	3.18	5.30	7.50	8.62	10.9	13.2	15.5	0.360	1.20	2.16	3.17	5.27	7.45	8.55	10.8	13.0	15.3	19.9	23.4
50.0	0.863	2.09	3.33	4.57	7.06	9.55	10.8	13.3	15.8	18.3	0.866	2.10	3.34	4.59	7.09	9.59	10.8	13.3	15.8	18.3	23.3	27.1
25.0	1.72	3.33	4.84	6.31	9.14	11.9	13.3	16.0	18.6	21.3	1.73	3.37	4.90	6.39	9.28	12.1	13.5	16.3	19.0	21.8	27.2	31.2
10.0	2.84	4.78	6.52	8.16	11.3	14.2	15.7	18.6	21.4	24.2	2.88	4.86	6.65	8.35	11.6	14.7	16.2	19.3	22.2	25.2	30.9	35.2
5.0	3.68	5.80	7.66	9.39	12.7	15.8	17.3	20.3	23.2	26.0	3.75	5.93	7.87	9.69	13.1	16.4	18.0	21.2	24.3	27.4	33.4	37.8
1.0	5.59	8.00	10.1	12.0	15.6	18.9	20.5	23.6	26.5	29.5	5.76	8.30	10.5	12.6	16.4	20.0	21.8	25.2	28.5	31.8	38.2	42.9
	0.25	1.0	1.5	2.5	4.0	X	6.5	X	10	X	0.25	1.0	1.5	2.5	4.0	X	6.5	X	10	X	15	X

Acceptable Quality Levels (tightened inspection)

Note: All values given in above table based on Poisson distribution as an approximation to the Binomial.

The sampling plan 1% AQL selected will be as follows:

$$N = 1000$$
$$n = 80$$

with AQL = 1%.

$$Ac\,2(\text{accept})$$
$$Re\,3(\text{reject})$$

The OC Curves and AOQL for this plan may be found in Tables 8.21a, b and Table 8.22, respectively. Table 8.23 indicates the comparison with the plan devised earlier and the MIL-STD.

One of the advantages of a standard is that everyone using it arrives at a similar plan under essentially the same conditions, whereas if all devised their own plans, many different plans would result.

The double sampling plan would be set at $N = 1000$.

$$n_1 = 50 \qquad Ac = 0 \qquad Re = 3$$
$$n_2 = 50 \qquad Ac = 3 \qquad Re = 4$$

The original plan devised previously was

$$n_1 = 82 \qquad Ac = 1 \qquad Re = 4$$
$$n_2 = 82 \qquad Ac = 3 \qquad Re = 4$$

The multiple sampling plan would be set at:

	MIL-STD				Original		
Sample No.	n	Ac	Re	Sample No.	n	Ac	Re
1	20	#	2	1	30	#	2
2	40	0	3	2	60	0	3
3	60	0	3	3	90	1	4
4	80	1	4	4	120	2	5
5	100	2	4	5	150	4	6
6	120	3	5	6	180	4	6
7	140	4	5	7	210	5	6

In devising inspection criteria the standard specifies critical, major and minor defect classifications. Many contracts specify AQL = 1% major, 2.5% minor, and 0% critical defects.

TABLE 8.22. Sampling Plans for Sample Size Code Letter: J

Acceptable Quality Levels (normal inspection) (top header)
Acceptable Quality Levels (tightened inspection) (bottom header)

Type of Sampling Plan	Cumulative Sample Size	Less than 0.15	0.15	0.25	X	0.40	0.65	1.0	1.5	2.5	X	4.0	X	6.5	X	10	X	15	Higher than 15
		Ac Re	Ac Re	Ac Re	Ac Re	Ac Re	Ac Re	Ac Re	Ac Re	Ac Re	Ac Re	Ac Re	Ac Re	Ac Re	Ac Re	Ac Re	Ac Re	Ac Re	Ac Re
Single	80	▽	0 1	*Use*	*Use*	*Letter H*	1 2	2 3	3 4	5 6	6 7	7 8	8 9	10 11	12 13	14 15	18 19	21 22	△
Double	50	▽	*	*Use*	*Use*	*Letter L*	0 2	0 3	1 4	2 5	3 7	3 7	5 9	6 10	7 11	9 14	11 16		△
	100			*Letter*	*Letter*	*K*	1 2	3 4	4 5	6 7	7 8	8 9	11 12	12 13	15 16	18 19	23 24	26 27	
Multiple	20	▽	*				# 2	# 2	0 2	0 3	1 4	0 4	0 4	0 5	0 6	1 7	1 8	2 9	△
	40						# 2	# 2	0 2	0 3	1 5	1 6	2 7	3 8	3 9	4 10	6 12	7 14	
	60						0 2	0 2	0 3	1 4	2 6	3 8	4 9	6 10	7 12	8 13	11 17	13 19	
	80						0 3	0 3	1 4	2 5	3 7	5 10	6 11	8 13	10 15	12 17	16 22	19 25	
	100						1 3	1 3	2 4	3 6	5 8	7 11	9 12	11 15	14 17	17 20	22 25	25 29	
	120						1 3	1 3	3 5	4 6	7 9	10 12	12 14	14 17	18 20	21 23	27 29	31 33	
	140						2 3	2 3	4 6	6 7	9 10	13 14	14 15	18 19	21 22	25 26	32 33	37 38	

Bottom (tightened inspection) AQL header row:

		Less than 0.25	0.25	0.40	X	0.65	1.0	1.5	2.5	X	4.0	X	6.5	X	10	X	15	Higher than 15

Legend:

△ = Use next preceding sample size code letter for which acceptance and rejection numbers are available.

▽ = Use next subsequent sample size code letter for which acceptance and rejection numbers are available.

Ac = Acceptance number

Re = Rejection number

* = Use single sampling plan above (or alternatively use letter M)

\# = Acceptance not permitted at this sample size.

235

TABLE 8.23.   Single Sampling Plan Comparing MIL-STD-105 to Original Plan
Original: N = 1000,   n = 82,   c = 2
MIL-STD 105 D N = 1000,   n = 80,   c = 2

True Percent Defective	Calculated Plan	MIL-STD Plan
.01	.95	.95
.02	.78	.80
.03	.54	.60
.04	.36	.38
.05	.22	.25
.06	.13	.14
.07	.08	.07
.08	.04	.04
.09	.02	.03
.10	.01	.01
AOQL	.0153	.017

8.11.   DODGE – ROMIG (D – R) SAMPLING PLANS

Using "canned sampling" plans saves a great deal of effort and standardizes the plans. When the user is thoroughly familiar with the construction of, and the limitations of, sampling plans the standardized plans may be used under most conditions. Very good approximations to the exact plan desired can be obtained in the D–R table.

Using the D–R approach all rejected lots must be screened and, for best results in selecting an appropriate plan, the consumer should know something about the producer's process average. This latter information may be estimated from samples drawn from the first lots submitted for inspection. The early estimates may then be substantiated by drawing continuing information from succeeding lots submitted for inspection.

There are four sets of tables:

1.   Single sampling—lot quality protection.
2.   Double sampling—lot quality protection.
3.   Single sampling—average quality protection.
4.   Double sampling—average quality protection.

Single Sampling — Lot Quality Protection.

These tables give lot quality protection if values are specified for lot tolerance percent defective (LTPD) which is the limiting percent defective in any lot and a consumer's risk, which expresses the probability of accepting a lot in which the true percent defective is equal to the lot tolerance. The tables then give for lot sizes from 1 to 100,000 values at n and c in combination to minimize inspection and best accommodate the relation between consumers and producers risk.

The tables demonstrate plans for lot tolerance values of .5%, 1.0%, 2.0%, 3.0%, 4.0%, 5.0%, 7.0%, and 10%, all with a consumer's risk of 10%.

For conditions of the single sample we derived earlier with $N = 1000$, lot tolerance of 5%, consumers risk of 10%, process average 1%, we would get from Table 8.24 the following:

$$n = 105, \quad c = 2, \quad \text{AOQL} = 1.2$$

Plan B gave $\quad n = 134, \quad c = 3, \quad \text{AOQL} = 1.4.$

Double Sampling Tables — Lot Quality Protection

These tables give lot quality protection if values are specified for LTPD and consumer risk for lot sizes 1 to 100,000 and proper combinations of pairs of n and c to minimize inspection and get the most efficient relationship between consumers and producers risk. The consumer's risk is now made up of the probability of accepting a lot that meets tolerance quality in the first sample plus the probability that the lot is accepted on the second sample if the lot does not pass inspection on the first sample.

Double sample tables have the same values of LTPD as single samples. The D–R sample is taken from Table 8.25 for $n = 1000$, LTPD = 5%, consumers risk = 10%, process average = 1.0%.

$$n_1 = 55 \qquad c_1 = 0$$
$$n_2 = 115 \qquad c_2 = 4 \qquad n_1 + n_2 = 170$$
$$\text{AOQL} = 1.4$$

The double sample plan devised earlier was

$$n_1 = 82 \qquad c_1 = 1$$
$$n_2 = 82 \qquad c_2 = 3 \qquad n_1 + n_2 = 164$$
$$\text{AOQL} = 1.4$$

Single Sampling — Average Quality Protection

Both single and double sampling plans for average quality protection assume sampling from a continuing, homogeneous process, under statistical control with a specified and maintained process average.

TABLE 8.24. Dodge–Romig Single Sampling Table for Lot Tolerance Percent Defective (LTPD) = 5.0%

Lot Size	Process Average 0 to 0.05%			Process Average 0.06 to 0.50%			Process Average 0.51 to 1.00%			Process Average 1.01 to 1.50%			Process Average 1.51 to 2.00%			Process Average 2.01 to 2.50%		
	n	c	AOQL %	n	c	AOQL %	n	c	AOQL %	n	c	AOQL %	n	c	AOQL %	n	c	AOQL %
1–30	All	0	0	All	0	0	All	0	0	All	0	0	All	0	0	All	0	0
31–50	30	0	0.49	30	0	0.49	30	0	0.49	30	0	0.49	30	0	0.49	30	0	0.49
51–100	37	0	0.63	37	0	0.63	37	0	0.63	37	0	0.63	37	0	0.63	37	0	0.63
101–200	40	0	0.74	40	0	0.74	40	0	0.74	40	0	0.74	40	0	0.74	40	0	0.74
201–300	43	0	0.74	43	0	0.74	70	1	0.92	70	1	0.92	95	2	0.99	95	2	0.99
301–400	44	0	0.74	44	0	0.74	70	1	0.99	100	2	1.0	120	3	1.1	145	4	1.1
401–500	45	0	0.75	75	1	0.95	100	2	1.1	100	2	1.1	125	3	1.2	150	4	1.2
501–600	45	0	0.76	75	1	0.98	100	2	1.1	125	3	1.2	150	4	1.3	175	5	1.3
601–800	45	0	0.77	75	1	1.0	100	2	1.2	130	3	1.2	175	5	1.4	200	6	1.4
801–1000	45	0	0.78	75	1	1.0	105	2	1.2	155	4	1.4	180	5	1.4	225	7	1.5
1,001–2000	45	0	0.80	75	1	1.0	130	3	1.4	180	5	1.6	230	7	1.7	280	9	1.8
2,001–3000	75	1	1.1	105	2	1.3	135	3	1.4	210	6	1.7	280	9	1.9	370	13	2.1
3,001–4000	75	1	1.1	105	2	1.3	160	4	1.5	210	6	1.7	305	10	2.0	420	15	2.2
4,001–5000	75	1	1.1	105	2	1.3	160	4	1.5	235	7	1.8	330	11	2.0	440	16	2.2
5,001–7000	75	1	1.1	105	2	1.3	185	5	1.7	260	8	1.9	350	12	2.2	490	18	2.4
7,001–10,000	75	1	1.1	105	2	1.3	185	5	1.7	260	8	1.9	380	13	2.2	535	20	2.5
10,001–20,000	75	1	1.1	135	3	1.4	210	6	1.8	285	9	2.0	425	15	2.3	610	23	2.6
20,001–50,000	75	1	1.1	135	3	1.4	235	7	1.9	305	10	2.1	470	17	2.4	700	27	2.7
50,000–100,000	75	1	1.1	160	4	1.6	235	7	1.9	355	12	2.2	515	19	2.5	770	30	2.8

n = sample size; c = acceptance. "All" indicates that each piece in the lot is to be inspected. AOQL = average outgoing quality limit.

Protection is provided if values are specified for AOQL. Then combinations of n and c give the minimum inspection for a process average of continuing uniform quality.

Both single and double sampling tables are designed for AOQL values of 0.1%, 0.25%, 0.5%, 0.75%, 1.0%, 1.5%, 2.0%, 2.5%, 3.0%, 4.0%, 5.0%, 7.0%, 10%.

LTPD (P_t) values at consumers risk of 10% are given for each plan. For the conditions of the plan devised earlier the D–R single sample plan $n = 1000$, $P_1 = .01$, $\alpha = .05$. Tables 8.26a and b give D–R tables for both 1% (A) and 1.5% (B) AOQL. For (A), AOQL = 1.0% the plan is

$$n = 120 \qquad c = 2$$
$$P_t = 4.3\%$$

For (B), AOQL = 1.5% the plan is

$$n = 85 \qquad c = 2$$
$$P_t = 6.2\%$$

The plan devised was

$$n = 134 \qquad c = 3$$
$$P_t = .05$$
$$\text{AOQL} = 1.4$$

Double Sampling — Average Quality Protection

The approach here is the same as for single sampling–average quality protection. For the same conditions as in the devised sample $n = 1000$, $P_1 = .01$, $\alpha = .05$, D–R Table 8.27a 1.0% AOQL gives a plan:

$$n = 125 \qquad c_1 = 1 \qquad n_1 + n_2 = 305$$
$$n_2 = 180 \qquad c_2 = 6$$
$$Pt = 3.5\%$$

Table 8.27b 1.5% AOQL gives a plan as follows (this is the plan we designed):

$$n_1 = 50 \qquad c_1 = 0 \qquad n_1 + n_2 = 155$$
$$n_2 = 105 \qquad c_2 = 4$$
$$Pt = 5.5\%$$
$$\text{AOQL} = 1.4\%$$

TABLE 8.25. Dodge–Romig Double Sampling Table for Lot Tolerance Percent Defective (LTPD) = 5.0%

Lot Size	Process Average 0 to 0.05%						Process Average 0.06 to 0.50%						Process Average 0.51 to 1.00%					
	Trial 1		Trial 2			AOQL	Trial 1		Trial 2			AOQL	Trial 1		Trial 2			AOQL
	n_1	c_1	n_2	n_1+n_2	c_2	in %	n_1	c_1	n_2	n_1+n_2	c_2	in %	n_1	c_1	n_2	n_1+n_2	c_2	in %
1–30	All	0	—	—	—	0	All	0	—	—	—	0	All	0	—	—	—	0
31–50	30	0	—	—	—	0.49	30	0	—	—	—	0.49	30	0	—	—	—	0.49
51–75	38	0	—	—	—	0.59	38	0	—	—	—	0.59	38	0	—	—	—	0.59
76–100	44	0	21	65	1	0.64	44	0	21	65	1	0.64	44	0	21	65	1	0.64
101–200	49	0	26	75	1	0.84	49	0	26	75	1	0.84	49	0	26	75	1	0.84
201–300	50	0	30	80	1	0.91	50	0	30	80	1	0.91	50	0	55	105	2	1.0
301–400	55	0	30	85	1	0.92	55	0	55	110	2	1.1	55	0	55	110	2	1.1
401–500	55	0	30	85	1	0.93	55	0	55	110	2	1.1	55	0	80	135	3	1.2
501–600	55	0	30	85	1	0.94	55	0	60	115	2	1.1	55	0	85	140	3	1.2
601–800	55	0	35	90	1	0.95	55	0	65	120	2	1.1	55	0	85	140	3	1.3
801–1,000	55	0	35	90	1	0.96	55	0	65	120	2	1.1	55	0	115	170	4	1.4
1,001–2,000	55	0	35	90	1	0.98	55	0	95	150	3	1.3	55	0	120	175	4	1.4
2,001–3,000	55	0	65	120	2	1.2	55	0	95	150	3	1.3	55	0	150	205	5	1.5
3,001–4,000	55	0	65	120	2	1.2	55	0	95	150	3	1.3	90	1	140	230	6	1.6
4,001–5,000	55	0	65	120	2	1.2	55	0	95	150	3	1.4	90	1	165	255	7	1.8
5,001–7,000	55	0	65	120	2	1.2	55	0	95	150	3	1.4	90	1	165	255	7	1.8
7,001–10,000	55	0	65	120	2	1.2	55	0	120	175	4	1.5	90	1	190	280	8	1.9
10,001–20,000	55	0	65	120	2	1.2	55	0	120	175	4	1.5	90	1	190	280	8	1.9
20,001–50,000	55	0	65	120	2	1.2	55	0	150	205	5	1.7	90	1	215	305	9	2.0
50,001–100,000	55	0	65	120	2	1.2	55	0	150	205	5	1.7	90	1	240	330	10	2.1

Process Average 1.01 to 1.50%, 1.51 to 2.00%, 2.01 to 2.50%

Lot Size	1.01–1.50% Trial 1 n_1	c_1	Trial 2 n_2	n_1+n_2	c_2	AOQL in %	1.51–2.00% Trial 1 n_1	c_1	Trial 2 n_2	n_1+n_2	c_2	AOQL in %	2.01–2.50% Trial 1 n_1	c_1	Trial 2 n_2	n_1+n_2	c_2	AOQL in %
1–30	All	0	—	—	—	0	All	0	—	—	—	0	All	0	—	—	—	0
31–50	30	0	—	—	—	0.49	30	0	—	—	—	0.49	30	0	—	—	—	0.49
51–75	38	0	—	—	—	0.59	38	0	—	—	—	0.59	38	0	—	—	—	0.59
76–100	44	0	21	65	1	0.64	44	0	21	65	1	0.64	44	0	21	65	1	0.64
101–200	49	0	51	100	2	0.91	49	0	51	100	2	0.91	49	0	51	100	2	0.91
201–300	50	0	55	105	2	1.0	50	0	80	130	3	1.1	50	0	100	150	4	1.1
301–400	55	0	80	135	3	1.1	55	0	100	155	4	1.2	85	1	105	190	6	1.3
401–500	55	0	105	160	4	1.3	85	1	120	205	6	1.4	85	1	140	225	7	1.4
501–600	55	0	110	165	4	1.3	85	1	145	230	7	1.4	85	1	165	250	8	1.5
601–800	90	1	125	215	6	1.5	90	1	170	260	8	1.5	120	2	185	305	10	1.6
801–1,000	90	1	150	240	7	1.5	90	1	200	290	9	1.6	120	2	210	330	11	1.7
1,001–2,000	90	1	185	275	8	1.7	120	2	225	345	11	1.9	175	4	260	435	15	2.0
2,001–3,000	120	2	180	300	9	1.9	150	3	270	420	14	2.1	205	5	375	580	21	2.3
3,001–4,000	120	2	210	330	10	2.0	150	3	295	445	15	2.3	230	6	420	650	24	2.4
4,001–5,000	120	2	255	375	12	2.1	150	3	345	495	17	2.3	255	7	445	700	26	2.5
5,001–7,000	120	2	260	380	12	2.1	150	3	370	520	18	2.3	255	7	495	750	28	2.6
7,001–10,000	120	2	285	405	13	2.1	175	4	370	545	19	2.4	280	8	540	820	31	2.7
10,001–20,000	120	2	310	430	14	2.2	175	4	420	595	21	2.4	280	8	660	940	36	2.8
20,001–50,000	120	2	335	455	15	2.2	205	5	485	690	25	2.5	305	9	745	1050	41	2.9
50,001–100,000	120	2	360	480	16	2.3	205	5	555	760	28	2.6	330	10	810	1140	45	3.0

Trial 1: n_1 = first sample size; c_1 = acceptance number for first sample. "All" indicates that each piece in the lot is to be inspected. Trial 2: n_2 = second sample size; c_2 = acceptance number for first and second samples combined. AOQL = average outgoing quality limit.

TABLE 8.26a. Dodge–Romig Single Sampling Table for Average Outgoing Quality Limit (AOQL) = 1.0%

Lot Sizes	Process Average 0 to 0.02%			Process Average 0.03 to 0.20%			Process Average 0.21 to 0.40%			Process Average 0.41 to 0.60%			Process Average 0.61 to 0.80%			Process Average 0.81 to 1.00%		
	n	c	$p_t\%$	n	c	$p_t\%$	n	c	$p_t\%$	n	c	$p_t\%$	n	c	$p_t\%$	n	c	$p_t\%$
1–25	All	0	—	All	0	—	All	0	—	All	0	—	All	0	—	All	0	—
26–50	22	0	7.7	22	0	7.7	22	0	7.7	22	0	7.7	22	0	7.7	22	0	7.7
51–100	27	0	7.1	27	0	7.1	27	0	7.1	27	0	7.1	27	0	7.1	27	0	7.1
101–200	32	0	6.4	32	0	6.4	32	0	6.4	32	0	6.4	32	0	6.4	32	0	6.4
201–300	33	0	6.3	33	0	6.3	33	0	6.3	33	0	6.3	33	0	6.3	65	1	5.0
301–400	34	0	6.1	34	0	6.1	34	0	6.1	70	1	4.6	70	1	4.6	70	1	4.6
401–500	35	0	6.1	35	0	6.1	35	0	6.1	70	1	4.7	70	1	4.7	70	1	4.7
501–600	35	0	6.1	35	0	6.1	75	1	4.4	75	1	4.4	75	1	4.4	75	1	4.4
601–800	35	0	6.2	35	0	6.2	75	1	4.4	75	1	4.4	75	1	4.4	120	2	4.2
801–1,000	35	0	6.3	35	0	6.3	80	1	4.4	80	1	4.4	120	2	4.3	120	2	4.3
1,001–2,000	36	0	6.2	80	1	4.5	80	1	4.5	130	2	4.0	130	2	4.0	180	3	3.7
2,001–3,000	36	0	6.2	80	1	4.6	80	1	4.6	130	2	4.0	185	3	3.6	235	4	3.3
3,001–4,000	36	0	6.2	80	1	4.7	135	2	3.9	135	2	3.9	185	3	3.6	295	5	3.1
4,001–5,000	36	0	6.2	85	1	4.6	135	2	3.9	190	3	3.5	245	4	3.2	300	5	3.1
5,001–7,000	37	0	6.1	85	1	4.6	135	2	3.9	190	3	3.5	305	5	3.0	420	7	2.8
7,001–10,000	37	0	6.2	85	1	4.6	135	2	3.9	245	4	3.2	310	5	3.0	430	7	2.7
10,001–20,000	85	1	4.6	135	2	3.9	195	3	3.4	250	4	3.2	435	7	2.7	635	10	2.4
20,001–50,000	85	1	4.6	135	2	3.9	255	4	3.1	380	6	2.8	575	9	2.5	990	15	2.1
50,001–100,000	85	1	4.6	135	2	3.9	255	4	3.1	445	7	2.6	790	12	2.3	1520	22	1.9

TABLE 8.26b. Dodge–Romig Single Sampling Table for Average Outgoing Quality Limit (AOQL) = 1.5%

Lot Size	Process Average 0 to 0.03%			Process Average 0.04 to 0.30%			Process Average 0.31 to 0.60%			Process Average 0.61 to 0.90%			Process Average 0.91 to 1.20%			Process Average 1.21 to 1.50%		
	n	c	$p_t\%$	n	c	$p_t\%$	n	c	$p_t\%$	n	c	$p_t\%$	n	c	$p_t\%$	n	c	$p_t\%$
1–15	All	0	—	All	0	—	All	0	—	All	0	—	All	0	—	All	0	—
16–50	16	0	11.6	16	0	11.6	16	0	11.6	16	0	11.6	16	0	11.6	16	0	11.6
51–100	20	0	9.8	20	0	9.8	20	0	9.8	20	0	9.8	20	0	9.8	20	0	9.8
101–200	22	0	9.5	22	0	9.5	22	0	9.5	22	0	9.5	22	0	9.5	44	1	8.2
201–300	23	0	9.2	23	0	9.2	23	0	9.2	47	1	7.9	47	1	7.9	47	1	7.9
301–400	23	0	9.3	23	0	9.3	49	1	7.8	49	1	7.8	49	1	7.8	49	1	7.8
401–500	23	0	9.4	23	0	9.4	50	1	7.7	50	1	7.7	50	1	7.7	50	1	7.7
501–600	24	0	9.0	24	0	9.0	50	1	7.7	50	1	7.7	50	1	7.7	50	1	7.7
601–800	24	0	9.1	24	0	9.1	50	1	7.8	50	1	7.8	80	2	6.4	80	2	6.4
801–1,000	24	0	9.1	55	1	7.0	55	1	7.0	85	2	6.2	85	2	6.2	85	2	6.2
1,001–2,000	24	0	9.1	55	1	7.0	55	1	7.0	85	2	6.2	120	3	5.4	155	4	5.0
2,001–3,000	24	0	9.2	55	1	7.1	90	2	5.9	125	3	5.3	160	4	4.9	200	5	4.6
3,001–4,000	24	0	9.2	55	1	7.1	90	2	5.9	125	3	5.3	165	4	4.8	240	5	4.4
4,001–5,000	24	0	9.2	55	1	7.1	90	2	5.9	125	3	5.3	205	5	4.6	280	7	4.2
5,001–7,000	24	0	9.2	55	1	7.1	90	2	5.9	165	4	4.8	205	5	4.6	325	8	4.0
7,001–10,000	24	0	9.2	55	1	7.1	130	3	5.2	165	4	4.8	250	6	4.6	390	10	3.8
10,001–20,000	55	1	7.1	90	2	5.9	130	3	5.2	210	5	4.4	340	8	3.8	515	12	3.4
20,001–50,000	55	1	7.1	90	2	5.9	170	4	4.7	295	7	4.0	480	11	3.5	860	19	3.0
50,001–100,000	55	1	7.1	130	3	5.2	210	5	4.4	340	8	3.8	625	14	3.3	1120	24	2.8

n = sample size; c = acceptance number. "All" indicates that each piece in the lot is to be inspected. p_t = lot tolerance percent defective with a consumer's risk (P_c) of 0.10.

TABLE 8.27a. Dodge–Romig Double Sampling Table for Average Outgoing Quality Limit (AOQL) = 1.0%

Lot Size	Process Average 0 to 0.02%						Process Average 0.03 to 0.20%						Process Average 0.21 to 0.40%					
	Trial 1		Trial 2			p_t %	Trial 1		Trial 2			p_t %	Trial 1		Trial 2			p_t %
	n_1	c_1	n_2	n_1+n_2	c_2		n_1	c_1	n_2	n_1+n_2	c_2		n_1	c_1	n_2	n_1+n_2	c_2	
1–25	All	0	—	—	—	—	All	0	—	—	—	—	All	0	—	—	—	—
26–50	22	0	—	—	—	7.7	22	0	—	—	—	7.7	22	0	—	—	—	7.7
51–100	33	0	17	50	1	6.9	33	0	17	50	1	6.9	33	0	17	50	1	6.9
101–200	43	0	22	65	1	5.8	43	0	22	65	1	5.8	43	0	22	65	1	5.8
201–300	47	0	28	75	1	5.5	47	0	28	75	1	5.5	47	0	28	75	1	5.5
301–400	49	0	31	80	1	5.4	49	0	31	80	1	5.4	55	0	60	115	2	4.8
401–500	50	0	30	80	1	5.4	50	0	30	80	1	5.4	55	0	65	120	2	4.7
501–600	50	0	30	80	1	5.4	50	0	30	80	1	5.4	60	0	65	125	2	4.6
601–800	50	0	35	85	1	5.3	60	0	70	130	2	4.5	60	0	70	130	2	4.5
801–1,000	55	0	30	85	1	5.2	60	0	75	135	2	4.4	60	0	75	135	2	4.4
1,001–2,000	55	0	35	90	1	5.1	65	0	75	140	2	4.3	75	0	120	195	3	3.8
2,001–3,000	65	0	80	145	2	4.2	65	0	80	145	2	4.2	75	0	125	200	3	3.7
3,001–4,000	70	0	80	150	2	4.1	70	0	80	150	2	4.1	80	0	175	255	4	3.5
4,001–5,000	70	0	80	150	2	4.1	70	0	80	150	2	4.1	80	0	180	260	4	3.4
5,001–7,000	70	0	80	150	2	4.1	75	0	125	200	3	3.7	80	0	180	260	4	3.4
7,001–10,000	70	0	80	150	2	4.1	80	0	125	205	3	3.6	85	0	180	265	4	3.3
10,001–20,000	70	0	80	150	2	4.1	80	0	130	210	3	3.6	90	0	230	320	5	3.2
20,001–50,000	75	0	80	155	2	4.0	80	0	135	215	3	3.6	95	0	300	395	6	2.9
50,001–100,000	75	0	80	155	2	4.0	85	0	180	265	4	3.3	170	1	380	550	8	2.6

244

Progress Average 0.41 to 0.60% | Process Average 0.61 to 0.80% | Process Average 0.81 to 1.00%

Lot Size	Trial 1 n_1	c_1	Trial 2 n_2	n_1+n_2	c_2	p_t %	Trial 1 n_1	c_1	Trial 2 n_2	n_1+n_2	c_2	p_t %	Trial 1 n_1	c_1	Trial 2 n_2	n_1+n_2	c_2	p_t %
1–25	All	0	—	—	—	—	All	0	—	—	—	—	all	0	—	—	—	—
26–50	22	0	—	—	—	7.7	22	0	—	—	—	7.7	22	0	—	—	—	7.7
51–100	33	0	17	50	1	6.9	33	0	17	50	1	6.9	33	0	17	50	1	6.9
101–200	43	0	22	65	1	5.8	43	0	22	65	1	5.8	47	0	43	90	2	5.4
201–300	55	0	50	105	2	4.9	55	0	50	105	2	4.9	55	0	50	105	2	4.9
301–400	55	0	60	115	2	4.8	55	0	60	115	2	4.8	60	0	80	140	3	4.5
401–500	55	0	65	120	2	4.7	60	0	95	155	3	4.3	60	0	95	155	3	4.3
501–600	60	0	65	125	2	4.6	65	0	100	165	3	4.2	65	0	100	165	3	4.2
601–800	65	0	105	170	3	4.1	65	0	105	170	3	4.1	70	0	140	210	4	3.9
801–1,000	65	0	110	175	3	4.0	70	0	150	220	4	3.8	125	1	180	305	6	3.5
1,001–2,000	80	0	165	245	4	3.7	135	1	200	335	6	3.3	140	1	245	385	7	3.2
2,001–3,000	80	0	170	250	4	3.6	150	1	265	415	7	3.0	215	2	355	570	10	2.8
3,001–4,000	85	0	220	305	5	3.3	160	1	330	490	8	2.8	225	2	455	680	12	2.7
4,001–5,000	145	1	225	370	6	3.1	225	2	375	600	10	2.7	240	2	595	835	14	2.5
5,001–7,000	155	1	285	440	7	2.9	235	2	440	675	11	2.6	310	3	665	975	16	2.4
7,001–10,000	165	1	355	520	8	2.7	250	2	585	835	13	2.4	385	4	785	1170	19	2.3
10,000–20,000	175	1	415	590	9	2.6	325	3	655	980	15	2.3	526	6	980	1500	24	2.2
20,001–50,000	250	2	490	740	11	2.4	340	3	910	1250	19	2.2	610	7	1410	2020	32	2.1
50,001–100,000	275	2	700	975	14	2.2	420	4	1050	1470	22	2.1	770	9	1850	2620	41	2.0

Trial 1: n_1 = first sample size; c_1 = acceptance number for first sample. "All" indicates that each piece in the lot is to be inspected. Trial 2: n_2 = second sample size; c_2 = acceptance number for first and second samples combined. p_t = lot tolerance percent defective with a consumer's risk (p_c) of 0.10.

TABLE 8.27b. Dodge–Romig Double Sampling Table for Average Outgoing Quality Limit (AOQL) = 1.5%

Lot Size	Process Average 0 to 0.03%						Process Average 0.04 to 0.30%						Process Average 0.31 to 0.60%					
	Trial 1		Trial 2			p_t	Trial 1		Trial 2			p_t	Trial 1		Trial 2			p_t
	n_1	c_1	n_2	n_1+n_2	c_2	%	n_1	c_1	n_2	n_1+n_2	c_2	%	n_1	c_1	n_2	n_1+n_2	c_2	%
1–15	All	0	—	—	—	—	All	0	—	—	—	—	All	0	—	—	—	—
16–50	16	0	—	—	—	11.6	16	0	—	—	—	11.6	16	0	—	—	—	11.6
51–75	23	0	11	34	1	10.5	23	0	11	34	1	10.5	23	0	11	34	1	10.5
76–100	26	0	14	40	1	9.4	26	0	14	40	1	9.4	26	0	14	40	1	9.4
101–200	31	0	18	49	1	8.4	31	0	18	49	1	8.4	31	0	18	49	1	8.4
201–300	33	0	22	55	1	8.0	33	0	22	55	1	8.0	38	0	37	75	2	7.0
301–400	34	0	21	55	1	7.9	34	0	21	55	1	7.9	39	0	41	80	2	6.9
401–500	35	0	20	55	1	7.8	35	0	20	55	1	7.8	39	0	46	85	2	6.9
501–600	35	0	20	55	1	7.8	40	0	45	85	2	6.8	40	0	45	85	2	6.8
601–800	35	0	20	55	1	7.8	41	0	49	90	2	6.7	46	0	74	120	3	6.0
801–1,000	36	0	19	55	1	7.8	42	0	48	90	2	6.5	47	0	78	125	3	5.9
1,001–2,000	44	0	51	95	2	6.3	44	0	51	95	2	6.3	49	0	81	130	3	5.7
2,001–3,000	45	0	50	95	2	6.2	45	0	50	95	2	6.2	55	0	110	165	4	5.3
3,001–4,000	45	0	50	95	2	6.2	50	0	85	135	3	5.5	55	0	115	170	4	5.2
4,001–5,000	45	0	50	95	2	6.2	50	0	85	135	3	5.5	55	0	120	175	4	5.1
5,001–7,000	46	0	54	100	2	6.1	50	0	90	140	3	5.4	60	0	155	215	5	4.7
7,001–10,000	46	0	54	100	2	6.1	50	0	90	140	3	5.4	60	0	160	220	5	4.6
10,001–20,000	46	0	54	100	2	6.1	50	0	90	140	3	5.4	60	0	165	225	5	4.5
20,001–50,000	47	0	53	100	2	6.1	55	0	125	180	4	5.0	65	0	195	260	6	4.4
50,001–100,000	47	0	53	100	2	6.1	55	0	130	185	4	4.9	115	1	235	350	8	4.0

	Process Average 0.61 to 0.90%						Process Average 0.91 to 1.20%						Process Average 1.21 to 1.50%					
	Trial 1		Trial 2			p_t %	Trial 1		Trial 2			p_t %	Trial 1		Trial 2			p_t %
Lot Size	n_1	c_1	n_2	n_1+n_2	c_2		n_1	c_1	n_2	n_1+n_2	c_2		n_1	c_1	n_2	n_1+n_2	c_2	
1–15	All	0	—	—	—	—	All	0	—	—	—	—	All	0	—	—	—	—
16–50	16	0	—	—	—	11.6	16	0	—	—	—	11.6	16	0	—	—	—	11.6
51–75	23	0	11	34	1	10.5	23	0	11	34	1	10.5	23	0	11	34	1	10.5
76–100	26	0	14	40	1	9.4	26	0	14	40	1	9.4	26	0	14	40	1	9.4
101–200	35	0	35	70	2	7.5	35	0	35	70	2	7.5	35	0	35	70	2	7.5
201–300	38	0	37	75	2	7.0	38	0	37	75	2	7.0	40	0	60	100	3	6.6
301–400	39	0	41	80	2	6.9	42	0	63	105	3	6.3	42	0	63	105	3	6.3
401–500	44	0	71	115	3	6.1	44	0	71	115	3	6.1	46	0	94	140	4	5.8
501–600	44	0	71	115	3	6.1	44	0	71	115	3	6.1	47	0	98	145	4	5.7
601–800	46	0	74	120	3	6.0	49	0	101	150	4	5.6	85	1	125	210	6	5.0
801–1,000	47	0	78	125	3	5.9	50	0	105	155	4	5.5	90	1	125	215	6	4.9
1,001–2,000	55	0	105	160	4	5.4	95	1	175	270	7	4.6	100	1	230	330	9	4.4
2,001–3,000	100	1	145	245	6	4.6	110	1	255	365	9	4.1	155	2	345	500	13	3.9
3,001–4,000	105	1	190	295	7	4.3	160	2	300	460	11	3.8	205	3	405	610	15	3.6
4,001–5,000	105	1	190	295	7	4.3	165	2	340	505	12	3.7	250	4	480	730	18	3.5
5,001–7,000	110	1	225	335	8	4.2	165	2	375	540	13	3.7	310	5	610	920	22	3.3
7,001–10,000	115	1	280	395	9	3.9	170	2	420	590	14	3.6	360	6	660	1020	24	3.2
10,001–20,000	120	1	315	435	10	3.8	210	3	420	630	15	3.6	415	7	835	1250	29	3.1
20,001–50,000	165	2	350	515	12	3.7	225	3	640	865	20	3.3	510	9	1130	1640	38	3.0
50,001–100,000	175	2	440	615	14	3.5	275	4	725	1000	23	3.2	570	10	1400	1970	45	2.9

Trial 1: n_1 = first sample size; c_1 = acceptance number for first sample. "All" indicates that each piece in the lot is to be inspected. Trial 2: n_2 = second sample size; c_2 = acceptance number for first and second samples combined. p_t = lot tolerance percent defective with a consumer's risk (P_c) of 0.10.

247

Thus, with much less effort the tables give us plans that were quite close to those we designed.

8.12. SAMPLING PLANS FOR CONTINUOUS PRODUCTION

In many continuous production processes there are no facilities for forming lots. In order to form inspection lots of material just produced, the product must be stored until it has been inspected. This in turn results in space, inventory, and materials handling cost. Further, if some of the items in a designated inspection lot have not yet completed the process and the defectives in the lot have exceeded the criteria for rejection, what happens to the balance of the lot?

Dodge (4) created the initial continuous sampling plan (now described as CSP-1) as follows:

(a) At start up, institute 100% inspection of production items (the usual and recommended procedure—but not mandatory) and continue until i items in succession have been inspected without finding a defect.

(b) When i items in succession have been found clear of defects, stop 100% inspection immediately and begin review of a sample to be selected one at a time from the production process in a random manner. (It is recommended that tables of random numbers or some mechanical device be used to locate each sample item selected.)

(c) When a defective is found revert to 100% inspection immediately and continue until i items are found clear of defectives. At this point revert to sample inspection as described in (b).

(d) All defects are to be removed and replaced with good items.

Selection of a plan may be undertaken as follows: combinations of i and f for a specific AOQL can be obtained from Figure 8.9. An example follows:

If we are inspecting a product with a process average of 1% and an AOQL of 1.5% and the sampling fraction is 1/25 or 0.04%, our plan will read as follows:

(a) Referring to Figure 8.9 for an AOQL of 1.5% and an f of .04, we get an i of approximately 120.

(b) Then inspect 100% until 120 consecutive items are clear of defects.

(c) Change to a sample of 1 in 25 selecting the items in a random manner.

(d) When a defect is found, revert to 100% inspection until 120 consecutive items have been shown to be clear of defects. At this point, return to sample inspection.

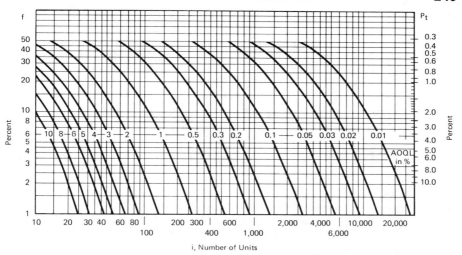

* Reproduced from the *Annals of Mathematical Statistics*, Vol. XIV (1943). p. 274.

p in percent = the value of percent defective, in a consecutive run of N = 1000 product units for which the probability of acceptance, P_a is 0.10 for a sample size of *f* percent.

FIGURE 8.9. Curves for determining values of *f* and *i* of a continuous sampling (CSP-1) plan for a given AOQL.

Some interesting information for evaluation of inspection costs and operation of the plan may be obtained by applying the following formulas: (all from Ref. 4.) For $i = 120$, $f = .04$, $p = .01$ (true percent defective in the process):

1. The average number of pieces screened after finding a defective $(\overline{R}_s)$ will be

$$\overline{R}_s = \frac{1 - Q^n}{P \times Q^n} = \frac{1 - (.99)^{120}}{(.01)(.99)^{120}} = \frac{1 - .299}{(.01)(.299)} = \frac{.701}{.00299} = \overline{R}_s = 234$$

2. The average number of items okayed until a defect is found $(\overline{R}_0)$ will be

$$\overline{R}_0 = \frac{1}{f \times p} = \frac{1}{(.04)(.01)} = \frac{1}{.0004} = 2500 \overline{R}_0 = 2500$$

3. The average percent of total items inspected $(\overline{R}_I)$ will be

$$\overline{R}_I = \frac{\overline{R}_s \pm (f \times \overline{R}_0)}{\overline{R}_s + \overline{R}_0} = 23.4 + \frac{(.04 \times 2500)}{234 + 2500} = .122$$

4. The average percent of items accepted ($\overline{R}_a$) will be (at $p = 1\%$)

$$\overline{R}_a = \frac{\overline{R}_0}{\overline{R}_s + \overline{R}_0} = \frac{2500}{234 + 2500} = \frac{2500}{2734} = 9.14$$

Using step (4) for selected values of p', a form of OC curve can be plotted for various true percent defective as follows (see Table 8.28 and Figure 8.10):

If we sampled 1/12, or 8.3%, with an AOQL of 1.5%, the i chosen from Figure 8.9 would be equal to 80. At $p' = 1\%$:

$\overline{R}_s = 138$ (average number of pieces screened after a defect)

$\overline{R}_0 = 1205$ (average number of pieces okayed until a delay is found)

$\overline{R}_I = 17.7\%$ (average % of total items inspected)

$p(\overline{R}_a) = 89.7\%$

The OC curve for AOQL = 1.5%, $f = .093$, $i = 80$, is shown in Table 8.29 and Figure 8.10.

Some "fiddling around" indicates a 5% sample (1/20) yields an $i = 100$. This plan gives a producer's risk (α) of about 8% and a consumer's risk (β) at about 11%. Thus we have somewhat the same condition specified in designing our original plan. The entire OC curve is shown in Table 8.29 and Figure 8.10.

There are modifications to Dodge CSP-1 plans suggested by Dodge and Torrey (3). One modification (CSP-2) suggests no return to 100% sampling

TABLE 8.28. Comparison of the OC Curves for Three Sample Plans All with AOQL of 1.5%

If We Sample: True Percent Defective in Lot	$P(\overline{R}_a)$		
	$(1/25) = .04$	$(1/20) = .05$	$(1/12) = .083$
	$f = .04, i = .120$	$f = .05, i = 100$	$f = .083, i = 80$
.01	91.4	92.0	90.7
.02	71.2	75.4	75.0
.03	40.6	50.2	53.5
.04	15.8	25.7	32.2
.05	4.8	10.8	17.3
.06	1.5	3.8	7.8
.07	.5	1.4	3.5
.08	.0	.4	1.2
.09	—	.2	.6
.10	—	—	.2

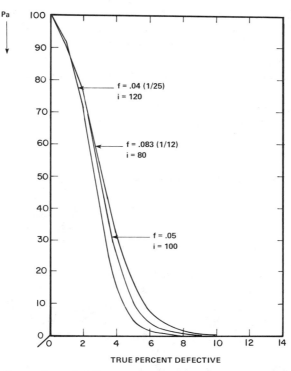

FIGURE 8.10. Comparison of OC curves for three continuous sampling plans—all with AOQL = 1.5%.

TABLE 8.29. **OC Curve for the Variable Sampling Plan $N = 19$, $k = 1.9471$**

$$(K - Z_p)\sqrt{n}$$
$$(1.9476 - Z_p)\sqrt{18}$$

P	Z_p	$(K - Z_p)\sqrt{n}$	P_a
.01	2.33	-1.625	.948
.02	2.05	$-.437$	.669
.03	1.88	.285	.388
.04	1.75	.836	.202
.05	1.65	1.260	.104
.06	1.55	1.685	.046
.07	1.48	1.982	.024
.08	1.41	2.279	.011
.09	1.34	2.576	.005
.10	1.28	2.830	.002

until two defectives have been found in successive sampling intervals (i) and CSP-3 suggests sampling four additional items after finding a defective. If a defective is found in the sample, return to 100% inspection is mandated immediately; otherwise follow rules for CSP-2.

Other approaches to designing continuous sampling plans may be found in Ref. 5.

No discussion of interim military standards is undertaken here since there is strong controversy on the subject. A good discussion of this will be found in Duncan (Ref. 2, pp. 367–371).

8.13. VARIABLES SAMPLING PLANS

When some variable is inspected by measurement, rejected if it is outside the tolerance limit, and merely counted as defective, the method of attributes (good or bad) has been applied. If the measurement is not recorded (which it very often is) and not used in analysis (which it very often is not) valuable information is discarded. Thus the effort expended to make and record the data may be wasted if emphasis is placed only on whether the item meets or does not meet the specifications.

Variable sampling plans make use of the data from all measurements made on the sample. Use of these plans encourages detailed analysis of characteristics of the variable, allows good estimates to be made of the true character of the product of different suppliers, and can provide a very nice picture of the extent to which the inspected product is meeting specifications (just outside, just inside, far out or in control). Further, if histograms are plotted, some idea of the mathematical model to which the particular family of measurements might belong may become known. A more important point is that in adopting variables sampling plans, a smaller sample size may be used with about the same protection as provided by the comparatively larger samples used in attribute plans, given the same acceptance criteria.

Of course variables sampling plans have some disadvantages. One is that there needs to be a separate analysis for each quality characteristic. This might result in different sample sizes (the maximum size sample could be used as a master sample from which to subsample for the smaller sizes with destructive tests made last) or a separate size plan invoked for each characteristic that could take advantage of smaller samples. Remember that if measurements are already being made and recorded there is only slight additional cost involved.

Another disadvantage, though the point is essentially theoretical, is that a lot might be rejected as being significantly different from specifications even though there were no defectives in the lot.

Further, the measurements of each characteristic must be distributed after the manner of the normal curve. Fortunately much industrial data is "mound shaped" and not asymmetrical and in analyzing measurements from data with a pattern such as this, Tchebychoff's inequality, and the Camp–Meidel variations thereto, may be invoked. Further, since sample means and standard deviations are involved, and measurements from these statistics follow the central limit theorem, some confidence may be generated, especially in the long run, in using estimates of the mean and standard deviation to make decisions to accept and reject. Further, the "subject matter" specialists in charge of operations might be given some credit for the development and use of "probability sense" in applying the use of this presently undervalued tool.

A stated disadvantage, that methods of calculation of mean and standard deviations must be explained and taught to inspection personnel, should no longer be much of a problem since the advent of pocket calculators, which have built-in programs to do this work, or with the availability of time-sharing computers which can be located at the inspection stations. Further the inspector today is rapidly becoming a professional as witnessed by the number of inspectors who have passed the rather stiff certification tests given by ASQC.

Design of a variables sampling plan may be approached as follows:

1. Assume the process is producing a quality characteristic that follows the "normal" curve in that the measurements of the characteristic product histograms which indicate the data are "mound shaped" and fairly symmetrical. Further, the process from which the samples are selected is in control as evidenced by $\overline{X}$ and R charts (see Chapter 16).

2. We need acceptance criteria and sample size n. Further, for this plan, we have only a one-sided lower specification (L_s) to consider. Set $P_1 = .01$, $P_2 = .05$, $\alpha = .05$, $\beta = .10$ as in the earlier examples.

3. By placing the lower specification in terms of a standard normal distribution $Z = (\overline{X} - L_s)/S$, the percent defective that will fall below this limit may be found by reference to the tabled values of the normal curve. If

$$L_s = 195$$

$$\overline{X} = 199$$

$$S = 2$$

Then

$$Z = \frac{195 - 199}{2} = -2.0 = .0228$$

as illustrated in Figure 8.11.

4. Using this criteria we can establish a procedure to accept a lot if $(\overline{X} - L_s)/S \geq K$, where K = acceptance criteria.

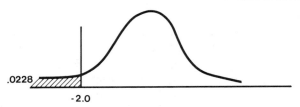

FIGURE 8.11. Percent defective based on normal distribution.

5. Developing this formula we can arrive at two values of K

a. $K = Z_{P_1} \dfrac{Z}{\sqrt{n}}$

b. $K = Z_{P_2} - \dfrac{Z}{\sqrt{n}}$

c. $n = \left(\dfrac{Z_\alpha + Z_\beta}{Z_{P_1} - Z_{P_2}} \right)^2$

where

Z_{P_1} = the Z value for the abscissa of the normal curve for P_1
Z_{P_2} = the Z value of the abscissa of the normal curve for P_2
Z_α = Z value for the abscissa of the normal curve for α
Z_β = Z value of the abscissa of the normal curve for β

6. Proceeding with our plan:

$$Z_{P_{.01}} = 2.327 \qquad \alpha_{.05} = 1.645$$
$$Z_{P_{.05}} = 1.645 \qquad \beta_{.10} = 1.282$$

(a) Sample size = $n = \left(\dfrac{1.645 + 1.282}{2.327 - 1.645} \right)^2 = 18.4 = \underline{\underline{18.}}$

(b) Lower limit for acceptance of the mean at P_1

$$P_1 = .01 = Z_{P_1} = \dfrac{X - 195}{2} = 2.327 = 199.7$$

(c) Lower limit for acceptance of the mean for P_2

$$Z_{P_2} = \dfrac{X - 195}{2} = 1.645 = 198.3$$

(d) Criteria for exactly α

$$K = 2.327 - \dfrac{1.645}{\sqrt{18}} = 1.939$$

Criteria for exactly β

$$K = 1.645 + \frac{1.282}{\sqrt{18}} = 1.645 + .302 = 1.947$$

(e) To get both we average

$$K = \frac{1.939 + 1.947}{2} = 1.943$$

(f) If we then take a sample of size 18, substitute into our formula

$$\frac{X - L_s}{S} \geq K$$

and realize a value greater than 1.641, we reject the lot.

7. An OC curve may be developed as follows:

(a) Obtain Z_P from the normal table for selected values of P'.
(b) Use $(K - Z)\sqrt{n}$ to obtain Z values for each selected P.
(c) Convert to P_a (positive) value of the area under the normal curve (see Table 8.29 and Figure 8.12).

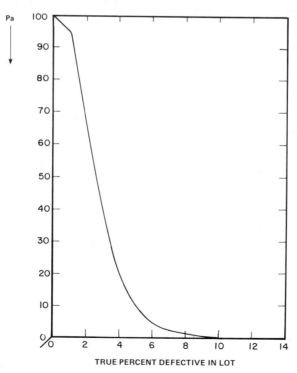

FIGURE 8.12. OC curve for variables sampling plan $n = 19$, $K = 1.9476$.

When there is no knowledge of the process standard deviation, a slightly different approach to designing a variables sampling plan must be used. The penalty for not having good information for sigma is an increase in sample size. Then to estimate n for a sampling plan where sigma is not known Ref. 4 gives the following formula for n:

$$K = \frac{Z_a Z_t + Z_\beta ZAQL}{Z_\alpha + Z_\beta}$$

For the conditions of

$$(AQL)P_1 = .01, \quad (p_i)P = .05, \quad \alpha = .05, \quad \beta = .10$$
$$Z_{P_1} = 2.33, \quad Z_{P_2} = 1.65, \quad \alpha_{.01} = 1.65, \quad \beta_{.05} = 1.28$$
$$K = \frac{(1.65)(1.65) + (1.28)(2.33)}{1.65 + 1.28} = 1.9471$$

and

$$n = \left(1 + \frac{K^2}{2}\right)\left(\frac{Z_\alpha + Z_\beta}{ZAQL - Z_1}\right)^2$$
$$= \left[1 + \frac{(1.9471)^2}{2}\right]\left(\frac{1.65 + 1.28}{2.33 - 1.65}\right)$$
$$(2.9)(18.6) = 53.8 = 54$$

The OC curve is calculated and plotted as in Table 8.30 and Figure 8.12.

TABLE 8.30.   OC Curve for Variables Sampling
$n = 54, P_1 = .01, P_2 = .05, \alpha = .05, B = .10, K = 1.9471$

P	Z_P	$\dfrac{K - Z_P}{\sqrt{\dfrac{1}{n} + \dfrac{K^2}{2n}}}$	P_a
.01	2.33	−1.6853	.954
.02	2.05	−.4529	.675
.03	1.88	+.2953	.384
.04	1.75	+.8675	.193
.05	1.65	+1.3077	.095
.06	1.55	+1.7478	.040
.07	1.48	+2.0559	.0198
.08	1.41	+2.3640	.009
.09	1.34	+2.6721	.004
.10	1.28	+2.9362	.002

REFERENCES

1. Molina, E. C., *Poisson's Exponential Binomial Limit*, Van Nostrand, New York, 1947.
2. Duncan, A. J., *Quality Control and Industrial Statistics*, 4th ed., Irwin, Homewood, IL, 1974.
3. Dodge, H. F., and Torrey, M. N., Additional continuous sampling inspection plans, *Ind. Qual. Control*, Vol. 7, No. 5, March 1951.
4. Dodge, H. F., A sampling inspection plan for continuous production, *Annals of Math. Statistics*, Vol. 14, No. 3, September 1943, pp. 264–279; and *Trans. of the ASME*, Vol. 66, No. 2, February 1944, pp. 127–133.
5. Dodge, H. F., Chain sampling inspection plans, *Ind. Qual. Control*, Vol. XI, No. 4, 1955.
6. Burstein, H., *Attribute Sampling: Tables and Explanations*, McGraw Hill, New York, 1971.
7. Dodge, H. F., and Romig, H. G., *Sampling Inspection Tables—Single & Double Sampling*, 2nd ed., Wiley, New York, 1959.
8. *Acceptance Sampling, A Symposium*, American Statistical Association, Washington, D.C., 1950.

9

COMPARISONS OF POPULATIONS

ANTHONY A. SALVIA

Behrend College
Pennsylvania State University
Erie, Pennsylvania

GARY E. MEEK

College of Business Administration
University of Akron
Akron, Ohio

9.1. INTRODUCTION

This chapter describes the fundamental types of inference employed in comparing two (or more) populations. There are several ways, not necessarily mutually exclusive, of subdividing this topic. One way is to consider the knowledge base the statistician begins with in any particular application. Obviously, there are essential differences in the following two situations:

1. A standard method of paint application has, historically, produced a number X of defects per square foot. X has, to a very close approximation, a Poisson distribution (see Chapter 3) with parameter $\lambda = 0.4$. A new method has been suggested which is not likely to

change the basic nature of the process, but is expected to reduce the average number of defects per square foot to some value $\lambda < 0.4$.

2.   There are in development, two proposed methods for paint application; what is required is to choose the method producing fewer average defects per square foot.

One difference, of course, is that in the first situation, a new method is being compared to a standard one, whereas in the second situation, two new methods are being compared. However, the principal difference between the situations lies in the knowledge (or assumption) base: in situation (1) we are given a random variable whose distribution is known, save for a constant (λ). In situation (2) we know nothing of the distributions involved, except that the random variables are discrete.

Most statistical applications in common use are from knowledge bases like (1). We might term the methods used in these applications "ordinary" statistical methods. In situations such as (2), an entirely different set of methods is employed. These are usually called *nonparametric* or *distribution-free* methods, and they are the subject matter of Chapter 11.

Another way of subdividing the body of methods and techniques of comparing populations is to focus on the number of random variables involved. *Univariate* methods are applicable in situations involving only one random variable; *multivariate* methods are appropriate when more than one random variable enters. Both types are interspersed throughout this book.

Still another subdivision could concentrate on the size of the experiment planned; if that size is not predetermined, but is permitted in some fashion to depend on the actual experimental results, we say that we are employing *sequential* methods (Chapter 17). By way of contrast, all other methods would then be termed *fixed* or *nonsequential*.

It is also possible to divide the methodology upon more or less philosophic grounds, which relate to the amount of weight the experimenter wishes to assign to his or her prior beliefs about the experimental matter. This line of division produces "classical methods" and "Bayesian methods."

This chapter will concentrate now on some rather simple "ordinary" techniques of inference which are of use in comparing populations.

Statistical populations are characterized by a set of one or more parameters; the values of these, if known, render determinable all the features of interest (statistical) of the population. Note that if these are known exactly, then we are not in an inferential situation since sampling can provide *no* new statistical information.

We begin with an example, using the first situation described. Let us suppose that, for some reason (probably cost) the experimenter decides to count the numbers of defects found in 10 samples, and upon doing so obtains the following information:

Defects Observed	Frequency
0	7
1	2
2	1
3 or more	none

A mathematical model for this data may be based on the probability function

$$f(x; \lambda) = \frac{e^{-\lambda}\lambda^x}{x!} \qquad x = 0, 1, 2, \ldots; \lambda > 0$$

which is the probability function of a Poisson random variable (Chapter 3). Prior to obtaining the sample values, we define

$$X_i = \text{number of defects in the } i\text{th square foot}$$
$$i = 1, 2, \ldots, 10$$

These variables have a joint probability function

$$f(x_1, x_2, \ldots, x_{10})$$

and *if they are independent,** we may write

$$f(x_1, x_2, \ldots, x_{10}) = f_1(x_1)f_2(x_2)\ldots f_{10}(x_{10})$$

We assume further that each $f_i(x_i)$ has the same form as $f(x; \lambda)$, and so arrive at

$$f(x_1, x_2, \ldots, x_{10}) = f(x_1; \lambda)f(x_2; \lambda)\ldots f(x_{10}; \lambda)$$
$$= \frac{e^{-10\lambda}\lambda^{\Sigma x_i}}{(x_1! x_2! \ldots x_{10}!)}$$

*In practice, one seldom really knows if the values are independent. If the conduct of the experiment is controlled in such a way as to make it difficult for any particular X_i to influence the others, independence is a reasonable assumption. In the example under discussion, if the promoter of the new method had observed seven or eight defects in the first panel, this may well have influenced his "random choice" of the remaining nine panels.

NOTE. When an experiment is conducted in such a way that it is reasonable to assume factorization into identical densities, as above, the data are said to have been obtained by *simple random sampling*. Often in the sequel we shall refer to a random sample $(X_1, X_2, \ldots, X_n)$; implicit in this reference is such a factorization.

The observed values were not uniquely specified in the short table above, but it is clear that 7 of the X's were found equal to zero, 2 were 1's, and 1 equaled 2. The probability function above then has the value

$$f(x_1, \ldots, x_{10}) = \frac{e^{-10\lambda}(\lambda)^4}{2}$$

Since the experiment was conducted to determine whether $\lambda < 0.4$, it is natural to ask how this value of f might shed some light on the question. Suppose we decide to graph f as a function of λ. Figure 9.1 is thus obtained.

Note that the maximum value in Figure 9.1 occurs at $\lambda = 0.4$. Since f is, in fact, equal to the probability of obtaining the particular values $X_1, X_2, \ldots, X_{10}$, which were actually observed, we see that the sample is more probable (likely) with $\lambda = 0.4$ than with any other value. In a sense, then, $\lambda = 0.4$ is a reasonable *estimate*, and a value of $\lambda < 0.4$ is less reasonable.

Had the sample appeared as:

Defects Observed	Frequency
0	9
1	1
2 or more	none

we would find f maximized at $\lambda = 0.1$; this would tend to support the hypothesis that $\lambda < 0.4$.

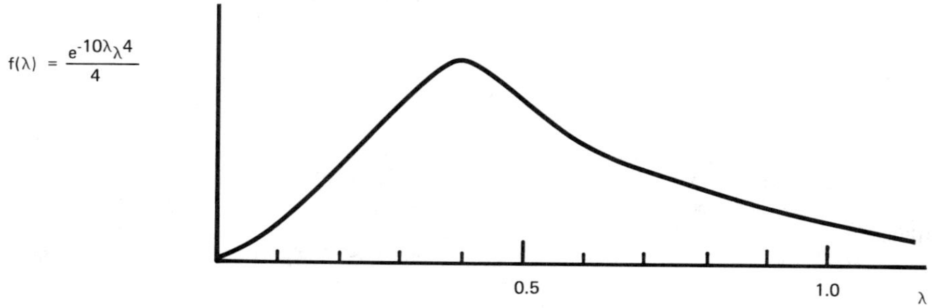

$f(\lambda) = \dfrac{e^{-10\lambda}\lambda^4}{4}$

FIGURE 9.1. f as a function of λ [$f(\lambda)$ versus λ].

At this point, the reader should recall some of the fundamental concepts of hypothesis testing; a discussion of these basic points is contained in Chapters 4 and 6.

Let us instead consider an example in which a "rough-and-ready" approach to population comparison is made. Suppose that two competing methods of synthesis of a particular product are under consideration. It is desirable to choose that method which produces a better yield, measured as percent by weight of final product. Due to random fluctuations—differences in temperature, humidity, and so on—the yield per batch is not constant, but may be considered to be a random variable with a normal distribution

$$f(x; \mu, \sigma^2) = \frac{1}{\sqrt{2\pi}\,\sigma} \exp\left[-\frac{\frac{1}{2}(x - \mu)^2}{\sigma^2}\right]$$

where the parameters μ and σ^2 are unknown. Examine Figures 9.2a, 9.2b, and 9.2c.

In Figure 9.2a, method A is clearly superior; it has, on the average, a higher yield, and is almost always better than B. In Figure 9.2b, most individuals would also prefer A; although the averages are the same, there is less fluctuation in A; the production process would be far more consistent. Even in Figure 9.2c, where A produces a smaller average yield, A may be preferred in practice because of the smaller fluctuations. Prior to any data gathering, we have no way to ascertain which of the above graphs comes closest to the true situation; it may also happen, of course, that the distributions are more or less identical.

The experimenter has two decisions to make: (1) what sample sizes should be taken, that is, how many of each type of batch should be measured, and (2) given the data, which may conveniently be represented as

$$X_1, X_2, \ldots, X_m \qquad \text{from method A}$$
$$Y_1, Y_2, \ldots, Y_n \qquad \text{from method B}$$

how is the "better" method chosen? A somewhat more mathematical way of posing this second question is to ask what function of the variables $\{X_i\}$ and $\{Y_j\}$ shall be used to decide upon a method?* We shall return to this example, after a brief aside to review the concepts of type I and type II errors.

---

*We mention in passing that the same questions need to be answered in comparing a *single* population with some standard, namely, "how big a sample?" and "what do we do with the numbers, once obtained?"

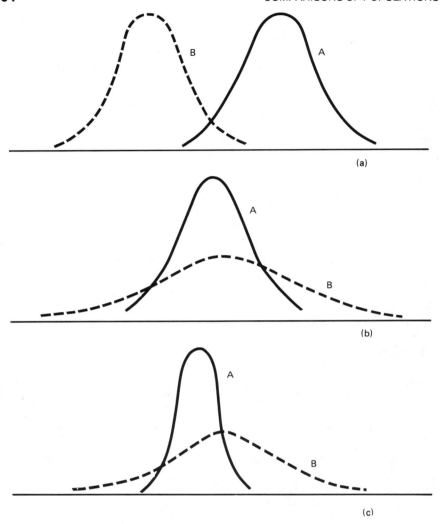

FIGURE 9.2. Comparison of methods A and B.

Type I Error

Suppose we hypothesize that the methods under study are identical; letting X and Y be the random variables associated with A and B, respectively, this hypothesis may be symbolized as

$$H_0: \quad X \text{ is normally distributed with parameters}$$
$$\mu_x, \sigma_x^2; \quad Y \text{ is normally distributed with parameters}$$
$$\mu_y, \sigma_y^2; \quad \text{and } \mu_x = \mu_y, \sigma_x^2 = \sigma_y^2$$

If H_0 is, in fact, true, there are no operational differences between the two methods, and the final choice may well be made on some other basis (cost, convenience, etc.).

After the sample sizes have been chosen and the data are observed, a decision about H_0 will be made. There is a set of only two possible decisions:

$$d_0 : H_0 \text{ is true}$$

$$d_1 : H_0 \text{ is false}$$

If decision d_1 is made, we *reject* the hypothesis H_0. Doing this when H_0 is true is an error; such an error is called a type I error. The probability that one will commit a type I error depends on the sample sizes, (possibly) the values of the unknown parameters, and the manner in which the decision emanates from the data. We use the symbol α to represent this probability.

$$\alpha = \text{probability of type I error}$$

$$= \text{probability } H_0 \text{ rejected when it is true}$$

Type II Error

A second possibility for an error is to accept H_0 (make decision d_0) when it is false. This is termed a type II error; the probability of its occurrence depends on whatever the "true" situation, say H_1, is. We generally denote this probability as β. In the example under consideration, β will vary as a function of μ_x, μ_y, σ_x^2, and σ_y^2; that is,

$$\beta = \beta\left(\mu_x, \sigma_x^2, \mu_y, \sigma_y^2\right)$$

This function is usually called the *operating characteristic* of the test procedure. Being a probability, it takes on values in the range $0 \le \beta \le 1$, as does its complement $1 - \beta$, which is known as the *power* of the test.

We return now to a discussion of our example. Let us suppose that it has been decided to sample five batches produced by each method; we will then have at our disposal the observed values of 10 random variables,

$$X_1, X_2, \ldots, X_5, Y_1, \ldots, Y_5$$

There are a multitude of choices of *test statistic* we may make; a test statistic here means a specific function of the random variables employed in arriving at one of the decisions d_0, d_1. For example, we may use

i. $X_1 - Y_1$

ii. $\max\limits_{i, j} |X_i - Y_j|$

iii. $\min\limits_{i,\,j}|X_i - Y_j|$

iv. $\text{median}\{X_j\} - \text{median}\{Y_j\}$

v. $|\overline{X} - \overline{Y}|$

In studying problems of this nature, we unfortunately lack the space to motivate properly the choice of any particular test statistic. We do stress the fact that this choice is contingent upon the assumptions we are willing to make about the populations being considered, the method (experimental design) used to collect the data, and the level of measurement that can be associated with the numerical values of the sample(s). Generally speaking, we shall simply present that particular statistic which is in one or more ways "optimal." In this instance, the choice is (v); that is, we may accept or reject H_0 on the basis of the value of the statistic T,

$$T = T(X_1, \ldots, Y_5) = |\overline{X} - \overline{Y}|$$

The choice we have just made leads to a t test for the difference between means. Similarly, a test for the equality of the variances σ_x^2 and σ_y^2 will ordinarily be based on the F distribution. In this chapter, our discussion is limited to those situations meeting the assumptions required for the application of the t and F distributions.

We now proceed to a taxonomy of methods having fairly wide use in the comparison of populations. The chart below should provide an idea of our direction.

Characteristic of Interest	Number of Populations	Assumptions	Test Statistic or Technique
1. Means	2	Normality; known variances	Z (normal deviate)
2. Means	2	Normality; variances assumed equal	t
3. Means	2	Normality	Various
4. Means	> 2	Normality; equal variances	Analysis of variance
5. Variances	2	Normality	F
6. Variances	> 2	Normality	Cochran's test or Bartlett's test

All of the above are parametric problems; there also exists a set of nonparametric counterparts. In the remainder of this chapter, we consider items 3, 4, and 6; items 1, 2, and 5 have been discussed earlier.

9.2. EQUALITY OF TWO MEANS — UNKNOWN AND UNEQUAL VARIANCES

This particular situation gives rise to what is known in the literature as the "Behrens–Fisher" problem. It has been studied by many noted statisticians and, to date, no *exact* test that can be easily applied has been developed. Many procedures that give approximate results (with respect to α) have been proposed. Among these are Fisher's (1) fiducial approach and Scheffe's (2) approach. Both are based on the construction of approximate confidence intervals for the difference of the means (or a transformation on the difference) with rejection occurring whenever the statistic falls outside the interval. Fisher's approach, though, has been openly questioned by Bartlett and Stein. Approximate tests have also been proposed by Wald (3) and Welch (4). We will consider Welch's test in detail here.

In addition to the approximate tests Wald developed an exact similar test, but Linnik (5) proved that under the given conditions, Wald's test cannot exist. Linnik (6) also proved that, if we only require measurability of a nonrandomized homogeneous test, then it exists for any size. Linnik's test is, at best, a difficult procedure to actually apply.

The most commonly used procedure is an approximation attributed originally to Welch. The basic approach is to use the statistic for the first situation (i.e., known variances) with s_1^2 and s_2^2 substituting for σ_1^2 and σ_2^2, respectively. The resulting statistic then has an approximate t distribution with the degrees of freedom being given by

$$\nu = \frac{\left(\left(s_1^2/n_1 + s_2^2/n_2\right)\right)^2}{\left\{s_1^2/[n_1(n_1-1)]\right\} + \left\{s_2^2/[n_2(n_2-1)]\right\}} - 2 \qquad (9.1)$$

where s_1^2 and s_2^2 are the usual unbiased estimates of σ_1^2 and σ_2^2, respectively. Since this computation will seldom yield an integer value, we generally round to the nearest integer. With this adjustment on the degrees of freedom, the null hypothesis, H_0: $\mu_1 = \mu_2$, is rejected if the absolute value of the statistic

$$\frac{\overline{X}_1 - \overline{X}_2}{\sqrt{s_1^2/n_1 + s_2^2/n_2}}$$

exceeds $t_{\nu,(\alpha/2)}$.

9.3. EQUALITY OF SEVERAL MEANS

In this section, we consider the basic statistical methods used in comparing the means of several populations, given available samples from each. The body of techniques which are used in attacking this problem is normally referred to as the "analysis of variance."

We shall introduce the principal idea by means of an example.

Example 9.1.   A manufacturing facility includes four production lines for a particular product. Management has expressed some concern that there has been considerable variation in product quality from line to line (a very common problem). You decide to collect samples from each line and measure a characteristic related to "quality" (high values = good quality). The following measurements are made.

Production Line

1	2	3	4
36	32	26	29
29	31	20	30
34	31	28	
		28	

The natural thing to do with the data at this point is to determine the average for each line:

$$\text{Line 1:} \qquad \frac{36 + 29 + 34}{3} = 33.0$$

$$\text{Line 2:} \qquad \frac{32 + 31 + 31}{3} = 31.3$$

$$\text{Line 3:} \qquad \frac{26 + 20 + 28 + 28}{4} = 25.5$$

$$\text{Line 4:} \qquad \frac{29 + 30}{2} = 29.5$$

We may also find it helpful to plot the data as in Figure 9.3.

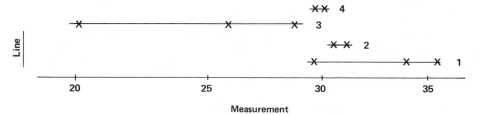

FIGURE 9.3.   Production line measurements of quality.

In Figure 9.3, we notice that two of the lines (1 and 3) have much more spread than the other two, and that, except for line 3, the data overlap from line to line. Presumably we would expect the *populations* from which the data are extracted to overlap as well.

The starting point for a detailed analysis is a *model equation* which the experimenter creates to describe the data. Let us write y_{ij} to represent the jth data item in the ith line, for example,

$$y_{12} = 29$$
$$y_{33} = 28$$

Further, write μ_i for the (unknown) population mean for line i. Then each observation may be written as

$$y_{ij} = \mu_i + \varepsilon_{ij} \tag{9.2}$$

Equation (9.2) states that the observation y_{ij} is equal to its mean value, plus an error term, ε_{ij}. The experimenter's interest is now, in statistical terms, to test whether the set of $\{\mu_i\}$ are all equal.

It turns out to be more convenient in the sequel to define the parameter

$$\mu = \frac{3\mu_1 + 3\mu_2 + 4\mu_3 + 2\mu_4}{12}$$

The overall mean, μ, is a weighted average of $\mu_1, \ldots, \mu_4$, the weights being equal to the sample size in each group of data. Let us define further the *group effects*.

$$A_i = \mu_i - \mu, \qquad i = 1, \ldots, 4 \tag{9.3}$$

It is easy to calculate that $\sum_{i=1}^{4} n_i A_i = 0$, where the $\{n_i\}$ are the group sample sizes. In terms of μ and the $\{A_i\}$, equation (9.2) may be rewritten as

$$y_{ij} = \mu + A_i + \varepsilon_{ij} \tag{9.4}$$

and the hypothesis $H_0: \mu_1 = \mu_2 = \mu_3 = \mu_4$ is equivalent to $H_0': A_1 = A_2 = A_3 = A_4 = 0$.

Consider now the quantity

$$SST = \sum_{i=1}^{4} \sum_{j=1}^{n_i} (y_{ij} - \bar{y})^2 \tag{9.5}$$

In equation (9.5), $\bar{y}$ is the sample mean of all the data. Except for division by its degrees of freedom, SST resembles a sample variance; in any event, it provides a measure of the total variability of the data. Note that this computation treats the data as if it were one large sample rather than four smaller ones. Now, letting $\bar{y}_i$ = the sample mean in group i, and omitting indices on the summations

$$SST = \Sigma\Sigma(y_{ij} - \bar{y})^2$$

$$= \Sigma\Sigma(y_{ij} - \bar{y}_i + \bar{y}_i - \bar{y})^2$$

$$= \Sigma\Sigma(y_{ij} - \bar{y}_i)^2 + \Sigma\Sigma(\bar{y}_i - \bar{y})^2$$

$$+ 2\Sigma\Sigma(y_{ij} - \bar{y}_i)(\bar{y}_i - \bar{y})$$

$$= SSW + SSB + C$$

SSW consists of the squared deviations of each observation from the average in its group; it measures variability *within* groups. SSB contains deviations of group means from the overall mean; it measures variability *between* groups. The cross product term $C = 0$.

What we have done is to separate (partition) the total sum of squares SST into two components,

$$SST = SSW + SSB \tag{9.6}$$

the first of which, SSW, measures what we might call the "natural" variation of the data, within group, and the second of which measures the variation between groups. It is this type of separation which lends the name analysis of variance to the technique.

We have previously encountered sums of squares of random variables with normal distributions; these led to chi-square distributions, and to proceed we needed to know the degrees of freedom associated with the sum. Consider the components of the relationship in equation (9.6):

SST contains (Σn_i) squared deviations, each measured from the same sample mean. It therefore possesses $(\Sigma n_i) - 1$ degrees of freedom (df).
$SSW = \Sigma\Sigma(y_{ij} - \bar{y}_i)^2$. In group i, there are n_i squared deviations, each taken from the same group mean; hence, group i contributes $n_i - 1$ df. Adding these for all groups we obtain $(\Sigma n_i) - 4$ df for SSW.
SSB contains four deviations, and hence 3 df.

Note that

$$(\Sigma n_i) - 1 = [(\Sigma n_i) - 4] + [4 - 1]$$

is an equation that relates degrees of freedom of the sums of squares in equation (9.6). There is a well-known theorem in mathematical statistics,

Cochran's theorem, which states essentially that, if a sum of squares (such as SST) can be partitioned in such a way that the degrees of freedom "balance," then the components of the partition (SSW and SSB) are statistically independent.

We shall always in the sequel have an implicit need for Cochran's theorem; the reason is that very shortly we shall form a ratio involving SSW and SSB and apply an F test. In order to have a true F distribution, the chi-square variables in the numerator and denominator must be independent.

The ratio in fact is, for this example,

$$F = \frac{SSB/3}{SSW/8}$$

and, if the hypothesis H_0' is true, F possesses the F distribution with 3 and 8 df, respectively. We usually summarize results of the analysis in a table such as the one below, where we have, for generality, let the number of groups be K.

Analysis of Variance

Source of Variation	Sum of Squares	df	Mean Square	F
Between groups	SSB	$K - 1$	$SSB/K - 1$	
Within groups	SSW	$(\Sigma n_i) - K$	$SSW/(\Sigma n_i - K)$	
Total	SST	$(\Sigma n_i) - 1$	—	

Large values of F throw suspicion on H_0', and we may test H_0' by the rule "reject if the calculated value of F exceeds the upper $100(1 - \alpha)\%$ point of F with $(K - 1)$ and $(\Sigma n_i) - K$ df" where, of course, α is the significance level (type I error) of the test.

Let us apply the F test to our data. We have

$$K = 4$$
$$n_1 = 3 \qquad \bar{y}_1 = 33.0$$
$$n_2 = 3 \qquad \bar{y}_2 = 31.3$$
$$n_3 = 4 \qquad \bar{y}_3 = 25.5$$
$$n_4 = 2 \qquad \bar{y}_4 = 29.5$$

The overall mean

$$\bar{y} = \frac{36 + 29 + \cdots + 30}{12} = 29.5$$

and

$$SST = \Sigma\Sigma(y_{ij} - \bar{y})^2 = (36 - 29.5)^2 + (29 - 29.5)^2 + \cdots (30 - 29.5)^2$$
$$= 181.00$$

The two terms on the right in equation (9.6) are

$$SSW = \Sigma\Sigma(y_{ij} - \bar{y}_j)^2$$
$$= (36 - 33.0)^2 + (29 - 33.0)^2 + (34 - 33.0)^2$$
$$+ (32 - 31.3)^2 + (31 - 31.3)^2 + (31 - 31.3)^2$$
$$+ (26 - 25.5)^2 + (20 - 25.5)^2 + (28 - 25.5)^2 + (28 - 25.5)^2$$
$$+ (29 - 29.5)^2 + (30 - 29.5)^2$$
$$= 70.17$$

and

$$SSB = \Sigma\Sigma(\bar{y}_i - \bar{y})^2 = \Sigma n_j(\bar{y}_j - \bar{y})^2$$
$$= 3(33 - 29.5)^2 + \cdots + 2(29.5 - 29.5)^2$$
$$= 110.47$$

Note that the value of 110.47 for SSB is slightly in error due to rounding $\bar{y}_2$ from 31.3̄3 to 31.3. If we do not round, the value would be 110.83 and

$$SST = 181.0 = 70.17 + 110.83 = SSW + SSB$$

verifying equation (9.6) for this example.

Summarizing these computations in the tabular format, we have

Source	SS	df	MS	F
Between lines	110.47	3	55.24	6.30
Within lines	70.17	8	8.77	
Total	181.00	11		

The hypotheses of interest here are

$$H_0: \mu_1 = \mu_2 = \mu_3 = \mu_4 \qquad H_1: \text{some } \mu_j \text{ differ}$$

or equivalently

$$H_0': A_j = 0 \text{ all } j \qquad H_1': \text{some } A_j \neq 0$$

Verbally, H_0 (and H_0') states that the quality is the same for all lines while H_A says that the quality differs between lines. To test this, we calculate the ratio MSB/MSW and reject $H_0(H_0')$ if it exceeds the appropriate value from an F table. Using an α of .05, the critical F value is $F_{3,8,.05} = 4.07$. Since our calculated F of 6.30 exceeds 4.07, we would reject H_0 and would conclude that the quality differs between lines.

It should be noted for Example 9.1 that for our comparison and resulting conclusion to be valid, the assumptions underlying the ANOVA procedure must be satisfied. We will address all of the assumptions in more detail later but one in particular is important at this juncture. This assumption is that the variances of the populations (lines) from which the samples were selected are all equal. In statistical jargon, this is referred to as homogeneity of variance. In the next section we provide a means of testing that assumption.

9.4. EQUALITY OF SEVERAL VARIANCES

Recall that one of the problems we wished to address earlier was a test for equality of several (more than two) variances. As noted in Figure 9.3, the four groups of data appeared to have different amounts of spread, and equal variances within groups are required for us to be able to employ ANOVA. It appears, then, that we should test that hypothesis even *before* we proceed to an ANOVA.

The tests most commonly employed for this situation are Hartley's test, Cochran's test, and Bartlett's test. We deal with Hartley's test first because it is the easiest one to apply.

The application of Hartley's test requires that the sample sizes are equal and that the populations are normal. The hypotheses to be tested are

$$H_0 : \sigma_1^2 = \sigma_2^2 = \cdots = \sigma_r^2 = \sigma^2 \qquad H_1 : \text{some } \sigma_j \neq \sigma^2$$

with the test statistic being based solely on the largest and smallest sample variances, denoted $\max s_j^2$ and $\min s_j^2$, respectively. This statistic is

$$H = \frac{\max s_j^2}{\min s_j^2} \tag{9.7}$$

with rejection of H_0 occurring for large values of H. That is, H_0 is rejected at level α if

$$H > H_{1-\alpha, r, n} \tag{9.8}$$

with the value for $H_{1-\alpha,\,r,\,n}$ being found in Table 9.1, which gives 95th and 99th percentiles for H where r represents the number of variances being compared and n is the common sample size.

Cochran developed a test which is similar to Hartley's based upon the same assumptions of equal sample sizes and normal populations. The difference in the two procedures is that Cochran compares the largest

TABLE 9.1. Percentiles of Hartley's *H* Statistic Distribution

Entry is H(1 − a; r, n) where P $\{H \leqslant H\,(1 - a;\, r,\, n)\}$ = 1 − a

$1 - a = .95$

n	2	3	4	5	6	7	8	9	10	11	12
3	39.0	87.5	142	202	266	333	403	475	550	626	704
4	15.4	27.8	39.2	50.7	62.0	72.9	83.5	93.9	104	114	124
5	9.60	15.5	20.6	25.2	29.5	33.6	37.5	41.1	44.6	48.0	51.4
6	7.15	10.8	13.7	16.3	18.7	20.8	22.9	24.7	26.5	28.2	29.9
7	5.82	8.38	10.4	12.1	13.7	15.0	16.3	17.5	18.6	19.7	20.7
8	4.99	6.94	8.44	9.70	10.8	11.8	12.7	13.5	14.3	15.1	15.8
9	4.43	6.00	7.18	8.12	9.03	9.78	10.5	11.1	11.7	12.2	12.7
10	4.03	5.34	6.31	7.11	7.80	8.41	8.95	9.45	9.91	10.3	10.7
11	3.72	4.85	5.67	6.34	6.92	7.42	7.87	8.28	8.66	9.01	9.34
13	3.28	4.16	4.79	5.30	5.72	6.09	6.42	6.72	7.00	7.25	7.48
16	2.86	3.54	4.01	4.37	4.68	4.95	5.19	5.40	5.59	5.77	5.93
21	2.46	2.95	3.29	3.54	3.76	3.94	4.10	4.24	4.37	4.49	4.59
31	2.07	2.40	2.61	2.78	2.91	3.02	3.12	3.21	3.29	3.36	3.39
61	1.67	1.85	1.96	2.04	2.11	2.17	2.22	2.26	2.30	2.33	2.36
∞	1.00	1.00	1.00	1.00	1.00	1.00	1.00	1.00	1.00	1.00	1.00

$1 - a = .99$

n	2	3	4	5	6	7	8	9	10	11	12
3	199	448	729	1,036	1,362	1,705	2,063	2,432	2,813	3,204	3,605
4	47.5	85	120	151	184	216	249	281	310	337	361
5	23.2	37	49	59	69	79	89	97	106	113	120
6	14.9	22	28	33	38	42	46	50	54	57	60
7	11.1	15.5	19.1	22	25	27	30	32	34	36	37
8	8.89	12.1	14.5	16.5	18.4	20	22	23	24	26	27
9	7.50	9.9	11.7	13.2	14.5	15.8	16.9	17.9	18.9	19.8	21
10	6.54	8.5	9.9	11.1	12.1	13.1	13.9	14.7	15.3	16.0	16.6
11	5.85	7.4	8.6	9.6	10.4	11.1	11.8	12.4	12.9	13.4	13.9
13	4.91	6.1	6.9	7.6	8.2	8.7	9.1	9.5	9.9	10.2	10.6
16	4.07	4.9	5.5	6.0	6.4	6.7	7.1	7.3	7.5	7.8	8.0
21	3.32	3.8	4.3	4.6	4.9	5.1	5.3	5.5	5.6	5.8	5.9
31	2.63	3.0	3.3	3.4	3.6	3.7	3.8	3.9	4.0	4.1	4.2
61	1.96	2.2	2.3	2.4	2.4	2.5	2.5	2.6	2.6	2.7	2.7
∞	1.00	1.0	1.0	1.0	1.0	1.0	1.0	1.0	1.0	1.0	1.0

sample variance to the sum of the sample variances. Thus, to test

$$H_0 : \sigma_1^2 = \sigma_2^2 = \cdots = \sigma_k^2 = \sigma^2 \qquad H_A : \text{some } \sigma_j^2 \neq \sigma^2$$

we calculate

$$C = \frac{\max s_j^2}{\sum\limits_{j=1}^{k} s_j^2} \tag{9.9}$$

with rejection of H_0 being implied if this ratio is too large; that is, reject H_0 if

$$C > C_{k,n,\alpha} \tag{9.10}$$

where $C_{k,n,\alpha}$ is found in Table 9.2. Table 9.2 gives critical values for C for α equal to .05 and .01, where k is the number of variances being compared and n is the common sample size.

Example 9.2. Let us assume that the four samples of Example 9.1 are all based on three observations with the sample variances being $s_1^2 = 13.00$, $s_2^2 = 0.33$, $s_3^2 = 14.33$, and $s_4^2 = 0.5$. The hypotheses of interest are

$$H_0 : \sigma_1^2 = \sigma_2^2 = \sigma_3^2 = \sigma_4^2 = \sigma^2 \qquad H_1 : \text{some } \sigma_j^2 \neq \sigma^2$$

We use both Hartley's and Cochran's procedure to test these hypotheses. For Hartley's test, we have

$$H = \frac{\max s_j^2}{\min s_j^2} = \frac{14.33}{0.33} = 43.42$$

and from Table 9.1, we obtain $H_{.95,4,3} = 142$. Thus, we cannot reject H_0 on the basis of these samples.

For Cochran's test we have $\max s_j^2 = 14.33$ and $\sum_{j=1}^{k} s_j^2 = 28.16$. Thus,

$$C = \frac{\max s_j^2}{\sum\limits_{i=1}^{k} s_j^2} = \frac{14.33}{28.16} = 0.51$$

In Table 9.2, we find $C_{.05,4,3} = 0.7679$. Again, we are unable to reject H_0 on the basis of these samples.

TABLE 9.2. Cochran's Table

Percentage points of the ratio $s^2{}_{max}. \Big/ \sum_{t=1}^{k} s^2{}_t.$ Upper 5% points

k / n	2	3	4	5	6	7	8	9	10	15	15	20
2	0.9985	0.9669	0.9065	0.8412	0.7808	0.7271	0.6798	0.6385	0.6020	0.5410	0.4709	0.3894
3	.9750	.8709	.7679	.6838	.6161	.5612	.5157	.4775	.4450	.3924	.3346	.2705
4	.9392	.7977	.6841	.5981	.5321	.4800	.4377	.4027	.3733	.3264	.2758	.2205
5	.9057	.7457	.6287	.5441	.4803	.4307	.3910	.3584	.3311	.2880	.2419	.1921
6	.8772	.7071	.5895	.5065	.4447	.3974	.3595	.3286	.3029	.2624	.2195	.1735
7	0.8534	0.6771	0.5598	0.4783	0.4184	0.3726	0.3362	0.3067	0.2823	0.2439	0.2034	0.1602
8	.8332	.6530	.5365	.4564	.3980	.3535	.3185	.2901	.2666	.2299	.1911	.1501
9	.8159	.6333	.5175	.4387	.3817	.3384	.3043	.2768	.2541	.2187	.1815	.1422
10	.8010	.6167	.5017	.4241	.3682	.3259	.2926	.2659	.2439	.2098	.1736	.1357
11	.7880	.6025	.4884	.4118	.3568	.3154	.2829	.2568	.2353	.2020	.1671	.1303
17	0.7341	0.5466	0.4366	0.3645	0.3135	0.2756	0.2462	0.2226	0.2032	0.1737	0.1429	0.1108
37	0.6602	0.4748	0.3720	0.3066	.2612	.2278	.2022	.1820	.1655	.1403	.1144	.0879
145	.5813	.4031	.3093	.2513	.2119	.1833	.1616	.1446	.1308	.1100	.0889	.0675
∞	.5000	.3333	.2500	.2000	.1667	.1429	.1250	.1111	.1000	.0833	.0667	.0500

Upper 1% points

k / n	2	3	4	5	6	7	8	9	10	12	15	20
2	0.9999	0.9933	0.9676	0.9279	0.8828	0.8376	0.7945	0.7544	0.7175	0.6528	0.5747	0.4799
3	.9950	.9423	.8643	.7885	.7218	.6644	.6152	.5727	.5358	.4751	.4069	.3297
4	.9794	.8831	.7814	.6957	.6258	.5685	.5209	.4810	.4469	.3919	.3317	.2654
5	.9586	.8335	.7212	.6320	.5635	.5080	.4627	.4251	.3934	.3428	.2882	.2288
6	.9373	.7933	.6761	.5875	.5195	.4659	.4226	.3870	.3572	.3099	.2593	.2048
7	0.9172	0.7606	0.6410	0.5531	0.4866	0.4347	0.3932	0.3592	0.3308	0.2861	0.2386	0.1877
8	.8988	.7335	.6129	.5259	.4608	.4105	.3704	.3378	.3106	.2680	.2228	.1748
9	.8823	.7107	.5897	.5037	.4401	.3911	.3522	.3207	.2945	.2535	.2104	.1646
10	.8674	.6912	.5702	.4854	.4229	.3751	.3373	.3067	.2813	.2419	.2002	.1567
11	.8539	.6743	.5536	.4697	.4084	.3616	.3248	.2950	.2704	.2320	.1918	.1501
17	0.7949	0.6059	0.4884	0.4094	0.3529	0.3105	0.2779	0.2514	0.2297	0.1961	0.1612	0.1248
37	.7067	.5153	.4057	.3351	.2858	.2494	.2214	.1992	.1811	.1535	.1251	.0960
145	.6062	.4230	.3251	.2644	.2229	.1929	.1700	.1521	.1376	.1157	.0934	.0709
∞	.5000	.3333	.2500	.2000	.1667	.1429	.1250	.1111	.1000	0833	.0667	.0500

$s^2{}_{max}$ is the largest in a set of k independent mean squares, $s^2{}_t$, each based on γ degrees of freedom.

Note that Hartley's test is particularly sensitive to a lack of homogeneity resulting from one relatively large variance and one relatively small with the other variances involved being essentially ignored. Cochran's test, on the other hand, is sensitive to a lack of homogeneity resulting from one variance being large in comparison to all others involved. Both tests assume equal sample sizes as well as normality. If the sample sizes differ than we may turn to Bartlett's procedure which also assumes normal populations but not equal sample sizes.

The hypotheses of interest remain as

$$H_0: \sigma_1^2 = \sigma_2^2 = \cdots = \sigma_r^2 = \sigma^2 \qquad H_1: \text{some } \sigma_j^2 \neq \sigma^2$$

To use Bartlett's test we are required to make a rather cumbersome transformation on the sample variances. His test statistic compares the MSE, as defined earlier with the weighted geometric mean of the s_j^2, denoted $GMSE$. If the s_j^2 are all equal, then $MSE = GMSE$, otherwise $GMSE < MSE$. Thus, under H_0, the ratio $MSE/GMSE$ has an expected value of 1. Bartlett showed that by transforming the log of this ratio, as given in equation (9.11) below, one obtains an approximate χ^2 distribution with $r - 1$ df when the n_j are large (generally $n_j \geq 5$ suffices). The appropriate form is

$$B = \frac{2.302585}{K}(n_T - r)(\log_{10}MSE - \log_{10}GMSE)$$

$$= \frac{2.302585}{K}\left[(n_T - r)\log_{10}MSE - \sum_{j=1}^{r}(n_j - 1)\log_{10}s_j^2\right] \quad (9.11)$$

where $n_T = \sum_{j=1}^{r}n_j$, r is the number of variances being compared, and

$$K = 1 + \frac{1}{3(r-1)}\left(\sum_{j=1}^{r}\frac{1}{n_j - 1} - \frac{1}{n_T - r}\right)$$

Note that K is always greater than 1. Thus, for the indicated hypothesis, H_0 is to be rejected if

$$B > \chi_{r-1,\alpha}^2 \quad (9.12)$$

where $\chi_{r-1,\alpha}^2$ is the $(1 - \alpha)$ 100th percentile for a chi-square distribution having $r - 1$ df and can be found in Appendix Table A.4.

Example 9.3. Since our data in Example 9.1 did not consist of equal sample sizes, Hartley's and Cochran's tests were actually inappropriate whereas Bartlett's is not, though it may be questionable due to small sample sizes. Realizing this shortcoming, we still apply it to illustrate the procedure.

Our hypotheses are

$$H_0: \sigma_1^2 = \sigma_2^2 = \sigma_3^2 = \sigma_4^2 = \sigma^2 \qquad H_1: \text{some } \sigma_j^2 \neq \sigma^2$$

with $n_1 = n_2 = 3$, $n_3 = 4$, $n_4 = 2$, and $s_1^2 = 13.00$, $s_2^2 = 0.33$, $s_3^2 = 14.33$, and $s_4^2 = 0.5$. Calculating K, we get

$$K = 1 + \frac{1}{3(4-1)}\left[\left(\frac{1}{3-1} + \frac{1}{3-1} + \frac{1}{4-1} + \frac{1}{2-1}\right) - \frac{1}{12-4}\right]$$

$$= 1.24537$$

Thus, we have

$$MSE = 8.77 \text{ (from Example 9.1) and}$$

$$\sum_{j=1}^{4} (n_j - 1)\log_{10}s_j^2 = (3 - 1)\log_{10}(13.00) + (3 - 1)\log_{10}$$
$$\times (0.33) + (4 - 1)\log_{10}(14.33) + (2 - 1)\log_{10}(0.5)$$
$$= 2(1.1139) + 2(-0.4815) + 3(1.1562) + 1(-0.301)$$
$$= 4.4326$$

Substituting these results in equation (9.11) yields

$$B = \frac{2.302585}{1.24537}\left[(12 - 4)\log_{10}(8.77) - 4.4326\right]$$
$$= \frac{2.302585}{1.24537}(3.1114) = 5.7527$$

For a chi-square distribution with 3 df, the 95th percentiles is $\chi_{3,.05}^2 = 7.81$. Since $5.7527 < 7.81$, we cannot reject H_0 and thus conclude that there is insufficient evidence to contradict the homogeneity of variance assumption.

It should be noted that all three of the above tests assume normality and are quite sensitive to violations of this assumption. In addition, the F test is robust with respect to the homogeneity assumption. That is, it is not affected appreciably by unequal variances, particularly if the sample sizes are equal, provided the differences are not excessively large.

9.5. DISCUSSION OF SINGLE-FACTOR MODELS

Let us pause to summarize what we have said so far about ANOVA, and to indicate what is coming. The experiment we have discussed in our example is typical of a *one-way fixed effects* experiment. The qualifier "one-way" refers to the fact that our experimental data were segregated using a single characteristic, namely, production line. The qualifier "fixed effects" refers to the fact that the hypotheses of interest were concerned only with the four production lines represented in the experiment. We often call the segregating variable a "factor," and its various values "levels." We might, then, equally well call our example a *single fixed factor* experiment.

Later in this book we shall wish to consider multifactor experiments; we shall also have need to consider experiments in which some levels of one or more factors are not present in the experimental data. For example, we may have had 10 production lines in operation but, because of cost, could include only 4 in our actual experiment. Under certain conditions, factors that have missing levels in experiments are termed "random" factors; multifactor experiments with both types of factors are called "mixed."

In all cases, it is best to begin with a statement of the model equation and the assumptions and hypotheses of interest. For one-way fixed experiments these usually look like

$$\text{Model:} \quad y_{ij} = \mu + A_i + \varepsilon_{ij} \qquad \begin{aligned} i &= 1, 2, \ldots, K \\ j &= 1, 2, \ldots, n_i \end{aligned}$$

Assumptions: The set $\{\varepsilon_{ij}\}$ are normally and independently distributed with means equal to zero and equal variances, σ_ε^2. This may be written in shorthand fashion as

$$\varepsilon_{ij} \sim \text{NID}(0, \sigma_\varepsilon^2)$$

Hypotheses: $H_0 : A_1 = A_2 = \cdots = A_K = 0$ versus H_1: some $A_i \neq 0$ are the usual hypotheses.

The actual experiment may be conceived as having several stages.

I. Write model; determine sample sizes and appropriate means for collection of the data.
II. Gather data and perform necessary assumption testing.
III. Do the analysis of variance if assumptions test out.
IV. Perform appropriate postanalyses.

We now proceed to consider each of these in detail.

Stage I

We have already discussed writing the model. While this may seem to be somewhat trite for single-factor experiments, it is absolutely essential for multifactor experiments, as considered in Chapters 12 and 13.

The determination of sample size is, as usual, a trade-off between the time and cost allotted for the experiment and the degree of precision of the desired results. Recall Chapter 4 in this instance. The determination begins with a fixing of the type I error. About how often would we be willing to risk stating that the groups are different, when in fact they are the same? The type II error is quite difficult to determine, since the number of possible alternatives to our hypothesis, H_0, is quite enormous.

Let us suppose that we have K groups; the thing that needs to be determined is, say, n, the number of observations per group. Even though it is not necessary to have equal group sizes, we adopt equal sizes as a convenience in determining the size of the experiment.

Now clearly, if any of the constants $A_1,\ldots,A_K$ are nonzero, and we *accept* H_0, we are committing a type II error. Presumably, the more these differ from zero, the more serious is the consequence of that error. From the way in which we structured the model, we know that

$$\Sigma n_i A_i = 0$$

or, if $n_1 = n_2 = \cdots = n_K = n$,

$$\Sigma A_i = 0,$$

and so we need to look at, say, ΣA_i^2 as a measure of "how big" the set $A_1,\ldots,A_K$ is.

At the same time, we might have, for example, a very large error variance σ_ε^2 relative to the A_i's, and so this measure ought to be *scaled*; this we accomplish by setting

$$\phi^2 = \frac{n\sum_1^K A_i^2}{K\sigma_\varepsilon^2}$$

Figure 9.4 shows graphs of power as a function of ϕ, n, K, and α. More specifically, we use $K(n-1)$ and $K-1$, the degrees of freedom for the two sums of squares in the experiment.

Example 9.4. Suppose $K = 4$. Then $df_1 = K - 1 = 3$. We wish to have a type II error of 10% or less (power of 90% or more) if any of the A_i's is as large as $\frac{1}{2}\sigma_\varepsilon$.

Take

$$A_1 = 0.5\sigma_\varepsilon, \qquad A_2 = -0.5\sigma_\varepsilon, \qquad A_3 = A_4 = 0$$

Then

$$\Sigma A_i^2 = \tfrac{1}{2}\sigma_\varepsilon^2 \quad \text{and} \quad \phi^2 = \frac{n \cdot \tfrac{1}{2}\sigma_\varepsilon^2}{4\sigma_\varepsilon^2} = \frac{n}{8}$$

If our type I error is .05 and we read over on the horizontal line $1 - \beta = .90$, we see it intersects each of the curves labeled df_2.

We try several values for n, until we strike a balance.

n	$\phi^2 = n/8$	df_2	ϕ from Graph	ϕ^2
2	0.25	6	2.65	
4	0.500	12	2.2	
10	1.250	30	2.0	
20	2.500	60	1.9	3.84

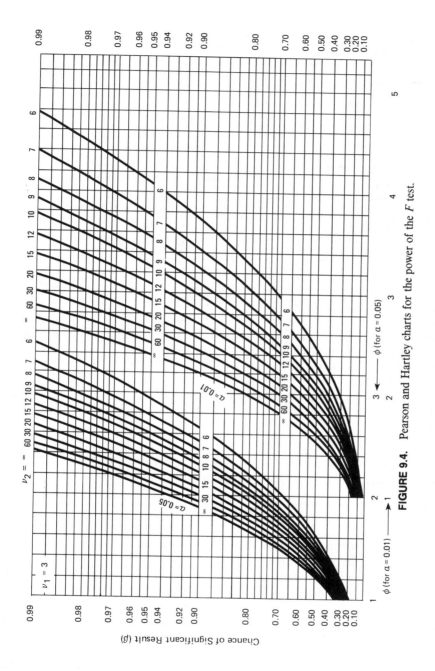

FIGURE 9.4. Pearson and Hartley charts for the power of the *F* test.

281

Clearly, we need $n > 20$ units in each group. If we reduce our power requirement to 0.50, $df_2 = 9$ (or $n = 3$) will do. If we decide upon three observations per group, we have a 50% chance of accepting H_0 if the A_i's differ from zero in any way such that

$$\Sigma A_i^2 \leq \frac{\sigma_\varepsilon^2}{2}$$

Generally speaking, increasing the power can only be accomplished at considerable cost when the number of groups is moderate; the above example shows that 50% power is relatively cheap; 90% is very costly. At this stage, you might want to spend a great deal of time determining *both n and K*, as well as deciding how big the A_i's can be and still be "tolerable."

The appropriate methods for collection of the data also need to be given some thought in the first stage. This means, among other things, that the experimenter needs to specify *who* collects the data, *how* it is to be recorded, and *when*. In our earlier example, suppose we discovered *after the fact*, that the production measurements from lines 1, 2, and 4 had been made by inspector A (a person with 10 years' experience), but line 3 was measured by a new apprentice, B, who just started working last week. In this event, we do not really know, and there is no way to find out, whether the observed differences are due to actual differences in the lines or in the inspectors.

Conceptually, we could model this situation as

$$y_{ij} = \mu + L_i + I_j + \varepsilon_{ij}$$

where we have used L for lines and I for inspectors. By ignoring the I_j terms in our analysis, we have in fact altered the error term from ε_{ij} to

$$\varepsilon_{ij} = I_j + \varepsilon_{ij}$$

This is only one of an infinite number of ways in which the lack of some foresight can invalidate an experiment.

In industrial applications, there are almost always a multitude of factors that *could* influence the data. Your job, in deciding who, how, and when the data are collected, is to either (1) control those extraneous factors, or (2) "average" them out. In the example we have been discussing, we could control the factor "inspector" by using only one inspector.

Here is an example in which we can "average out" the influence of a nuisance factor. Suppose we are attempting to compare two different methods of making a product by looking at the resulting viscosities. We decide to obtain four measurements on the viscosity for each method, and the total determination time takes 1 hour. On the chart below we have the average ambient temperature for the laboratory space where we are measuring the viscosity.

Time of day	8 A.M.	9	10	11	1	2	3	4
Temperature, °F	68	73	73	75	78	78	78	78

Clearly, since viscosity is temperature dependent, it would be wise not to make all the "method 1" determinations in the morning, leaving "method 2" to the afternoon. To do so would introduce a second factor, temperature, which we do not want to analyze. If we can *control*, that is, keep the temperature constant, so much the better.

Otherwise, we ought to *balance* by taking some of each method in the morning as well as the afternoon, for example,

$$1 \quad 2 \quad 2 \quad 1 \quad \quad 2 \quad 1 \quad 2 \quad 1$$

This should result in any differences due to temperature being averaged "out."

This process of assigning a different order is usually called *randomization*; most texts dealing with principles of experimental design will discuss and illustrate further the need for randomization. In general, the rule is quite simple: if for any reason you suspect that the *order* in which the data are observed will influence the results, use a random order.

A second example should suffice to drive this point home. Suppose a synthetic material is produced in a continuous strip 18 in. wide. We wish to test three different coatings to be applied to the material (coatings A, B, and C). A 6 × 6 in. test piece is the minimum size needed, and we decide to test five pieces for each coating. Figure 9.5 shows where we would cut a 30-in. length of the material.

The experimental conditions will be completely specified if we can label each piece as A, B, or C. One possibility is

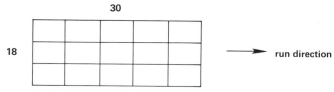

FIGURE 9.5. Experimental design of coatings A, B, and C as applied to test substrate.

If, however, there are any "machine effects" that exist in the direction transverse to the run, those effects will be buried in our data. To be safe, we assign the treatments (coatings) at random to the pieces. This is most easily accomplished by the use of a table of random numbers, as given in Table 9.3.

Suppose we start in the 21st row and the 16th column of paired digits. The entry there is 0.03. We will divide by 3 and consider the remainder, using this correspondence

$$0 \leftrightarrow \text{coating A}$$
$$1 \leftrightarrow \text{coating B}$$
$$2 \leftrightarrow \text{coating C}$$

and assign coatings as they occur from left to right and top to bottom.

	$\dfrac{3}{3}$	has a remainder	$0 \leftrightarrow A$

(next number down is) $\dfrac{73}{3}$ — $1 \leftrightarrow B$

$\dfrac{67}{3}$ — $1 \leftrightarrow B$

$\dfrac{35}{3}$ — $2 \leftrightarrow C$

$\dfrac{6}{3}$ — $0 \leftrightarrow A$

The first row than appears as

A   B   B   C   A

and we continue in this way until all 15 test pieces have been assigned a coating; obviously this could take more than 15 random numbers, since we need to end up with exactly five each of A, B, C. The complete layout if we continue down the column is

A	B	B	C	A
A	A	B	C	B
A	C	C	B	C

To summarize this rather lengthy discussion of stage I:

1.   Consider power in determining sample sizes.
2.   *Before* any data are gathered, consider all the factors you are not experimenting with. Control these by (1) holding them constant, or (2) randomization.

TABLE 9.3. Random Digits

```
8827453568    4217377248    2495206720    2017168112    1808544608
3163368288    6771200320    4605186432    2145441488    0852123376
6509292928    9784592768    5904044224    2642608768    5041245440
5405520320    9310618112    4345846720    9647445376    6444313792
7726936768    0200608214    0875678312    3149511232    1323423920
7454329984    1634068784    7033283392    2899511360    0889101016
1976160592    4141926336    2987933376    6570059328    2683603936
8476500416    2067174208    5031046592    2939419168    6872209792
9393789568    0049860980    4572305088    0082615473    2653819680
0102953700    7539961152    3252276096    2320012448    5526913088


8506892096    2398911040    9289076352    5351579136    6559663552
1963677568    4060682112    3707132672    7290077824    1925657664
0407428688    6645002752    8664238208    1414147600    6172497920
6149936896    9282970624    9189843712    8519211520    9058076032
1826571776    9803518592    0644211776    4533380160    1664351616
7678281408    1653483296    4692246528    2796940224    9799015168
0987936704    8903665536    0486826760    1130086976    3472770336
3229757120    2120099136    5319921536    0752330640    2409031776
5029854528    6688536000    9559689088    2722220160    2199248064
7931624960    6957654848    0018014677    3949648448    1412294384


7375441600    8113566656    5906870784    0349033644    5952707648
8722083968    0725953744    9646023168    7390187136    4997370944
3242450016    4549369344    9193632768    6722655616    5905700224
8750359808    2885287584    6108548992    3537057024    1994336016
6444603456    9102477184    3643529600    0691272800    5927429120
8049433408    0704816816    3461662112    3900380448    7481390848
0039197616    6988792448    7014618048    5750222464    9041247616
5986018944    4730745792    1156469568    0771851472    4525116928
5894052160    2163804192    7068476480    4409313728    9546811648
7226528576    5531182208    9712091648    1051866288    0694955536


6927327936    5040356928    4289203744    7201012800    6745452160
8518065856    5374753920    7863916800    6816289856    8923906048
3038119840    7593257792    0106397150    9854122112    5483463232
3704340928    3977352672    2219083072    3245263488    1998127120
4391003520    9819682432    2238167872    7395617216    5728724672
1187167520    0887652600    2154023456    8080161152    4307488512
9915713152    5629571136    6044421568    9288764416    3043302912
5489275840    0274190644    7467957184    1558737648    1211538288
5180731456    3890347872    5388786752    6652901312    3405244864
3532910976    7996668352    9395813248    9849554432    0080177667


2118666720    0529741092    5246587392    9498683264    8038592576
9455474944    1438618016    9093593472    5786803712    4239972960
7718744320    6168599680    7338003904    8986584192    9778566016
8847682432    4569541376    5622997696    9992531584    5054832064
3429422848    8446759424    9956924416    4085322208    4450763904
5705700928    3388957376    8639630336    4436893312    3428124992
9678940800    3628598400    3828830208    2655089568    5575999360
1495927136    8954353536    2164635872    4675298816    2241599304
8045382080    4715303616    1562257584    9767318528    4597882752
5773776576    9656708352    4602578432    1336028288    9629003264
```

Stage II

At this point, we collect our data. Before the actual ANOVA it may be desirable to test for the validity of the model and the necessary assumptions. There are four principal assumptions built into the model:

A(1)   The effects of the factors we are investigating are additive.

A(2)   The random error terms, ε_{ij}, are statistically independent.

A(3)   The error terms are normally distributed.

A(4)   The error variance is constant within each group.

Assumption A(4) is the easiest to deal with; we simply use either Hartley's test, Cochran's test, or Bartlett's test for the equality of the variances within each group. If we can accept the null hypothesis $H_0 : \sigma_1^2 = \sigma_2^2 = \cdots = \sigma_K^2 \ (= \sigma_\varepsilon^2)$ the analysis may proceed. Otherwise, we may wish to consider a transformation of the data to produce nearly equal variances. Several possibilities are discussed in Section 9.7.

NOTE.   There is a certain amount of caution with which one should approach data transformations, however. Suppose that the experimenter is satisfied that A(1) through A(3) hold for the data, but discovers that the groups have unequal variances. By transforming we may equalize the variances, but then the earlier assumptions, particularly the first and third, will not hold. A further difficulty lies in interpretation of the results; the original measurements y_{ij} were, presumably, measured in units which made sense, for example, calories per gram. How is one to interpret physically something such as $\cos[1/(1 + y_{ij})]$? Clearly, we can "play" a little too much with the data.

Very often, if it appears that departures from the assumptions are quite severe, it is advisable to abandon the standard ANOVA procedures. If we do so, we can usually find another method of analysis which is less heavily dependent on the assumptions, for example, a nonparametric ANOVA based on the ranks of the data.

If the variances are unequal and we decide to proceed with the analysis anyway, it is important to be aware of how our results may be influenced by unequal variances. A discussion of the effects of violating the various assumptions is given in Section 9.7.

In considering assumption A(1) we ought to consider alternatives to an additive model. Recall that our model was, ignoring subscripts,

$$y = \mu + A + \varepsilon$$

It is entirely possible that the correct model is, for example,

$$y = \mu A \varepsilon$$

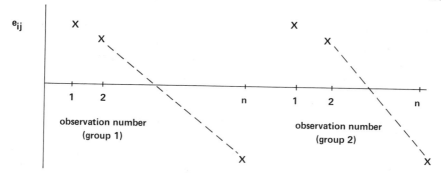

FIGURE 9.6. Residual error terms.

or perhaps

$$y = \mu^A \varepsilon$$

or, indeed, any function of μ, A, and ε. The additive model is, of course, the simplest one we can adopt; in practice it works reasonably well. It may be possible, if the experimenter strongly suspects that a nonadditive model is correct, to transform to an additive model. In the first model above we have, for example,

$$\log y = \log \mu + \log A + \log \varepsilon$$

while in the second model no such simple transformation can be found.* Note that the transformed model needs to satisfy the other assumptions; in the above model $y = \mu A \varepsilon$, we do not *want* ε to be normally distributed; we want its logarithm to be so.

In practice there is not much we can do even to detect nonadditivity in simple experiments of this type. In regression analysis, however, it is usually possible (see Chapter 10) to test the adequacy of the assumed linear model. Nonadditivity will tend to inflate the error term and thus we will usually understate the statistical significance of the F test.

The exact treatment of nonadditive models would take us too far afield here. A few references on the subject are given at the end of the chapter.

Independence of the $\{\varepsilon_{ij}\}$, assumption A(2), is rather critical. Suppose we had conducted an experiment in which two groups were compared. Afterwards, we decided to plot the residual error terms; we estimate ε_{ij} by $e_{ij} = y_{ij} - \bar{y}_i$. If the plot appeared as in Figure 9.6, we would have reason to be distressed.

---

*However, if we recall that A is a group effect and μ is the overall mean ($\mu_i = \mu + A_i$ in our original model) we could write an equivalent model $y_{ij} = (e^{\mu_i}) \cdot \varepsilon_{ij}$; then $\ln y_{ij} = \mu_i + \ln \varepsilon_{ij}$ is linear.

Something is plainly amiss here; in each group the first observation has the largest positive error, and the errors decrease regularly through the last observation. What probably happened is that an unsuspected factor was in operation which produced the trend lines in the two groups. If the observations were numbered in the order in which they were acquired, there may well be a "time" variable operating.

Very often, a lack of independence such as illustrated here is the result of one or more factors overlooked in the planning of the experiment. Occasionally, we ought to suspect human errors as well. It may be that a well-meaning individual assigned to gather the data feels, say, that the first few observations are higher than they should be. He might "edit" subsequent data to compensate for this. Unfortunately, this type of action would render the experiment meaningless, by and large.

To check the assumption of independence *completely* is not possible. If one looks at any set of numbers long enough—in particular, if one examines the residuals closely enough—some pattern will emerge. Expressed another way, we can probably devise, for a given set of data, a test that would reject the hypothesis that the data are independent. Usually, however, it is sufficient to plot the residuals in one or more ways and examine the plot. The text by Neter and Wasserman (7) contains an excellent treatment of residual plotting, which can be used to assess not only independence, but normality and constancy of variance as well.

Finally, assumption A(3), normality, is perhaps the least critical of the assumptions we make in ANOVA. From a theoretical point of view, we need the normality assumption to be able to assert that the ratio MS_A/MS_E follows an F distribution. Considerable research has indicated that, provided departure from normality is not too severe, the ratio behaves in a manner that is close enough to the F distribution that little damage is done in using the F tables. A complete discussion of the consequences of violating this assumption (and the others as well) is contained in Scheffe's book (2).

Stage III

The actual ANOVA was described earlier in this chapter. The computational steps are purely mechanical, and most computer installations have software available to generate the analysis of variance table.

Stage IV

In many ways, this final step is the most interesting and fruitful for the experimenter. The F ratio was calculated in stage III; it will turn out to be either "statistically significant" or "statistically nonsignificant" relative to the type I (α) error the experimenter preselected. Before proceeding, it is pertinent to comment upon the meaning of "significance" here. To say that our results are statistically significant means merely that the result of some rather complex arithmetic produced a number higher than a predetermined cutoff value. *Whether this is of any practical importance is up to the experimenter to determine.* Consider for the moment the algebraic expression for the calculated F ratio:

$$F = \frac{MS_A}{MS_E} = \frac{SS_A/(K-1)}{SS_E/K(n-1)}$$

$$= \frac{K(n-1)SS_A}{(K-1)SS_E}$$

If K is fixed, then a very large value of n will most likely produce a very large F, leading us to conclude that there are significant differences among groups. While the conclusion is proper, we arrived at it by conducting (via our choice of large n) a very "sensitive" experiment, that is, one that would distinguish very small actual differences. If, for instance, one is comparing five resins and finds, with 100 observations for each, that F is barely significant, in all likelihood the actual differences are so small that, in application, one would select the least expensive rather than the "statistically" best.

In order to do *any* postanalyses, the experiment needs to result in findings that are *both* statistically significant *and* practically important. To analyze further when these conditions do not hold is at best counterproductive.

Often an experimenter embarks upon a study anticipating significant results, and finds there are none. It is a temptation to backtrack at this point and say, for example, "Well, if we only consider groups 2 and 5, they would probably come out different," or similar things. To the author, this is very much like betting on a toss of a coin, losing, and then asking for "best two out of three." Once you have gone carefully through the design of the experiment, monitored the data gathering, and checked the assumptions, you really ought to live with the results. It *is* true that, had you analyzed only two groups, you may have come to a different conclusion. The point, however, is that such a statement refers to an experiment you did not perform. The one you *did* perform said the groups were alike—or at least that they are sufficiently close to be considered alike. About the most you can say is that the experiment you completed contains some hints about how you ought to plan your next experiment. It is, in the author's opinion, a serious misuse of statistical techniques to "massage" the data over and over until some positive result occurs. One approach to statistical analysis which ought to be avoided is this "shotgun" approach, which says "gather data, and then do whatever you can think of until something good happens." It is far better to plan a series of experiments, the first for more or less general directions (typically this would have many groups and only a few observations in each), and subsequent ones for specific goals.

At this point, we descend from the soapbox and assume that there is legitimate reason to conduct further analysis of the data. We have *statistically significant* results, and there is the potential for valuable practical conclusions to be drawn.

9.6. A POSTERIORI ANALYSES

The statistics available for comparing groups consist of (i) the individual group means, $\overline{Y}_1, \overline{Y}_2, \ldots, \overline{Y}_K$, based on $n_1, n_2, \ldots, n_K$ observations, respectively, and (ii) the value of MS_E, which is an estimate of σ_ε^2, based on $N - K$ df ($N = \Sigma n_i$). To test for differences between two particular groups (say groups 1 and 2) we may develop a t test similar to that in Chapter 6, in the following manner. Under the assumptions A(1)–A(4), we find

$$\overline{Y}_1 \text{ is normal,} \qquad \text{mean} = \mu_1, \qquad \text{variance} = \frac{\sigma_\varepsilon^2}{n_1}$$

$$\overline{Y}_2 \text{ is normal,} \qquad \text{mean} = \mu_2, \qquad \text{variance} = \frac{\sigma_\varepsilon^2}{n_2}$$

Therefore $\overline{Y}_1 - \overline{Y}_2$ is normal with mean $\mu_1 - \mu_2$ and variance $\sigma_\varepsilon^2(1/n_1 + 1/n_2)$. Then

$$Z = \left\{ (\overline{Y}_1 - \overline{Y}_2) - (\mu_1 - \mu_2) \div \sqrt{\sigma_\varepsilon^2 \left(\frac{1}{n_1} + \frac{1}{n_2} \right)} \right\}$$

is a standard normal variable. If we replace σ_ε^2 by its estimate MS_E we obtain Student's t with $(N - K)$ df. To test the hypothesis $\mu_1 - \mu_2 = 0$, we calculate

$$t = \frac{\overline{Y}_1 - \overline{Y}_2}{\sqrt{MS_E} \sqrt{\dfrac{1}{n_1} + \dfrac{1}{n_2}}}$$

and compare the calculated t with the appropriate percentile of Student's t with $(N - K)$ df.

Example 9.5. Let us assume, for illustrative purposes, that we had concluded that the line means in Example 9.1 are different and that we wish to compare lines 1 and 3. To this end we have

$$MSE = 8.77, \quad n_1 = 3, \quad \bar{y}_1 = 33, \quad n_3 = 4, \quad \bar{y}_3 = 25.5$$

and want to test the hypothesis set

$$H_0 : \mu_1 = \mu_3 \qquad H_1 : \mu_1 \neq \mu_3$$

The appropriate decision rule, for $\alpha = 0.05$, is to reject H_0 if the absolute value of the calculated t exceeds $t_{N-K, \alpha/2} = t_{8, 0.025} = 2.306$.

For the data above,

$$|t_{calc}| = \left| \frac{33 - 25.5}{\sqrt{8.77} \sqrt{\frac{1}{3} + \frac{1}{4}}} \right|$$

$$= \left| \frac{7.5}{2.262} \right| = 3.316$$

Thus, since postanalysis was indicated, we would reject H_0 and conclude that lines 1 and 3 differ.

Two points should be stressed with respect to this example. First, postanalysis was indicated by the results of Example 9.1 and, secondly, we are letting the *data indicate* the comparison to be made. Thus, the stated α of .05 is not really correct.

If we denote by $t_{N-K, \alpha/2}$, the value which the calculated t must exceed (in absolute value, since usually these comparisons are two sided), we can state for any two groups i and j how large a difference would cause us to conclude that i and j differ; this is simply

$$D_{ij}(\max) = \sqrt{MS_E} \sqrt{\frac{1}{n_i} + \frac{1}{n_j}} \, t_{N-K, \alpha/2} \tag{9.13}$$

and, if all groups are the same size, n,

$$D(\max) = \left(\sqrt{\frac{2MS_E}{n}} \right) t_{K(n-1), \alpha/2} \tag{9.14}$$

Simple comparisons of this kind very often are all that need be done in the way of postanalysis. We make two observations before continuing.

1. Strictly speaking, the pairs of groups we wish to compare should be selected prior to gathering any data. Otherwise, we are letting the data lead us (cf. earlier comments).
2. Error probabilities can very easily become intolerable if we make too many comparisons. The basic problem is that, each time we

make a comparison and draw a conclusion, we risk either a type I or type II error; the probability of a long series of error-free conclusions decreases rapidly (recall the multiplication law for probabilities) as the number of conclusions increases. In experiments with only three or four groups, this is seldom a real problem, and the above method is usually sufficient.

Student's t may also be used to compare subsets of the set of groups. Suppose, for example, that prior to performing the experiment, the experimenter feels that groups 1 and 2 are probably alike, as are 3, 4, and 5, but the set $\{1, 2\}$ may be different from the set $\{3, 4, 5\}$. The weighted average for the first set is $(n_1\mu_1 + n_2\mu_2)/(n_1 + n_2)$ and for the second $(n_3\mu_3 + n_4\mu_4 + n_5\mu_5)/(n_3 + n_4 + n_5)$, and our hypothesis is that these are equal. We consider the variable

$$U = \frac{n_1\overline{Y}_1 + n_2\overline{Y}_2}{n_1 + n_2} - \frac{n_3\overline{Y}_3 + n_4\overline{Y}_4 + n_5\overline{Y}_5}{n_3 + n_4 + n_5}$$

Under the hypothesis just stated, U will have a normal distribution with mean zero. Its variance will be

$$\sigma_U^2 = \left\{ \frac{n_1^2}{(n_1 + n_2)^2}\left(\frac{1}{n_1}\right) + \frac{n_2^2}{(n_1 + n_2)^2}\left(\frac{1}{n_2}\right) + \frac{n_3^2}{(n_3 + n_4 - n_5)^2}\left(\frac{1}{n_3}\right) \right.$$
$$\left. + \cdots + \frac{n_5^2}{(n_3 + n_4 + n_5)^2}\left(\frac{1}{n_5}\right) \right\} \cdot \sigma_\varepsilon^2$$

We replace σ_ε^2 by its estimate MSE to conduct a t test.

Note that U was defined as a linear combination of the group means,

$$U = \Sigma a_i \overline{Y}_i$$

and the sum of the coefficients in that combination, Σa_i, is zero. Such linear combinations are called *contrasts*; in general, if

$$U = \Sigma C_i \overline{Y}_i, \quad \text{with} \quad \Sigma C_i = 0$$

then

$$E(U) = \Sigma C_i \mu_i \tag{9.15}$$

$$\text{var}(U) = \left(\Sigma \frac{C_i^2}{n_i}\right)\sigma_\varepsilon^2 \tag{9.16}$$

We can obviously define any contrast U (or set of contrasts) we choose prior to the experiment, and employ Student's t to test $H_0: E(U) = C_0$, where C_0 is some particular value of interest. Usually $C_0 = 0$.

Numerous other techniques are employed in practice in the postanalysis of ANOVA data. Among these we mention: (1) Newman–Keuls analysis, used to rank groups in ascending order (this is a variation of Student's t, employing a "Studentized range" statistic); (2) Scheffe's method, used to determine confidence regions for all possible contrasts; and (3) Tukey's method, used to perform all t tests for differences "simultaneously."

In more complex experiments, particularly experiments in which K is rather large, the extra labor required to employ these techniques is worthwhile. The books by Guenther (8) and Scheffe (2) describe them in considerable detail, and should be consulted as needed.

Random Effects Model

The model equation we have heretofore employed in ANOVA is

$$y_{ij} = \mu + A_i + \varepsilon_{ij}$$

The setting in which we developed this model led to an interpretation of the A_i as "group effects." Our interest in conducting the experiment was founded in a curiosity about the different groups; which was best, were they all alike, and so on.

Another type of experimentation which is, at least mechanically, quite similar, is based on a different viewpoint. That viewpoint is that *the groups from which we collect data are merely a random selection from a larger set of groups from which we might have chosen.* This viewpoint often arises in the following setting. Suppose that we manufacture a product in well-defined "batches," for example, 100-lb drums. We wish to measure or control some chemical property of the product. From experience we recognize that the measurement of the property will vary, within any given batch, from sample to sample. There will also be variation overall from one batch to another. If we identify "batches" with "groups," (1) would appear to be a reasonable model; μ is the average value of the measurement, A_i the effect for batch i, and ε_{ij} the error associated with the jth determination in the ith batch. There is, however, one important difference. We cannot (unless we continue the experiment indefinitely) sample from *every* batch. Furthermore, our interest usually centers on determining the overall variability in the product, rather than specific comparisons of some fixed number of batches.

To account for these differences, we introduce a slightly different model. Suppose we have a large population of batches and we select, at random, K of these. In batch i we make n_i determinations y_{ij}. We write

$$y_{ij} = \mu + \alpha_i + \varepsilon_{ij} \qquad (9.17)$$

In this equation, μ and ε_{ij} have the same interpretation as before, but we write α_i, rather than A_i, to represent the random additive component which results from our having included that particular batch in the experiment. That is, α_i is a random variable chosen from a population of α's, each α being associated with its own batch. The usual assumptions about the $\{\alpha_i\}$ are that (i) they are normally distributed with zero means, (ii) they possess constant variance σ_A^2, (iii) they are uncorrelated with one another and with the $\{\varepsilon_{ij}\}$.

The purposes of the ANOVA are now (i) to estimate σ_ε^2, (ii) to estimate σ_A^2, and, usually, (iii) to estimate $\text{var}(y_{ij}) = \sigma_A^2 + \sigma_\varepsilon^2$, the variability of the individual measurements. To distinguish the model from our earlier one, we term it a random effects or type II model (the earlier one is then called fixed effects or type I).

For single-factor experiments the computations in the ANOVA table are identical in both models. Formally, we may consider the F ratio that is calculated to be a test statistic for the hypothesis $H_0 : \sigma_A^2 = 0$. In practice we would certainly expect to reject this hypothesis, since in most testing situations like the one we have been describing the variation between batches is considerably larger than the variation within batches.

When the number of observations per batch is constant ($n_1 = n_2 = \cdots = n_K = n$) our best estimates of σ_ε^2 and σ_A^2 are

$$\hat\sigma_\varepsilon^2 = MS_E$$
$$\hat\sigma_A^2 = (MS_A - MS_E)/n$$

The second formula results since MSA includes both error variation and batch-to-batch variation.

If the second of these is negative, we replace it by zero and conclude that batch-to-batch variation is essentially zero. The best estimate of $\text{var}(Y_{ij})$ is the sum $MS_E + (MS_A - MS_E)/n$. Confidence limits for σ_ε^2 and for the ratio $\sigma_A^2/\sigma_\varepsilon^2$ are discussed in a number of the references.

9.7. VIOLATIONS OF ASSUMPTIONS AND POSSIBLE CORRECTIONS

As a final note, we briefly discuss what effects violations of assumptions have on the analysis and corresponding conclusions. To check for possible violations of assumptions, you may use residual plots and formal tests when the appropriate information is available. We recommend that you consult either Draper and Smith (9) or Neter and Wasserman (7) for a thorough discussion of these methods.

Recall that our basic assumptions underlying the ANOVA procedure are:

1. Constant (homogeneity) variances.
2. Independence of error terms.
3. Normal distributions for the error terms.

Our discussion of the effects of departures and possible corrections is a summary of the more complete presentation in Chapter 15 of Neter and Wasserman (7).

The criticality of the assumption of homogeneity of variances is dependent on whether or not we have equal sample sizes. It turns out that if the sample sizes are equal, the procedures utilizing the F test are only slightly affected with the true α being only slightly larger than the indicated one when the variances differ. With respect to posterior analyses, multiple comparisons are only mildly affected if the sample sizes are equal, but individual (pairwise) comparisons may be affected in such a way that the actual confidence levels can differ significantly from the stated ones. If it is determined that the variances are not constant, various transformations are available to help to achieve homogeneity. The appropriate transformation is dependent on the type of nonconstancy. Four of the more common transformations are summarized below.

i.   If $\sigma_j^2 = K\mu_j$, then consider either

$$Y' = \sqrt{Y} \quad \text{or} \quad Y' = \sqrt{Y} + \sqrt{Y + 1}$$

ii. If $\sigma_j = K\mu_j$, then consider the logarithmic transformation,

$$Y' = \log Y$$

iii. If $\sigma_j = K\mu_j^2$, then the reciprocal transformation may be useful; that is,

$$Y' = \frac{1}{Y}$$

iv. If the response is a proportion, that is, $Y_{ij} = X_{ij}/n_{ij}$, then consider using

$$Y_{ij}' = 2 \arcsin\sqrt{Y_{ij}}$$

If the random effects model is being used, lack of homogeneity has a much greater effect on inferences even when the sample sizes are equal.

If the error terms are dependent, inferences based on the ANOVA, procedures may be affected seriously. This is true for both the fixed effects and random effects models. The best way to avoid problems of this type is in the initial design of the experiment.

In terms of the normality assumption slight departures from normality have little effect with respect to the fixed effects model. Surprisingly, it turns out that peakedness of the distribution (deviations in either direction from the normal) is more important than skewness. Usually the true value for α is slightly higher than the stated value if the distributions are non-normal.

For the random effects model we are interested in variances rather than means, and, hence, lack of normality is more critical. The actual confidence levels of any statements about the variances may deviate significantly from those stated.

9.8. SUMMARY

In this chapter, we have presented a summary of the classical two-sample procedures for comparing means and variances. In addition, the classical Behrens–Fisher problem (testing two means when the variances are unknown and unequal) was discussed briefly. For the Behrens–Fisher situation, no usable exact procedure exists but some approximate techniques are available; namely, an approximation due to Welch (4) which adjusts the degrees of freedom for the t statistic.

The majority of the chapter deals with comparisons involving more than two means with a section on the comparison of several variances. For making inferences about the equality of a set of means from several populations, we generally use a procedure known as analysis of variance (ANOVA). This technique evaluates the hypothesis of equal means in a fixed effects model by comparing the variability between the corresponding sample means to that which is attributed to error. These comparisons are based on an F test which assumes: (i) random samples; (ii) underlying normal distributions; (iii) constant variances; and (iv) independence of the error terms.

Tests for evaluating the assumption of equal variances are given in Section 9.4. All of these assume normal distributions with Hartley's and Cochran's adding an assumption of equal sample sizes.

The basic model that we have considered assumes only one factor and can be expressed as

$$y_{ij} = \mu + A_i + \varepsilon_{ij}$$

when the effects are fixed; that is, the levels in the experiment are the only ones of interest. For random effects (the levels used represent only a sample of all possible ones), the model becomes

$$y_{ij} = \mu + \alpha_i + \varepsilon_{ij}$$

where α_i is now a random variable. The primary differences in the two models lie in the fact that for fixed effects, we are interested in isolating where differences are occurring while in the random case, our objective is to estimate the contributions to the overall variability. Multiple-factor models, utilizing the same basic approach, are discussed in the chapter on experimental designs.

If the ANOVA F test leads to rejection of the null hypothesis, we usually proceed to a posteriori analysis. For fixed effects, this implies using one of the multiple comparison procedures discussed in Section 9.6, such as Newman–Keuls. Posterior analysis in the random effects model attempts to estimate the various components of variation.

The chapter concluded with a brief discussion of the effects on our inferences of departures from the assumptions listed above. More detailed presentations of these effects can be found in the references.

REFERENCES

1. Fisher, R. A., *Contributions to Mathematical Statistics*, Wiley, New York, 1950.
2. Scheffe, H., *The Analysis of Variance*, Wiley, New York, 1959.
3. Wald, A., *Selected Papers in Statistics and Probability*, McGraw-Hill, New York, 1955.
4. Welch, B. L., *Biometrika*, Vol. 29, 1937, pp. 350–362.
5. Linnick, Y. V., *Sankyā*, Series A, Vol. 25, Pt. 1, 1963, pp. 377–380.
6. Linnick, Y. V., *Soviet Mathematics, Doklady*, Vol. 4, Pts. 1 and 2, 1963, pp. 359–362, 580–582, 644–646, 1360–1361; Vol. 5, Pt. 1, 1964, pp. 570–572.
7. Neter, J., and Wasserman, W., *Applied Linear Statistical Models*, Irwin, Homewood, IL, 1974.
8. Guenther, W. C., *Analysis of Variance*, Prentice-Hall, Englewood Cliffs, NJ, 1964.
9. Draper, N. R., and Smith, H., *Applied Regression Analysis*, Wiley, New York, 1981.
10. Davies, O. L., *Design and Analysis of Industrial Experiments*, 2nd ed., Oliver and Boyd, London, 1963.
11. Kendall, M. G., and Stuart, A., *The Advanced Theory of Statistics*, Vol. 2, Chas.-Griffin & Co., London, 1946.
12. Lehman, E. L., *Testing Statistical Hypotheses*, Wiley, New York, 1959.

10

CORRELATION AND REGRESSION

H. EARL HILL

Lord Corporation
Erie, Pennsylvania

10.1. CORRELATION COEFFICIENT

Regression analysis is a powerful statistical method used to deal with the general problem of finding a relationship and its statistical meaning between two or more variables. The technique is a prime tool in measuring the degree of association between variables and thus estimating the effect of one variable on another, relating the values from one test method to another, and so on. Very frequently, a cause/effect relationship is suspected and a dependent (observed response) variable (Y) is being estimated from an independent variable (X) at either selected or controlled levels. In other cases, a dependent, random variable is being measured as a function of an independent variable. Analysis of such data is called regression analysis, and the line obtained is referred to as the line of regression, or as commonly determined the "least-squares" line. Broadly speaking, correlation attempts to determine how well a linear or other equation describes or explains the relationships among variables.

Correlation is concerned with five general areas:

1. Finding a quantitative measure of the degree of relationship between the variables independent of the terms in which the variables were expressed.

2. Determining a regression equation that allows estimation of one variable from another.

3. Determining a measure of divergence of actual values from computed values, also referred to as standard error of estimate and similar to a standard deviation.
4. Tests of the significance of the parameters of the regression equation.
5. Estimation of the confidence limits of the parameters of the regression equation.

When only two variables are concerned, we refer to this as simple correlation and regression analysis. Analysis of greater than two variables is correspondingly referred to as multiple correlation and regression. The first step in investigating the degree of association between two variables is to construct a scatter diagram of the sets (X_i, Y_i) of data. If they follow closely along a straight line with positive slope, a high positive correlation exists between the two variables (for negative slope, a high negative correlation). Numerically, correlation varies from $+1.0$ for a perfect direct relationship, to -1.0, for a perfect inverse relationship. For a strictly random pattern, there is zero correlation (no clear linear relationship exists between X and Y). The value of r is a measure of the degree to which changes in the level of one variable are associated with changes in the level of the other variable. We distinguish between the sample correlation coefficient r, and the population coefficient ρ. The sample correlation coefficient r is dimensionless and for nonzero values, implies that there is an association, however, not necessarily cause/effect, between the values of X_i and Y_i. A high correlation coefficient merely indicates a linear relationship between the two variables.

The correlation coefficient can be mathematically derived from various equations, using many of the same values determined during the process of calculation of the least-squares line, to be detailed later. The correlation coefficient is introduced at this time rather than subsequent to the treatment of regression for the least-squares line, since among other things it may be used in alternate formulas for calculating the reliability of parameter estimates for the best-fit line. The correlation coefficient r is calculated by the following ratio:

$$r = \frac{\Sigma xy}{\left(\Sigma x^2 \Sigma y^2\right)^{1/2}} \tag{10.1}$$

where

$$x = X_i - \overline{X} \quad \text{and} \quad y = Y_i - \overline{Y}$$

For calculating purposes, the denominator terms in equation (10.1) are determined by

$$\Sigma\left(X_i - \overline{X}\right)^2 = \Sigma X_i^2 - \frac{\left(\Sigma X_i\right)^2}{n} \tag{10.2}$$

and

$$\Sigma(Y_i - \overline{Y})^2 = \Sigma Y_i^2 - \frac{(\Sigma Y_i)^2}{n} \tag{10.3}$$

and for the numerator

$$\Sigma(X_i - \overline{X})(Y_i - \overline{Y}) = \Sigma X_i Y_i - \frac{\Sigma X_i Y_i}{n} \tag{10.4}$$

Equations for calculations involving grouped data and coded data are available (1).

The correlation coefficient r_{xy} measures association between X and Y. The regression line slope coefficient b_1 measures the size of the change in Y which can be predicted when a unit change is made in X. Although providing different interpretations, they are closely related.

In this regard, it should be noted that

$$(n-1)s_y^2 = \Sigma(Y_i - \overline{Y})^2$$
$$(n-1)s_x^2 = \Sigma(X_i - \overline{X})^2$$

and that

$$b_1 = \frac{s_y}{s_x}r_{xy}$$

and also that

$$s_{y\cdot x}^2 = s_y^2(1 - r^2)$$

where $s_{y\cdot x}$ is the standard error of estimate of Y on X, s_y is the variance of Y, and so on.

The significance of the correlation coefficient r in indicating the extent to which two variables are related is as follows. If r is found to be nearly $+1$, it means that as one variable increases the other does likewise, and correspondingly for a value of r nearly -1, one value increases while the other decreases. In practice we will generally not get a value of $r = 1$ in view of the error contribution. The question to be answered is then, for a certain determined value of r for a small sample size, what is the probability of obtaining a value of r equal to this purely by chance observation? The answer is obtained by Appendix Table A.6 for a given number of n pairs of observations, degrees of freedom $\nu = (n-2)$. Common practice is to set a significance level of $P = 0.05$ as a rule of thumb and if the table value is exceeded by the calculated value of r, then it is reasonably certain 1 out of 20 times that a true correlation exists.

Too much significance should not be attached to r; two values of, say, 0.4 and 0.8, mean that there are two positive correlations, one stronger than the other but not necessarily twice as good. However, for $r = 0.8$ it can be said that $(100r^2)\%$ or 64% of the variation in the random variable Y is accounted for by differences in the variable X. To claim good correlation exists between two methods, a value of at least 0.9 for r is required, since $r^2 = 0.81$ indicates that only 81% of the variation in Y can be ascribed to X. If a value of 0.7 or less is obtained for r, the degree of association is considered to be inadequate for assuming that the two procedures are measuring the same property. It should be noted that a high value for r does not guarantee that values of Y can be predicted precisely from values of X. The value of r is strongly dependent on sample size. Even moderately large values of r with small sample sizes cannot necessarily be considered as real (2).

The correlation coefficient serves well to measure the correlation between X and Y if the true regression curves are lines, that is if r is the linear correlation coefficient. However, for nonlinear cases, for example, where the regression curves are of higher powers such as quadratic, r may not be at all applicable. For these cases, for example, a value of $r = 0$ implies a lack of linearity but not necessarily a lack of any kind of association. Furthermore, a significant value of r does not necessarily imply that there is a cause/effect relationship between X and Y. A suspected casual relationship needs further investigation by scientific experiment to confirm (or deny) the cause/effect mechanism.

10.2. THE LEAST-SQUARES REGRESSION LINE

For purposes of illustration, we assume that we have two test methods of measuring gloss (2). We want to find out whether the two methods are comparable. The results of 10 panels measured by each of the two methods are given in Table 10.1, for the (X_i, Y_i) pairs of measurements. The line we are seeking to represent the data in Figure 10.1 is called the regression line, the least-squares line, or the best-fit line. In this case, where the line will be used to estimate the value of method B, it is referred to as the line of regression of Y on X. Here we are considering that method A is the independent variable and method B is the dependent variable. Regression analysis fits a line or function to data by minimizing the sum of the squares of the vertical distances from the data points to the fitted line. This is a measure of effectiveness of the best *fit* between the line and the data. However, it must be kept in mind that we do not know if this line is the *best* or only line representing the data.

This best-fit line is the result of an equation developed via regression analysis to express the (linear) relationship between the two variables. A

TABLE 10.1. Gloss of Test Samples as Measured by Two Different Test Methods

Sample	(X_i) Method A	(Y_i) Method B
1	20	40
2	10	30
3	40	60
4	30	49
5	25	45
6	15	35
7	60	75
8	55	75
9	50	70
10	40	58

straight line plot is a consequence of a linear regression model. However, the relationship is based on tests for fit of the data to the linear model. The factors influencing the location of the line of regression (2) are shown in Figure 10.2.

If the variables are linearly related, then the response variable can be said to be a function of the input variable and the line of regression for the population of X's and Y's for the preceding case is

$$Y = \beta_0 + \beta_1 X \qquad (10.5)$$

For any single point X_i, Y_i there will be an error contribution where

$$Y_i = \beta_0 + \beta_1 X_i + \varepsilon \qquad (10.6)$$

The regression equation for the sample data is

$$\hat{Y} = b_0 + b_1 X_i \qquad (10.7)$$

The sample coefficients are defined as the slope b_1 and the intercept b_0. The value of b_1 is obtained from the normal equation (3)

$$b_1 = \frac{n\Sigma x_i y_i - (\Sigma x_i)(\Sigma y_i)}{n\Sigma x_i^2 - (\Sigma x_i)^2} \qquad (10.8)$$

and the value of b_0 by

$$b_0 = \frac{\Sigma y_i}{n} - b_1 \frac{\Sigma x_i}{n} = \bar{y} - b_1 \bar{x} \qquad (10.9)$$

Using the data from Table 10.2 the coefficient values by substitution are

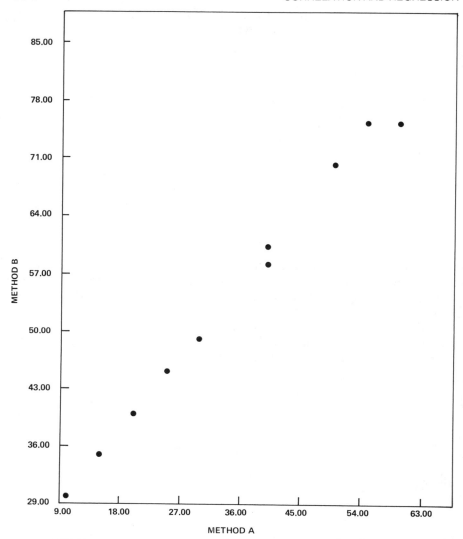

FIGURE 10.1. Specular gloss as measured by two different test methods.

$b_1 = 0.9499$ and $b_0 = 20.929$. The calculated least-squares regression line is then $\hat{Y} = 20.929 + 0.9499X$ (see Figure 10.3).

The correlation coefficient is $r = 0.9965$ from equation (10.1), suggesting a strong positive linear correlation between the two methods of gloss measurement. The significance of the value of r for this case is tested by setting up the null hypothesis that its value is actually zero and that the value calculated occurred by chance alone. The number of degrees of freedom ν for the data are $10 - 2$ or 8 since the data were used to estimate two parameters in the regression equation. Appendix Table A.6 is used to

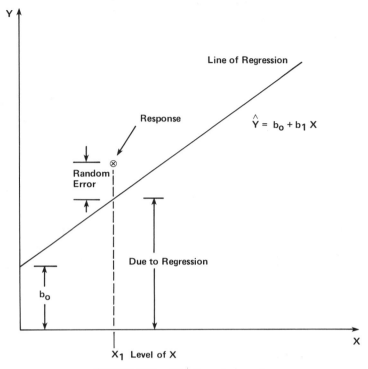

FIGURE 10.2. The line of regression.

determine the significance of r. Since the value of r with $v = 8$ at the 10, 5, 2, and 1% significance levels is not nearly exceeded by any of the table values, there is a very high indication that methods A and B are correlated.

Confidence Limits

Using the calculated regression line, values of the response variable $\hat{Y}$ may be predicted for any desired value of the input variable X_0.

Just as the sample standard deviation measures the variation of scatter about the arithmetic mean, the sample standard error of estimate is a measure of variation or scatter about a regression line. It is merely the root mean square of the Y deviations about the fitted curve, and as such is a measure of the precision of a least-squares fit. The Y deviations, d ($d = Y$ actual minus Y theoretical), are often called the *residuals* around the regression line, and $s_{y \cdot x}$ is referred to as the standard deviation of residuals:

$$s_{y \cdot x} = \frac{\Sigma d^2}{n - 2}$$

The standard error of estimate for small samples calculated from the

TABLE 10.2. Paint Gloss Rating by Two Different Rating Scales

	Gloss Method A X_i	Gloss Method B Y_i	X_iY_i	X_i^2	Y_i^2	x^2	xy	y^2
	20	40	800	400	1,600	210.25	198.65	187.69
	10	30	300	100	900	600.25	580.65	561.69
	40	60	2,400	1,600	3,600	30.25	34.65	39.69
	30	49	1,470	900	2,401	20.25	21.15	22.09
	25	45	1,125	625	2,025	90.25	82.65	75.69
	15	35	525	225	1,225	380.25	364.25	349.69
	60	75	4,500	3,600	5,625	650.25	543.15	453.69
	55	75	4,125	3,025	5,625	420.25	436.65	453.69
	50	70	3,500	2,500	4,900	240.25	252.65	265.69
	40	58	2,320	1,600	3,364	30.25	23.65	18.49
Sum	345	537	21,065	14,575	31,265	$\Sigma x^2 = 2,672.5$	$\Sigma xy = 2,538.5$	$\Sigma y^2 = 2,428.1$
Mean	$\overline{X}_i = 34.5$	$\overline{Y}_i = 53.7$						

$$\overline{X_iY_i} = 2,106.5 \qquad \overline{X_i^2} = 1,457.5 \qquad \overline{Y_i^2} = 3,126.5$$

$$(\overline{X}_i)^2 = 1,190.25 \qquad \overline{X}_i\overline{Y}_i = 1,852.65 \qquad (\overline{Y}_i)^2 = 2,883.69$$

$$\Sigma(X_i - \overline{X})^2 = 2,672.5 \qquad \Sigma(Y_i - \overline{Y})^2 = 2,428.1 \qquad n = 10$$

$$x = X_i - \overline{X}$$
$$y = Y_i - \overline{Y}$$

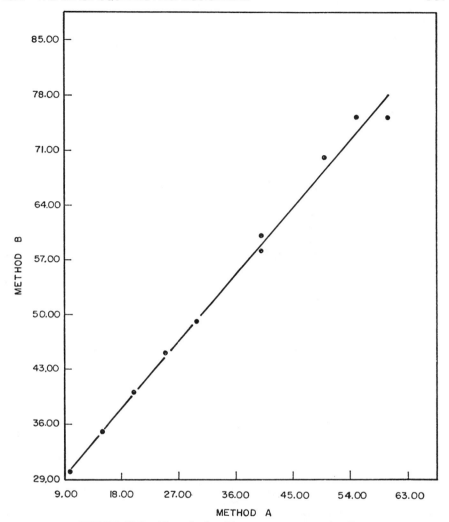

FIGURE 10.3. The calculated least-squares regression line.

regression line of Y on X is given by

$$(n - 2)s_{y \cdot x}^2 = \Sigma(Y_i - \hat{Y}_i)^2 \qquad (10.10)$$

or

$$s_{y \cdot x} = \sqrt{\frac{\Sigma(Y_i - \hat{Y}_i)^2}{n - 2}} \qquad (10.11)$$

Also

$$\Sigma(Y_i - \hat{Y}_i)^2 = \Sigma Y_i^2 - b_0 \Sigma Y_i - b_1 \Sigma X_i Y_i \tag{10.12}$$

and

$$\Sigma(X_i - \overline{X})^2 = \Sigma X_i^2 - \frac{(\Sigma X_i)^2}{n} \tag{10.13}$$

The sample standard deviation of Y is given by

$$s_y = \sqrt{\frac{\Sigma Y^2 - (\Sigma Y)^2/n}{n - 1}} \tag{10.14}$$

An alternate expression for the standard error of estimate, useful for checking purposes, is

$$s_{y \cdot x} = s_y \sqrt{(1 - r^2)\frac{n - 1}{n - 2}} \tag{10.15}$$

Using equations (10.11) and (10.12), the value of $s_{y \cdot x}$ for the gloss data is 1.4526. The confidence limits of estimate for single values of x_0 are given by

$$\hat{Y} \pm t_{n-2, \, \alpha/2} s_{y \cdot x} \sqrt{\frac{1}{n} + \frac{(X_0 - \overline{X})^2}{\Sigma(X_i - \overline{X})^2}} \tag{10.16}$$

To place confidence limits, using Student's t for small samples, on the estimated Y at the 5% level ($\alpha = 5$) with $\overline{X} = 34.5$ at, for example, $X_0 = 10$, substitution is made in equation (10.16). We then find that $\hat{Y} + 1.904$ defines the confidence limits at $X_0 = 10$. By substituting in the "best-fit" equation for the line we find that, again at $X_0 = 10$, our $\hat{Y}$ is 30.428, and the true value of $\hat{Y}$ (at an $\alpha = 5$ risk level) lies between 28.524 and 32.332. Other values of the confidence limit for the response variable $\hat{Y}$ are determined in a similar manner. For example, for an X_0 of 60, the $\hat{Y}$ is $75.965 - 79.875$. Calculating a number of these values falling about the best-fit line and connecting them by a smooth curve then defines the confidence bands as shown in Figure 10.4. Ideally, the complete detailed operation is done by computer. However, where one is unavailable, the approximate confidence band can be easily found by determining the confidence limits at the mean and those close to the extremes of the X_0 values.

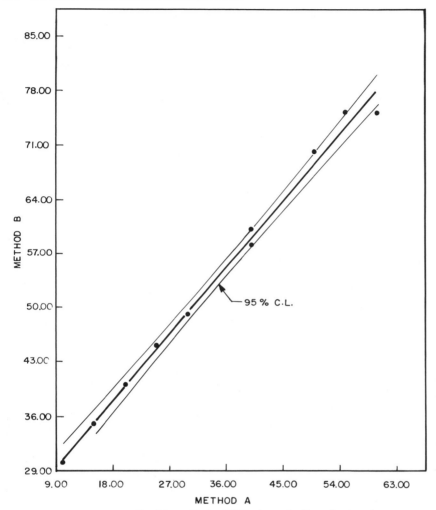

FIGURE 10.4. Confidence limits (95%) about the line of regression.

In this section, as well as in the following sections, the distinction between small and large sample size should be especially noted particularly in referring to equations for calculating the various sample parameters and coefficients. This is also important in the relationships between coefficients such as that given in equation (10.15). For example, where n is large, equation (10.15) reduces to $s_{y \cdot x}^2 = s_y^2(1 - r^2)$. Furthermore, the confidence limits of estimate are based on Student's t, whereas where n is large, bands one, two, and three times the standard error, (approached as a limit) measured on each side of the line of regression, will contain 68%, 95%, and 99.7% of the sample points.

Standard Error of the Slope, Confidence Limits on b_1

To place confidence limits about the slope, we first calculate the standard error (effectively the estimated standard deviation) of the slope using Equation (10.17).

$$s_{b_1} = \frac{s_{y \cdot x}}{\left\{ \Sigma \left(X_i - \overline{X} \right)^2 \right\}^{1/2}} = \frac{s_{y \cdot x}}{\left\{ \Sigma X_i^2 - \left(\Sigma X_i^2 / n \right) \right\}^{1/2}} \tag{10.17}$$

An alternate equation for the standard error of the slope, useful for checking purposes, is

$$s_{b_1} = \frac{b_1 \sqrt{\dfrac{1}{r^2} - 1}}{\sqrt{n - 2}} \tag{10.18}$$

The value of s_{b_1} from equation (10.17) is found to be 0.0280994. The 95% confidence limits ($\alpha/2 = 0.025$, $\nu = 8$, $t = 2.306$) about the true line of regression slope coefficient β_1 are given by equation (10.19)

$$\beta_1 = b_1 \pm t_{n-2, \alpha/2} s_{b_1} \tag{10.19}$$

and again, for the data in Table 10.2, by substitution and for a value of $s_{b_1} = 0.0281$

$$\beta_1 = 0.94985 \pm 2.306(0.0281) \quad \text{or} \quad 0.8851 \leq \beta_1 \leq 1.015$$

These values are substituted for the slope parameter in the regression equation and, by substitution of selected $X_i Y_i$ values, used to arrive at values for plotting the confidence limit (of the slope) lines.

It is also possible, if desired, to test the significance of the value for β_1. This is done by testing the null hypothesis that the slope could be zero, or that there is no relationship between the two methods. The calculated value for Student's t (b_1/s_{b_1}) is 33.8035 which is larger than $t = 2.306$ (one-sided). Therefore, the hypothesis that a linear relationship between Y and X might *not* exist is rejected at this confidence level.

Standard Error of the Intercept; Confidence Limits on b_0

Confidence limits about the intercept b_0 are calculated similarly to those of the slope. The standard error of the intercept is first determined using equation 10.20

$$s_{b_0} = s_{y \cdot x} \sqrt{\frac{\Sigma X_i^2}{n \Sigma X_i^2 - \left(\Sigma X_i \right)^2}} \tag{10.20}$$

or as an estimating formula for a cross-check

$$s_{b_0} = \sqrt{\frac{s_{y \cdot x}^2}{n} + s^2 b_1 (\overline{X})^2} \tag{10.21}$$

The value of s_{b_0} is found to be 1.07275. The 95% confidence limits ($\alpha/2 = 0.025$, $\nu = 8$, $t = 2.306$) for the intercept with the Y axis are given by equation (10.22)

$$\beta_0 = b_0 \pm t_{n-2, \alpha/2} s_{b_0} \tag{10.22}$$

and for the same example, by substitution

$$\beta_0 = 20.929 \pm 2.306(1.07275)$$

or $18.455 \leq \beta_0 \leq 23.403$.

These values may be substituted for the value of the intercept in the regression equation and used to arrive at the plot of the confidence limit of the intercept for the line of regression.

As with the slope, it is possible to perform a t test for the significance of the value for β_0. With the hypothesis H_0: $\beta_0 = 0$,

$$t = b_0/s_{b_0}$$
$$= 20.929/1.07275$$
$$= 19.51$$

Since this value is greater than the critical value for t, the hypothesis that the intercept is zero is rejected at this confidence level.

10.3. ANOVA FOR REGRESSION

The line of best fit was defined as the line for which the sum of the squares of the deviations of the Y values were a minimum. This definition of the line of best fit agrees with the concepts of ANOVA or the analysis of variance. The sum of squares (SS) can be used as a test of the extent to which the calculated line really represents the data. ANOVA is a short-cut least-squares analysis which can be used because of the balanced method of data collection (2).

The ANOVA values provide the means to test whether the model fits the data and to determine if a real relationship exists between the variables. Aside from this, the ANOVA values also serve as alternate methods of calculating values for r^2 (and thus r), and cross-checks on the t- and F-test values. Details of ANOVA for regression are given in Refs. 2 and 3.

The steps required for ANOVA are given in Table 10.3. The first test to run is the one for lack of fit. To do this, the MS (mean square) for lack of fit is compared to the MS for error. The null hypothesis is that the variance due to error is equal to or greater than the variance due to lack of fit. The alternative hypothesis is that the variance due to lack of fit is greater than the variance due to error.

To test these hypotheses we find an F ratio of the corresponding MS terms at an α risk of 0.05, with the corresponding $\nu_1 = 7$ and $\nu_2 = 1$. The value of F is $2.126/2.000 = 1.06$. The table value for F critical is 237. Therefore, the null hypothesis is accepted and we conclude that the data give us no reason to reject the linear model for this situation.

The regression coefficient is then tested for significance (3). To do this, again from the ANOVA table, the MS for regression is compared to the MS for the residual (also the value of $s_{y \cdot x}^2$).

The null hypothesis being tested is $H_0: \beta_1 = 0$. Again we find the F ratio (MS_R/s^2) and compare it to the table value with $\nu_1 = 1$ and $\nu_2 = 8$. The value of F critical, at $\alpha = 0.05$, is 5.32. Since the calculated F (1143) exceeds the critical value from the table, we reject the hypothesis that $\beta_1 = 0$. In other words, we entertain the hypothesis that the value of the response variable Y is genuinely related to the level of the input variable X. In the particular case of fitting a straight line, this F test for regression is exactly the same as the t test for the regression coefficient ($\beta_1 = 0$) given earlier. This is because (3, 4) the ratio $MS_R/s^2 = t^2$ and for $H_0: \beta_1 = 0$

$$t = \frac{b_1 - \beta_1}{s_{b_1}}$$

and therefore

$$t = \frac{b_1}{s_{b_1}}$$

Conclusions for the significance of the correlation coefficient will also be identical with F tests for the significance of the regression coefficient (3). However, when there are more regression coefficients, the overall F test for regression, an extension of the one given here, does not correspond to the t test of a coefficient. However, tests for individual coefficients can be made either in t or $t^2 = F$ form by a similar argument. The F form is frequently seen in computer programs.

From the foregoing, it follows (3) that the percentage of the total variation explained by the regression equation can be obtained by the value $r^2 = SS(b_1)/SS_T$. Thus r^2 approximates the fraction of the total variation in the values of Y accounted for by the line of regression, and is $2411.21/2428.09 = 0.993$, or 99.3%. Thus the regression equation obtained explains 99.3% of the total variation.

TABLE 10.3. Summary ANOVA for Regression

Steps	Source	Method	SS	$df(v)$	ms	F-ratio
1	Crude SS (variability of Y about the origin)	ΣY_i^2	0.312650E 05	10	0.312650E 04	
2	SS due to b_0 (sum of squares of $\overline{Y}$ about the origin)	$\dfrac{(\Sigma Y_i)^2}{n}$	0.288369E 05	1	0.288369E 05	
3	Total SS (total sum of squares of Y about $\overline{Y}$)	Step 1 – Step 2	0.252810E 04	9	0.269789E 03	
4	SS due to b_1 [sum of squares due to relationship of X and Y (regression)]	$\dfrac{b_1(n\Sigma X_i Y_i - \Sigma X_i \Sigma Y_i)}{n}$	0.241121E 04	1	0.241121E 04	1142.67
5	SS residual (sum of squares still unexplained)	Step 3 – Step 4	0.168811E 02	8	0.211014E 01	
6	SS error (sum of squares due to replication)	$\Sigma(Y_i - \overline{Y}_i)^2$	0.200000E 01	1	0.200000E 01	
7	SS lack of fit (sum of squares still unexplained)	Step 5 – Step 6	0.148811E 02	7	0.212588E 01	1.06

10.4. COEFFICIENT OF DETERMINATION

Some further considerations of the degree of relationship between variables are now in order. Correlation seeks to determine the problem of how well a linear or other equation describes the relationship between variables. The values of variables satisfying an equation range from those of perfect correlation through those showing some degree of correlation to those showing no relationship. Numerically, the values range from ± 1.0 to 0.0, and are dimensionless. A value of zero indicates no correlation at all. It can correspond to one of two conditions. One, data points are scattered over an area symmetrical with respect to both the X and Y directions, or two, the least-squares line is horizontal. With two variables which, when plotted, lie about a straight line, we refer to simple linear correlation.

The form of the relationships is shown in Figures 10.5a, 10.5b, and 10.5c. The degree of this relationship is measured by the sample correlation coefficient r. The square of r is termed the sample coefficient of determination. In theory, r^2 explains the relative amount of variation in the estimating equation.

The value of r is most generally given by

$$r = \frac{\Sigma(x_i - \bar{x})(y_i - \bar{y})}{\sqrt{\left[\Sigma(x_i - \bar{x})^2\right]\left[\Sigma(y_i - \bar{y})^2\right]}} \tag{10.23}$$

or alternatively by

$$r = b_1 \sqrt{\frac{\Sigma(X_i - \bar{X})^2}{\Sigma(Y_i - \bar{Y})^2}} \tag{10.24}$$

Two problems are commonly encountered with the correlation coefficient. The first of these is determining whether or not its value differs significantly from zero. The second is the comparison of a coefficient

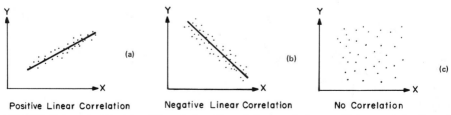

Positive Linear Correlation   Negative Linear Correlation   No Correlation

FIGURE 10.5 **(a)**, **(b)**, **(c)**. The relationships of the correlation coefficient and the line of regression.

derived from a linear equation, to one derived from a transformed equation, that is, one where the data have been transformed to a linear relationship, for example, by taking logarithms of one or more of the variables. In most cases, it is obvious by inspection that one equation gives superior correlation to another. However, where the r values are close to each other, the values of r are conventionally tested by a t test to determine whether the improvement is real or fortuitous. This is done by setting up the negative hypothesis that the difference is not real and disproving it at a selected confidence level (5).

The coefficient of determination, r^2, is that proportion of the total variability in the dependent variable that is accounted for by the regression equation in the independent variable(s). For the straight line case only, the value of $r = r^2$; for cases other than the straight line $r \doteq \sqrt{r^2}$. It essentially acts as a measure of the goodness of fit.

The total variation is given by $\Sigma(Y_i - \overline{Y})^2$ where $\overline{Y}$ is the mean Y value. Furthermore,

$$\Sigma\left(Y_i - \overline{Y}\right)^2 = \text{unexplained} - \text{explained variation}$$

or

$$\Sigma\left(Y_i - \overline{Y}\right)^2 = \Sigma(Y_i - \hat{Y})^2 + \Sigma(\hat{Y} - \overline{Y})^2 \tag{10.25}$$

The ratio of the explained variation to the total variation gives the value of r^2, that is,

$$r^2 = \frac{\Sigma(\hat{Y} - \overline{Y})^2}{\Sigma\left(Y_i - \overline{Y}\right)^2}$$

If the total variation of the dependent variable is all assigned to the effect of the independent variable there is zero unassigned variation, and $r^2 = 1.0$. If the total variation is all unassigned there is zero assigned variation, and $r^2 = 0.0$. For intermediate cases, the value of r^2 lies between 0 and 1.

The value of r^2 is thus equal to the fraction of the total variation of the Y data points explained by the least-squares line. Its graphical relationship to the line of regression is shown in Figure 10.6 (adapted from Refs. 1 and 6). The physical significance is that $E = $ explained deviation from the mean, $U = $ unexplained deviation from the mean, and $T = $ total deviation from the mean.

The meaning of r^2 is such that a value of $r = 0.6$ indicates that about 36% of all the variation in the data is attributable to a relationship between the variables. Likewise $r = 0.95$ means that about 90% of the total variation can be related to a relationship between the variables. The balance of variation is associated with error or other sources of variation.

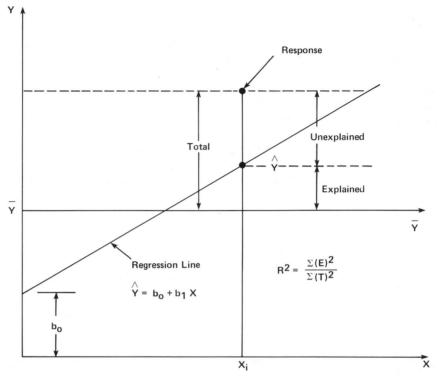

FIGURE 10.6. The graphical relationship of the coefficient of determination and the least-squares best-fit line.

For a given set of data, if, for example, the value of r^2 for a power law equation is higher than that of a least-squares line, then we consider that the power law curve gives a better fit for the data. However, the reader should consult the excellent article by Hahn (7), and for further aspects and details those books by Daniel and Wood (8), Draper and Smith (3), and the article by Anscombe (9), for detailed interpretation and the limitations of the coefficient of determination.

10.5. LINEARIZING TRANSFORMATIONS

There are many cases where simple linear regression analysis does not serve to produce a line which adequately expresses the relationship among the variables. In other words, the relationship between the two variables being correlated is nonlinear. In much chemical work, this relationship is in fact a curvilinear one. A simple scatter plot of the data will frequently indicate that such is indeed the case. This "nonfit" (to a straight line) may be confirmed by a low value of the coefficient of determination. Where this is the case the data may be transformed by taking logarithms of one or both variables, squaring, reciprocals, or any other appropriate step so that the relationship between the transformed variables can be expressed as a straight line. As mentioned previously, if it is suspected (say from a rough plot of the data) that the relationship is, for example, better described by a logarithmic relationship, the two correlation coefficients may be tested to determine whether the degree of improvement is or is not real.

As an illustration, suppose it is suggested from some preliminary plotting that the relationship between the viscosities of ball mill pastes in Krebs units (Y_i) and their true viscosities in poise (X_i) is linear when log X is plotted against Y. That is, $Y = b_0 + b_1(\log X)$ where X in the ordinary sense is replaced by log X. The values of the correlation coefficients for the linear (X_i plotted against Y_i) and logarithmic relationship are first determined, then tested against the hypothesis that the difference between them is not real (5), thus indicating at a chosen confidence level whether the transformation brings about a true improvement in describing the relationship.

The various possible transformations that allow the linear correlation and regression techniques to be extended to cover the cases of curvilinear relationship are given in numerous texts (10, 11). Use of these transformations has one main object, namely to yield a relationship in a linear form. The transformed data will not necessarily satisfy certain assumptions (for example, the assumption that the variability of Y given X is the same for all X) which are theoretically necessary for operations such as estimating the true value of Y associated with a particular value of X, determining confidence interval estimates for the least-squares line as a whole, and so on. However, for many purposes and within the range of the data considered, the transformations often serve well in providing practical relationships.

It is perhaps more helpful or useful to view these transformations from the somewhat converse aspect of determining, empirically, the form of a relationship between variables, a procedure that is frequently the case in actual practice, rather than using transformations to convert a known relationship to a linear form. For example, if a scatter diagram of log Y versus X plots as a straight line, then the relationship is highly likely to be $Y = ab^X$ (exponential), or $\log Y = \log a + (\log b)X$, where $y = \log Y$, $a = \log a$, and $b = \log b$ for $y = a + bX$. In certain instances the nature of the data itself, such as the viscosity of a resin in solution (Y_i) observed at a series of temperatures (X_i), would automatically suggest that a plot of the reciprocal of temperature versus viscosity be tried first as a linearizing step. In other cases, curved plots of the data on an x–y coordinate system may suggest, for example, that where the curve passes through the origin or is asymptotic to both axes it may be a power function. In contrast, exponential curves do not pass through the origin and are always asymptotic to the x axis, for $y = ae^{bx}$.

Some of the more useful transformations $(5, 10)$ (for $Y = a + bX$) are: (1) plot Y versus $1/X$ for $Y = a + b/X$; (2) plot X versus $1/Y$ for $Y = 1/(a + bx)$; (3) plot $\log Y$ versus $\log X$ for $Y = aX^b(\log Y = \log a + b \log X)$, the geometric or power function; (4) plot $\ln Y$ versus X for $Y = ae^{bx}(\ln Y = \ln a + bX)$ the exponential function, or for $Y = ab^x$.

An example of the application of $Y = aX^b$ would be, from (12), $\mu = KP^m$ or $\log \mu = \log K + m \log P$. For this, $\mu =$ effective viscosity in centipoise for a TiO_2 dispersion in alkyd resin, K (consistency coefficient) $= 54.51$ to 193.8, $m = 0.636$ to 0.904, and $P = ndz$, where $n =$ rpm, $d =$ diameter of the rotor, and $z =$ number of teeth in a gear mill. This relationship would

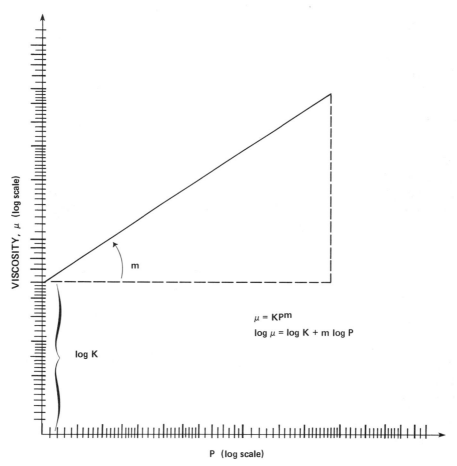

FIGURE 10.7. Viscosity of a pigment dispersion in alkyd resin.

be expressed graphically as in Figure 10.7. This same relationship is also encountered (13) in the rheology of power law fluids where the shear stress–shear rate relationship takes the form $\tau = k\gamma^n$.

The equation for $Y = ae^{bx}$ is that for the exponential trend which is, for example, quite useful to project short-term trends in sales, earnings, or production with a fitted least-squares line. Of course, as for $Y = ab^x$, conversions of the straight line constants to the original constants must be carried out where these are transformed, as for example for $Y = ae^{bx}$, $a = \log a = b_0$, to arrive at the equation of the line. Confidence limits on the estimate of Y, confidence limits about the best-fit transformed line slope and intercept parameters, tests of significance for the slope and intercept, and so on, are determined as before.

An example of a curvilinear relationship is given in Table 10.4. There, the viscosity of a given weight mixture of a higher and a lower molecular weight vinyl resin in a glycol ether solvent at 10% total resin solids by weight is tabulated. The area of interest for the series of mixtures was restricted to about 35–60 parts due to viscosity constraints and solubility considerations. When viscosity was plotted against the amount of the higher molecular weight resin it appeared that a curved line might fit the data points better than a straight line (see Fig. 10.8).

Expressed as a linearizing transformation, the equation for the least-squares line would be $\log \hat{Y}_i = b_1 X_i + b_0$, where Y_i is the viscosity, X_i is the weight mixture, and b_1 and b_0 are empirical constants to be determined from experimental data. These constants will necessarily differ from those of other equations such as $\hat{Y} = b_1 X_1 + b_0$. The procedure for fitting the relationship is to take logs of all the observed Y values and use the procedures of Section 10.2 to determine b_0 and b_1. Where equations are nonlinear in their coefficients, the constants of the original equation are determined by substituting the calculated coefficient values in, for example, $b_0 = \log a$, and taking the required antilogs. All cases are somewhat different, for example, in the present instance the value of the intercept will be the intercept on the $\log Y_i$ axis. Also, placing confidence limits on the

TABLE 10.4. Viscosity of Resin Mixtures

(X_i) Parts Higher MW Component	(Y_i) Viscosity, cP
35	40
40	60
45	95
50	200
55	235
60	305

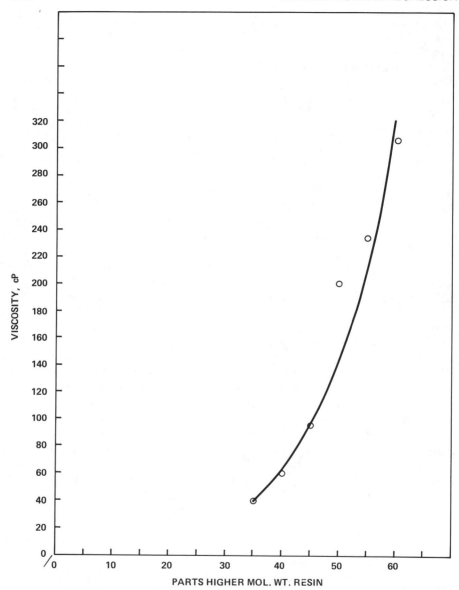

FIGURE 10.8.   Solution viscosity as a function of resin composition.

intercept at various values of X_i necessitates taking the antilog of log Y as solved for during the substitution process using the appropriate expressions from Section 10.2. These conversions are discussed in detail in Ref. 10 and also in Refs. 3 and 14.

Inspection of the viscosity data indicated that a straight line fit was possible with log viscosity (Y_i) versus the series of mixtures (X_i). The equation for this line was determined by a least-squares fit and found to be $\log Y = 0.0372 X + 0.3178$ with a value of r^2 of 0.966. Using this calculated regression line, the value of the response variable Y for any assigned value of the variable X_0 can be predicted. As before, confidence limits for this prediction should be determined, as well as those for the regression coefficients themselves. Particular attention should be paid to the equations used to calculate these confidence limits, in that small sample sizes are generally involved, rather than large ones, and the limits for small sample sizes thus calculated will be different from those of large sample sizes. The worth of these predictions is inversely related to the uncertainty in the data and the true relationship between the variables is probably best expressed as a band within which the estimates probably fall. One final point—where two possible linearizing expressions differ in r^2 by a rather small amount, but where both are close to 1.0, for practical purposes it would generally be advisable to select the simpler relationship for expressing the relationship between the variables.

All of the previous discussion has been devoted to the use of the least-squares method for arriving at a relationship between two variables. For those wanting to dig deeper, other methods are available for linear regression. In fact, as discussed by King (15), a fitted *line* may give estimates closer to the known population values than do conventional calculations. The best-fit line may be better suited than a least-squares line since it is much less influenced by "wild," "outlying," or "spurious" values. This could be the case where the data were obtained through careless sampling or from a poorly controlled process.

10.6. RANK CORRELATION

Quite often we are confronted with the problem of assessing the relative quality of factors or effects which are difficult or impractical to express quantitatively, such as odor, the degree of yellowing, general appearance, and so on. In this case, ranking methods are quite widely used. These frequently concern subjective evaluations between two or more persons, frequently at different competence levels, for any pairs of observations. Rank correlation has recently been used (16) to evaluate Florida exposure tests and to compare accelerated weathering to natural weathering.

A coefficient that measures the correlation between assigned ranks has been devised by Spearman (17). This coefficient of rank correlation is r_s. It is similar to the linear correlation coefficient r discussed previously. For subjective evaluations between persons, if the differences in assigned ranks of observations are small, r_s will be close to $+1$. This value indicates a degree of agreement between the judges. Complete agreement would make all differences zero, giving a value of $r_s = +1$. Complete disagreement between the judges would give $r_s = -1$. A value of 0 would indicate no relationship. Values of 0.90 to 0.95 indicate excellent agreement. At lower levels, the principal value of r_s is in comparing two or more sets of data to another set. The value of r_s is determined by

$$r_s = 1 - \frac{6\Sigma d_i^2}{n(n^2 - 1)} \tag{10.26}$$

where d_i are the differences between the ratings and n is the number of items ranked.

The problem can be illustrated (18) as follows: Consider the ranking of the odors of 10 paints on a relative odor scale, with 0 being assigned to the paint with the least objectionable odor and 10 to the paint with the most objectionable odor. In your laboratory you have three chemists who have a good "nose" for odor, from their association with various and sundry organic chemicals. They would have more of a tendency to give an objective evaluation of odor—which is what is wanted in this particular test. We are interested in how well these three individuals agree among themselves. The results of the ranking are shown in Table 10.5.

The resulting rank correlation coefficients are:

$$\begin{array}{ll}
\text{For A + B} & r_s = -0.212 \\
\text{For B + C} & r_s = -0.297 \\
\text{For A + C} & r_s = +0.636
\end{array}$$

By inspection, it can be seen that, because of their higher correlation, chemists A and C have the nearest approach to common judgment. This value can be tested for significance (2) by determining the minimum

TABLE 10.5. Ranking Table — Paint Odor Tests

Paint No.	Chemists			A Versus B		B Versus C		A Versus C	
	A	B	C	d_i	d_i^2	d_i	d_i^2	d_i	d_i^2
1	1	3	6	2	4	3	9	5	25
2	6	5	4	1	1	1	1	2	4
3	5	8	9	3	9	1	1	4	16
4	10	4	8	6	34	4	16	2	4
5	3	7	1	4	16	6	36	2	4
6	2	10	2	8	64	8	64	0	0
7	4	2	3	2	4	1	1	1	1
8	9	1	10	8	64	9	81	1	1
9	7	6	5	1	1	1	1	2	4
10	8	9	7	1	1	2	4	1	1
					200		214		60

required value for the significance of r_s. This is done by use of the following:

$$\frac{t_\alpha}{\sqrt{t_\alpha^2 + n - 2}}$$

where t_α is the table value of Student's t_ν for $n - 2$ at an α risk of 0.05 (5% level). Using the value of t_α of 1.86 we calculate a value of 0.549. Since the A + C value of 0.636 exceeds this minimum value, we conclude that A + C have some degree of statistical agreement between them.

An alternate method of test (19) is by use of

$$t_s = r_s \frac{(n - 2)^{1/2}}{1 - r_s^2}$$

By substitution, t_s is found to be 2.33. If $t_s >$ critical $t(\nu = n - 2)$ for the one-tailed test, we can say that there is evidence of agreement between A and C. Since this is so ($t_\alpha = 1.86$) we conclude that there is agreement between these two judges. Further details, for example, problems dealing with proportions or percentages, pairs of quantitative observations, the handling of ties, and so on, concerning the use of Spearman's test are given in Refs. 19 and 20.

REFERENCES

Text References

1. Longly-Cook, L. H., *Statistical Problems*, Barnes & Noble, New York, 1971.
2. Rickmers, A. D., and Todd, H. N., *Statistics: An Introduction*, McGraw-Hill, New York, 1967.
3. Draper, N. R., and Smith, H., *Applied Regression Analysis*, Wiley, New York, 1966.
4. Middlebrooks, E. J., *Statistical Calculations—How to Solve Statistical Problems*, Ann Arbor Science Publishers, Ann Arbor, 1976.
5. Gore, W. L., *Statistical Methods for Chemical Experimentation*, Interscience, New York, 1956.
6. Perry, C. C., *Interpreting least-squares lines*, *Machine Design*, Vol. 33, No. 14, June 8, 1961, pp. 177–184.
7. Hahn, G. J., The coefficient of determination exposed!, *Chemical Technology*, Vol. 3, No. 10, October 1973, p. 609.
8. Daniel, C., and Wood, F. S., *Fitting Equations to Data*, Wiley, New York, 1971.
9. Anscombe, F. J., Graphs in statistical analysis, *Amer. Statist.* Vol. 27, No. 1, 1973, p. 17.
10. Natrella, M. G., *Experimental statistics*, Chapter 5, N.B.S. Handbook 91, U.S. Government Printing Office, Washington, D.C., 1966.

11. Freund, J. E., Livermore, P. E., and Miller, I., *Manual of Experimental Statistics*, Chapter IID, Prentice-Hall Inc., Englewood Cliffs, NJ, 1960.

12. Leibzon, L. N., and Mikheenkova, E. A., Determination of the shear rate, shear stress, and effective viscosity during grinding of pigment pastes, (*USSR*). *Khim. Khim. Tekhnol., Sint. Issled. Plenoobrazuyushchikh Veschestv Pigm.*, *1975*, *71-3* (Russ.), Edited by Ermilov, P. I. Yarosl. Politekh. Inst.: Yaroslavl., USSR.

13. Hyman, W. A., *Ind. Eng. Chem., Fundam.*, Vol. 15, No. 3, 1976, p. 215.

14. Heller, H., Choosing the right formula for calculator curve fitting, *Chemical Engineering*, February 13, 1978, p. 119.

15. King, J. R., TEAM, *TEAM Easy Analysis Methods*, Vol. 4, No. 1, 1977.

16. Grossman, G. W., *J. Ctngs. Tech.*, Vol. 49, No. 633, 1977, p. 45.

17. Spearman, C., *American Journal of Psychology*, 1904, pp. 72–101.

18. Prane, J. W., *Pt. Varnish Prod.*, October, 1956, pp. 1–42.

19. Mack, C., *Essentials of Statistics for Scientists and Technologists*, Plenum/Rosetta Ed., Plenum, New York, New York, 1975.

20. Langley, R., *Practical Statistics Simply Explained*, Dover, New York, 1971.

General References

Acton, F. S., *Analysis of Straight-Line Data*, Dover New York, 1966.

Ball, W. E., and Johnson, R. C., Solve equations graphically, *Chemical Engineering*, November, 1957, pp. 272–281.

Bennett, C. A., and Franklin, N. L., *Statistical Analysis in Chemistry and the Chemical Industry*, Wiley, New York, 1954.

Box, G. E. P., Use and abuse of regression, *Technometrics*, Vol. 8, No. 4, November, 1966, pp. 625–630.

Daniel, C., and Wood, F. S., *Fitting Equations to Data*, Wiley, New York, 1971.

Dixon, W. J., and Massey, F. J., *Introduction to Statistical Analysis*, 2nd ed., McGraw-Hill, New York, 1957.

Hahn, G. J., Regression for prediction versus regression for control, *Chemical Technology*, Vol. 4, No. 9, September 1974, pp. 574–576.

Hahn, G. J., Meeker, W. Q., and Feder, P. I., The evaluation and comparison of experimental designs for fitting regression relationships, *Journal of Quality Technology*, Vol. 8, No. 3, July 1976, pp. 140–157.

Hahn, G. J., and Shapiro, S. S., The use and misuse of regression analysis, *Industrial Quality Control*, Vol. 23, No. 4, October 1966, pp. 184–189.

Hoel, P. G., Testing goodness of fit, *Introduction to Mathematical Statistics*, 4th ed., Wiley, New York, 1965, pp. 244–259.

Johnson, N. L., and Leone, F. C., *Statistics and Experimental Design in Engineering and the Physical Sciences*, Vols. 1 and 2, Wiley, New York, 1964.

Neter, J., and Wasserman, W., *Applied Linear Statistical Models*, Irwin, Homewood, IL, 1974.

Spiegel, M. R., *Schaum's Outline of Theory and Problems of Probability and Statistics*, McGraw-Hill, New York, 1975.

Statistical methods: New tools for solving paint problems, *Official Digest*, February 1959.

Youden, W. J., *Statistical Methods for Chemists*, Wiley, New York, 1964.

Young, H. D., *Statistical Treatment of Experimental Data*, McGraw-Hill, New York, 1962.

11

NONPARAMETRIC STATISTICS

GARY E. MEEK

College of Business Administration
University of Akron
Akron, Ohio

In the previous chapters a variety of analytical methods were introduced which have become known as classical statistical procedures. As each was discussed, the assumptions underlying the application and use, and hence the limitations, of the particular test were indicated. A basic assumption of all the tests, other than those relating to the binomial, was either that the population(s) being sampled could be reasonably well approximated with the normal distribution or that a large sample was selected. The second assumption inherent in all procedures other than the binomial was that the data or numbers that were collected were "good" measurements. By "good" it is meant that for any two observations it is possible to compare them, to determine which is numerically larger, and to specify precisely how much larger, and that these comparisons are logical and meaningful.

In numerous practical situations the normality assumption is suspect and it may be impossible, for any number of reasons, to take a large sample. Second, the data that are observed will often not be of the level indicated above and the classical procedures will no longer be applicable even for large samples. Thus it is imperative that procedures with less restrictive assumptions be available. These methods are called "distribution-free" since they make no specific assumptions about the type of family to which the population distribution belongs.

These tests are also more commonly called nonparametric. This term is somewhat of a misnomer since it implies that no hypothesis is made about the value of a parameter in the probability function when, in fact, the null hypothesis may concern the value of a parameter. Generally, the only assumption for many of the nonparametric tests is that the sampled population be continuous and occasionally the additional requirements of symmetry and, in the multisample case, of identical shapes may be made.

Because nonparametric tests do not use the actual magnitudes of the observations, they do not test for parameters computed from them in the same way that classical procedures test for equal means or variances. Generally nonparametric procedures will test for values which can be computed from characteristics of the observations, such as frequency or position in an array. In some cases the distribution-free test will also be a test for the classical parameters. For example, if the populations are assumed to be symmetric then a test for equal medians is also a test for equality of means. Nonparametric tests are of particular value in the coatings industry, and have been briefly introduced in Chapter 10. Nonparametric tests have been applied to numerous situations in coatings such as evaluating whiteness of panels, odor of paints, and weathering tests.

11.1. LEVELS OF MEASUREMENT

As indicated previously, the level of measurement which is attained in the sample data influences which statistical procedure is applicable. Thus it is necessary to define what these levels are and as each procedure is introduced, the level of measurement required by it will be specified. Only the three levels of measurement required by the various procedures of this and the previous chapters will be discussed.

The simplest or weakest level of measurement that can be obtained is called *nominal*. The measuring instrument in this case simply classifies the objects in the sample into categories. That is, it may assign each item a number but the numerical value is actually little more than a name, hence, the term nominal. An example of nominal data would be the visual identification of a set of finishes with each finish being identified as either glossy or flat. There is no inherent quantitative difference between the words glossy and flat and thus with most nominal data the variable used is the frequency or number of observations belonging to each category.

The next step up in the hierarchy of levels of measurement is the *ordinal* scale. As is implied by the name, data satisfying this level are able to be ordered or ranked in some way. The ranking need not be accomplished using actual numerical values but may take the form of preferences; for example, in comparing shades of a paint color, a person may indicate which is least preferable assigning 1 to the least and n to the most preferable. This ranking procedure will be adhered to in this chapter but the order may be reversed without affecting the results. Though numerical values may designate the observations, they serve only to identify the relative positions of the observations when rated from "worst" to "best." With strictly ordinal data it is meaningless to compare two observations in terms of the distance between them.

The highest level of measurement that will be considered is that obtained on an *interval* scale. Data satisfying an interval scale imply that not only are the observations able to be ranked but also that an exact difference between any two observations can be obtained and is meaningful. The interval level of measurement does not assume a natural origin, that is, a natural zero, but it is necessary that a one-unit change on the scale corresponds to a one-unit change in the object being measured. An example of the interval scale would be time as measured in years or days, since the selection of an origin is arbitrary. The previous test procedures based on the normal, t, χ^2, and F distributions all assumed at least an interval level of measurement.

Note that data that satisfy one level of measurement will also satisfy any level of measurement which is less sophisticated. That is, observations on an interval scale can obviously be ranked and hence satisfy the requirements of the ordinal scale. It is imperative that one be familiar with the manner in which numerical data are obtained. In many situations if only the numbers are considered, the data may appear to be interval; whereas, if their manner of origin is considered, the scale of measurement may be ordinal at best and possibly no better than nominal.

11.2. COMPARISON OF DISTRIBUTION-FREE AND CLASSICAL PROCEDURES

Only a few of the points of comparison of nonparametric versus the previous classical tests are considered. Both types of procedures have areas of application in which one is superior to the other. Whether a nonparametric test or its classical counterpart should be used is dependent on the level of measurement and how well the underlying assumptions of the test are satisfied by the particular situation under consideration, as well as the ability of the user to determine the degree of satisfaction. In general, the comparisons are quite favorable for the nonparametric tests.

In terms of the derivation of corresponding test statistics and their related distributions the nonparametric tests are more easily derived since most of them rely only on simple combinatorial (counting) techniques. Thus, most nonparametric tests arise via a logical process that can be easily understood by the user. Also, the user is more capable of understanding the underlying assumptions of the test and is less likely to misuse it. At the other extreme the classical tests require a sophisticated level of mathematics to understand their derivations. Since most users are not conversant in the higher levels of mathematics they are reduced to following cookbook procedures and to applying overgeneralized rules of thumb to justify the assumptions. Nonparametric techniques generally have an advantage in both computational ease and speed. In most situations all that is necessary are simple arithmetic operations and ranking when the sample size is less than 30. As the sample size increases, tests based on ranking procedures become more cumbersome and time consuming than their classical counterparts.

The major argument in opposition to distribution-free tests is based on an alleged lack of efficiency in the mathematical sense. This type of efficiency is known as asymptotic relative efficiency (ARE).

Definition 11.1. If T_1 and T_2 are consistent tests of a null hypothesis H_0 against an alternative H_1, at significance level α, then the asymptotic relative efficiency of T_1 to T_2 is the limiting value of the ratio n_1/n_2, where n_1 is the sample size required by test T_1 for the power of test T_1 to equal the power of test T_2 based on n_2 observations, as n_2 approaches infinity and H_1 approaches H_0.

Definition 11.2. A test is said to be consistent for a specific alternative H_1 if the probability of rejecting the null hypothesis approaches 1 as the sample size approaches infinity.

The claim of lower ARE for nonparametric tests is often based on a comparison of the nonparametric test to the classical test under conditions satisfying the assumptions of the classical. Under those conditions, nonparametric tests usually have relative efficiencies slightly less than 1.0 for small samples with the efficiency decreasing as the sample size increases. The ARE is obtained by letting the sample size become infinite at which point nonparametric tests ordinarily have their lowest efficiencies relative to the classical counterpart. Under conditions that violate some of the classical assumptions the nonparametric test is often superior to the most efficient classical test. Since ARE is a large sample property it may not have much importance for small- or even moderate-sized samples which is the recommended arena for nonparametric tests. When samples are small (say $n \simeq 15$) nonparametric tests are easier, faster, and almost as efficient even if the classical assumptions are met. At these sample sizes deviations from the classical assumptions are most critical and generally very difficult to detect. Thus, unless it is known in advance that the classical assumptions are justified, it will generally be wiser to use a nonparametric test.

In terms of applicability, the nonparametric tests are obviously superior. Being based on fewer and less elaborate assumptions, they can be applied in all situations where the more restrictive classical assumptions are satisfied as well as a much larger class of situations where they are not.

11.3. ONE-SAMPLE SIGN TEST

One of the simplest, safest, and most easily understood tests is the sign test. It is based on the binomial distribution and as such it has readily available tables and is intuitively appealing. It is not the most powerful test but its efficiency is surprisingly high considering the simplistic nature of the information used. This information is merely the algebraic sign of the difference between two numerical values.

The hypothesis to be tested is that the median of the population is equal to some specified value, say Me_0, that is, the hypotheses are

$$H_0: Me = Me_0 \qquad H_1: \neq Me_0 \qquad (11.1)$$

Another way of stating the null hypothesis is, since the median is the fiftieth percentile, $P(X > Me_0) = P(X < Me_0) = .5$, or, by subtracting the median from each value, $P(X - Me_0 > 0) = P(X - Me_0 < 0) = .5$. The latter statement leads directly to a statement based on the sign of the difference, that is, P (difference is positive) = P (difference is negative) = .5. This will be the case if the null hypothesis is true and the following assumptions are satisfied.

Assumptions of the Sign Test

i. The observations are from an underlying continuous distribution or $P(X = Me_0) = 0$. The basic assumption is still satisfied if the population is discrete but the median is not an observable value.

ii. The observations constitute a random sample.

iii. The observations are independent which is satisfied by randomness if the population is infinite or by sampling with replacement if it is not.

iv. At least nominal data are available. Under these assumptions we may restate the hypotheses in equation (11.1) as

$$H_0: p_+ = .5 \qquad H_1: p_+ \neq .5 \qquad (11.2)$$

where $p_+ = P(X - Me_0 > 0)$. Thus, intuitively, the test is based on the observed number of plus signs. If the null hypothesis is true then the number of +'s and the number of −'s should be the same, differing only because of random variation. The logical rejection region corresponds to situations in which there are either too many or too few +'s. If the true median is less than that specified in H_0, we should observe more −'s than +'s and if it is greater than Me_0 we should observe more +'s than −'s. (See Figure 11.1). As usual, the measure of too many or too few is the probability of occurrence when H_0 is true.

In Chapter 6 we based our tests on standardized statistics realizing that an extreme value of the statistic corresponded to a very small probability

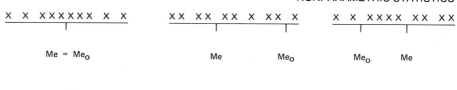

FIGURE 11.1. Illustration of possible configurations of data points observed when the (a) true median equals Me_0, (b) true median is less than Me_0, (c) true median is greater than Me_0. Note that when Me_0 is subtracted from each observation there will be (a) 5 +'s, 5 −'s, (b) 1 +, 9 −'s, (c) 8 +'s, 2 −'s.

of occurrence. When the calculated value was more extreme than the tabulated value corresponding to α, it implied that the probability of occurrence was less than α and H_0 was rejected. For the sign test we shall actually calculate the probability of occurrence and compare it to a stated level of significance. The procedure will be as follows:

1. State the hypothesis and the nominal α to be used for testing.
2. Select a random sample from the population under consideration.
3. Subtract the hypothesized value, Me_0, from every observation recording only the sign of the difference.
4. Count the number of plus signs, say n_+, and the number of minus signs, say n_-.

Depending on whether the hypotheses are one-sided or two-sided the following rules or test criteria arise.

5(a). If the hypotheses are two-sided, that is,

$$H_0: Me = Me_0 \qquad H_1: Me \neq Me_0$$

then we reject H_0 at the nominal α if either

$$P\left(X \leq n_+ | p = \tfrac{1}{2}, n = n_+ + n_- \right) < \alpha/2 \qquad (11.3)$$

or

$$P\left(X \geq n_+ | p = \tfrac{1}{2}, n = n_+ + n_- \right) < \alpha/2 \qquad (11.4)$$

5(b). If the hypotheses are of the form

$$H_0: Me \leq Me_0 \qquad H_1: Me > Me_0$$

the null hypothesis is rejected at α if

$$P\left(X \geq n_+ | p = \tfrac{1}{2}, n = n_+ + n_- \right) < \alpha \qquad (11.5)$$

5(c). For the other one-sided hypotheses,

$$H_0: Me \geq Me_0 \qquad H_1: Me < Me_0$$

H_0 is rejected if

$$P\left(X \leq n_+ | p = \tfrac{1}{2}, n = n_+ + n_-\right) < \alpha \qquad (11.6)$$

To be able to calculate the probabilities indicated above it is necessary that the distribution of the number of positive differences be known. Under the assumptions, we have a finite number of independent trials from a very large dichotomous population with $p_+ = p_- = \tfrac{1}{2}$. We recognize this as a binomial situation. Since any observation has a 50% chance of falling above the median, p equals $\tfrac{1}{2}$. Theoretically no observations should equal Me_0 and n should be the sample size. Practically, due to imprecise measurements, values will occasionally equal Me_0 resulting in a difference of 0. When this occurs the 0 differences are discarded and the sample size reduced correspondingly to $n = n_+ + n_-$. Thus the probabilities indicated in equations (11.3) and (11.6) would be determined according to the binomial distribution as

$$P\left(X \leq n_+ | p = \frac{1}{2}, n = n_+ n_-\right) = \sum_{x=0}^{n_+} C_x^n \left(\frac{1}{2}\right)^x \left(\frac{1}{2}\right)^{n-x} \qquad (11.7)$$

For equations (11.4) and (11.5) they would be

$$P\left(X \geq n_+ | p = \frac{1}{2}, n = n_+ + n_-\right) = \sum_{x=n_+}^{n} C_x^n \left(\frac{1}{2}\right)^x \left(\frac{1}{2}\right)^{n-x} \qquad (11.8)$$

You will recall that, for appropriate values of n, these probabilities can be found by use of the binomial table.*

Example 11.1. An experimental enamel is being evaluated as to its ability to retain its gloss under various weathering conditions. Sixteen steel sheets were finished with the enamel and then subjected to a variety of accelerated weathering conditions. Each sheet was then visually compared to a sheet coated with the standard enamel which had been subjected to the same conditions. The experimental sheets were labeled as being either glossier or less glossy than the standard. The results of comparisons are

$$M \quad M \quad M \quad L \quad M \quad L \quad L \quad - \quad L \quad M \quad M \quad L \quad M \quad M \quad M \quad L$$

*See Chemical Rubber Co., *Handbook of Tables for Probability & Statistics,* 2nd ed., W. H. Beyer, ed., Table III.2, "Cumulative Terms, Binomial Distribution," pp. 194–205.

where M indicates the experimental is glossier, L indicates the experimental is less glossy, and — indicates an inability to differentiate.

Letting Me_0 represent the median glossiness of the standard enamel, the indicated hypotheses are

$$H_0: Me \le Me_0 \qquad H_1: Me > Me_0$$

with rejection of H_0 occurring at $\alpha = .05$ if

$$P(X \ge n_M | n = 15, p = .5) < .05$$

For the data $n_M = 9$, $n_L = 6$ and

$$P(X \ge 9 | n = 15, p = .5) = .3036 \not< .05$$

Since .3036 is not less than .05 the null hypothesis cannot be rejected. The resulting conclusion is that the $9M$'s (or $+$'s) is a reasonable occurrence due to chance variation alone.

In Example 11.1 we used a nominal α of .05. This is only nominal since the distribution of $+$'s is discrete and the stated α will not generally correspond to one of the possible values which the variable can assume. The actual or exact α for Example 11.1 is .3036. That is, if we decide to reject H_0 in Example 11.1 then .3036 is the probability of doing so by chance alone.

NOTE. If it can be assumed that the population with which we are concerned has a symmetric distribution then the mean and median are identical. Thus the test of the hypothesis $H_0: Me = Me_0$ is equivalent to testing $H_0: \mu = \mu_0$ and the sign test provides an alternative to the one-sample t test discussed in Chapter 6 but with much less restrictive assumptions.

When the sample size increases to the extent that n cannot be found in the binomial table the probability is calculated using the normal approximation to the binomial.

11.4. ONE-SAMPLE WILCOXON SIGNED-RANK TEST

A test which is similar to the sign test but which has somewhat more restrictive assumptions is the Wilcoxon signed-rank test for population location. The test is based on ranking the differences of the observations minus the hypothesized median (or mean). The test is surprisingly powerful, having an ARE of .955 relative to the one-sample t test even when the underlying population is normal and in most cases an ARE greater than 1.00 when compared to the t with non-normal populations. In most situations it is either the most powerful test or a close second. These assumptions are given below.

Assumptions of Wilcoxon's Signed-Rank Test

   i. The underlying distribution is continuous and symmetric. (Continuity may be relaxed if the population is either infinite or very large.)

  ii. The sample data are of at least an ordinal scale.

 iii. The observations constitute a random sample and are independent of each other.

Under the above assumptions we can test hypotheses about the population median or mean since the distribution is assumed to be symmetric. The basic hypotheses for the two-sided case are

$$H_0: Me = Me_0 \qquad H_1: Me \neq Me_0$$

As with the sign test we state a recommended procedure to be used. The steps to be followed are:

1. State the hypotheses and select the nominal α at which they are to be tested.
2. Select a random sample.
3. Subtract the hypothesized value, Me_0, from every observation recording both the sign and the difference.
4. Rank the absolute values of the differences from low to high assigning 1 to the lowest nonzero absolute difference, 2 to the next lowest, up to n for the highest (discard zero differences).
5. Calculate T_+ and T_- where

$$T_+ = \text{sum of the ranks of the positive differences, and}$$

$$T_- = \text{sum of the ranks of the negative differences} \qquad (11.9)$$

The decision rules for the two-sided and various one-sided alternatives are based on the magnitudes of T_+ and T_-. If Me_0 is the true median then the high and low ranks should be evenly divided among the positive and negative differences, differing only due to random variation, and T_+ and T_- should be approximately equal. If $Me < Me_0$ then there should be more negative differences and correspondingly more of the high ranks should be assigned to them. Thus T_- should be greater than T_+ since their sum must equal $n(n + 1)/2$ which is the sum of the first n integers. In this case n is the number of nonzero differences. Conversely, by the same reasoning, if $Me > Me_0$ then T_+ should exceed T_-. Table 11.1 gives critical values for the smaller sum corresponding to nominal α's of .05, .025, .01, and .005. Depending on the stated alternative the following decision rules arise.

6(a). If the alternative is two-sided, that is, H_1: $Me \neq Me_0$, then we reject H_0 at the stated α if

$$T = \min(T_+, T_-) \leq T_{n,\alpha/2} \qquad (11.10)$$

6(b). For H_0: $Me \geq Me_0$, H_1: $Me < Me_0$, T_+ would be expected to be smaller if H_1 is true and we reject H_0 at the stated α if

$$T_+ \leq T_{n,\alpha} \qquad (11.11)$$

6(c). For H_0: $Me \leq Me_0$, H_1: $Me > Me_0$ we expect T_- to be smaller if H_1 is true and reject H_0 at the stated α if

$$T_- \leq T_{n,\alpha} \qquad (11.12)$$

TABLE 11.1. Lower-Tail Critical Values for Wilcoxon's Signed-Rank Statistic

n	Nominal .05	.025	.01	.005	n	Nominal .05	.025	.01	.005
5	0	—	—	—	23	83	73	62	54
6	2	0	—	—	24	91	81	69	61
7	3	2	0	—	25	100	89	76	68
8	5	3	1	0	26	110	98	84	75
9	8	5	3	1	27	119	107	92	83
10	10	8	5	3	28	130	116	101	91
11	13	10	7	5	29	140	126	110	100
12	17	13	9	7	30	151	137	120	109
13	21	17	12	9	31	163	147	130	118
14	25	21	15	12	32	175	159	140	128
15	30	25	19	15	33	187	170	151	138
16	35	29	23	19	34	200	182	162	148
17	41	34	27	23	35	213	195	173	159
18	47	40	32	27	36	227	208	185	171
19	53	46	37	32	37	241	221	198	182
20	60	52	43	37	38	256	235	211	194
21	67	58	49	42	39	271	249	224	207
22	75	65	55	48	40	286	264	238	220

Table 11.1 is indexed by n, the total number of $+$'s and $-$'s, and by a nominal α. The value of n identifies the row and α identifies the column in which the critical value T_n will be found. The probability is only nominal since T is a discrete random variable. The true α for a test that rejects H_0 will be somewhat less than the stated α.

Example 11.2.   A particular type of paint is supposed to have an average adhesion reading of at least 900 after a 15-day drying period when applied to previously painted surfaces. Twelve painted surfaces exhibiting marked deterioration due to weathering were repainted with this type of paint. After the prescribed time lapse adhesion readings were taken for each surface. The readings are as follows: 921, 902, 886, 838, 895, 864, 890, 918, 856, 890, 913, 886. The question that arises is are the values below 900 due to an inferior adhesion property in the paint? Selecting a significance level of .05, the decision rule for the hypotheses,

$$H_0: \mu \geq 900 \qquad H_1: \mu < 900$$

is to reject H_0 if $T_+ \leq T_{12,.05} = 17$. The differences, $d_i = x_i - 900$, are

d: 21   2 -14 -62 -5 -36 -10 18 -44 -10 13 -14

Taking the absolute values of the d_i and ranking them from low to high gives (circled values correspond to positive differences)

$R|d_i|$:   ⑨   ①   6.5   12   2   10   3.5   ⑧   11   3.5   ⑤   6.5

Thus,
$$T_+ = 9 + 1 + 8 + 5 = 23$$

Since $23 \nleq 17$ we cannot reject H_0 and must conclude that the deviations appear to be due to chance.

Observe that in Example 11.2 there were two absolute differences with a value of 14 and two with a value of 10. Theoretically ties cannot occur but in practice they do, either from sampling with replacement from a finite population or from using an imprecise measuring instrument on a continuous population. The general practice for handling ties in ranking procedures is to assign the average of the ranks which would have been assigned to each of the tied values. Our two 14's should have received the ranks 6 and 7 but to avoid biasing the statistic with an arbitrary assignment they were each assigned a rank of 6.5. The next value, 18, then received the rank of 8 and so on.

NOTE.   If there are any doubts about the symmetry of the population, then the sign test is much preferred over the signed-rank test. This is also the case if there are serious questions about the precision of the measurements.

For situations in which the sample size is too large for Table 11.1 we may use the normal approximation to the distribution of T. It can be shown, using arithmetic series, that the mean and variance of T are

$$E(T) = \frac{n(n + 1)}{4} \quad \text{and} \quad \sigma_T^2 = \frac{n(n + 1)(2n + 1)}{24}$$

Thus, for large samples, the null hypothesis is rejected if

$$Z = \frac{T - \dfrac{n(n + 1)}{4}}{\sqrt{\dfrac{n(n + 1)(2n + 1)}{24}}} \tag{11.13}$$

is less than the appropriate z value from the standard normal table. Whether one uses T_+, T_-, or $\min(T_+, T_-)$ depends on the hypotheses being tested. For two-sided tests the critical z will be $-z_{\alpha/2}$ while for both one-sided tests it is $-z_\alpha$. The approximation should not be used if extremely small values of α are selected.

11.5. RUNS TEST

An assumption of all the test procedures has been that a random sample was selected. The runs test provides a method for testing the validity of this assumption. The test does not have high power but it is extremely simple and easy to use even for large samples. However, the efficiency may be quite adequate for most situations.

Assumptions of the Runs Test

    i.   A random sample has been selected.
   ii.   The observations are able to be dichotomized.
  iii.   At least nominal level data are available.

The runs test specifically tests whether or not the order of occurrence of the observations follows a random pattern. The procedure is quite simple and is given below.

1.   State the hypothesis to be tested and the level of significance. Ordinarily the hypotheses will be H_0: random pattern of occurrence versus H_1: nonrandom pattern.

2. List the observations in order of occurrence.
3. Indicate to which of the two classes each observation belongs.
4. Count the number of runs, R, in the data where a run is defined to be an unbroken sequence of elements of the same type.

Nonrandomness in the data can be indicated in two ways. There can be too few runs, that is, elements of the same type tend to be grouped together, resulting in long runs, or too many runs, in which case an element of one type tends to be followed by an element of the opposite type resulting in very short runs and cycling. Thus, the decision rule is given by

5. Reject H_0 at α if either

$$R \leq R_{n_1, n_2, \alpha/2} \quad \text{or} \quad R \geq R_{n_1, n_2, 1-\alpha/2} \qquad (11.14)$$

Table 11.2 gives upper and lower critical values for R at significance levels of .10, .05, and .025. The table is based upon combinatorial formulas and thus is indexed by n_1, the number of elements of one type, and n_2, the number of elements of the other type, where $n_1 \leq n_2$. It is irrelevant which is identified as type 1 and which as type 2.

Example 11.3. When all components are in control a particular coating material should have an average viscosity reading of 4.5. Quality control samples each batch produced, and takes a viscosity reading on each sample. The last 14 batches all had viscosity readings within the required limits but it is also of interest to check for any patterns that might be developing in the process. The viscosity readings for the 14 batches are:

Batch:	1	2	3	4	5	6	7	8	9
Reading:	4.40	4.45	4.35	4.35	4.45	4.55	4.55	4.60	4.65
Batch	10	11	12	13	14				
Reading:	4.70	4.60	4.65	4.55	4.60				

Choosing an α of .10 the hypotheses are

H_0: random occurrence H_1: nonrandom occurrence

with rejection of H_0 occurring if either

$$R \leq R_{n_1, n_2, .05} \quad \text{or} \quad R \geq R_{n_1, n_2, .95}$$

To determine R, n_1, and n_2 each observation is compared to the standard of 4.5 and identified as being either above or below the average value. (We assume symmetry.) The identifications are given below.

$$\underline{B \quad B \quad B \quad B \quad B} \quad \underline{A \quad A \quad A \quad A \quad A \quad A \quad A \quad A \quad A}$$

TABLE 11.2. Critical Values for the Total Number of Runs (R). R_{n_1, n_2}, Such That $P(R \leq R_{n_1, n_2, \alpha}) \leq \alpha$ and $R_{N_1, n_2, 1 - \alpha}$ Such That $P(R \geq R_{n_1, n_2, 1 - \alpha}) \leq \alpha$

(n_1, n_2)	$\alpha = .025$	$\alpha = .05$	$\alpha = .10$	$1 - \alpha = .90$	$1 - \alpha = .95$	$1 - \alpha = .975$
(2, 3)	—	—	—	5	—	—
(2, 4)	—	—	—	—	—	—
(2, 5)	—	—	2	—	—	—
(2, 6)	—	—	2	—	—	—
(2, 7)	—	—	2	—	—	—
(2, 8)	—	2	2	—	—	—
(2, 9)	—	2	2	—	—	—
(2, 10)	—	2	2	—	—	—
(3, 3)	—	—	2	6	—	—
(3, 4)	—	—	2	7	7	—
(3, 5)	—	2	2	7	—	—
(3, 6)	2	2	2	—	—	—
(3, 7)	2	2	3	—	—	—
(3, 8)	2	2	3	—	—	—
(3, 9)	2	2	3	—	—	—
(3, 10)	2	3	3	—	—	—
(4, 4)	—	2	2	8	8	—
(4, 5)	2	2	3	8	9	9
(4, 6)	2	3	3	9	9	9
(4, 7)	2	3	3	9	9	—
(4, 8)	3	3	3	9	—	—
(4, 9)	3	3	4	9	—	—
(4, 10)	3	3	4	—	—	—
(5, 5)	2	3	3	9	9	10
(5, 6)	3	3	3	9	10	10
(5, 7)	3	3	4	10	11	11
(5, 8)	3	3	4	10	11	11
(5, 9)	3	4	4	10	11	—
(5, 10)	3	4	5	11	11	—
(6, 6)	3	3	4	10	11	11
(6, 7)	3	4	4	11	11	12
(6, 8)	3	4	5	11	12	12
(6, 9)	4	4	5	11	12	13
(6, 10)	4	5	5	12	12	13
(7, 7)	4	4	5	11	12	12
(7, 8)	4	4	5	12	13	13
(7, 9)	4	5	5	12	13	13
(7, 10)	5	5	6	13	13	14
(8, 8)	4	5	6	12	13	14
(8, 9)	5	5	6	13	14	14
(8, 10)	5	6	6	13	14	15
(9, 9)	5	6	6	14	14	15
(9, 10)	5	6	7	14	15	16
(10, 10)	6	6	7	15	16	16

Each underlined set is a run. We have that $R = 2$, $n_1 = 5$, $n_2 = 9$ with $R_{5,9,.05} = 4$ and $R_{5,9,.95} = 11$. Since $R = 2 < 4$ we *reject* H_0 and conclude that there appears to be a pattern (in fact a trend) developing in the viscosities of the batches.

The runs test can also be used to test for a trend or a cyclical pattern in the data over time. For alternatives of this type the median of the sample is determined and each observation is identified as being above or below the median value with observations equal to the median being ignored. If an upward trend is present, we would expect the initial values to be below the median and later ones to be above the median.

If a downward trend is present, these should be reversed. In either case we would expect a small number of runs. A cyclical pattern would be indicated by short alternating sequences of values above and below the median. A situation of this type would be indicated by a large number of runs. These three situations are illustrated in Figures 11.2a, 11.2b, and 11.2c respectively.

As indicated in Figure 11.2, trend and cyclical alternatives lead to one-sided tests. The hypotheses and decision rules are formalized in the following statements. For trend alternatives the hypotheses are

$$H_0: \text{random pattern} \qquad H_1: \text{trend pattern}$$

and we reject H_0 at α if

$$R \le R_{n_A, n_B, \alpha} \tag{11.15}$$

where n_A is the number of observations above the median and n_B is the number of observations below the median. For the cyclical alternative we have

$$H_0: \text{random pattern} \qquad H_1: \text{cyclical pattern}$$

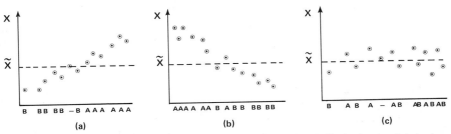

FIGURE 11.2. Illustrations of possible trend, (a) and (b), and cyclical patterns, (c), in data with corresponding numbers of runs: (a) 2; (b) 4; (c) 11.

with H_0 being rejected at α if

$$R \geq R_{n_A, n_B, 1-\alpha} \qquad (11.16)$$

The mean and variance of R are given by

$$\mu_R = \frac{2n_1 n_2}{n_1 + n_2} + 1 \quad \text{and} \quad \sigma_R^2 = \frac{2n_1 n_2(2n_1 n_2 - n_1 - n_2)}{(n_1 + n_2)^2(n_1 + n_2 - 1)}$$

As n_1 and n_2 approach infinity with their ratio remaining constant, the distribution of R approaches a normal distribution. Thus for sample sizes too large for n_1, n_2, to be included in Table 11.2 we calculate

$$Z = \frac{R - \mu_R}{\sigma_R} \qquad (11.17)$$

and reject for (a) the two-sided alternative if the absolute value exceeds $+z_{\alpha/2}$, (b) trend alternatives if $Z < -z_\alpha$, and (c) cyclical alternatives if $Z > +z_\alpha$.

11.6. SIGN TEST FOR PAIRED DATA

The sign test of Section 11.3 can also be used as a test for identical location parameters with paired samples. As such, it provides a nonparametric alternative to the paired t test of Chapter 6. For this application the basic assumptions of the sign test are altered as follows.

Assumptions for the Sign Test for Paired Data

  i.   The underlying populations are continuous.
 ii.   The observations constitute a random sample of matched pairs.
iii.   The pairs are independent of each other.
 iv.   At least nominal data are available; that is, within each pair it is possible to classify each member of the pair into one of two mutually exclusive categories.

Under these assumptions we are able to test for identical population medians (or means if we can assume symmetry). The procedure consists of subtracting the second observation in each pair from the first and recording the sign of the difference. The tests that follow will correspond to the differences being taken in this direction. If the direction is reversed the decision rules corresponding to the one-sided alternatives must be altered to compensate for the reversal. The logical derivation then follows the same line of reasoning as was presented in Section 11.3.

For hypotheses of the form

$$H_0: Me_1 = Me_2 \qquad H_1: Me_1 \neq Me_2$$

we reject H_0 at the specified α if either

$$P\left(X \leq n_+ | p = \tfrac{1}{2}, n = n_+ + n_- \right) < \frac{\alpha}{2}$$

or if

$$P\left(X \geq n_+ | p = \tfrac{1}{2}, n = n_+ + n_- \right) < \frac{\alpha}{2}$$

If the hypotheses are

$$H_0: Me_1 \leq Me_2 \qquad H_1: Me_1 > Me_2$$

we reject H_0 if

$$P\left(X \geq n_+ | p = \tfrac{1}{2}, n = n_+ + n_- \right) < \alpha$$

The other possible hypotheses are

$$H_0: Me_1 \geq Me_2 \qquad H_1: Me_1 < Me_2$$

and H_0 is rejected if

$$P\left(X \leq n_+ | p = \tfrac{1}{2}, n = n_+ + n_- \right) < \alpha$$

Example 11.4. To evaluate two types of rust inhibitors 20 steel rods were selected and half of each rod was coated with type I and the other half with type II. The rods were then subjected to various soil, weather, and climactic conditions for 1 year. At that time the rods were inspected visually and the two halves of each rod compared. The hypotheses of interest are

$$H_0: Me = Me \qquad H_1: Me \neq Me$$

and α was selected to be .10.

The results of the inspection with the more resistant appearing half identified by type of inhibitor are

Inhibitor: II II I II II I II I I II II II I II
II II II I II II

Letting II correspond to "a minus" and I to "a plus" we have $n_+ = 6$, $n_- = 14$ with H_0 being rejected if either

$$P(X \leq 6|20, .5) < .05$$

or

$$P(X \geq 6|20, .5) < .05$$

Using the binomial tables these probabilities are

$$P(X \leq 6|20, .5) = .0577 \quad \text{and} \quad P(X \geq 6|20, .5) = .9793$$

Since $.0577 \nless .05$, H_0 cannot be rejected at a significance level of .10. Therefore it cannot be concluded that the inhibitors differ in effectiveness.

Situations in which this procedure is applicable are identical to those for the paired t. The ideal experiment would be of the before–after type. Another way of satisfying the paired assumption is to match the experimental subjects on pertinent variables before treatment. The discussion and words of caution pertaining to the sign test in Section 11.3 are also applicable here.

11.7. SIGNED-RANK TEST FOR PAIRED DATA

If it is logical to add the assumptions of ordinal data and symmetric populations to those of the previous section, then we can use the signed-rank test of Section 11.4 to test the hypothesis of identical locations, that is, of no treatment effect. It is assumed that the experimental design randomly assigns treatments within a pair. The ARE of the test relative to the paired t test under the assumption of normality is .955. As the sample size decreases, the corresponding efficiency increases toward 1.00.

The procedure for the test is to calculate $X_{i1} - X_{i2}$ for each pair, where X_{i1} is the first observation in the ith pair and X_{i2} is the second, and then to apply the steps indicated in Section 11.4. Once this is accomplished the following decision rules are applied to the appropriate hypotheses. For the two-sided alternative,

$$H_0: Me_1 = Me_2 \qquad H_1: Me_1 \neq Me_2$$

we reject H_0 at α if

$$T = \min(T_+, T_-) \leq T_{n, \alpha/2}$$

where n = number of nonzero differences.

For hypotheses of the form

$$H_0: Me_1 \le Me_2 \qquad H_1: Me_1 > Me_2$$

H_0 is rejected at α if

$$T_- \le T_{n,\alpha}$$

The remaining possibility is

$$H_0: Me_1 \ge Me_2 \qquad H_1: Me_1 < Me_2$$

and we reject H_0 at α if

$$T_+ \le T_{n,\alpha}$$

As with the one-sample signed-rank test, zero differences are discarded while tied differences are handled by the average rank procedure.

Example 11.5. To evaluate different photoinitiator replacements, experiments were conducted to determine scrub resistance. A secondary aim of the experiment was to try to determine if the percent gloss loss differed between two trials of scrub testing. The same UV cured clear coating was used with each of 16 different photoinitiator materials.

The scrub resistance was determined by curing a 3-mil drawdown of each formulation with UV radiation. Each was then subjected to two scrubbing trials. Each trial comprised 150 cycles on a Gardner scrub machine using 10 g of standard soap solution and a hog bristle brush. The initial and final gloss readings of each drawdown were recorded using a 60° gloss meter. The data in Table 11.3 represent the percent gloss loss for each trial.

To determine if differences in percent gloss loss between the two trials are indicated the signed-rank test will be used. The hypotheses are

$$H_0: Me_1 = Me_2 \qquad H_1: Me_1 \ne Me_2$$

In words, the null hypothesis is that there is no difference other than random variation while H_1 indicates that there are other differences present. For an α of .1, H_0 will be rejected if $\min(T_+, T_-) \le T_{n,.05}(n = n_+ + n_-)$. The computations for determining T_+, T_- are given in Table 11.4.

As we see from Table 11.4 there are 15 nonzero differences and $\min(T_+, T_-) = \min(54, 66) = 54$.

Table 11.1 gives $T_{15,.05} = 30$. On the basis of these data H_0 cannot be rejected and we conclude that the differences in the two trials appear to be due to random variation.

TABLE 11.3.   Percent Gloss Loss of Two Scrubbing Trials in Evaluating Photoinitiators

(Sample) Drawdown	Percent Gloss Loss	
	Trial 1	Trial 2
A	6.7	8.6
B	3.2	3.2
C	4.3	12.2
D	6.4	4.4
E	7.4	4.4
F	4.3	6.8
G	18.1	13.0
H	8.8	7.4
I	4.3	6.7
J	6.4	5.4
K	3.3	7.4
L	8.5	10.9
M	14.9	5.4
N	4.2	3.2
O	3.3	2.1
P	8.7	3.2

Note that in the example above the zero difference was discarded. Also the two 1's each received a rank of 1.5, the average of the first two ranks, while the two 2.4's received the rank of 7.5, the average of the seventh and eighth ranks.

Comments relating to the power and applicability of the procedure are identical to those stated in Section 11.4. The large sample approximation of Section 11.4 can also be applied in this situation. If the symmetry assumption of the signed-rank test is believed to be unreasonable for the

TABLE 11.4.   Computations for Example 11.5

Sample	A	B	C	D	E	F	G	H	I
Difference	1.9	0	−7.9	2.0	3.0	−2.5	5.1	1.4	−2.4
Rank of Difference	5	—	14	5	10	9	12	4	7.5
Negative Ranks	5		14			9			7.5

Sample	J	K	L	M	N	O	P
Difference	1.0	−4.1	−2.4	9.5	1.0	1.2	5.5
Rank of Difference	1.5	11	7.5	15	1.5	3	13
Negative Ranks		11	7.5				

Sum of negative ranks = 54; sum of positive ranks = 66

data in Example 11.5 then the sign test should be used. Otherwise, the stated α is only approximate at best.

11.8. MANN–WHITNEY *U* TEST FOR IDENTICAL DISTRIBUTIONS

Sections 11.6 and 11.7 provided us with procedures for handling paired data, that is, the subjects were matched on certain attributes before experimentation. In practice, matching is often impossible to accomplish and we must resort to selecting two independent random samples. When this is true the Mann–Whitney *U* test provides an alternative to the *t* test given in Chapter 6 for two independent samples. The *U* test is a variation of a test known as the Wilcoxon rank-sum test and the computational form of its statistic will use the rank sums. In reality, the procedure tests for any differences in the underlying distributions but it is particularly sensitive to differences in location, that is, in medians or means.

As with all statistical procedures there are certain assumptions inherent in the *U* test. These are listed below.

Assumptions of the Mann–Whitney *U* Test

  i. Both samples have been randomly selected.
 ii. The populations are infinite or else sampling is with replacement. Thus, the samples are independent.
iii. The populations are identical except ·for possible differences in location.
 iv. The data are at least ordinal.

In the second assumption, the condition that the population distributions are continuous is often imposed. The third assumption may be relaxed but it is noted that it is possible to reject H_0 with near certainty, even though it is true, for extreme violations of this condition when small samples are used. Most examples illustrating this point are contrived and would not be expected to occur in practical situations.

Letting x_i, $i = 1, 2, \ldots, n$, and y_j, $j = 1, 2, \ldots, m$, denote the two sets of sample observations where $n \leq m$, the test statistic, U_X, is defined to be

$$U_X = \sum_{j=1}^{m} u_j \qquad (11.18)$$

where u_j equals the number of x's which are smaller than Y_j. Alternatively, we could use

$$U_Y = \sum_{i=1}^{n} u_i$$

$$Y\ XY\ \ X\ X\ Y\ Y\ X\ \ X\ Y\ Y\ X\ Y\ X$$
$$\overline{Me_X = Me_Y}$$

(a)

$$X\ X\ XX\ \ X\ X\ Y\ X\ Y\ Y\ \ Y\ Y\ \ Y\ Y$$
$$\overline{Me_XMe_Y}$$

(b)

$$YY\ Y\ \ YY\ \ Y\ X\ X\ Y\ X\ X\ \ X\ X\ X\ X$$
$$\overline{Me_YMe_X}$$

(c)

FIGURE 11.3. Possible configurations of X's and Y's where
(a) $Me_X = Me_Y$ and $U_X = 23$, $U_Y = 26$
(b) $Me_X < Me_Y$ and $U_X = 55$, $U_Y = 1$
(c) $Me_X > Me_Y$ and $U_X = 2$, $U_Y = 54$

where u_i is the number of y's which are smaller than x_i. The determination of the u_j's (or u_i's) is facilitated by first ranking the combined samples from low to high and identifying each element in the array as to its origin. Then, if H_0 is true and the assumptions are satisfied, all possible configurations of nX's and mY's are equally likely with each having probability $1/C_n^{n+m}$. Under H_0 the X's and Y's would be expected to be interspersed and U_X should be approximately equal to U_Y, differing only because of unequal sample sizes and random variation (see Figure 11.3a). If the median of the X distribution is less than that of the Y distribution, then the X values would be expected to be grouped at the low end of the array resulting in a disproportionately large value of U_X (see Figure 11.3b). Conversely, if the Y's are smaller than the X's we expect U_X to be disproportionately small (see Figure 11.3c). Note that $U_X + U_Y = nm$.

An alternative method of calculating U_X and U_Y is to replace the observation with their ranks in the overall array and to calculate T_X and T_Y where

$$T_X = \text{sum of ranks associated with } X \text{ values}$$

and

$$T_Y = \text{sum of ranks associated with } Y \text{ values}$$

Then

$$U_X = nm + \frac{n(n+1)}{2} - T_X \qquad (11.19)$$

and

$$U_Y = nm + \frac{m(m+1)}{2} - T_Y \qquad (11.20)$$

Summarizing, the overall procedure is as follows.

1. State the hypotheses to be tested and select an α.
2. Select random samples from each of the populations being compared.
3. Rank all the values as if they constitute one large sample of size $n + m$.
4. Calculate U_X and U_Y.
5. (a) If the hypotheses in (1) are

$$H_0: Me_X = Me_Y \qquad H_1: Me_X \neq Me_Y$$

we reject H_0 at the nominal α if

$$U = \min(U_X, U_Y) \leq U_{n,m,\alpha/2} \qquad (11.21)$$

(b) For

$$H_0: Me_X \geq Me_Y \qquad H_1: Me_X < Me_Y$$

reject H_0 if

$$U_Y \leq U_{n,m,\alpha} \qquad (11.22)$$

(c) If the hypotheses are

$$H_0: Me_X \leq Me_Y \qquad H_1: Me_X > Me_Y$$

H_0 is rejected if

$$U_X \leq U_{n,m,\alpha} \qquad (11.23)$$

Critical values for U are given in Table 11.5. The table is indexed by n, the smaller sample size, m, the larger sample size, and α, the nominal probability of being less than or equal to U. The exact probability will be less than or equal to that stated at the top of the column.

Example 11.6. In a coating process the density of the coating material is of prime consideration. Two coating materials are being considered for use.

TABLE 11.5. Lower-Tail Critical Values for the Mann–Whitney U Statistic. $U_{n1, n2, \alpha}$ Such That $P(U \leq U_{n1, n2, \alpha}) \leq \alpha$

$(n_1 \leq n_2)$ n_1, n_2	$\alpha = .10$	$\alpha = .05$	$\alpha = .025$	$\alpha = .01$	$\alpha = .005$
(2, 3)	0	—	—	—	—
(3, 3)	1	0	—	—	—
(2, 4)	0	—	—	—	—
(3, 4)	1	0	—	—	—
(4, 4)	3	1	0	—	—
(2, 5)	1	0	—	—	—
(3, 5)	2	1	0	—	—
(4, 5)	4	2	1	0	—
(5, 5)	5	4	2	1	0
(2, 6)	1	0	—	—	—
(3, 6)	3	2	1	—	—
(4, 6)	5	3	2	1	0
(5, 6)	7	5	3	2	1
(6, 6)	9	7	5	3	2
(2, 7)	1	0	—	—	—
(3, 7)	4	2	1	0	—
(4, 7)	6	4	3	1	0
(5, 7)	8	6	5	3	1
(6, 7)	11	8	6	4	3
(7, 7)	13	11	8	6	4
(2, 8)	2	1	0	—	—
(3, 8)	5	3	2	0	—
(4, 8)	7	5	4	2	1
(5, 8)	10	8	6	4	2
(6, 8)	13	10	8	6	4
(7, 8)	16	13	10	7	6
(8, 8)	19	15	13	9	7
(1, 9)	0	—	—	—	—
(2, 9)	2	1	0	—	—
(3, 9)	5	4	2	1	0
(4, 9)	9	6	4	3	1
(5, 9)	12	9	7	5	3
(6, 9)	15	12	10	7	5
(7, 9)	18	15	12	9	7
(8, 9)	22	18	15	11	9
(9, 9)	25	20	17	14	11
(1, 10)	0	—	—	—	—
(2, 10)	3	1	0	—	—
(3, 10)	6	4	3	1	0
(4, 10)	10	7	5	3	2
(5, 10)	13	11	8	6	4
(6, 10)	17	14	11	8	6
(7, 10)	21	17	14	11	9
(8, 10)	24	20	17	13	11
(9, 10)	28	24	20	16	13
(10, 10)	32	27	23	19	16

The company would prefer material X since it is less expensive, unless its density is significantly lower than that of material Y. Thus they have stated the hypotheses to be tested as

$$H_0: Me_X \geq Me_Y \qquad H_1: Me_X < Me_Y$$

and selected a significance level of .01. Accordingly, H_0 is to be rejected if

$$U_Y \leq U_{n,m,.01}$$

To make the evaluation, random samples of eight observations on X and seven on Y were taken with the following densities being recorded.

$$X: \quad 3.28 \quad 3.51 \quad 2.98 \quad 3.16 \quad 3.34 \quad 2.95 \quad 3.17 \quad 3.40$$
$$Y: \quad 3.15 \quad 3.22 \quad 3.24 \quad 3.06 \quad 2.99 \quad 3.12 \quad 3.18$$

Ranking the data in ascending order gives (with the material identified):

Observation:	2.95	2.98	2.99	3.06	3.12	3.15	3.16	3.17
Material:	X	X	Y	Y	Y	Y	X	X
Observation:	3.18	3.22	3.24	3.28	3.34	3.40	3.51	
Material:	Y	Y	Y	X	X	X	X	

The ranks associated with the Y's are 3, 4, 5, 6, 9, 10, and 11 giving $T_Y = 48$ and $U_Y = (7)(8) + 7(7 + 1)/2 - 48 = 56 + 28 - 48 = 36$.

From Table 11.5 we find $U_{7,8,.01} = 7$. Since $36 \not\leq 7$ we cannot reject H_0 and conclude that the density for material X appears to be acceptable.

If tied values occur, use the previous average rank procedure. For sample sizes outside the range of Table 11.5 use the Z statistic given by

$$Z = \frac{U - \mu_U}{\sigma_U}$$

with rejection of H_0 indicated for values of Z less than $-z_{\alpha/2}$ in the two-sided case and for values of Z less than $-z_\alpha$ in both one-sided cases. The mean and variance of U are given by

$$\mu_U = \frac{nm}{2}$$

and

$$\sigma_U^2 = \frac{nm(n + m + 1)}{12}$$

The approximation will be good for both n and m greater than or equal to 25 and standard values of α. If n is considerably less than m and α is quite small, the approximation is a poor one.

The ARE of the U test relative to the comparable t test under normal conditions is .955 and has been shown to be never less than .864. For rectangular distributions the U test is just as powerful as the t test giving an ARE of 1.00 and in many situations has an ARE greater than 1.00, possibly approaching infinity.

NOTE. As mentioned at the beginning of this section, the U test is really a test of identical distributions. Thus, for it to be purely a test of location differences, the assumptions of equal variances and common shapes must be included with those stated earlier. If this is not the case, the test may become less sensitive to differences in location; for example, if the two populations are skewed in opposite directions over the same range of possible values.

11.9. SIEGEL–TUKEY TEST FOR DISPERSION

For the Mann–Whitney U test we assigned ranks to the data in the pooled sample corresponding to each observation's position in the array. Utilizing a different procedure for ranking, insofar as it remains independent of whether the observation is an X or a Y, will not affect the distribution of the rank sums, T_X and T_Y. Thus, the distribution of U_X and U_Y is also independent of the method used in assigning ranks. The Siegel–Tukey test utilizes this fact to assign the ranks in such a way that the U statistic becomes sensitive to differences in scale parameters, that is, in variances. As such, it provides an alternative to the F test for equal variances as presented in Chapter 6.

The assumptions of the test are similar to those of the Mann–Whitney U test. The only change in the assumptions is that the populations have distributions which may differ in variance but are otherwise identical. This assumption implies that the populations have the same median and mean. If this assumption is violated, the results of the test may be questionable. Thus in using this test rather than an F test we substitute the assumption of identical locations for that of normal populations and interval data.

The hypotheses we will be testing are the equality of variances and the one-sided alternatives. Except for the method of assigning ranks to the data the procedure will correspond to that of Section 11.8. We will first place the data of the combined samples into an array. Next we rank the data from both ends toward the middle as follows:

1. Assign a rank of 1 to the smallest value and ranks of 2 and 3 to the largest and next largest values, respectively.

2. Assign ranks of 4 and 5 to the second and third smallest values, respectively, and ranks of 6 and 7 to the third and fourth largest values, respectively, and continue the pattern until all $n + m$ ranks have been assigned.

After the ranks have been assigned, T_X, T_Y, U_X, and U_Y are calculated as before.

If the null hypothesis of equal variances is true, we would expect the two sets of observations to be interspersed throughout the array, yielding values for T_X and T_Y, and hence U_X and U_Y, of approximately the same magnitude. If $\sigma_X^2 < \sigma_Y^2$ the X values would be expected to be grouped in the middle of the array and to receive the high ranks. Thus T_X would be disproportionately large with U_X correspondingly small. Conversely, for $\sigma_X^2 > \sigma_Y^2$, the Y values become clustered in the middle with T_Y becoming large and U_Y becoming small. These situations are illustrated in Figure 11.4.

As an example of the ranking procedure we will assign ranks to the configurations just given. Ranking from both ends toward the middle we obtain

$$1 \quad 4 \quad 5 \quad 8 \quad 9 \quad 12 \quad 13 \quad 14 \quad 11 \quad 10 \quad 7 \quad 6 \quad 3 \quad 2$$

for all three configurations. The corresponding values for T_X and T_Y are

$$(a) \ T_X = 50, \quad T_Y = 55; \quad (b) \ T_X = 63, \quad T_Y = 42;$$
$$(c) \ T_X = 34, \quad T_Y = 71$$

From the above considerations we construct the tests which are appropriate to the hypotheses of interest. If we are interested in only

$$\underline{X \quad Y \quad Y \quad X \quad X \quad Y \quad X \quad Y \quad Y \quad X \quad Y \quad X \quad X \quad Y}$$
(a)

$$\underline{Y \quad Y \quad X \quad Y \quad X \quad X \quad X \quad Y \quad X \quad Y \quad X \quad X \quad Y \quad Y}$$
(b)

$$\underline{X \quad X \quad X \quad Y \quad Y \quad X \quad Y \quad Y \quad Y \quad Y \quad X \quad Y \quad X \quad X}$$
(c)

FIGURE 11.4. These are the possible configurations resulting when $n = m = 7$ and (a) $\sigma_X^2 = \sigma_Y^2$, (b) $\sigma_X^2 < \sigma_Y^2$, and (c) $\sigma_X^2 > \sigma_Y^2$.

detecting a difference, our hypotheses become

$$H_0: \sigma_X^2 = \sigma_Y^2 \qquad H_1: \sigma_X^2 \neq \sigma_Y^2$$

and we reject H_0 at the stated α if

$$U = \min(U_X, U_Y) \leq U_{n, m, \alpha/2} \qquad (11.24)$$

If the indicated hypotheses are

$$H_0: \sigma_X^2 \geq \sigma_Y^2 \qquad H_1: \sigma_X^2 < \sigma_Y^2$$

we reject H_0 if

$$U_X \leq U_{n, m, \alpha} \qquad (11.25)$$

For the remaining one-sided case

$$H_0: \sigma_X^2 \leq \sigma_Y^2 \qquad H_1: \sigma_X^2 > \sigma_Y^2$$

H_0 is rejected if

$$U_Y \leq U_{n, m, \alpha} \qquad (11.26)$$

The following example illustrates the total procedure used in testing one of the hypotheses.

Example 11.7. Refer to Example 11.6. Another consideration in the choice between the coating materials is the variability of the material density. It is important that the density from batch to batch be fairly uniform. Thus, before making a final decision the company has decided to test the variability of the materials. Since X is preferred over Y unless it is inferior the hypotheses are

$$H_0: \sigma_X^2 \leq \sigma_Y^2 \qquad H_1: \sigma_X^2 > \sigma_Y^2$$

Using α of .10 the test is to reject H_0 if

$$U_Y \leq U_{n, m, \alpha} = U_{7, 8, .10} = 16$$

From Example 11.6 the ranked data are

Observation:	2.95	2.98	2.99	3.06	3.12	3.15	3.16	3.17
Material:	X	X	Y	Y	Y	Y	X	X
Rank:	1	4	5	8	9	12	13	15
Observation:	3.18	3.22	3.24	3.28	3.34	3.40	3.51	
Materials:	Y	Y	Y	X	X	X	X	
Rank:	14	11	10	7	6	3	2	

Thus,

$$T_Y = 5 + 8 + 9 + 12 + 14 + 11 + 10 = 69$$

and

$$U_Y = 7(8) + 7(7 + 1)/2 - 69 = 15$$

Since $15 < 16$ we reject H_0 on the basis of these data and conclude that the variability between batches of the X material appears to be significantly higher than that of the Y material.

For sample sizes outside the range of Table 11.5 use the normal approximation with $Z = (U - \mu_U)/\sigma_U$ where

$$\mu_U = \frac{nm}{2} \quad \text{and} \quad \sigma_U^2 = \frac{nm(n + m + 1)}{12}$$

The decision rule corresponding to the two-sided alternative rejects H_0 if $Z < -z_{\alpha/2}$, and that corresponding to the one-sided alternatives will be to reject H_0 in both cases if $Z < -z_\alpha$.

For situations involving tied values use the average rank procedure. Note that in this case, though, you may not be averaging consecutive values. For example, if in Example 11.7 the third and fourth values in the ranked data had both been 2.99 then the ranks 5 and 8 would have been averaged to give 6.5.

11.10. KRUSKAL–WALLIS TEST

In Section 11.8 we discussed the Mann–Whitney U test for unequal location parameters using two independent random samples. The Kruskal–Wallis test is a direct extension of the U test to cover situations involving c independent random samples. Thus, it provides a distribution-free alternative to the F test used in one-way analysis of variance. As is the case with the U test the Kruskal–Wallis procedure is a test of identical distributions and is particularly sensitive to location differences. For $c = 2$ the test is identical to the rank-sum form of the U test.

Except for the obvious extension to cover c populations the assumptions of the procedure are identical to those listed in Section 11.8. For ease in reference they are restated below with the indicated changes.

Assumptions of the Kruskal–Wallis Test

i. Independent random samples of size n_j are selected from each of the c populations of interest (or are selected before application of the c treatments).

ii.   The populations are infinite or sampling is with replacement.
iii.   The underlying distributions are identical with respect to shape and scale.
iv.   No ties occur between observations in different samples.
v.   The level of measurement is at least ordinal.

Under the above assumptions the hypotheses to be tested are

$$H_0: \quad Me_1 = Me_2 = \cdots = Me_c$$
$$H_1: \quad \text{some } Me_j \text{ differ} \tag{11.27}$$

Then if H_0 is true, the distribution of the Kruskal–Wallis statistic, H, can be determined by considering all possible permutations of the ranks of the n observations in the combined sample. To calculate H we first rank all observations from 1 to n by increasing order of magnitude and calculate T_j, the sum of the ranks corresponding to the observations in the jth sample, for all $j = 1, 2, \ldots, c$. Then H is determined by the equation

$$H = \frac{12}{n(n+1)} \sum_{j=1}^{n} n_j \left(\overline{R}_j - \frac{n+1}{2} \right)^2 \tag{11.28}$$

where

$$n_j = \text{the number of observations in the } j \text{th sample}$$

$$n = \text{the total number of observations}$$

$$= \sum_{j=1}^{n} n_j \quad \text{and} \quad \overline{R}_j = \frac{T_j}{n_j}$$

The quantity $(n + 1)/2$ in the square term of equation (11.28) is the average of the ranks for the combined sample. If H_0 is true, each $\overline{R}_j$ has an expected value equal to $(n + 1)/2$. Thus, since H will equal zero only when all $\overline{R}_j = (n + 1)/2$, a value of zero for H indicates perfect agreement between the null hypothesis and the sample data. Conversely, large values for H will occur only when some of the $\overline{R}_j$'s differ significantly from $(n + 1)/2$ which happens if some samples receive a disproportionate share of either the high ranks or the low ranks. The occurrence of this would indicate that the observations corresponding to the aforesaid samples are grouped at the extremes of the array which implies rejection of the null hypothesis of equal medians. In summary, our rejection region will correspond only to large values of the statistic H.

Equation (11.28) can be shown to be algebraically equivalent to the following:

$$H = -3(n + 1) + \frac{12}{n(n+1)} \sum_{j=1}^{n} \frac{T_j^2}{n_j} \tag{11.29}$$

This is known as the computational form for H and involves fewer arithmetic operations.

Tables for the exact distribution of the H statistic are limited and exist only for very small sample sizes. For cases with $c = 3$ and all $n_j = 5$ critical values corresponding to an α of .1, .05, or .01 can be found in *Owen's Handbook of Statistical Tables* (1962). In other situations or if Owen's tables are not available, it has been proven that the distribution of H can be approximated with a χ^2 distribution having $c - 1$ df. As the sample sizes increase, the approximation improves and becomes exact in the limiting case.

As usual, we summarize the above procedure and the test in the following steps.

1. State the hypotheses to be tested and select α.
2. Select random samples of size n_j from each of the c populations under consideration, or select a random sample of size n from a single population and randomly assign n_j items to receive the jth treatment, $j = 1, 2, \ldots, c$.
3. Construct an array of the combined set of observations and replace each observation with its corresponding position in the array.
4. Calculate H using equation (11.29).
5. For hypotheses of the form,

$$H_0: \quad Me_1 = Me_2 = \cdots = Me_c$$
$$H_1: \quad \text{some } Me_j \text{ differ}$$

reject H_0 at α if

$$H > \chi^2_{c-1, \alpha} \tag{11.30}$$

where

$$\chi^2_{c-1, \alpha}$$

is found in Table A.4 of the Appendix. We illustrate the above procedure with an example.

Example 11.8. In attempting to determine if weathering properties of a particular housepaint differ by area, six locations were randomly selected from each of the following areas of the United States: Northeast, Southeast, Northwest, and Southwest. One house in each location was randomly selected and painted with the housepaint in question. The same preparation and painting crew was used for every house. The houses were examined periodically to check for evidence of requiring repainting. The number of months until it was determined that repainting was necessary are given below. One house in each of the Northeast and Southwest was damaged by fire, one in the Northwest incurred flood damage, and another in the Northeast was demolished for a parking lot. The sample sizes are reduced accordingly.

Northwest	Southeast	Northeast	Southwest
18.3	17.2	14.1	16.7
17.6	15.3	15.2	15.9
20.1	16.9	16.5	17.4
19.5	14.5	15.7	15.8
19.0	15.1		18.0
	16.4		

Using $\alpha = .05$, can it be concluded that the weathering properties of the housepaint differ by geographical area? The hypotheses indicated by the question are

$$H_0: Me_{NW} = Me_{SE} = Me_{NE} = Me_{SW} \qquad H_1: \text{some differ}$$

with α specified at .05. Thus we will reject H_0 if $H > \chi^2_{4-1, .05} = 7.815$. Placing the data into an array we have, with areas identified.

14.1	14.5	15.1	15.2	15.3	15.7	15.8	15.9	16.4	16.5
NE	SE	SE	NE	SE	NE	SW	SW	SE	NE
16.7	16.9	17.2	17.4	17.6	18.0	18.3	19.0	19.5	20.1
SW	SE	SE	SW	NW	SW	NW	NW	NW	NW

Replacing each observation in the original table with its position in the array we obtain the table of ranks and the corresponding rank sum for each sample.

	Northwest	Southeast	Northeast	Southwest
	17	13	1	11
	15	5	4	8
	20	12	10	14
	19	2	6	7
	18	3		16
		9		
T_j:	89	44	21	56

Squaring the rank sums gives

$$T_{NW}^2 = 7921, \qquad T_{SE}^2 = 1936, \qquad T_{NE}^2 = 441, \qquad T_{SW}^2 = 3136,$$

with $n = n_j = 5 + 6 + 4 + 5 = 20$. Then, using equation (11.29),

$$\begin{aligned}
H &= -3(n+1) + \frac{12}{n(n+1)} \sum_{j=1}^{c} \frac{T_j^2}{n_j} \\
&= -3(21) + \frac{12}{20(21)} \frac{7921}{5} + \frac{1936}{6} + \frac{441}{4} + \frac{3136}{5} \\
&= -63 + \frac{12}{420}(1584.20 + 322.67 + 110.25 + 627.2) \\
&= -63 + \frac{12}{420}(2644.32) = -63 + 75.552 = 12.552
\end{aligned}$$

Since $12.552 > 7.815$ we reject H_0 and conclude that the weathering property appears to differ by geographical area.

Our assumptions were such that, theoretically, ties should not occur. In practice, they do. If there are ties in the data, we will use the average rank procedure. Ties within the same sample will not reduce the value of H nor its distribution but those between samples will reduce the value of H, altering its distribution, and will make the test less sensitive to actual population differences. If numerous between sample ties occur, it is recommended that the user dichotomize the combined sample in terms of each observation's being above or below the overall median and construct a $2 \times c$ contingency table. If the sample sizes are large enough so that all $E_{ij} \geq 5$, the hypotheses may be tested using the χ^2 procedure of Chapter 9.

The ARE's of the Kruskall–Wallis test relative to the F test for a one-way analysis of variance are identical to those of the Mann–Whitney U test with respect to the t test for two independent samples. In specific situations the ARE can be as high as infinity but it is never lower than .864. If the populations are normally distributed the ARE is .955 and if the distributions are uniform it is 1.000.

The Kruskal–Wallis test can be adjusted to test H_0 versus one-sided alternatives and has been generalized to handle factorial designs testing both main effects and interactions. For a thorough treatment of the appropriate changes in the procedure, see Bradley's *Distribution-Free Statistical Tests* (1968).

11.11. FRIEDMAN'S TEST

In using the sign test for paired data (Section 11.6) we calculated the within pair differences and counted the resulting number of positive differences.

Equivalently, we could have ranked the data within each pair, assigning a 1 to the smaller value and a 2 to the larger value, and used as a statistic the squared difference between the rank sums of the two treatments. Friedman's test uses a direct extension of the latter procedure to two or more treatments. It is equivalent to the sign test when only two treatments are being considered.

The procedure for Friedman's test consists of ranking the observations within each row (matching set) and adding the ranks associated with each treatment. Then the test statistic, denoted by S, is calculated by comparing the rank sums for each treatment with the average rank sum. The statistic is

$$S = \sum_{j=1}^{c} \left(T_j - \frac{n(c+1)}{2} \right)^2 \tag{11.31}$$

and will have a random distribution; that is, determined by considering the results associated with all possible permutations of the c ranks within each matched set, provided the following assumptions are satisfied.

Assumptions for Friedman's Test

   i.   There are available n sets of c objects each with the elements in each set being similar before treatments.

  ii.   The assignment of treatments to objects within a set is random, or if the same unit receives all treatments, that there is no carryover effect. The latter part of the assumption can be approximately satisfied by administering the treatments in a random order.

 iii.   There are no ties within sets (rows) after treatments.

 iv.   The level of measurement within each row is at least ordinal. If the null hypothesis of no treatment effects, $H_0: Me_1 = Me_2 = \cdots = Me_c$ is true, then the ranks within each row should occur randomly. Hence, the treatment rank sums, T_j, would also be random and should equal the average rank sum. This average is given by $n(c+1)/2$ since the sum of the ranks, 1 to c, within each row is $c(c+1)/2$, the number of rows equals n, and the total, $nc(c+1)/2$, is divided by the number of treatments, c. If $S = 0$, T_j equals $n(c+1)/2$ for all j and there is perfect agreement with the null hypothesis. Conversely, if, for example, one treatment yields appreciably higher responses, say treatment 1, then it will receive all of the high ranks within each row. The high ranks will be c's and the corresponding sum would equal nc. The resultant value for S would be larger than should be expected if H_0 was true. Thus, the null hypothesis is rejected for larger values of S.

Equation (11.31) is algebraically equivalent to

$$S = \sum_{j=1}^{c} T_j^2 - \frac{c}{4}[n(c+1)]^2 \qquad (11.32)$$

where
   c = the number of treatments
   n = the number of matched sets
   $T_j, j = 1, 2, \ldots, c,$ = the sum of the ranks assigned to subjects
      receiving treatment j

Equation (11.32) is computationally easier than equation (11.31) and is recommended.

We summarize the overall procedure for performing Friedman's test in the outline below.

1. State the hypotheses of interest and select α.

2. Select a sample of n sets consisting of c matched elements in each.

3. Randomly assign a treatment to each element in each set and record the results, taking care to maintain the matches.

4. Within each set, rank the results in ascending order, replacing each observation with its rank, and calculate S using equation (11.32).

5. For the hypotheses

$$H_0: Me_1 = Me_2 = \cdots = Me_c \qquad H_1: \text{some } Me_j \text{ differ}$$

reject H_0 if

$$S \geq S_{c,n,\alpha} \qquad (11.33)$$

where values of $S_{c,n,\alpha}$ are given in Table 11.6 for selected values of c and n at nominal α's of .10, .05, .01, and .005.

Example 11.9. In evaluating four painting methods seven 4×8 ft sheets of plywood were selected and each cut into four 4×2 ft sections. Each section was prepared for painting in the same manner. Using seven different colors of paint each section of the same sheet was painted a given color but different methods were used for each section. After the required drying time three judges were asked to rate the sections within a color as to

TABLE 11.6. Critical Values for Friedman's Statistic.
$S_{c,n,\alpha}$ Such That $P(S \geq S_{c,n,\alpha}) \leq \alpha$

(c, n)	$\alpha = .10$	$\alpha = .05$	$\alpha = .01$	$\alpha = .005$
(3, 3)	18	18	—	—
(3, 4)	24	26	32	—
(3, 5)	26	32	42	50
(3, 6)	32	42	54	72
(3, 7)	38	50	62	86
(3, 8)	42	50	72	98
(3, 9)	50	56	78	114
(3, 10)	50	62	96	122
(3, 11)	54	72	104	146
(3, 12)	62	74	114	150
(3, 13)	62	78	122	168
(3, 14)	72	86	126	186
(3, 15)	74	96	134	194
(4, 2)	20	20	—	—
(4, 3)	33	37	45	—
(4, 4)	42	52	64	74
(4, 5)	53	65	83	105
(4, 6)	64	76	102	128
(4, 7)	75	91	121	161
(4, 8)	84	102	138	184

their appearances with a four indicating the best and one the worst. The ratings then were averaged and the results are recorded in the table below.

Color	Method 1	Method 2	Method 3	Method 4
1	1.33	4.00	2.67	3.00
2	2.33	4.00	2.00	1.67
3	1.67	3.33	2.33	2.67
4	2.67	3.67	1.33	2.67
5	3.00	4.00	2.00	1.00
6	1.00	3.67	2.00	3.33
7	2.00	3.33	1.00	3.67

The hypotheses to be tested at $\alpha = .05$ are

$$H_0: Me_1 = Me_2 = Me_3 = Me_4 \qquad H_1: \text{some } Me_j \text{ differ}$$

The corresponding decision rule implies rejection of H_0 if

$$S \geq S_{c,n,\alpha} = S_{4,7,.05} = 91$$

To calculate the value of S for the data we first rank the observations within colors and replace the observed value with its corresponding rank. The ranks under each method are then added to obtain the T_j's. This is accomplished in the following table.

Color	Method 1	Method 2	Method 3	Method 4
1	1	4	2	3
2	3	4	2	1
3	1	4	2	3
4	3	4	1	2
5	3	4	2	1
6	1	4	2	3
7	2	3	1	4
T_j:	14	27	12	17

Substituting into equation (11.32) we obtain

$$S = \sum_{j=1}^{c} T_j - \frac{c}{4}[n(c+1)]^2$$

$$= (14^2 + 27^2 + 12^2 + 17^2) - \frac{4}{4}[7(5)]^2$$

$$= (196 + 729 + 144 + 289) - (35)^2$$

$$= 1358 - 1225 = 133$$

Since $133 > 91$, we reject H_0 and conclude that the methods appear to differ in the final appearance.

For values of c and n outside the range of Table 11.6 the statistic $12S/nc(c+1)$ is approximately distributed as a chi-square variable with $c - 1$ df. In such cases an approximate test is obtained by rejecting the null hypothesis if

$$\frac{12S}{nc(c+1)} > \chi^2_{c-1, \alpha} \tag{11.34}$$

Friedman's test provides an alternative to the F test used in the randomized block design of Chapter 13. When compared to this F test, with the condition of normality satisfied, its ARE varies depending on the number of treatments being compared. For $c = 2$ the ARE is 0.637, the same as the sign test, and it increases with the number of treatments to a maximum value of 0.955. If the underlying distributions are non-normal, the actual ARE may be either above 0.955 or below 0.637 or between the two.

11.12. SPEARMAN'S RANK CORRELATION

Spearman believed that the assumptions required by Pearson's product moment correlation, namely, a bivariate normal distribution and interval measurement for both X and Y, are often unreasonable. Therefore, he derived an alternative procedure, with much less restrictive assumptions, by replacing the actual values in Pearson's formula, given below, with their corresponding ranks within each individual set of values.

$$r = \frac{\Sigma XY - \dfrac{(\Sigma X)(\Sigma Y)}{n}}{\sqrt{\left(\Sigma X^2 - \dfrac{(\Sigma X)^2}{n}\right)\left(\Sigma Y^2 - \dfrac{(\Sigma Y)^2}{n}\right)}}$$

Utilization of this equation assumes that the available data are an interval level of measurement. (The reader will recall that values for ρ are restricted to the interval -1 to $+1$, inclusive, with either -1 or $+1$ indicating a perfect linear relationship and 0 indicating no linear relationship.) Since r is calculated from sample data it is only an estimate of ρ. Using r to test if ρ differs from 0 requires that X and Y have a bivariate normal distribution. If the assumption of joint normality is satisfied, we are able to use the t distribution for constructing the test.

Spearman's procedure was derived by replacing the actual values in Pearson's formula with their corresponding ranks within each individual set of values. The distribution of Spearman's correlation coefficient, denoted r_s, that results from these substitutions can be derived if the following conditions are satisfied.

Assumptions for Testing Spearman's Coefficient

   i.   A random sample of n units is selected.

  ii.   A measurement of each variable, X, Y, is obtained for each unit in the sample.

 iii.   The underlying distribution of each variable is continuous or at least infinite (to eliminate ties).

 iv.   The data are at least ordinal.

The ARE of the test based upon r_s has a value of 0.912 when the assumption of normality is satisfied and a value of 1.000 if the distribution is uniform. For other distributions the ARE will vary.

The procedure for calculating r_s is summarized in the following steps. We assume a random sample of n pairs of observations has already been obtained.

1. Rank the n observations corresponding to X among themselves, replacing each value with its rank, R_{X_i}.
2. Rank the Y_i's among themselves, replacing each with its rank, R_{Y_i}.

Then r_s is calculated using the equation

$$r_s = \frac{\Sigma R_X R_Y - \dfrac{(\Sigma R_X)(\Sigma R_Y)}{n}}{\sqrt{\left(\Sigma R_X^2 - \dfrac{(\Sigma R_X)^2}{n}\right)\left(\Sigma R_Y^2 - \dfrac{(\Sigma R_Y)^2}{n}\right)}}$$

which can be simplified to the form

$$r_s = \frac{\Sigma R_X R_Y - \dfrac{n(n+1)^2}{4}}{\sqrt{\left(\Sigma R_X^2 - \dfrac{n(n+1)^2}{4}\right)\left(\Sigma R_Y^2 - \dfrac{n(n+1)^2}{4}\right)}} \tag{11.35}$$

If there are no ties in either the X's or the Y's, the denominator will reduce to

$$\Sigma R_X^2 - \frac{n(n+1)^2}{4} = \frac{n(n^2-1)}{12}$$

since ΣR_X^2 and ΣR_Y^2 will both equal $n(n+1)(2n+1)/6$, which is the sum of the squares of the first n integers.

An alternative method, which simplifies the arithmetic considerably, is available for calculating r_s. If there are no ties in the data, the resulting value will be identical to that obtained using Equation (11.35). When the number of ties is small and they are handled by the average rank procedure, the two methods yield slightly different values. If there are an excessive number of ties, equation (11.35) should be used. The alternate procedure is to calculate the differences between corresponding X and Y ranks, that is, $d_i = R_{X_i} - R_{Y_i}$, and then to use

$$r_s = 1 - \frac{6\Sigma d_i^2}{n(n^2-1)} \tag{11.36}$$

Spearman's rank correlation coefficient has a maximum value of 1, occurring when a perfect positive relationship exists. This can be seen from equation (11.36) since, in such a situation, $R_{X_i} = R_{Y_i}$ and $d_i = 0$ for all i. Conversely, if there is a perfect negative relation the smaller X values will correspond to the larger Y values resulting in $R_{X_i} = n - R_{Y_i} + 1$. That is, if $R_{X_i} = i$, then $R_{Y_i} = n - i + 1$ and $d_i = 2i - (n + 1)$, in which case it can be shown that

$$\Sigma d_i = \frac{n(n^2 - 1)}{3}$$

with a resultant r_s of -1. If there is no relation between X and Y, there should be no pattern between their respective ranks and the sum of the d_i^2 would be expected to be half-way between the extremes; that is, to equal $n(n^2 - 1)/6$ giving an r_s of zero. Thus r_s is also restricted to being within the interval from -1 to $+1$ inclusive and it would be expected to be close to the true value of ρ.

Utilizing the considerations above, we are able to test for a relationship between X and Y. To do so we:

1. State the hypotheses of interest and select α.
2. Select a random sample of n pairs of observations on X and Y.
3. Separately rank the X_i's and Y_i's assigning a rank of 1 to the smallest X and also to the smallest Y, and so on.
4. Calculate r_s using equation (11.36) [(11.35) if there is an excessive number of ties].
5. The decision rules for the appropriate hypotheses are given below.

**Hypotheses and Corresponding Decision Rules
Based on Spearman's Rank Correlation Coefficient.**

	Hypothesis Set		Decision Rule for a Nominal α
i.	$H_0: \rho \leq 0$	$H_1: \rho > 0$	Reject H_0 if $r_s > r_{n,\alpha}$
ii.	$H_0: \rho \geq 0$	$H_1: \rho < 0$	Reject H_0 if $r_s < -r_{n,\alpha}$
iii.	$H_0: \rho = 0$	$H_1: \rho \neq 0$	Reject H_0 if $|r_s| > r_{n,\alpha/2}$.

Critical values for r_s are given in Table 11.7. The table is indexed by n, the number of pairs of observations, and α, the nominal level of significance. The number, $r_{n,\alpha}$, in the body of the table is the largest value possible such that $P(r \geq r_{n,\alpha}) \leq \alpha$ if no relationship exists. Since the distribution of r_s is

TABLE 11.7.   Upper-Tail Critical Values for Spearman's Rank Correlation Coefficient.
$r_{n,\alpha}$ **Such That** $P(r \geq r_{n,\alpha}) \leq \alpha$

n	$\alpha = .05$	$\alpha = .025$	$\alpha = .01$	$\alpha = .005$
5	.900	—	—	—
6	.829	.886	.943	—
7	.714	.786	.893	—
8	.643	.738	.833	.881
9	.600	.683	.783	.833
10	.564	.648	.745	.794
11	.523	.623	.736	.818
12	.497	.591	.703	.780
13	.475	.566	.673	.745
15	.441	.525	.623	.689
16	.425	.507	.601	.666
17	.412	.490	.582	.645
18	.399	.476	.564	.625
19	.388	.462	.549	.608
20	.377	.450	.534	.591
21	.368	.438	.521	.576
22	.359	.428	.508	.562
23	.351	.418	.496	.549
24	.343	.409	.485	.537
25	.336	.400	.475	.526
26	.329	.392	.465	.515
27	.323	.385	.456	.505
28	.317	.377	.448	.496
29	.311	.370	.440	.487
30	.305	.364	.432	.478

symmetric about zero, lower tail critical values are obtained by merely multiplying the corresponding upper tail value by a negative one (-1).

The following example illustrates the computations required for determining the rank correlation coefficient and the test procedure. Equation (11.36) is used for the computations.

Example 11.10.   To attempt to determine if the drying time of a type of oil-base paint increases with the amount of pigment suspending agent used, the following hypotheses were formulated:

$$H_0: \rho \leq 0 \qquad H_1: \rho > 0$$

An α of .05 was selected and the following data were collected in a controlled environment and calculations completed.

Paint	Grams of Agent (X)	Drying Time (Y), hours	R_X	R_Y	d_i	d_i^2
1	.5	8.2	1	2	-1	1
2	1.0	7.6	2	1	1	1
3	1.5	9.1	3	5	-2	4
4	2.0	8.7	4	3	1	1
5	2.5	8.9	5	4	1	1
6	3.0	9.3	6	6	0	0
7	3.5	9.8	7	9	-2	4
8	4.0	9.5	8	7	1	1
9	4.5	9.7	9	8	1	1
10	5.0	10.1	10	10	0	0
						14

$$r_s = 1 - \frac{6\Sigma d_i^2}{n(n^2 - 1)} = 1 - \frac{6(14)}{10(99)}$$

$$= 1 - .0848 = .9152$$

From Table 11.7 we find that $r_{10,.05} = 0.564$. Since the alternative indicates a positive relationship, we reject H_0 if $r_s \geq 0.564$. The calculated value is 0.9152 which implies rejection of H_0 since $0.9152 > 0.564$. It appears that the drying time increases with increasing amounts of the suspending agent.

For values of $n > 30$ the t distribution may be used to obtain an approximate test. In such situations, the statistic

$$\frac{r_s\sqrt{n - 2}}{\sqrt{1 - r_s^2}}$$

will have an approximate t_{n-2} distribution.

11.13. SUMMARY

The procedures of this chapter are alternatives to the classical tests of Chapters 6, 9, and 10. A wealth of other nonparametric procedures exist for various types of situations. For a more extensive and complete discussion of other nonparametric alternatives, consult Bradley or Gibbons; both are listed in the references. For nonparametric interval estimation, the procedures of this chapter can be altered to yield exact distribution-free confidence intervals. Specific procedures are also given in the references.

REFERENCES

1. Anderson, L. B., Nonparametric statistics provide comparisons between distributions, *Chemical Engineering*, July 8, 1963, pp. 139–144.

2. Bradley, J. V., *Distribution-free statistical tests*, Prentice-Hall, Englewood Cliffs, NJ, 1968.

3. Baker, R. A., Subjective panel testing, *Ind. Qual. Control*, Vol. 19, No. 3, Sept. 1962, pp. 22–28.

4. Fraser, D. A. S., *Nonparametric Methods in Statistics*, Wiley, New York, 1957.

5. Gibbons, J. D., *Nonparametric Statistical Inference*, McGraw-Hill, New York, 1971.

6. Majek, J., *Nonparametric Statistics*, Charles University, Prague, Czechoslovakia, 1969.

7. Kendall, M. G., *Rank Correlation Methods*, 3rd ed., Hafner Publishing Co., New York, 1962.

8. Lehmann, E. L., *Nonparametrics: Statistical Methods Based on Ranks*, University of California, Berkeley, 1975.

9. Meek, G. E., *Distribution-Free Confidence Interval for the Coefficients in a Linear Model*, Dissertation, 1970.

10. Meek, G. E., and Turner, S. J., *Statistical Analysis for Business Decisions*, Houghton-Mifflin Co., Hopewell, NJ, 1983.

11. Olds, E. G., The five percent significance levels of sums of squares of rank differences and a correction, *Annals of Math. Stat.*, Vol. 20, 1949, pp. 117–118.

12. Quenouille, M. H., *Rapid Statistical Calculations*, Hafner Publishing Co., New York, 1959.

13. Savage, I. R., *Bibliography of Nonparametric Statistics*, Harvard University Press, Cambridge, MA, 1962.

14. Siegel, S., *Nonparametric Statistics for the Behavioral Sciences*, McGraw-Hill, New York, 1956.

15. Teegarden, K. L., Critical sum d^2 values for rank order correlations, *Ind. Qual. Control*, Vol. 16, No. 11, May, 1960, pp. 48–49.

16. Walsh, J. E., *Handbook of Nonparametric Statistics*, D. Van Nostrand, Princeton, NJ, 1962.

17. Wilcoxon, F., and Wilcox, R. A., *Some Rapid Approximate Statistical Procedures*, Lederle Laboratories, Pearl River, NY, 1964.

DESIGN AND OPTIMIZATION

ROGER E. ECKERT

School of Chemical Engineering
Purdue University
West Lafayette, Indiana

Should statistical design be used for all experiments? If not, what is the alternative? Experiments must be run and someone must decide the conditions. It is this decision that is the design. A strategy for deciding the conditions for a series of experiments must be decided.

The question might be asked: how much formal education have employees had in selecting the correct conditions under which to perform a series of tests to answer a technical question? In most organizations, the answer is little on the average and perhaps none for the majority of employees. Frequently, decisions such as these are regarded as just "common sense." Many individuals say, "I've planned many experiments and obtained useful results" (with a frequently convenient lapse of memory for those experiments which did not give useful results or which could not be unraveled). On the positive side, most science really is an extension of our common sense. However, without some formal schooling in physics, chemistry, and mathematics, it is extremely difficult to "go it alone." However, the backlog of information passed on to us by previous workers enables us to apply known principles to our research, development, and production, and to concentrate on the new aspects.

So it is with statistics. The simple situations for designing experiments are "common sense." The formal methods extend our common sense in a manner that is no more or less obvious than the principles of chemistry and physics. It is a rare person who can develop these principles alone and know the consequences without recourse to the formal structure of design of experiments. This structure originated in the field of statistics. Statistics deals with the prediction of useful results in the presence of random error. When random error is large, the design of experiments to obtain information is particularly essential. Thus, it was logical for statisticians to extend their formal treatment of random error to the design of experiments. The usefulness of designed experiments is not limited to cases of large random error or even any error for that matter. There are reasons to use designed experiments even when results are precise.

However, many experiments have error as large as the magnitude of the desired effects that are to be evaluated. Therefore, this error, as well as the design, is a matter of concern.

12.1. DEFINITIONS

As with all fields of specialization, there is a technical vocabulary that promotes communication once the meanings are understood. The following five definitions are necessary for our purpose:

1. Factor—One of the independent variables that can be set to a desired value (e.g., input temperature, chemical concentration, or feed rate). There are quantitative factors that have numerical values and qualitative factors such as solvent or type of additive.
2. Level—The numerical value or qualitative feature of each factor.
3. Treatment—Specific combinations of the levels, one from each factor, which are tested.
4. Response—The numerical (usually) result of an observation made with a particular treatment combination.
5. Factorial experiment—In a complete factorial design of an experiment, all combinations of all levels of the factors are tested. If any are repeated, all must be repeated to retain the balance of the factorial experiment.

12.2. FACTORIAL DESIGNS

One-Factor Case

First we will consider a descriptive experimental design which is simple and in agreement with "common sense" selection of treatment conditions.

The object is to determine the yield of a chemical reaction as a function of temperature. Over the temperature range of interest, 60–100 °C, we must select a number of temperature levels and perform chemical reactions to measure the yield. Here temperature is the factor and yield the response. If five temperature levels equally spaced at 10 °C intervals are a good common sense compromise between cost and information, then the design of this experiment is shown in Figure 12.1(1, a)*. Crosses represent the temperatures selected. The results of these five experiments are given in Figure 12.1(1, b). At low temperature, the reaction is incomplete and at high temperature leads to undesired secondary or side reactions yielding by-products. The highest yield of 38% is attained at 75 ° C, the maximum of the curve. Although the number of levels of temperature selected here are arbitrary, it truly depends on the precision needed and, of course, the complexity of the curve. If random error of a response is large, precision can be increased by repeating the experiment. A flat response needs less data than a steep-peak response. Since the complexity of the response curve is seldom known in advance, sequential experimentation is often used and will be discussed later. Although the experiment designed for this purpose was described in common sense terms, it is in reality a factorial experiment with five levels of one factor.

Two Factor Case

The response of an experiment almost always depends on more than one factor. What became of the other factors affecting our yields in the chemical reaction we are considering? Of course, the levels of other factors were held constant during the runs described. But has this met our objective of determining the dependence of yield upon temperature? In Figure 12.1(2, a) a second factor, ratio of the two reactants fed to the reactor, has been introduced. The two dimensions of the "table top" are used to display the levels of the factors and vertical height shows the yield at any combination of factor levels. Ratio was held constant at 0.4 mole/mole in determining the results shown in Figure 12.1(1, b). This same information can be seen in Figure 12.1(2, b) as the curve on the front face of the cube. If other levels of ratio such as 0.6 and 0.8 are chosen and held constant, while the yield is again determined as it depends on temperature only, curves represented on planes of constant ratio parallel to the front plane will result. The effect of temperature is then seen to be different for different ratios. In fact, we can determine the effect of ratio on yield by holding temperature constant. However, the more inclusive objective should now be more obvious and that is to define the dependence of yield jointly on the levels of temperature and ratio. In the table-top plane of the two factors (Figure 12.1(2, a), we see each "treatment" to be defined by the intersection of the temperature and ratio used. If five levels of ratio are selected, the factorial experiment is a two-factor, five levels each, and the number of individual runs required is 5 × 5 or 25. The criteria for choosing the number of levels are again complexity of the response curve (now a surface as shown in Figure 12.1(2, b), precision needed, and cost or effort required.

*Numbers refer to rows, letters refer to columns in the figure.

To emphasize the experimental design and simplify the two-dimensional representation of the three-dimensional Figure 12.1(2, b), a contour graph of this same information is used in the figure immediately below Figure 12.1(2, b). The table-top, two-factor plan is shown in the row labeled "2 (contour)" as the two dimensions of the paper. The third dimension, height or yield, is represented by projecting intersections of horizontal planes at

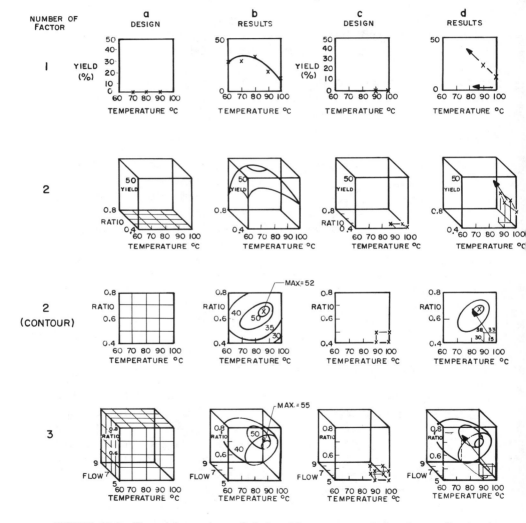

FIGURE 12.1. Factorial experimental design. The rows are numbered according to the number of factors under study. The two-factor case is displayed in two ways, as three-dimensional graphs and as contours. Columns a and b designate the experimental design and results, respectively, when five levels of each factor are used. Columns c and d give the corresponding graphs when only two-levels of each factor are used. Arrows represent path of steepest ascent.

fixed values of yield with the response surface of Figure 12.1(2, b) downward onto the table-top plane. This will be recognized as the contours as displayed on topological maps. The advantage of this method of presentation is that the design of the experiment is emphasized and accurately shown. The maximum yield of 52% attained at a temperature of 85°C and ratio of 0.7 is also readily apparent.

Three-Factor Case

Let us extend our objective to describe the yield dependence of the reaction on three factors. The three dimensions of a cube are now needed to display the levels of the three factors. Thus, the front contour plane of Figure 12.1(3, b) is identical with the contour graph of Figure 12.1(2, b). The third dimension, "into the page," is used for different levels of the third factor, flow rate. In the runs previously described, flow rate was held constant at 5 liters/hour. However, the yield contours change with flow rate. Selecting five levels of flow rate and performing a factorial experiment gives 125 (5 × 5 × 5) yield values which are represented as points of intersection of planes on Figure 12.1(3, a). The contours of yield can be formed resulting in curved surfaces, one for each yield. These resemble squashed headlights in a somewhat concentric arrangement as shown in Figure 12.1(3, b).

More Than Three Factors

Many experimental problems have more than three factors which affect the response and need to be studied. Visual displays in our three-dimensional world are cumbersome but the factorial design is readily extended mathematically in principle to any number of factors. Statistical methods such as analysis of variance and regression enable the effects to be judged for significance (for example, is the effect of temperature larger than we would obtain from random error alone?) and the magnitude described (yield versus temperature is a curved relationship with a maximum of 38% at 75°C).

The number of individual treatments needed in a factorial experiment with n factors each at five levels is 5^n. This is a large number and completely unreasonable for most situations involving three or more factors. The 125 treatments required by the three-factor experiment with five levels each might seem to be completely unwieldy. Perhaps even the 25 treatments of the two-factor case seem troublesome. The apparent problem for the engineer or paint chemist might be stated as follows: "This type of design is of no value for the work I do. I need an answer in a short time and a design of that size can't possibly be run within the time available." The solution of this problem of the minimum size factorial experiment will next be presented. We will consider how much information we can obtain from it and what information is lost by not performing the large factorial designs.

12.3. MINIMUM SIZE FACTORIAL DESIGNS

Two level factorial experiments, 2^n, have the following properties:

1. The minimum number of levels to obtain any measure of the effect of a factor on the response is two.
2. When two levels of each of the factors are chosen for a designed experiment, the smallest number of treatments possible for a factorial experiment is required.
3. In these experiments, effects are estimated with the greatest possible precision, since every response is used to estimate the effect of each factor on the response.

These properties will now be illustrated by examples with several numbers of factors.

One-Factor Case

Again consider the effect of temperature upon yield, other factors being held constant. The two-level factorial design is identical with the "common sense" plan we have been using. Two temperatures are selected [Figure 12.1(1, c)] and the effect of temperature on yield is the difference in the measured yield at high temperature minus that at low temperature [Figure 12.1(1, d)]. Thus, the effect of varying temperature from 90 to 100 °C is $(15 - 30) = -15$ (15% less yield). A further strategy of designing experiments follows from these two results. Higher yield is more likely to be in the direction of increasing slope of the yield/temperature graph if the temperature values are not too widely separated. (A maximum could be missed if they are.) Therefore, if the object is to reach high yield, select temperature level(s) beyond the one that gave the higher yield of the first pair of results. This is the simplest use of the "path of steepest ascent." The reduction in the number of treatments from, say the originally proposed five, would be unimportant for this one factor case, but can be substantial when more factors are explored, as will be shown.

Two-Factor Case

In Figure 12.1(2, c), a two-level factorial design in temperature and reactant ratio is shown with treatments at the corners of a square. It is necessary to select two ratios as well as two temperatures to design this experiment. The effect of each factor can be estimated as the average increase in yield at the high level over the yield at the low level of that factor [Figure 12.1(2, d), contour]. Thus, the effect of a temperature change from 90 to 100°C is

$$\frac{(15 - 30) + (33 - 38)}{2} = -10\% \ (10\% \ \text{decrease})$$

The effect of a ratio change from 0.4 to 0.5 is

$$\frac{(38 - 30) + (33 - 15)}{2} = +13\% \ (13\% \ \text{increase})$$

If the purpose of the study is to attain high yield, the direction of decreasing temperature and increasing ratio would be the preferred conditions. The factor that shows the larger effect, ratio, should be emphasized and the "path of steepest ascent" is the one in which the change in level for each factor is proportional to the magnitude of the effect of the factor.

This procedure is equivalent to fitting a plane for the response [Figure 12.1(2, d)] and following the gradient (steepest direction) in the plane. In this case, the "path" favors the larger effect due to increasing the ratio by 0.1 over decreasing the temperature by 10°C. These intervals are the scale units shown on Figures 12.1(2, c), 12.1(2, d). Sequential treatments (or runs) are then performed at suitable intervals along the path until the value of the response starts to decrease. As can be seen, this path leads toward the maximum yield, but does not necessarily intersect the exact maximum.

Since the selection of the "path of steepest ascent" depends on fitting a plane to experimental data, reasons for the path not leading to the exact maximum can be either experimental error and/or the true response is not planar. Virtually all experimental results are subject to some error. Fitting a plane is only a convenient approximation to the response and the plane must depart from the true curved response as the maximum is reached. However, more detailed study of the region of the maximum can be performed, if justified, using factorial experiments. The maximum response along the path is the place to center such a design.

Three-Factor Case

Figure 12.1(3, c) shows a two-level design in each of the factors temperature, ratio, and flow rate. The treatments are defined by the eight corners of the cube. Thus the eight treatments consist of the four previously described in the two-factor case at a flow rate of 5, and four corresponding treatments with the same levels of temperature and ratio but with a flow rate of 6.

Again, the effect of each factor is estimated as the average increase in yield at the high level of that factor over the yield at the corresponding low level. Since there are four comparisons to average, the effects of temperature and ratio will be altered from the values presented under the two-factor case. In fact, the effect of temperature, for example, need not be the same as all other conditions, but usually such an average is useful in planning further experiments. The "path of steepest ascent" is drawn as a three-dimensional vector, the components of which are again proportional to the effects of each of the factors (Figure 12.1(3, d)). This path appears similar to the two-factor case when viewed from the front plane, but is actually going back into the plane of paper. The double circle represents the true three-factor maximum of 55%. It is also behind the two-factor maximum of 52%. The "path of steepest ascent" does not necessarily intersect the true maximum for reasons discussed with the two-factor case. Again, planning another factorial experiment centered around the maximum experimental yield obtained from conducting treatments along the path will closely locate the maximum.

12.4. EXAMPLE OF AN EXPERIMENTAL DESIGN

Fast-Dry Alkyd for Transportation Finishes

There are a number of criteria or specifications which a satisfactory finish must meet. We will assume that the standard alkyd resin to be studied in this example meets all the test criteria except drying time. It has a minimum dry to stencil of 3 to 4 hours and must meet a 2-hour maximum dry for this application. Factors that affect or are thought to affect drying time are considered and the three listed below selected for experimentation in a two-level factorial design. The levels of the ingredients must be chosen to avoid such undesired phenomena as gelation, while attempting to meet the drying use-test criteria.

	Levels	
Factor	Low	High
1. Amount of soybean oil	20%	40%
2. Type of terminator	Benzoic acid	t-Butyl benzoic acid
3. Excess hydroxyl	15%	25%

From each of these eight resins of the factorial design, a paint sample will be made in a standard manner. These represent all possible combinations of the levels of the three factors listed above, designated as a $2 \times 2 \times 2 = 2^3 = 8$ treatments. Each treatment can be represented as the corner of a cube in three-dimensional space. The properties of each finish are to be evaluated. In addition to dry to stencil, impact resistance, durability, and so on, will be experimentally determined.

While the main objective of the experiment is to determine how to change the resin formulation to reduce the drying time, it is necessary to measure other responses to ensure that they do not change adversely with improved drying. If one of the eight finishes called for by this design meets the drying specification as well as being satisfactory in the other responses, the immediate problem is solved. However, if the drying time is not reduced to 2 hours, the guidance for further lowering of this response will come from comparison of the average results at the high and low level of each factor. The direction to alter each factor for shorter drying time to stencil will thus be established. This information can be combined into a "path of steepest ascent" (descent in this case) and further treatments can be performed along this path.

In paint formulation problems, a compromise of test properties is often necessary. The information on the several responses that must be compromised is well provided by this experimental design procedure. The proper compromise within the factorial is readily shown by comparing the analysis of the important responses. If the compromise must be made outside the two-level factorial experiment originally designed, then the direction is clearly indicated by comparing the paths for these responses. Sometimes the criteria cannot be met, at least with the study of the factors selected. This fact can then be established clearly from the results of the experiments described.

In summary, the advantages of statistical design and optimization are:

1. Good design requires the minimum number of runs to reach a desired objective.

2. It also provides the maximum protection against being misled by random error.

3. Results apply over a wider range of conditions (if this is consistent with the objectives).

4. Insight is gained into the multivariable system which reinforces scientific or technical thought.

5. Communications are clearer because there is preciseness and a framework to present objectives, plan of attack, and results.

BIBLIOGRAPHY

Bennett, C. A., and Franklin, N. L., *Statistical Analysis in Chemistry and the Chemical Industry*, Wiley, New York, 1954.

Blank, L., *Statistical Procedures for Engineering Management and Science*, McGraw-Hill, New York, 1980.

Box, G. E. P., Hunter, W. G., and Hunter, J. S., *Statistics for Experimenters*, Wiley, New York, 1978.

Davies, O. L. ed., *Statistical Methods in Research and Production*, Hafner Press, New York, 1957.

Davies, O. L., ed., *Design and Analysis of Industrial Experiments*, Hafner Press, New York, 1956.

Dixon, W. J., and Massey, F. J., *Introduction to Statistical Analysis*, McGraw-Hill, New York, 1969.

Draper, N. R., and Smith, H., *Applied Regression Analysis*, Wiley, New York, 1981.

Guttman, I., Wilks, S. S., Hunter, J. S., *Introductory Engineering Statistics*, Wiley, New York, 1982.

Hicks, C. R., *Fundamental Concepts in the Design of Experiments*, Holt, Rinehart and Winston, New York, 1964.

Himmelblau, D. M., *Process Analysis by Statistical Methods*, Wiley, New York, 1970.

Hultquist, R. A., *Introduction to Statistics*, Holt, Rinehart and Winston, New York, 1969.

Mendenhall, W., *Introduction to Linear Models and the Design and Analysis of Experiments*, Wadsworth, Belmont, CA, 1968.

Miller, I., and Freund, J. E., *Probability and Statistics for Engineers*, Prentice-Hall, Englewood Cliffs, NJ, 1965.

Natrella, M. G., *Experimental Statistics*, NBS Handbook 91, 1963.

Ostle, B., *Statistics in Research*, Iowa State University Press, 1963.

Peng, K. C., *The Design and Analysis of Scientific Experiments*, Addison-Wesley, Boston, 1967.

Snedecor, G. W., and Cochran, W. G., *Statistical Methods*, Iowa State University Press, 1967.

Volk, W., *Applied Statistics for Engineers*, McGraw-Hill, New York, 1969.

13

EXPERIMENTAL DESIGN AND ANALYSIS

THOMAS D. MURPHY

American Cyanamid Company
Consumer Products Research Division
Clifton, New Jersey

13.1. BASIC DESIGN PRINCIPLES

Benefits of Experimental Design

Experimental design is an important tool to aid the experimenter in coping with the complexities of technical investigation. The more useful techniques can be mastered quickly, and are summarized in this chapter. A comprehensive survey of experimental design is given in Box et al. (1). A few of the benefits of experimental design are discussed in this chapter.

Experimental design is an organized approach to the collection of information. Because of the balanced nature of good designs, the experimental results are easily interpreted, and the conclusions are often evident without extensive statistical analysis. Unfortunately, results from a haphazard experimental program are difficult to extract, even when extensive statistical techniques are applied.

In most scientific work, many variables can influence the outcome of an experiment and these variables can often interact in complex ways. Good designs allow for estimation and interpretation of these interactions, which can be accomplished only through study of more than one variable at a time. Accounting for interactions gives more certainty that optimal conditions can be found. Ignoring the presence of interactions and studying only one variable at a time will usually result in more experimentation than necessary.

379

An assessment of the precision of the experimental results is another benefit of good experimental design, which considers variability from several sources: experimental, sampling, and measurement processes. A statement of the reliability of the results from an experimental program lends higher credibility to the stated conclusions.

The Experimental Run

The structure of the experimental run, defined as a single experiment, is illustrated in Figure 13.1. The outcome of a run is a *response value* or observation made on a physical *experimental unit*. The value of the response will vary depending on the settings, or *levels*, of one or more experimental variables, or *factors*, which are under the direct control of the experimenter. Examples are as follows:

Hardness of a paint film using different vehicles
    Response: hardness
    Experimental unit: paint film on a substrate
    Factor: vehicle formula
Pigment strength versus batch reaction conditions
    Response: strength of a pigment in lab test
    Experimental unit: batch of chemicals in reaction
    Factors: coupling temperature and stirring rate
Rat-feeding toxicity test
    Response: percent survival of rats in dose group
    Experimental unit: dose group of rats
    Factor: concentration of chemical in diet feed to dose group

It is not unusual to have multiple responses from a single experimental run (e.g., a battery of physical tests run on a plastic strip). It is highly

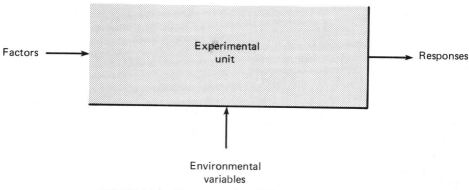

FIGURE 13.1. Representation of the experimental run.

desirable to have multiple factors in an experimental run. An *experimental design* comprises a group of experimental runs with a predefined objective.

In addition to the factors under the control of the experimenter, there exist other variables, termed environmental factors, which may also influence the outcome of a run. Some of these environmental factors can be identified and others will remain unknown. Examples of these are: (1) impurity level in a raw material which affects the cure rate of a formulation, (2) particle size of a powder in a drum which affects the settling rate of a suspension, (3) technique of operators tending a batch, and (4) ambient conditions, such as temperature or humidity. A good experimental design will consider the impact of environmental factors and plan for them.

An efficient experimental program will frequently consist of several experimental designs run sequentially. At each stage, the results of the preceding experimental design will determine the next design. Each design is kept to a size consistent with the objective. The sequential approach of many small designs (usually 4–16 runs each) allows for factors to be dropped and added and for factor levels to be changed as new information comes to light.

Response Variables

The response variables constitute the observation values or data obtained from the experimental run. If there are multiple responses, "trade-off" situations may occur, lending to compromises in arriving at an optimal response level. Although there are multivariate statistical techniques which allow for treatment of multiple responses, for example, Kramer (2), this chapter will consider only the separate evaluation of responses.

Responses can be classified as quantitative, qualitative, or quantal types. *Quantitative* responses are those which can be measured on a continuous numerical scale, such as Rockwell hardness, yield, or light absorbance, and these responses are the easiest to work with. *Qualitative* responses are those which are categorized in nature, such as pencil hardness of a film, odor, or appearance. If the categories can be ordered, then the category levels should be put in correspondence with a numerical scale. If possible, the intervals of the numerical scale should reflect the size of the qualitative differences between categories. Finally, the *quantal* responses comprise only two values, such as yes/no, pass/fail, or survived/died. These can be transformed to numerical values only through combining the results of several experimental units, yielding a percent responding value (i.e., percent survival).

For most responses a separate measurement step exists which is in addition to the experimental process. It is essential that the experimenter consider both the experimental variability and the measurement variability. If an aliquot, or portion, of the experimental unit is taken for measurement, then the sampling variability must also be considered. The overall variability is the total of experimental, sampling, and measurement variability and is usually expressed in terms of the standard deviation σ. The magnitude of σ will influence the number of experimental runs which are necessary in a design to attain the experimental objectives.

Factor Variables

The variables under the direct control of the experimenter during an experimental run are termed *factors*. The level of a factor is its value, or setting, during the experimental run. A combination of factor levels defines the conditions of an experimental run. For example, in the study of a pigment batch process the two factors might be reaction temperature and solvent type. In a particular experimental run the factor levels might be 70°C and water, respectively.

The number of factors in an experimental design will depend on the current stage of the experimental program. At the start there may be a large number of potential factors. As knowledge of the system increases, only the most important factors will remain. The range of levels under study for a given factor may also change as the experimental program proceeds.

Factor levels can be measured on either continuous or categorical scales. The continuous levels are the easiest to work with, examples being temperature, time, concentration, or stirring rate. Factors measured on a categorical scale include solvent type, catalyst type, or material of construction.

For two or more factors having continuous levels, functional combinations of factors may be more interpretable than the original factors. For example, the experimenter can work with the concentrations of two ingredients as (1) concentration of A and concentration of B or (2) ratio of A/B and total concentration $A + B$. Either approach may be appropriate, based on the experimenter's definition of the problem.

Experimental Units

An experimental design is composed of a definite number of experimental runs, each of which requires a distinct experimental unit. Ideally, these experimental units should be homogeneous, or indistinguishable from each other, but in practice this condition is rarely met. Examples are rubber, textile, or plastic specimens cut from a sheet in which there are gradients (e.g., thickness) in a horizontal, vertical, or radial direction. In formulation work, different batches of raw material may be used over the course of an experimental design. In abrasion tests, several test units may be run in parallel.

Differences in experimental units may affect the response value and be confounded with (or indistinguishable from) the effects of the factors of interest. To minimize any bias, the experimental units should be assigned to the experimental runs in a random manner. This can be done through random number tables, cards, dice, or by drawing numbers from a container. The experimental unit is assigned a number and this number is randomly assigned to an experimental run.

If the differences among experimental units can be identified then the experimental units can be categorized into groups, or *blocks*, of homogeneous units. This creates another factor in the design known as *blocking factor*. This process separates the differences in experimental units from other sources of experimental variability, reducing total variability of the responses. The minimum number of homogenous experimental units is termed the *block size*, and will be an important consideration in an experimental design where blocking variables are identified.

Environmental Factors

Closely related to the inhomogeneity of experimental units are the environmental factors, or extraneous variables not of direct interest in the investigation but influential to the response in an experimental run. If such factors can be controlled, they can be held at a constant level during the experimental design, reducing both systematic and random variability. If these factors can be identified but not controlled, they can be treated as *blocking factors*, with resultant decrease in experimental error. If they can be neither controlled nor identified, then randomizing the order of carrying out the experimental runs is the best insurance against confounding environmental factor effects with factor effects of interest.

13.2. MATHEMATICAL MODELS

Uses of Models with Experimental Design

A mathematical model relating the response value to the levels of one or more factors is an indispensable aid in the interpretation of results from an experimental design. To make this process more effective the design should be set up with a potential mathematical model in mind. Interpretation will be more economical in presentation when considered within the framework of a mathematical model.

When the response values and factor levels are continuous in scale, simple empirical models can be used as descriptive tools. First- or second-degree polynomials are usually sufficient; on occasion third-degree polynomials can be used. Transformations of either the response or factor scales can be performed, as will be discussed in Chapter 14. Qualitative factors are more difficult to handle, and for these, models dealing with comparisons of responses between factor levels can be used. Finally, in the more advanced stages of investigation, mechanistic models derived from first principles may be considered.

Single-Factor Empirical Models

For the single-factor case, the simplest empirical model is the first-order polynomial:

$$Y = b_0 + b_1 X_1 \qquad (13.1)$$

where

$$Y = \text{response value}$$

$$X_1 = \text{level of the factor } X_1$$

$$b_0 = \text{intercept parameter}$$

$$b_1 = \text{slope parameter for factor } X_1$$

as shown in Figure 13.2a. In this model b_1 describes the change in Y for a given change in X_1 over a limited range, and is useful as a preliminary description of this relationship. Unless the experimental factor levels include (or are close to) zero, the intercept parameter b_0 may be of little interest, since b_0 estimates the response at zero factor level.

The simplest experimental design to estimate the parameters b_0 and b_1 requires only two levels of the factor X_1, as shown in Figure 13.2b. For generality, the design levels of factor X_1 are coded as -1 (low level) and $+1$ (high level). The corresponding response values are Y_1 and Y_2, respectively. Under this coding, the intercept b_0 is the average of the responses Y_1 and Y_2, and predicts the response at $X_1 = 0$. The main effect of the factor X_1 is defined as the difference in response $(Y_2 - Y_1)$ between the high and low values of the factor. Under this coding, the slope b_1 is one-half the value of the main effect. To summarize:

$$b_0 = (Y_2 + Y_1)/2$$

$$b_1 = (Y_2 - Y_1)/2 \qquad (13.2)$$

If it is desired to state the model in terms of the original scale of X_1, the model can be "decoded" as given below.

First order model: $Y = b_0 + b_1 X_1$

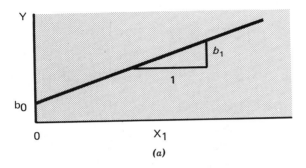

Estimation of First Order Effect

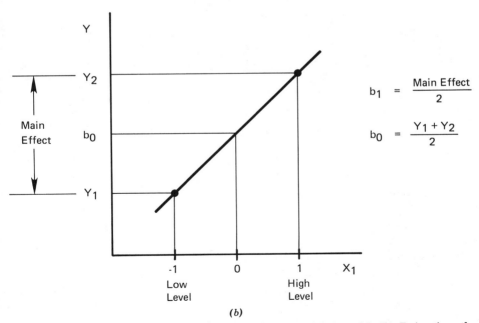

$$b_1 = \frac{\text{Main Effect}}{2}$$

$$b_0 = \frac{Y_1 + Y_2}{2}$$

(b)

FIGURE 13.2. First-order model—single factor. (a) Empirical model. (b) Estimation of first-order effect.

385

Let

$$T_1 = \text{factor value at } X_1 = -1$$

$$T_2 = \text{factor value at } X_1 = +1$$

$$\overline{T} = \frac{T_1 + T_2}{2}$$

$$W = \frac{T_2 - T_1}{2}$$

Then

$$Y = \left[b_0 - \frac{b_1 \overline{T}}{W} \right] + \left[\frac{b_1}{W} \right] \cdot T \tag{13.3}$$

The second-order polynomial is more general, covering situations where the response attains a maximum or minimum over the experimental range of the factor

$$Y = b_0 + b_1 X_1 + b_{11} X_1^2 \tag{13.4}$$

where Y, X_1, b_0, b_1 are as in equation (13.2) and $b_{11} = $ curvature parameter as shown in Figure 13.3a. The estimated optimum response occurs at the factor value $X_1 = -b_1/2b_{11}$.

The simplest experimental design to estimate the parameters b_0, b_1, and b_{11} requires three levels of the factor X_1, as shown in Figure 13.3b. The third level of factor X_1 is set halfway between the low and high levels and has the coded value 0 (midlevel). The parameters are estimated as follows:

$$b_0 = Y_2$$

$$b_1 = (Y_3 - Y_1)/2$$

$$b_{11} = (Y_3 + Y_1)/2 - Y_2 \tag{13.5}$$

The estimate b_{11}, termed the *curvature* effect, measures the deviation of the second-order model prediction from the first-order model prediction.

Two-Factor Empirical Models

For two factors, the simplest empirical model is the first-order or additive model as shown in Figure 13.4a:

$$Y = b_0 + b_1 X_1 + b_2 X_2$$

Second order model: $Y = b_0 + b_1 X_1 + b_{11} X_1^2$

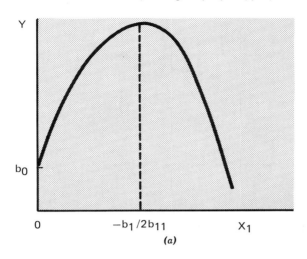

(a)

Estimation of Second Order Effect

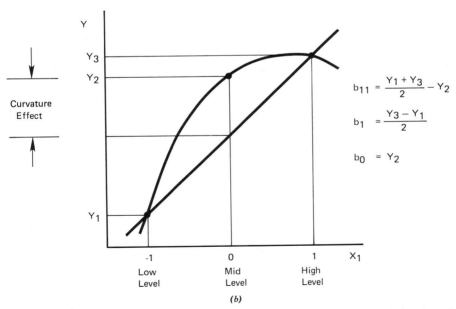

$$b_{11} = \frac{Y_1 + Y_3}{2} - Y_2$$

$$b_1 = \frac{Y_3 - Y_1}{2}$$

$$b_0 = Y_2$$

(b)

FIGURE 13.3. Second-order model—single factor. (*a*) Empirical model. (*b*) Estimation of second-order effect.

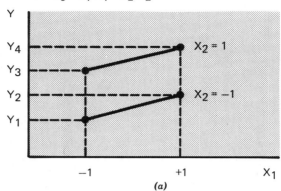

Two Factor Additive Model:

$$Y = b_0 + b_1 X_1 + b_2 X_2$$

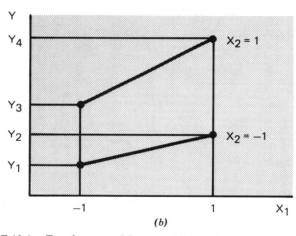

Two Factor Interactive Model:

$$Y = b_0 + b_1 X_1 + b_2 X_2 + b_{12} X_1 X_2$$

FIGURE 13.4. Two-factor models. (a) Additive model. (b) Interactive model.

where

Y, b_0, b_1, X_1 are as in equation (13.2)

b_2 = slope parameter for factor X_2

X_2 = level of factor X_2 (13.6)

This model states that the main effect of each factor is constant, or that the factors act independently on the response, over the range of the factors studied. In Figure 13.4a, the slope of the Y versus X_1 relationship is the same, regardless of the level of X_2.

The main effects for this model can be estimated using only three experimental runs (say, responses Y_1, Y_2, and Y_3), which would constitute a one-variable-at-a-time approach. The fourth response (say, Y_4) would be predicted on the basis of parallel lines.

The interactive model for two factors as shown in Figure 13.4b is:

$$Y = b_0 + b_1 X_1 + b_2 X_2 + b_{12} X_1 X_2 \qquad (13.7)$$

where

$$b_{12} = \text{interaction parameter of } X_1 \text{ and } X_2$$

This model adds the second-order cross-product term to the model and states that the slope of the Y versus X_1 relationship changes continually with the level of X_2:

$$Y = \underbrace{[b_0 + b_2 X_2]}_{\text{intercept}} + \underbrace{[b_1 + b_{12} X_2]}_{\text{slope}} X_1$$

The effect of b_{12} is to change the slope of (Y versus X_1) as X_2 changes in value. When $b_{12} = 0$, the slope of (Y versus X_1) remains constant and the additive model holds. A positive interaction ($b_{12} > 0$) means that the slope of (Y versus X_1) increases as X_2 increases in value.

The interaction concept is symmetric in X_1 and X_2 and we could just as well have interpreted the interaction in terms of Y versus X_2:

$$Y = \underbrace{[b_0 + b_1 X_1]}_{\text{intercept}} + \underbrace{[b_2 + b_{12} X_1]}_{\text{slope}} X_2$$

A positive interaction implies that the (Y versus X_2) relationship increases as the level of X_1 increases.

The estimation of the four terms of the interactive model requires four distinct experimental runs, for example, the four combinations of (X_1, X_2): $(-1, -1)$, $(-1, 1)$, $(1, -1)$, and $(1, 1)$ as shown in Figure 13.4b. Estimation of the model parameters for two or more factors will be deferred to a later section.

The full second-order model in two factors is

$$Y = b_0 + b_1 X_1 + b_2 X_2 + b_{11} X_1^2 + b_{12} X_1 X_2 + b_{22} X_2^2 \qquad (13.8)$$

where

$$b_{22} = \text{curvature effect of factor } X_2$$

Experimental design and parameter estimation for the full second-order models will be discussed in Chapter 14.

Three-or-More-Factor Models

The three-factor additive model is

$$Y = b_0 + b_1 X_1 + b_2 X_2 + b_3 X_3 \tag{13.9}$$

and is first-order in all factors. The graphical picture is shown in Figure 13.5a, where the slope of the Y versus X_1 relationship is constant over all values of X_2 and X_3.

When two-factor interactions are admitted to the model, we have

$$Y = b_0 + b_1 X_1 + b_2 X_2 + b_3 X_3 + b_{12} X_1 X_2 + b_{13} X_1 X_3 + b_{23} X_2 X_3 \tag{13.10}$$

where b_{12}, b_{13}, and b_{23} are the interaction parameters. A possible configuration of Y versus X_1 relationships is shown in Figure 13.5b. This model can be extended to the full second-order model by the addition of the three curvature terms: $b_{11} X_1^2$, $b_{22} X_2^2$, $b_{33} X_3^2$.

An additional effect often considered is the three-factor interaction $b_{123} X_1 X_2 X_3$, which is a third-order polynomial term. This may be interpreted as a change in the level of X_3 affecting the $X_1 X_2$ interaction in the Y versus X_1 and X_2 relationship:

$$Y = \underbrace{[b_0 + b_3 X_3]}_{\text{intercept}} + \underbrace{[b_1 + b_{13} X_3] X_1}_{\text{main effect } X_1} + \underbrace{[b_2 + b_{23} X_3] X_2}_{\text{main effect } X_2}$$
$$+ \underbrace{[b_{12} + b_{123} X_3] X_1 X_2}_{\text{interaction } X_1 X_2}$$

as shown in Figure 13.5c. A comparison of Figures 13.5b and 13.5c shows that the change of slope of Y versus X_1 with change in X_2 is unchanged in X_3 in Figure 13.5b, but is changed in X_2 in Figure 13.5c.

While meaningful three factor interactions can exist, the three-factor interaction is not often included in empirical models due to the difficulty of interpretation. Existence of three-factor interactions can be indicative of extreme nonlinearity of relationships which can be as effectively approximated by the full second-order model. For more than three factors, the interaction terms of higher order become even more difficult to interpret and are not considered in the model.

Qualitative Factor Models

For factors having qualitative or categorical levels the Y versus X_1 relationship cannot be depicted as a plot on a continuous scale of X_1. A compara-

Three Factor Additive Model

$$Y = b_0 + b_1 X_1 + b_2 X_2 + b_3 X_3$$

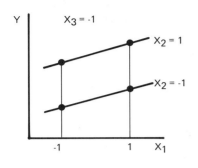 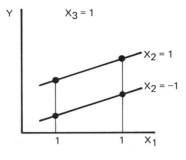

(a)

Three Factor Interactive Model

$$Y = b_0 + b_1 X_1 + b_2 X_2 + b_3 X_3$$
$$+ b_{12} X_1 X_2 + b_{13} X_1 X_3 + b_{23} X_2 X_3$$

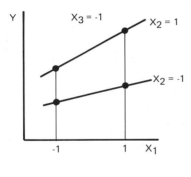 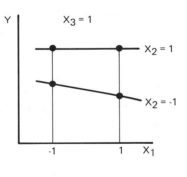

(b)

Three Factor Interactive Model

$$Y = b_0 + b_1 X_1 + b_2 X_2 + b_3 X_3$$
$$+ b_{12} X_1 X_2 + b_{13} X_1 X_3 + b_{23} X_2 X_3 + b_{123} X_1 X_2 X_3$$

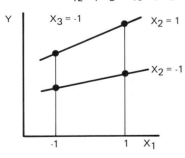 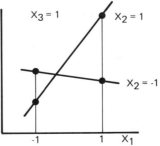

(c)

FIGURE 13.5. Three-factor models. (a) Additive model. (b) Two-factor interactions. (c) Three-factor interaction.

tive model for n levels of X_1 is

$$Y = \mu + A_i \qquad (13.11)$$

where

$$\mu = \text{mean response over all levels}$$
$$A_i = \text{differential effect of level } i$$
$$\text{from the mean } \mu$$
$$\Sigma A_i = 0$$

Usually, the differences between two levels $(A_i - A_j)$ are of more interest than the A_i or μ estimates themselves. For the special case of two levels, $(A_1 - A_2)$ is simply the main effect for that factor, and the qualitative factor can be combined with quantitative factors in a design. The interpretation of the main effect of a qualitative factor should not be that of a slope, but as a comparison between two levels or categories of the factor.

Mechanistic Models

Although empirical models are adequate for description of response-factor relationships, a mechanistic model may be desirable to develop a deeper understanding of the process under consideration. These models, developed from first principles, are usually nonlinear in the parameters (b's). For example, in the first-order chemical reaction $A \rightarrow B$, where Y, the concentration of compound B, varies as a function of time, X_1 is

$$Y = b_0 \exp(b_1 X_1) \qquad (13.12)$$

A second-order empirical model would fit data from this model quite well in the region of experimentation, but would diverge markedly from it outside the region. A good experimental design must not only be able to obtain good estimates for the model parameters, but also discriminate between two rival mechanistic models. Such designs are outside the scope of this book, but are covered in Ref. 1.

13.3. 2^k FACTORIAL DESIGNS

Uses and Limitations

Two-level factorial designs in k factors, or 2^k *factorial designs*, are experimental designs in which: (1) each factor is set at one of two levels (low or high), and (2) all 2^k combinations of factor levels are run. These designs are useful for estimation of parameters for interactive empirical models. The two levels of each factor are coded as -1 and $+1$ (or simply $-$ and $+$ in tables of factorial designs).

The two-factor design is the 2^2 design shown in Figure 13.6a. Each factor, X_1 and X_2, is run at two levels and all four combinations of levels are run. Figure 13.6a plots the design layout in the *factor space* ($X_1 - X_2$ plane) and also gives the design in tabular form. In the table, the experimental runs (rows) are defined by the settings of X_1 and X_2 (columns).

A practical example of a 2^2 design is depicted in Figure 13.6b, a study of mixing time and temperature on product color. The temperature levels were chosen at 50 and 70°C and the times were set at 0.5 hour and 2.0 hours. The 2^2 factorial experimental design comprised the four runs: (50°C, 0.5 hour), (70°C, 0.5 hour), (50°C, 2.0 hours) and (70°C, 2.0 hours).

The three-factor design is a 2^3 factorial design involving an additional factor X_3 also run at two levels. The design layout forms a cube in factor space (Figure 13.7a) and the table shows all eight runs (rows) as combinations of the three factors (columns).

To illustrate the 2^3 design, suppose a third factor, stirring rate, was added to the mixing study in the example just preceding. The levels

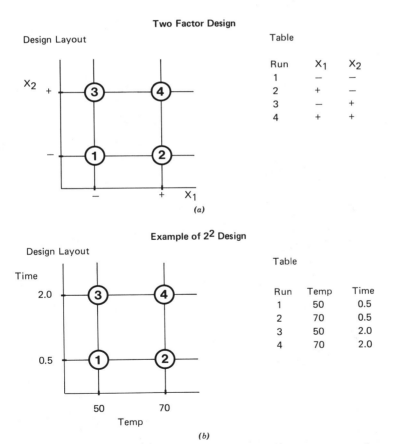

Two Factor Design

Design Layout

Table

Run	X_1	X_2
1	−	−
2	+	−
3	−	+
4	+	+

(a)

Example of 2^2 Design

Design Layout

Table

Run	Temp	Time
1	50	0.5
2	70	0.5
3	50	2.0
4	70	2.0

(b)

FIGURE 13.6. 2^2 factorial design. (a) Two-factor design. (b) Example of 2^2 design.

Three Factor Design

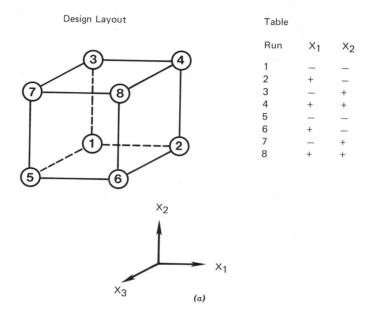

Design Layout

Table

Run	X_1	X_2	X_3
1	−	−	−
2	+	−	−
3	−	+	−
4	+	+	−
5	−	−	+
6	+	−	+
7	−	+	+
8	+	+	+

(a)

Example 2^3 Design

Design Layout

Table

Run	Temp	Time	Stir
1	50	0.5	16
2	70	0.5	16
3	50	2.0	16
4	70	2.0	16
5	50	0.5	28
6	70	0.5	28
7	50	2.0	28
8	70	2.0	28

(b)

FIGURE 13.7. 2^3 factorial design. (a) Three-factor design. (b) Example of 2^3 design.

selected were 16 and 28 rpm stirrer speed, resulting in the 2^3 design shown in Figure 13.7*b*.

Advantages of the 2^k factorial designs are that relatively few runs per factor are required but they cover the factor space in a uniform manner. The limitations of 2^k designs are that the number of runs increase geometrically as the number of factors increase and that 2^k designs cannot estimate curvature.

Selection of Factors and Levels

A complete layout of 2^k designs for up to five factors is shown in Table 13.1. The number of factors (k) determines the number of runs (2^k) for the

TABLE 13.1. Layout of 2^k Factorial Designs

Design	Run	Factor X_1	X_2	X_3	X_4	X_5
	1	−	−	−	−	−
	2	+	−	−	−	−
	3	−	+	−	−	−
2^2	4	+	+	−	−	−
	5	−	−	+	−	−
	6	+	−	+	−	−
	7	−	+	+	−	−
2^3	8	+	+	+	−	−
	9	−	−	−	+	−
	10	+	−	−	+	−
	11	−	+	−	+	−
	12	+	+	−	+	−
	13	−	−	+	+	−
	14	+	−	+	+	−
	15	−	+	+	+	−
2^4	16	+	+	+	+	−
	17	−	−	−	−	+
	18	+	−	−	−	+
	19	−	+	−	−	+
	20	+	+	−	−	+
	21	−	−	+	−	+
	22	+	−	+	−	+
	23	−	+	+	−	+
	24	+	+	+	−	+
	25	−	−	−	+	+
	26	+	−	−	+	+
	27	−	+	−	+	+
	28	+	+	−	+	+
	29	−	−	+	+	+
	30	+	−	+	+	+
	31	−	+	+	+	+
2^5	32	+	+	+	+	+

design. Although this table could be easily extended for $k > 5$, such large designs are uncommon in industry and are actually unnecessary.

The levels of each factor must then be selected and assigned to the − or + positions in the table. Each run (horizontal row) consists of a group of factor settings, which determines the actual design conditions.

Factor levels must be set far enough apart to ensure that a meaningful effect can be detected, but not so far apart that the experimental condition is not safe or feasible. It is the experimenter's responsibility to select the factor levels, based on scientific judgment and experience with similar processes. When the design table is completed, each run condition should also be checked for feasibility, safety, or any other important criteria.

Use of Center Points with 2^k Designs

One limitation of the 2^k designs is their inability to estimate curvature effects. As noted under single-factor empirical models, curvature estimation requires that three levels be run on each factor with other factor levels held constant, and this would lead to designs with a prohibitive number of runs. Efficient designs for full second-order models will be presented in Chapter 14.

In the early stages of an investigation the primary interest is to define important main effects and interactions. Nevertheless, it may be desirable to obtain some crude indication of curvature effects and additionally to explore the inner region of the design space. Both objectives can be accomplished by adding a *center point* to the design. The center point is defined as an experimental run with all factors set at values midway between their lower and upper levels (i.e., *midpoints*). The response for the center point run can be compared with the responses in the 2^k design to check for overall curvature (curvature effects due to one or more factors).

For two-factor designs the center point allows estimation of b_c (overall curvature) in the following model:

$$Y = b_0 + b_1 X_1 + b_2 X_2 + b_{12} X_1 X_2 + b_c \left(X_1^2 + X_2^2 \right) \qquad (13.13)$$

If the overall curvature term is important, then the individual curvature terms b_{11} and b_{22} could be estimated with additional experimentation. Following the 2^2 design example of 2^k factorial designs,

	−	+
Temperature	50°C	70°C
Stir Time	.05 hour	2.0 hours

the center point would be run at 60°C and 1.25 hours.

The center point concept is easily extended to a higher number of factors. In the 2^3 design example where the third factor was stirrer speed (16 rpm, 28 rpm), the center point would be run at 60°C, 1.25 hours, and 22 rpm.

In cases where one or more of the factors are qualitative, it will not usually be possible to attain a meaningful center point level. Suppose, in the example above, that the third factor was the material of construction of the mix tank instead of stir rate. If the levels of the third factor were carbon steel and stainless steel (as "low" and "high" levels) there would be no meaningful midpoint level for this factor. In this instance, two center point runs would be defined: (60°C, 1.25 hours, carbon steel) and (60°C, 1.25 hours, stainless steel). This would allow overall curvature estimates on the temperature and time factors for each level of material of construction.

Replication and Precision

The precision of a parameter estimate depends on the response variation σ and on the number of experimental runs used in the estimate. The precision of estimates of main effects, interaction effects, and overall curvature must be sufficient to detect meaningful differences. In a 2^k design, response data from all of the runs are used to estimate main effects and interactions. Therefore, the experimenter must set the total number of runs, N, that are required to achieve the desired precision.

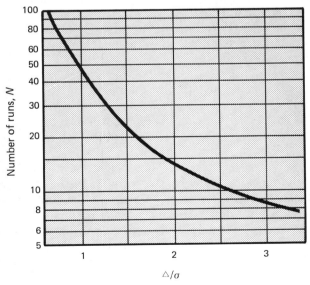

FIGURE 13.8. Design size as function of precision.

The first step is to determine the minimum change in response, Δ, that is of interest to detect with high assurance. Next an estimated value of σ should be obtained from a pilot experiment or prior experience with similar systems. The required number of runs N can be found from the relationship between N and Δ/σ in Figure 13.8.

For example, if we wished to detect an effect equal to twice the response error ($\Delta/\sigma = 2$), a value of $N = 14$ would be required. If a three-factor design is contemplated ($2^3 = 8$ runs) then this design would be replicated twice to achieve a total of 16 runs, which will meet the desired objective.

To qualify as a replicate, a run must be an entire repeat of the experiment and not just a part. For example, if a batch is prepared and analyzed in duplicate, the two results do not represent repeat runs, but only represent repeated measurements on the same run.

Blocking and Randomization

As discussed under experimental units and environmental variables, non-homogeneity of experimental units and the presence of environmental variables both give rise to blocking factors which can contribute to increased experimental error or biased estimates of effects. If a blocking factor can be identified, it should be added to the design as an additional factor. If blocking factors cannot be identified, the experimental units must be assigned to the runs by a random process.

When working blocking factors into a design, block size is of high importance. The simplest case occurs when block size is $\geq N/2$ (half the number of experimental runs or greater). Then only two levels of the blocking variable are required, and the blocking variable becomes an additional factor in the design.

For example, suppose a 2^3 factorial design is to be replicated twice, for a total of 16 runs. Suppose also that a batch of raw material will make 8–10 runs, and that raw material batches have been identified as a blocking factor. In this case two raw material batches will be required, eight runs per batch, and the blocking factor can be added as factor X_4 in a 2^4 factorial design.

If block size is less than $N/2$, then more than one factor must be allocated to the blocking variable. This will be discussed in Chapter 14.

When blocking variables cannot be identified, a random assignment of experimental units to runs will ensure that no biased (unconscious or otherwise) assignment is made. If the design is carried out over time, the order of experimentation should also be randomized. If the measurement process is carried out separately, then the testing order should also be performed in a random order.

13.4. ANALYSIS OF 2^k FACTORIALS

Estimation of Main Effects and Interactions

An introduction to the estimation of main effects, interactions, and curvature effects was presented under two-factor empirical models. A general procedure known as the *method of contrasts* will now be given for calculation of factor main effects and two-factor interactions for 2^k factorial designs. A 2^2 factorial will be used as a running example to illustrate the techniques. Table 13.2 gives a layout for the analysis of 2^2, 2^3, and 2^4 designs and this table can be extended to any larger 2^k factorial design.

Factor main effects are calculated as the difference between the average responses at the high and low factor levels. For a 2^2 design (Figure 13.6a), the main effect of X_1 is

$$\frac{Y_2 + Y_4}{2} - \frac{Y_1 + Y_3}{2} \qquad (13.14)$$

where Y_i is the response for run i. Similarly, the main effect for factor X_2 is

$$\frac{Y_3 + Y_4}{2} - \frac{Y_1 + Y_2}{2} \qquad (13.15)$$

The two-factor interaction effect for X_1 and X_2 is the difference between the average responses at the high and low levels of the cross product $X_1 X_2$.

TABLE 13.2. Analysis of 2^i Factorial Designs

Design	Run	X_1	X_2	X_1X_2	X_3	X_1X_3	X_2X_3	X_4	X_1X_4	X_2X_4	X_3X_4
	1	−	−	+	−	+	+	−	+	+	+
	2	+	−	−	−	−	+	−	−	+	+
	3	−	+	−	−	+	−	−	+	−	+
2^2	4	+	+	+	−	−	−	−	−	−	+
	5	−	−	+	+	−	−	−	+	+	−
	6	+	−	−	+	+	−	−	−	+	−
	7	−	+	−	+	−	+	−	+	−	−
2^3	8	+	+	+	+	+	+	−	−	−	−
	9	−	−	+	−	+	+	+	−	−	−
	10	+	−	−	−	−	+	+	+	−	−
	11	−	+	−	−	+	−	+	−	+	−
	12	+	+	+	−	−	−	+	+	+	−
	13	−	−	+	+	−	−	+	−	−	+
	14	+	−	−	+	+	−	+	+	−	+
	15	−	+	−	+	−	+	+	−	+	+
2^4	16	+	+	+	+	+	+	+	+	+	+

In Table 13.2, the $X_1 X_2$ column is the algebraic cross product of columns X_1 and X_2, which can be verified by multiplying the row entries in Table 13.2. Runs 1 and 4 correspond to the $+$ or high level of $X_1 X_2$ and runs 2 and 3 correspond to the $-$ or low level of $X_1 X_2$. Therefore, the estimate of the two-factor interaction $X_1 X_2$ is

$$\frac{Y_1 + Y_4}{2} - \frac{Y_2 + Y_3}{2} \qquad (13.16)$$

The interaction estimate can be rationalized by reference to Figure 13.4. If the model is additive (equal slopes of Y versus X_1 at both values of X_2), then $Y_4 - Y_3 = Y_2 - Y_1$ or $Y_4 - Y_3 - Y_2 + Y_1 = 0$. If the model is interactive then the interaction effect is proportional to the increase in slope of Y versus X_1 between the low level of X_2 and its high level.

To estimate the parameters of the interactive model the following calculations can be made:

$$b_0 = (Y_4 + Y_3 + Y_2 + Y_1)/4 = \text{average response}$$
$$b_1 = (Y_4 - Y_3 + Y_2 - Y_1)/4 = X_1 \text{ main effect}/2$$
$$b_2 = (Y_4 + Y_3 - Y_2 - Y_1)/4 = X_2 \text{ main effect}/2$$
$$b_{12} = (Y_4 - Y_3 - Y_2 + Y_1)/4 = X_1 X_2 \text{ interaction}/2 \qquad (13.17)$$

The factor 2 enters into the relationship between the b's and the effects since $\Delta X = 2$ under the $-1, +1$ factor coding.

Example 13.1. A designed study was made to determine the effect of mixing time and temperature on the yellowness index of a paint formulation.

$$Y = \text{yellowness index}$$
$$X_1 = \text{mixing temperature, 50 and 70°F}$$
$$X_2 = \text{mixing time, 0.5 and 2 hours}$$

The design gave the following results:

Run	Temperature	Time	Yellowness
1	50	0.5	6.0
2	70	0.5	8.0
3	50	2.0	7.5
4	70	2.0	8.5

To estimate the effects, use the portion of Table 13.2 for the 2^2 design and calculate average response for high and low levels of each factor and interaction as below:

Run	X_1	X_2	X_1X_2	Y
1	−	−	+	6.0
2	+	−	−	8.0
3	−	+	−	7.5
4	+	+	+	8.5
Average + level	8.25	8.0	7.25	
Average − level	6.75	7.0	7.25	
Difference	1.5	1.0	−0.5	

The model parameters are [from equation (13.1)]:

$$b_0 = (8.5 + 7.5 + 8.0 + 6.0)/4 = 7.5$$
$$b_1 = 1.5/2 = 0.75$$
$$b_2 = 1.0/2 = 0.5$$
$$b_{12} = -0.5/2 = -0.25$$

resulting in the model equation:

$$Y = 7.5 + 0.75X_1 + 0.5X_2 - 0.25X_1X_2$$

where $X_1 = \pm 1$ and $X_2 = \pm 1$ (coded).

A graphical summary is given in Figure 13.9. The temperature effect on yellowness is less at the higher stir time than at the lower stir time. Generally higher times and temperatures increase yellowness in this formulation over the ranges studied.

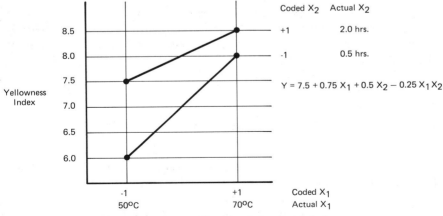

FIGURE 13.9. Graphical analysis of a 2^2 design.

Estimation of Precision

Main effects and interactions are calculated from response data that are subject to experimental error. In order to assess the reliability of these estimates it is first necessary to estimate the experimental error σ, which is best done through replicated measurements.

The sample standard deviation s, calculated from replicates, is an estimate of σ (see Chapter 1). For the special case of duplicate measurements $s = \sqrt{d^2/2}$ where d = difference in response between the duplicates, and this estimate has 1 degree of freedom (df).

Several estimates of σ: s_1, s_2, s_3, and so on, may be pooled together to form a single estimate:

$$s = \sqrt{\frac{\sum\limits_{i=1}^{p} \nu_i s_i^2}{\sum\limits_{i=1}^{p} \nu_i}} \tag{13.18}$$

where

$$\nu_i = \text{the degrees of freedom corresponding to } s_i$$

$$p = \text{the number of estimates}$$

This estimate has ν df where

$$\nu = \sum_{i=1}^{p} \nu_i \tag{13.19}$$

The precision of a main effect or interaction may be stated in the form of a confidence interval, or range of values of an effect that can be supported by the data at the desired confidence level. Confidence intervals have been discussed in Chapter 7.

The standard error of an effect for a 2^k design is

$$\frac{s}{\sqrt{N/4}} = \frac{2s}{\sqrt{N}}$$

where N is the number of runs in the factorial design. The confidence interval for an effect is

$$(\text{effect}) \pm t(\text{std error of effect})$$

or

$$(\text{effect}) \pm 2ts/\sqrt{N}$$

where t is the value of Student's t corresponding to the desired confidence level and degrees of freedom for s (Appendix Table A.3). The above confidence interval estimate holds for both main effects and two factor interactions.

Example 13.2. Returning to the example of the previous section, Example 13.1, suppose duplicate experiments were carried out at each run condition resulting in the following data:

Run	X_1 Temperature	Y_2 Time	Yellowness Results	(Y) Average	s_i^2	ν_i
1	70	0.5	5.8, 6.2	6.0	0.08	1
2	80	0.5	8.3, 7.7	8.0	0.18	1
3	70	2.0	7.5, 7.5	7.5	0.00	1
4	80	2.0	8.4, 8.6	8.5	0.02	1

The pooled standard deviation

$$s^2 = \frac{\Sigma s_i^2}{4} = \frac{0.28}{4} = 0.07$$

$$s = 0.26 \text{ with 4 df}$$

The value of t for 95% confidence and 4 df is 2.776. The standard error of effect is

$$\frac{2s}{\sqrt{N}} = \frac{2(0.26)}{\sqrt{8}} = 0.18$$

since $N = 8$ runs in the design. The 95% confidence intervals on the effects are:

Temperature main effect $\qquad 1.5 \pm (2.776)(0.18) = 1.5 \pm 0.5$

Time main effect $\qquad 1.0 \pm (2.776)(0.18) = 1.0 \pm 0.5$

Time + temperature interaction $\qquad -0.5 \pm (2.776)(0.18) = -0.5 \pm 0.5$

Graphical Analysis

A good visual summary of the data can be accomplished by plots of Y versus one factor at various levels of the other factors. This has been shown in Figure 13.9 for our two-factor example. Illustrations of three- and four-factor examples will be shown in later sections.

Another graphical technique for three factors is to plot the response averages at each experimental condition on the corners of a cube. For four factors, two cubes can be shown side by side. This technique will also be illustrated in later sections.

Estimation of Overall Curvature

Inclusion of one or more center points in a 2^k factorial design allows estimation of overall curvature as well as gaining response information on the interior of the experimental space. Furthermore, the center point may be replicated to gain precision information without destroying the balance of the design as would happen with partial replication of the 2^k design. The number of center points will not affect the estimation of main effects or interactions or the precision of these estimates.

The estimate of overall curvature is simply the difference between the average response of the 2^k factorial points, $(\overline{X}_f)$ and the average response of the center points $(\overline{X}_c)$. The precision of the overall curvature effect can be stated in the form of a confidence interval:

$$\left(\overline{X}_c - \overline{X}_f\right) \pm ts\sqrt{\frac{1}{N} + \frac{1}{C}} \tag{13.20}$$

where

$$N = \text{number of runs in the } 2^k \text{ factorial design}$$
$$C = \text{number of center point runs}$$

with t and s as defined in the preceding section.

Example 13.3. In the preceding study of yellowness, three center point runs were made at 60°C and 1.25 hours, giving yellowness values of 7.5, 8.0, and 7.6. Then $\overline{X}_c = 7.7$, $s_c^2 = 0.07$, $s_c = 0.26$, with 2 df. Pooling the estimate s_c with the previous estimate of σ from the duplicate factorial runs:

$$s^2 = \frac{(2)(0.07) + 4(0.07)}{6} = 0.07 \qquad s = 0.26 \quad \text{with} \quad 6 \text{ df}$$

For 95% confidence and 6 df, $t = 2.447$.

$$\text{curvature effect} = \overline{X}_c - \overline{X}_f = 7.7 - 7.5 = 0.2$$

$$95\% \text{ confidence interval} = 0.2 \pm (2.447)(0.26)\sqrt{\tfrac{1}{8} + \tfrac{1}{3}}$$

$$= 0.2 \pm 0.4$$

Example of a 2^3 Factorial Design

Example 13.4. A new product is being manufactured by an air oxidation reaction and the objective is to increase product yield. Experimental variables selected were reaction temperature, reaction time, and stir rate. The following factor levels were chosen by the experimenter:

	−	+
X_1 temperature	50°C	70°C
X_2 time	0.5 hour	2.0 h°
X_3 stir rate	16 rpm	28 rpm

It was desired to have sufficient precision to detect a 6% yield with high assurance. From similar studies the yield standard deviation was approximately 3% yield. Since $\Delta/\sigma = 2$, the number of runs required (from Figure 13.8) was $N = 14$. The final design was a 2^3 factorial design replicated twice, resulting in $N = 16$. The order of the runs was conducted randomly. The design and the results are shown in Table 13.3.

Since each run was duplicated, a precision estimate (s) with 1 df could be calculated for each run, and these values could be pooled together to form an estimate (s) with 8 df. These calculations are also shown in Table 13.3.

The calculation of main effects and interactions are shown in Table 13.4, which was copied from the pertinent columns of Table 13.2 with addition of a column of average response values for each run condition. Also shown were the ± values for the 95% confidence intervals, which apply to all estimates of main effects and interactions. The 95% confidence intervals were: [effect, % yield ± 3.3% yield].

TABLE 13.3. 2^3 **Design and Response Values**

Run	X_1 Temp	X_2 Time	X_3 Stir	Yield (Y) Results	Avg.	$s^2 = d^2/2$
1	50	0.5	16	28,21	24.5	24.5
2	70	0.5	16	43,43	43.0	0.0
3	50	2.0	16	30,30	30.0	0.0
4	70	2.0	16	47,47	47.0	0.0
5	50	0.5	28	59,66	62.5	24.5
6	70	0.5	28	88,86	87.0	2.0
7	50	2.0	28	65,60	62.5	12.5
8	70	2.0	28	81,83	82.0	2.0
						65.5

$s^2 = 8.19$ $s = 2.86$ with 8 df

TABLE 13.4. 2^3 Design — Estimation of Main Effects and Interactions

Run	Avg. Yield	X_1	X_2	X_1X_2	X_3	X_1X_3	X_2X_3	$X_1X_2X_3$
1	24.5	−	−	+	−	+	+	−
2	43.0	+	−	−	−	−	+	+
3	30.0	−	+	−	−	+	−	+
4	47.0	+	+	+	−	−	−	−
5	62.5	−	−	+	+	−	−	+
6	87.0	+	−	−	+	+	−	−
7	62.5	−	+	−	+	−	+	−
8	82.0	+	+	+	+	+	+	+
Average + level		64.8	55.4	54.0	73.5	55.9	53.0	54.4
Average − level		44.8	54.2	55.6	36.1	53.7	56.6	55.3
Diff (effect)		20.0	1.2	-1.6	37.4	2.2	-3.6	-0.9

95% Confidence Interval: $t = 2.306$ (8 df)

$$\pm\ 2ts/\sqrt{N}\ =\ \frac{\pm\ 2\ (2.306)\ (2.86)}{\sqrt{16}}\ =\ \pm\ 3.3$$

The three-factor interaction, though not given in Table 13.2, was also calculated. The $X_1X_2X_3$ column was generated by cross multiplying columns X_1X_2 with X_3.

The dominant effects were stir rate and temperature, each having a positive effect on the yield (yield increases with temperature and stir rate in the ranges studied). The most important interaction was time-stir rate. The time main effect and the other interactions were smaller and their confidence interval included zero (i.e., strong possibility of no effect).

Figure 13.10 shows a plot of yield versus time at various levels of temperature and stir rate. At the highest stir rates, time effect is zero or negative, whereas yield increased with time at the lower stir rates.

Figure 13.11 plots the average yields on the corners of a cube. Points joined from left to right show the temperature effect, which is always positive. Points joined from back to front show the stir rate effect which is also always positive. Points joined up and down show the time effect which is irregular.

2^4 Design with Blocking Factor

Example 13.5.   A development program was started to improve the color of a resin currently being manufactured. Color was measured after application of the resin to a standard substrate. Three factors related to the batch

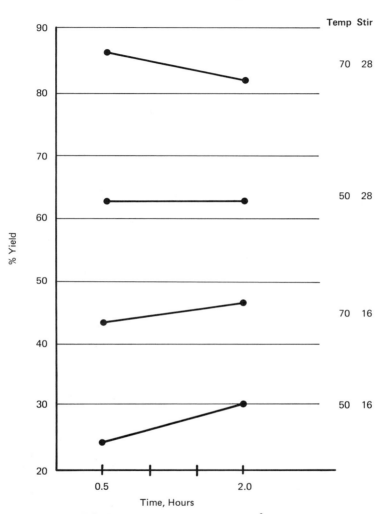

FIGURE 13.10. Y versus X plot—2^3 design.

resin process were selected for study together with their levels:

		−	+
X_1	pH	3	5
X_2	Aldehyde, xs	2.0%	5.0%
X_3	Catalyst	$1x$	$2x$

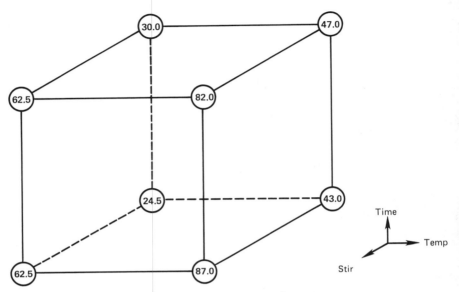

Cube Plot — 2³ Design

FIGURE 13.11. Cube plot—2³ design.

TABLE 13.5. 2⁴ Design and Response Values

Run	X_1 pH	X_2 Aldehyde (XS)	X_3 Catalyst	X_4 Pilot Unit	Y Color Value
1	3	2	1X	A	70.6
2	5	2	1X	A	63.1
3	3	5	1X	A	66.5
4	5	5	1X	A	35.2
5	3	2	2X	A	40.9
6	5	2	2X	A	26.4
7	3	5	2X	A	29.1
8	5	5	2X	A	23.0
9	3	2	1X	B	71.6
10	5	2	1X	B	57.4
11	3	5	1X	B	63.8
12	5	5	1X	B	39.9
13	3	2	2X	B	44.9
14	5	2	2X	B	22.8
15	3	5	2X	B	34.6
16	5	5	2X	B	19.1
17	4	3.5	1.5X	A	46.4
18	4	3.5	1.5X	A	41.4
19	4	3.5	1.5X	B	42.7
20	4	3.5	1.5X	B	44.8

408

Past experience with the process indicated that the color variability, expressed as a standard deviation, was about 5 color units ($\sigma = 5$). It was deemed necessary to detect an decrease of 8–10 color units ($\Delta = 8$–10) with high probability. From Figure 13.8, for Δ/σ of 1.6–2.0, the number of runs necessary for the design would be 14–17.

To expedite the development program, it was suggested by the research director that the batches be made in two available pilot units, each having a slightly different configuration (size, agitator type, and tank baffling). It was decided to identify the pilot unit as a blocking factor (X_4) and add this factor to the design. The final design was therefore a 2^4 design ($N = 16$), which met all criteria related to precision and blocking.

In order to check for overall curvature, four center points were added to the factorial design, and these were split equally between the two pilot units (two runs each). The center point conditions were pH = 4, excess aldehyde = 3.5%, and catalyst = 1.5x. The resulting design is given in Table 13.5.

TABLE 13.6. 2^4 Design — Estimation of Main Effects and Interactions

Run	X_1	X_2	X_1X_2	X_3	X_1X_3	X_2X_3	X_4	X_1X_4	X_2X_4	X_3X_4	Y
1	−	−	+	−	+	+	−	+	+	+	70.6
2	+	−	−	−	−	+	−	−	+	+	63.1
3	−	+	−	−	+	−	−	+	−	+	66.5
4	+	+	+	−	−	−	−	−	−	+	35.2
5	−	−	+	+	−	−	−	+	+	−	40.9
6	+	−	−	+	+	−	−	−	+	−	26.4
7	−	+	−	+	−	+	−	+	−	−	29.1
8	+	+	+	+	+	+	−	−	−	−	23.0
9	−	−	+	−	+	+	+	−	−	−	71.6
10	+	−	−	−	−	+	+	+	−	−	57.4
11	−	+	−	−	+	−	+	−	+	−	63.8
12	+	+	+	−	−	−	+	+	+	−	39.9
13	−	−	+	+	−	−	+	−	−	+	44.9
14	+	−	−	+	+	−	+	+	−	+	22.8
15	−	+	−	+	−	+	+	−	+	+	34.6
16	+	+	+	+	+	+	+	+	+	+	19.1
Avg. +	35.9	38.9	43.2	30.1	45.5	46.1	44.4	43.3	44.8	44.6	$\overline{X}_F = 44.3$
Avg. −	52.8	49.7	45.5	58.5	43.1	42.6	44.3	45.3	43.8	44.0	
Effect	-16.9	-10.8	-2.3	-28.4	2.4	3.5	0.1	-2.0	1.0	0.6	

Centerpoints: Unit 1 46.4, 41.4 $\overline{X}_C = 43.8$
Unit 2 42.7, 44.8   S = 2.2 (3 df)

95% Confidence Interval for Effects: t = 3.181 (3 df)

$$\pm\, 2ts/\sqrt{N} \;=\; \frac{\pm (2)\,(3.181)\,(2.2)}{\sqrt{16}} \;=\; \pm\, 3.5$$

95% Confidence Interval for Curvature $(\overline{X}_C - \overline{X}_F) \pm ts\sqrt{\dfrac{1}{N} + \dfrac{1}{C}}$

$$-0.5 \pm 3.9$$

Y vs X Plot — 2⁴ Design

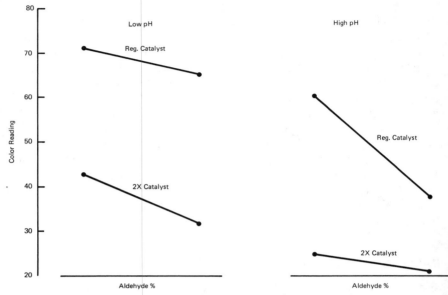

FIGURE 13.12. Y versus X plot—2^4 design.

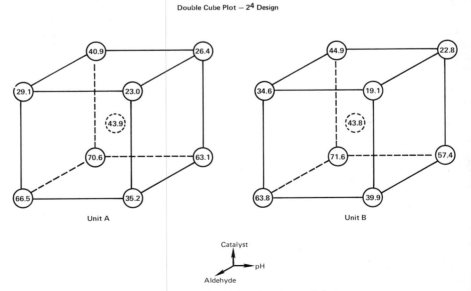

Double Cube Plot — 2⁴ Design

FIGURE 13.13. Double cube plot—2^4 design.

The 10 runs assigned to each pilot unit were then run in random order, and color determined on each batch of resin made. The color results are also shown in Table 13.5.

The main effects and interactions were calculated as shown in Table 13.6, using the layout from Table 13.2. Increase in pH, excess aldehyde, and catalyst level all decreased the color level. The results for catalyst and aldehyde level are shown in Figure 13.12. The effect of the blocking factor (pilot unit) was negligible. All of the interactions were low in magnitude, compared with the main effects.

Since there was no blocking factor effect, all center point responses were pooled together for calculation of center point mean $\overline{X}_c$ and an estimate of experimental standard deviation s. The 95% confidence intervals for the main effects and interactions were [effect ± 3.5 color units].

The average of all 16 factorial responses was $\overline{X}_F = 44.3$, which was 0.7 color unit higher than the center point average of $\overline{X}_c = 43.8$ (see Figure 13.13). The 95% confidence interval for curvature was -0.5 ± 3.9 color units, which was not significant at 95% confidence.

The investigators concluded that lower pH, lower excess aldehyde, and lower catalyst usage all increased the color, and that these factors were additive. The next stage of the investigation (if it were conducted) would involve probing the upper limits of these factors still further.

13.5. FRACTIONAL 2^k FACTORIAL DESIGNS

Resolution and Confounding

As discussed in Section 13.3, a disadvantage of 2^k factorial designs is the geometric increase in the number of runs as the number of factors increases. Since main effects and two-factor interactions are chiefly of interest, and not higher order interactions, a subset or fraction of the full 2^k design can be run by sacrificing the estimation of higher order interactions.

A simple example of a fractional 2^k factorial is shown in Figure 13.14. Either set of four runs constituting a one-half fraction (one-half replicate) of a 2^3 factorial design can be run at the sacrifice of estimating the three-factor interaction.

An important feature of fractional 2^k factorial designs is *confounding*, or the problem of identical estimates for two different effects. For design 2 in Figure 13.14 it can be seen that the X_1 column is identical with the $X_2 X_3$ column, thus the X_1 main effect is confounded with the $X_2 X_3$ interaction. Similarly, X_2 main effect is confounded with $X_1 X_3$ interaction and X_3 main effect is confounded with $X_1 X_2$ interaction. Design 1 has the same confounding pattern except that the main effects are confounded with the same interactions but of opposite sign.

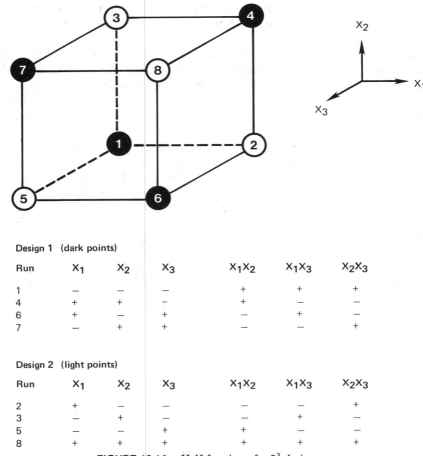

Design 1 (dark points)

Run	X_1	X_2	X_3	X_1X_2	X_1X_3	X_2X_3
1	−	−	−	+	+	+
4	+	+	−	+	−	−
6	+	−	+	−	+	−
7	−	+	+	−	−	+

Design 2 (light points)

Run	X_1	X_2	X_3	X_1X_2	X_1X_3	X_2X_3
2	+	−	−	−	−	+
3	−	+	−	−	+	−
5	−	−	+	+	−	−
8	+	+	+	+	+	+

FIGURE 13.14. Half-fraction of a 2^3 design.

Fractional factorial designs can be classified by their *resolution*, defined as follows:

Resolution   I   Main effects nonestimable
Resolution   II   Main effects confounded with main effects
Resolution   III   Main effects confounded with two-factor interactions
Resolution   IV   Two-factor interactions confounded with each other
Resolution   V   No confounding of main effects or two-factor interactions with each other

The design just discussed, the half replicate of a 2^3, written 2^{3-1} (or $\frac{1}{2}2^3$), is a resolution III design. A rotation that fully describes the fractional

TABLE 13.7.   Eight-Run Fractional Factorial Designs

Run	X_1	X_2	X_3	X_4	X_5	X_6	X_7
1	−	−	−	−	+	+	+
2	+	−	−	+	+	−	−
3	−	+	−	+	−	+	−
4	+	+	−	−	−	−	+
5	−	−	+	+	−	−	+
6	+	−	+	−	−	+	−
7	−	+	+	−	+	−	−
8	+	+	+	+	+	+	+
Estimates	1	2	3	4	5	6	7
of main	45	35	25	15	23	13	12
effects and	36	46	16	26	14	24	34
interactions	27	17	47	37	67	57	56

factorial design is 2_r^{k-p} where 2^{-p} is the fraction of the 2^k design and r denotes the resolution. Since $\frac{1}{2} = 2^{-1}$, the design shown in Figure 13.14 is a 2_{III}^{3-1} design.

To facilitate the use of fractional factorial designs, tables of such designs in 8 runs (Table 13.7) and 16 runs (Table 13.8) are given. The 8-run screening design can accommodate up to seven variables in the study. As factors are added, the resolution of the designs decreases (confounding increases). To illustrate the use of these tables a detailed summary of the 8-run designs (Table 13.7) will be given, with confounding patterns shown

TABLE 13.8.   Sixteen-Run Fractional Factorial Designs

Run	X_1	X_2	X_3	X_4	X_5	X_6	X_7	X_8	X_9	X_{10}	X_{11}	X_{12}	X_{13}	X_{14}	X_{15}
1	−	−	−	−	−	−	−	−	+	+	+	+	+	+	+
2	+	−	−	−	+	−	+	+	−	−	−	−	+	+	+
3	−	+	−	−	+	+	+	−	−	−	+	+	−	−	+
4	+	+	−	−	−	+	−	+	+	+	−	−	−	−	+
5	−	−	+	−	+	+	−	+	−	+	−	+	−	+	−
6	+	−	+	−	−	+	+	−	+	−	+	−	−	+	−
7	−	+	+	−	−	−	+	+	+	−	−	+	+	−	+
8	+	+	+	−	+	−	−	−	−	+	+	−	+	−	−
9	−	−	−	+	−	+	+	+	−	+	+	−	+	−	−
10	+	−	−	+	+	+	−	−	+	−	−	+	+	−	−
11	−	+	−	+	+	−	−	+	+	−	+	−	−	+	−
12	+	+	−	+	−	−	+	−	−	+	−	+	−	+	−
13	−	−	+	+	+	−	+	−	+	+	−	−	−	−	+
14	+	−	+	+	−	−	−	+	−	−	+	+	−	−	+
15	−	+	+	+	−	+	−	+	−	−	−	−	+	+	+
16	+	+	+	+	+	+	+	+	+	+	+	+	+	+	+
Estimates	1	2	3	4	5	6	7	8	16	12	13	14	23	24	34
of main									45	35	25	56	15	36	26
effect and									37	47	67	27	46	17	57
interactions									78	68	48	38	78	58	18

at the bottom of the table:

Three factors	A complete factorial or 2^3 with all main effects and interactions separately estimable.
Four factors	A one-half fraction of a 2^4 design or 2_{IV}^{4-1}, which is resolution IV because groups of two interactions are confounded. Main effects are separately estimable from interactions.
Five factors	A one-quarter fraction of a 2^5 design or 2_{III}^{5-2}, which is resolution III. Main effects $X_1 - X_4$ are confounded with interactions X_4X_5, X_3X_5, X_2X_5, and X_1X_5, respectively. X_5 is confounded with two interactions.
Six factors	A one-eighth fraction of a 2^6 design, or 2_{III}^{6-3}, all main effects are confounded with two interactions.
Seven factors	A one-sixteenth fraction of a 2^7 design or 2_{III}^{7-3}. All main effects are confounded with three interactions. This is known as a "saturated" design.

The 16-run table (Table 13.8) is complete for up to four factors, is resolution IV for five to eight factors, and is resolution III for over eight factors. The confounding pattern is shown for up to eight factors only.

Both Tables 13.7 and 13.8 represent one possible fraction out of many possible subsets of 2^7 or 2^{15} factorials, respectively. In some instances one or more set of run conditions may be undesirable from a safety or feasibility viewpoint. These designs may be changed by reversing the $-$ and $+$ signs of any column or any number of columns (the *entire* column must be switched, however).

Replication, Blocking, and Randomization

The amount of replication necessary to ensure precision in a fractional 2^k design is equivalent to that for the parent 2^k design. The previous relationship of N to Δ/σ applies (Figure 13.8). If N exceeds the size of the chosen design, however, replication is not recommended. Instead it is better to move up to the larger-sized fraction, as from an 8- (Table 13.7) to a 16-run (Table 13.8) design.

The same equivalence applies to blocking when blocking factors can be identified. One or more factors in the design may be designated as blocking factors, which may reduce the maximum number of factors that can be studied with a particular design.

As with the full factorial, randomization of assignment of experimental units to runs and time order in which the run is made applies equally to the fractional design.

Analysis of Fractional Factorial Designs

The estimation of main effects follows the same procedure as the full factorial except that Tables 13.3 and 13.4 are used to match columns with main effects and interactions instead of Table 13.2. For designs of resolution III and IV, a column estimate may represent either a single main effect or interaction, or group of interactions, or a main effect plus a group of interactions.

Precision usually cannot be estimated from a fractional factorial as there is no replication. To estimate the experimental error it is best to run replicated center points. Such points can also be used to estimate overall curvature in the usual way (comparison of the factorial run average with the center point average). If no replicates have been run, and there is no estimate of σ available from other sources, then the *half-normal* plot technique due to Daniel (3) can be used to assess the relative importance of effects. In this approach, the magnitudes (ignoring sign) of each main effect and interaction are ordered from high to low. The magnitude order (on a special half-normal scale) is plotted against the absolute values of the effects. If no effects are outstanding the points in the half-normal plot will fall on a straight line through the origin. Effects that are significantly larger than the rest will fall off to the right of the straight line.

Fractional Factorial Design (Eight Runs)

Example 13.6. A development study was initiated to study a new catalyst in a batch pigment-making process. The yield from the old catalyst was 60–65% and caused problems in effluent treatment. The objective was to detect a real yield increase of at least 10%. Since the process standard deviation in yield was 3% ($\Delta/\sigma = 3.3$), about eight experimental runs would be required.

Four-factor variables were selected for study:

	Low	High
Reactant ratio B/A	1.0	1.5
Modifier level, %	0.02	0.06
Catalyst level (compared with current level)	$1x$	$1.5x$
Reaction temperature	40°F	60°F

TABLE 13.9. 2_{IV}^{4-1} **Design and Response Values**

Run	X_1 Reactant	X_2 Modifier	X_3 Catalyst	X_4 Temp.	Y Yield
1	1.0	0.02	1X	40	70.7
2	1.5	0.02	1X	60	71.7
3	1.0	0.06	1X	60	86.0
4	1.5	0.06	1X	40	74.7
5	1.0	0.02	1.5X	60	73.2
6	1.5	0.02	1.5X	40	69.0
7	1.0	0.06	1.5X	40	65.2
8	1.5	0.06	1.5X	60	86.8
9	1.25	0.04	1.25X	50	75.6
10	1.25	0.04	1.25X	50	78.2
11	1.25	0.04	1.25X	50	79.4

These factors were assigned to the first four columns of an eight-run fractional factorial design (Table 13.7). This design, being resolution IV, allowed the estimation of the four-factor main effects plus the estimation of three groups of two two-factor interactions. In addition three center points were run to estimate overall curvature and experimental variability. The final design is shown in Table 13.9, together with the product yields obtained.

TABLE 13.10. 2_{IV}^{4-1} **Design — Estimation of Main Effects and Interactions**

Run	Yield	X_1	X_2	X_3	X_4	X_1X_4 X_2X_3	X_1X_3 X_2X_4	X_1X_2 X_3X_4
1	70.7	−	−	−	−	+	+	+
2	71.1	+	−	−	+	+	−	−
3	86.0	−	+	−	+	−	+	−
4	74.7	+	+	−	−	−	−	+
5	73.2	−	−	+	+	−	−	+
6	69.0	+	−	+	−	−	+	−
7	65.2	−	+	+	−	+	−	−
8	86.8	+	+	+	+	+	+	+
Avg. +		75.4	78.2	73.6	79.3	73.4	78.1	76.4
Avg. −		73.8	71.0	75.6	69.9	75.7	71.0	72.8
Diff (Effect)		1.6	7.2	-2.0	9.4	-2.3	7.1	3.6

$\overline{X}_F = 74.6$ $\overline{X}_C = 77.7$ $S = 1.94$ with 2 df from centerpoints

95% Confidence Intervals:

On main effects and interactions: $\pm 2\,ts/\sqrt{N} = \pm 2\,(4.303)\,(1.94)/\sqrt{8}$
$$= \pm 5.7$$

On Curvature:

$$\overline{X}_C - \overline{X}_F \pm ts\sqrt{\frac{1}{N} + \frac{1}{C}}$$

$$77.7 - 74.6 \pm (4.303)\,(1.94)\sqrt{\frac{1}{8} + \frac{1}{3}}$$

$$3.1 \pm 5.6$$

The analysis of this design is shown in Table 13.10. The factors having the largest main effects were X_2 (modifier level) and X_4 (temperature). A comparably large interaction effect was either $X_1 X_3$ or $X_2 X_4$, of which the latter was selected as most likely, since its component main effects were the largest. (This may not always be a good assumption, however.)

The experimental standard deviation was calculated from the triplicate center points as 1.94% yield with 2 df, and the 95% confidence intervals were calculated for the main effects and interactions as follows:

X_1	Reactant ratio	$1.6 \pm 5.7\%$ yield
X_2	Modifier level	$7.2 \pm 5.7\%$ yield
X_3	Catalyst level	$-2.0 \pm 5.7\%$ yield
X_4	Temperature	$9.4 \pm 5.7\%$ yield

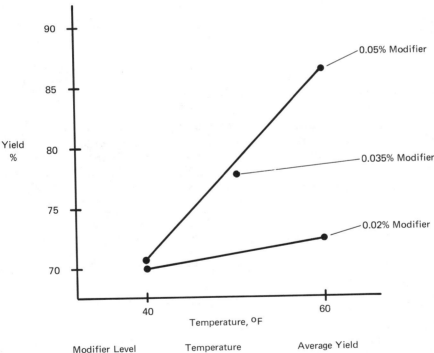

Modifier Level	Temperature	Average Yield
0.02	40	70.0
0.02	60	72.2
0.035	50	77.7 (Center Point)
0.05	40	70.0
0.05	60	86.4

FIGURE 13.15. Y versus X plot—2_{IV}^{4-2} design.

$X_1 X_4$	and $X_2 X_3$	$-2.3 \pm 5.7\%$ yield
$X_1 X_3$	and $X_2 X_4$	$7.1 \pm 5.7\%$ yield
$X_1 X_2$	and $X_3 X_4$	$3.6 \pm 5.7\%$ yield

From the above, only X_2, X_4 main effects and the (more likely) $X_2 X_4$ interaction did not include 0% yield in their 95% confidence intervals, hence their effects were statistically significant at 95% confidence.

The nature of the modifier–temperature interaction can be shown in the Y versus X plot in Figure 13.15. At low modifier levels there was little temperature effect on yield, but at high modifier levels, the increase was 16.4%.

Overall curvature was calculated as $\overline{X}_c - \overline{X}_f = 3.1\%$ yield. The 95% confidence interval or curvature was $3.1 \pm 5.6\%$ yield which also included the value zero and was not statistically significant at 95% confidence.

It was decided that the investigation should go ahead with the new catalyst for further investigation at the current catalyst level and at the stoichiometric B/A ratio, which were the best economic levels of each factor. Increased temperature and modifier levels would be investigated in the next phase.

2_{III}^{7-4} Saturated Design

Example 13.7. Recalling the example of Section 13.4, in which a 2^3 design was run to study the effect of stir rate, time, and temperature on product yield, we now extend the 2^3 design to study seven factors, four more than the original three. In this example, the response was product color and the seven factors and their levels were chosen as follows:

Factor	Level	
	Low	High
X_1 stir temperature	50°F	70°F
X_2 stir time	0.5	2.0 hours
X_3 stir rate	16	28 rpm
X_4 alcohol type	E	P (ethanol or propanol)
X_5 nitrogen blanket	No	Yes
X_6 pH	5	6
X_7 kettle material of construction	316 stainless steel	Carbon steel

(For X_1, X_2, X_3 high levels: as before)

TABLE 13.11. 2_{III}^{7-4} **Design and Response Values**

Run	X_1 Temp	X_2 Time	X_3 Stir	X_4 Solvent	X_5 Wash	X_6 pH	X_7 Matl	Y Color
1	50	0.5	16	A	Y	6	CS	69
2	70	2.0	16	W	Y	5	SS	52
3	50	0.5	16	W	N	6	SS	60
4	70	2.0	16	A	N	5	CS	83
5	50	0.5	28	W	N	5	CS	71
6	70	2.0	28	A	N	6	SS	50
7	50	0.5	28	A	Y	5	SS	59
8	70	2.0	28	W	Y	6	CS	88

The design appears in Table 13.11 with factor levels replacing the − and + values in Table 13.7. Table 13.11 also shows the colors resulting from each run.

This design is termed a "saturated" design since no more factors can be added without creating a resolution II design (confounding of main effects with each other). Such designs are useful for screening factor variables early in the investigational program. If many interactions are thought to be present, then this resolution III design will not be able to estimate them, since all interactions are confounded with main effects.

The estimates of main effects and interactions are shown in Table 13.12. The factors with the largest effect on color were the kettle material of construction and the stirring time.

Since no replicates were run, a half-normal plot was made to roughly ascertain the significance of the factor effects. As shown in Figure 13.16, the two factors with the largest effects appear to be well to the right of a straight line drawn through the five factors with smaller effects. As a working hypothesis, only stir time and material of construction are seen to

TABLE 13.12. 2_{III}^{7-4} **Design — Estimation of Main Effects and Interactions**

Run	Y Color	X_1	X_2	X_3	X_4	X_5	X_6	X_7
1	69	−	−	−	−	+	+	+
2	52	+	−	−	+	+	−	−
3	60	−	+	−	+	−	+	−
4	83	+	+	−	−	−	−	+
5	71	−	−	+	+	−	−	+
6	50	+	−	+	−	−	+	−
7	59	−	+	+	−	+	−	−
8	88	+	+	+	+	+	+	+
		68.2	72.5	67.0	67.8	67.0	66.8	77.8
		64.8	60.5	66.0	65.2	66.0	66.2	55.2
		3.4	12.0	1.0	2.6	1.0	0.6	22.6

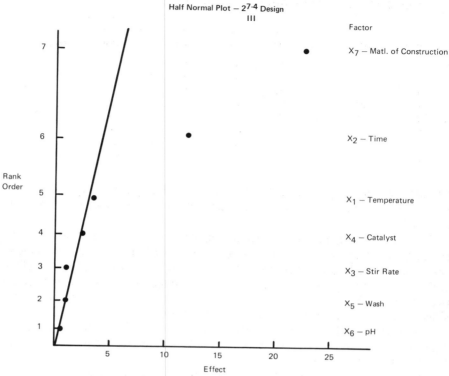

FIGURE 13.16. Half-normal plot—2_{III}^{7-4} design.

influence color, and this experiment can then be reanalyzed as a 2^2 factorial replicated twice.

The reanalysis is summarized in Figure 13.17, showing the color values for each combination of levels for the two factors and the average color. Although the two values are not truly duplicates, their agreement is reasonably good, tending to confirm the working hypothesis that the influence of the other five factors is small. The Y versus X plot shows an increase in color with time, with the time effect slightly higher for carbon steel than for 316 stainless steel.

2_{IV}^{8-4} Fractional Factorial Design

Example 13.8. The 16-run fractional factorial design is a useful resolution IV design for from five to eight factors. This example will illustrate this design for seven factors plus a blocking factor, for a total of eight factors in all.

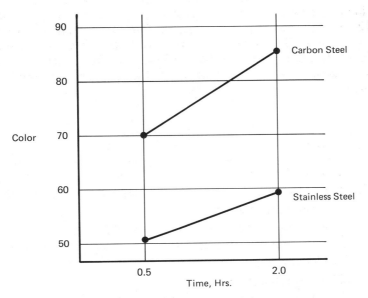

		Color	
Time	Matl. Const.	Values	Avg.
0.5	SS	50, 52	51.0
0.5	CS	69, 71	70.0
2.0	SS	59, 60	59.5
2.0	CS	83, 88	85.5

FIGURE 13.17. Y versus X plot—2_{III}^{7-4} design.

TABLE 13.13. 2_{IV}^{8-4} Design

Run No.	Acid Conc	Acid Vol	$\frac{B}{A}$	$\frac{Oxid}{A}$	$\frac{Mod}{A}$	Add Time	RX Temp	Chem
1	10	180	1.00	0.2	.002	2	70	W
2	70	180	1.00	0.2	.007	2	100	Z
3	10	550	1.00	0.2	.007	5	100	W
4	70	550	1.00	0.2	.002	5	70	Z
5	10	180	1.15	0.2	.007	5	70	Z
6	70	180	1.15	0.2	.002	5	100	W
7	10	550	1.15	0.2	.002	2	100	Z
8	70	550	1.15	0.2	.007	2	70	W
9	10	180	1.00	0.6	.002	5	100	Z
10	70	180	1.00	0.6	.007	5	70	W
11	10	550	1.00	0.6	.007	2	70	Z
12	70	550	1.00	0.6	.002	2	100	W
13	10	180	1.15	0.6	.007	2	100	W
14	70	180	1.15	0.6	.002	2	70	Z
15	10	550	1.15	0.6	.002	5	70	W
16	70	550	1.15	0.6	.007	5	100	Z

This investigation involved the maximization of yield for a chemical reaction of the type $A + B \rightarrow C$, and the response was the yield of product C. The reaction took place in an acid medium with an oxidizer and modifier added. To expedite the program, two chemists conducted the bench work, each running eight batches. Therefore, chemists were identified as a blocking factor. The eight factors and their levels were as follows:

Factor	Level	
	High	Low
X_1 acid concentration	10	70%
X_2 acid volume	180	550 ml
X_3 molar ratio of B/A	1	1.15
X_4 molar ratio of oxidizer/A	0.2	0.6
X_5 molar ratio of modifier/A	0.002	0.007
X_6 addition time of B	2	5 hours
X_7 reaction temperature	70	100°F
X_8 chemist	W	Z

The complete design as taken from Table 13.8 appears in Table 13.13. Run order was randomized within chemists.

TABLE 13.14. 2_{IV}^{8-4} **Design — Response Values and Estimation of Effects**

		Main Effects								Interactions						
Run	Yield	Acid Conc X_1	Acid Vol X_2	$\frac{B}{A}$ X_3	Oxid $\frac{}{A}$ X_4	Mod $\frac{}{A}$ X_5	Add Time X_6	RX Temp X_7	Chem X_8	X_1X_6 X_4X_5 X_3X_7	X_1X_2 X_3X_5 X_4X_7	X_1X_3 X_2X_5 X_6X_7	X_1X_4 X_5X_6 X_2X_7	X_2X_3 X_4X_6 X_1X_5	X_2X_4 X_3X_6 X_1X_7	X_3X_4 X_2X_6 X_5X_7
1	48	−	−	−	−	−	−	−	−	+	+	+	+	+	+	+
2	44	+	−	−	−	+	−	+	+	−	−	−	−	+	+	+
3	56	−	+	−	−	+	+	+	−	−	−	+	+	−	−	+
4	52	+	+	−	−	−	+	−	+	+	+	−	−	−	−	+
5	49	−	−	+	−	+	+	−	+	−	+	−	+	−	+	−
6	49	+	−	+	−	−	+	+	−	+	−	+	−	−	+	−
7	52	−	+	+	−	−	−	+	+	+	−	−	+	+	−	−
8	50	+	+	+	−	+	−	−	−	−	+	+	−	+	−	−
9	65	−	−	−	+	−	+	+	+	−	+	+	−	+	−	−
10	58	+	−	−	+	+	+	−	−	+	−	−	+	+	−	−
11	45	−	+	−	+	+	−	−	+	+	−	+	−	−	+	−
12	80	+	+	−	+	−	−	+	−	−	+	−	+	−	+	−
13	56	−	−	+	+	+	−	+	−	+	+	−	−	−	−	+
14	64	+	−	+	+	−	−	−	+	−	−	+	+	−	−	+
15	42	−	+	+	+	−	+	−	−	−	−	−	−	+	+	+
16	78	+	+	+	+	+	+	+	+	+	+	+	+	+	+	+
Avg. +		59.4	56.9	55.0	61.0	54.5	56.1	60.0	56.1	54.8	59.8	56.9	60.6	54.6	54.4	55.0
Avg. −		51.6	54.1	56.0	50.0	56.5	54.9	51.0	54.9	56.2	51.2	54.1	50.4	56.4	56.6	56.0
Effect		7.8	2.8	-1.0	11.0	-2.0	1.2	9.0	1.2	-1.4	8.6	2.8	10.2	-1.8	-2.2	-1.0

The estimation of main effects and interactions from the resulting yields are shown in Table 13.14. The factors with the largest main effects were acid concentration (X_1), oxidizer ratio (X_4), and reaction temperature (X_7). All other factor main effects, including the blocking factor effect, were small in comparison to the largest three. Two interaction groups also had large effects:

$$(X_1 X_2 \quad \text{and} \quad X_3 X_5 \quad \text{and} \quad X_4 X_7)$$

$$\text{and} \quad (X_1 X_4 \quad \text{and} \quad X_5 X_6 \quad \text{and} \quad X_2 X_7)$$

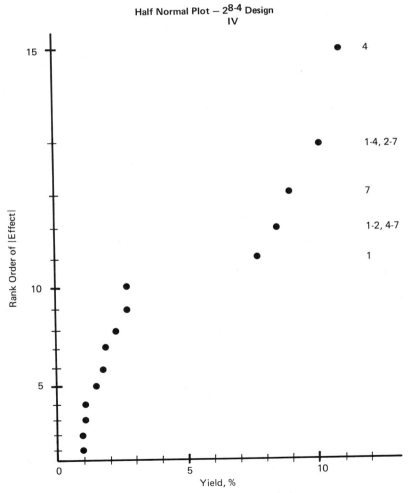

FIGURE 13.18.   Half-normal plot—2_{IV}^{8-4} design.

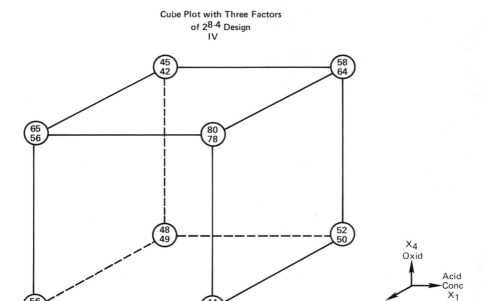

FIGURE 13.19. Cube plot for three factors of 2_{IV}^{8-4} design.

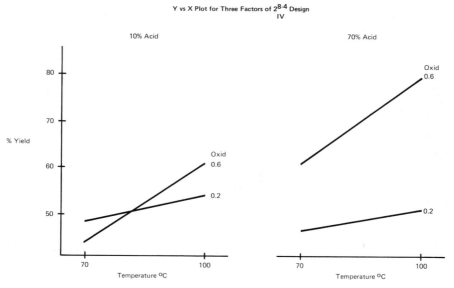

FIGURE 13.20. Y versus X plot for three factors of 2_{IV}^{8-4} design.

The most likely interactions, corresponding with pairs of factors with significant main effects, were $X_4 X_7$ in the first group and $X_1 X_4$ in the second group.

Since no replicates were available for estimation of experimental variation, a half-normal plot was created to assess the relative importance of the effects, shown in Figure 13.18. In this plot, the 5 largest effects clearly stand out from the 10 smaller effects. This helps to establish a working hypothesis that factors X_1, X_4, and X_7 are the important contributors, and that they are interactive.

This experiment was reanalyzed as a 2^3 factorial with two "replicates" per run condition. The response data are plotted on a cube in Figure 13.19. All "duplicates" agree reasonably well, confirming the working hypothesis. A Y versus X plot shown in Figure 13.20 indicates the nature of the interactions, and shows that the higher acid concentration and higher oxidizer/reactant ratio at the higher temperature is the most promising area for yield maximization. The next stage of the investigation would be concerned with even higher levels of all three factors.

REFERENCES

1. Box, G. E. P., Hunter, W. G., and Hunter, J. S., *Statistics for Experimenters*, Wiley, New York, 1978.
2. Kramer, C. Y., *A First Course in Methods of Multivariate Analysis*, Virginia Polytechnic Institute and State University, Blacksburg, VA, 1972.
3. Daniel, C., Use of half-normal plot in interpreting factorial two-level experiments, *Technometrics*, Vol. 1, 1959, p. 149.

14

RESPONSE SURFACES

GEORGE C. DERRINGER

Battelle
Columbus Laboratories
Columbus, Ohio

A typical industrial problem is to investigate the nature of the relationship between a response variable, Y, and independent or controllable variables, $X_1, X_2, X_3, \ldots, X_k$. The solution to the problem is often represented by isopleth diagrams commonly known as contour plots, an example of a contour plot is shown in Figure 14.1. Such plots enable predictions to be made of response level at settings of the independent variables not actually measured but lying within the boundaries of the X variables. Such predictions are called interpolations and the collection of tools used to accomplish such interpolations is known as response surface methodology. The sequence of operations commonly involved in response surface methodology is

1. Selection of experimental design.
2. Collection of data.
3. Fitting of regression equation to collected data.
4. Diagnosis of fitted equation to determine adequacy of fit.
5. Transformations of data if necessary to achieve adequate fit.
6. Construction of contour plots from regression equation to illustrate effects and simplify interpolation.

The fitted regression equation is the primary output from a response surface study. This equation is almost always a polynomial of no higher than third order. For example, for two independent variables, X_1 and X_2, the linear (first-order), quadratic (second-order), and cubic (third-order)

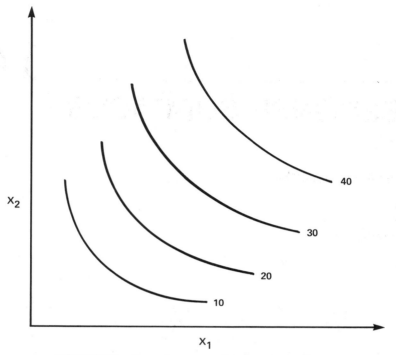

FIGURE 14.1.   Example contour plot (response surface).

equations are as follows:

$$Y = b_0 + b_1 X_1 + b_2 X_2 \tag{14.1}$$

$$Y = b_0 + b_1 X_1 + b_2 X_2 + b_{11} X_1^2 + b_{22} X_2^2 + b_{12} X_1 X_2 \tag{14.2}$$

$$\begin{aligned}
Y = &\, b_0 + b_1 X_1 + b_2 X_2 + b_{11} X_1^2 + b_{22} X_2^2 \\
&+ b_{12} X_1 X_2 + b_{111} X_1^3 + b_{222} X_2^3 + b_{112} X_1^2 X_2 \\
&+ b_{122} X_1 X_2^2 + b_{123} X_1 X_2 X_3
\end{aligned} \tag{14.3}$$

where Y is the response variable, X_1 and X_2 are the independent variables, and the b_i, b_{ij}, b_{ijk} are fitted regression coefficients. The reader should note the notation for the coefficients as it is relatively standard in response surface work. The complexity of the relationship between Y and the X's will dictate the order of equation required. Fortunately, experimental designs are available which enable sequential data collection and equation fitting so that the investigator can go through the sequence, linear $\rightarrow$ quadratic $\rightarrow$ cubic in a systematic, sequential manner. Wide experience, however, has shown that quadratic relationships are usually adequate and so most response surface designs in the literature are for fitting a quadratic equation. Of course a linear equation can always be fitted to data collected according to a second-order design and a quadratic equation to that from a third-order design. The reverse, however, is not true.

14.1. USE OF CATALOGED RESPONSE SURFACE DESIGNS

A collection of commonly used response surface designs is given in the appendix to this chapter. For each design the individual rows represent experiments and the columns represent the settings for each independent variable. To use these designs, we must first decide what range of independent variables will be used. Suppose we are going to use design C, and X_1 will represent reaction temperature which we decided to vary between 30 and 80 °C, and X_2 will represent reaction time which will range between 30 and 90 min. The first thing we must do is to express temperature, T, and time, t, in the standardized experimental design units. The scales are laid out as follows:

$-\sqrt{2}$	-1	0	$+1$	$\sqrt{2}$	X_i	Design units
30		55		80	T	Temperature, °C
30		60		90	t	Time, min

Obviously for $T = 30$, $X_1 = -\sqrt{2}$; $T = 55$, $X_1 = 0$, and so on. Therefore a little algebra will give

$$T = 72.68 \quad \text{for} \quad X_1 = +1$$
$$T = 37.32 \quad \text{for} \quad X_1 = -1$$
$$t = 81.21 \quad \text{for} \quad X_1 = +1$$
$$t = 38.79 \quad \text{for} \quad X_1 = -1$$

Of course we can make things a little easier by setting the T variable as follows:

$T-$	35	55	75	$T+$	T_t
	-1	0	$+1$		X_1

Clearly

$$T_+ = 55 + \frac{1.414}{1.0}(|75 - 55|) = 83.3 \quad \text{and}$$

$$T_- = 55 - \frac{1.414}{1.0}(|35 - 55|) = 26.7$$

Either assignment results in a linear relationship between T and X. The same argument applies for t.

Collection of Data and Randomization

After the most appropriate design is selected and the correspondence between actual and design variables established, one further thing must be done before actually carrying out the experiments—randomization. For valid interpretation of the regression analysis the individual experiments must be run in random order. This can be achieved in many ways including coin tossing, drawing numbers from a hat, selecting numbers from a random number table, and so on. For example, suppose we select a design with 10 experiments. To use the second method above one would write the numbers 1 through 10 on slips of paper, deposit them in a container, mix them up, and then withdraw them one at a time. The order of withdrawal then would be the order in which the experiments are run. For example, if the slips came out in the order 3, 7, 8, 2, 4, 1, 5, 6, 10, 9, row (experiment) 3 of the design would be run first, row 7 second, and so on. If complete randomization will, for one reason or another, significantly increase the cost of the experimental program then its advisability should be considered carefully. A rigid insistence on randomization without regard to the possible harmful consequences of such randomization (that is, increased cost, increased chance of an error in identification, and so on) is unrealistic. On the other hand, failure to randomize experiments because it is a nuisance is equally unrealistic. All statistical experimental designs should be randomized to the extent possible without incurring an unreasonable increase in experimental cost.

14.2. BLOCKING OF RESPONSE SURFACE DESIGNS

It is not uncommon in response surface studies, especially those involving a large number of runs, to have one or more of the following complications:

1. Several days or even weeks are required to run all of the experiments in the design.
2. Personnel shifts must be made during the course of the design. For example, all of the testing cannot be done by the same person.

3. Insufficient materials are on hand to complete the entire design.
4. In coating studies, outdoor or accelerated weathering exposure studies cannot be carried out under identical conditions between blocks.
5. It is desirable to evaluate preliminary results before a commitment is made to a larger study.

If one or more of the above constraints is anticipated, a response surface design can be selected which is run in two or more parts called blocks. If the design has the property that the blocking is orthogonal,* any change in conditions between blocks (i.e., different lots of material, testing on different days, and so on) will not affect the estimated coefficients of the fitted polynomial. Designs C1, G, and J are typical examples of response surface designs in orthogonal blocks.

The easiest way to visualize the block effect(s) is to think of it as another variable in the design and in the response equation. For the two-variable, second-order design in two blocks (C1), the corresponding equation fitted to the data would be

$$Y = b_0 + b_1 X_1 + b_2 X_2 + b_{11} X_1^2 + b_{22} X_2^2 + b_{12} X_1 X_2 + B_1 Z \quad (14.4)$$

where B_1 is the blocking coefficient and Z is the blocking variable. For designs containing P blocks the fitted equation will contain $P - 1$ blocking variables. If the regression analysis on the blocked data is run ignoring blocks, the b_i and b_{ij} will be unchanged provided the design has the property of orthogonal blocking. Thus the first- and second-order derivatives are independent of changes in conditions between blocks.

Since the concept of orthogonal blocking is often difficult to understand, an example will be helpful. Consider the 2^2 factorial design and assume the response given in the Y column.

			Variables		
Run	X_1	X_2	Z (Block Variable)	Y	Y'
1	-1	-1	$+1$	15	20
2	-1	$+1$	-1	35	35
3	$+1$	-1	-1	25	25
4	$+1$	$+1$	$+1$	45	50

The Z column splits the four runs into two orthogonal blocks assuming that the $X_1 X_2$ interaction is zero. For example, runs 1 and 4 constitute one block and runs 2 and 3 the other.

*All of the blocked designs in the appendix are orthogonal or nearly orthogonal.

To illustrate the concept of orthogonal blocking, a hypothetical block effect of 5 was added to the responses for the $Z = +1$ block and the results shown in the Y' column. Linear equations were then fitted to Y and Y' with and without the block variable. The results are as follows:

$$Y = 30 + 5X_1 + 10X_2 \tag{14.5}$$

$$Y' = 30 + 5X_1 + 10X_2 + 5Z \tag{14.6}$$

Note that the X_1 and X_2 coefficients are identical in both equations. In other words, a change in conditions between the two blocks did not change the estimation of the linear relationship between Y and X_1 and X_2.

Usually the block effect B has no meaningful interpretation. For example, if the blocks represent two widely separated days, the block effect could be attributed to any number of causative factors such as differences in temperature, humidity, operator, testing instrument, and so on. In other words, B cannot be assigned a causal effect so it is, in itself, not particularly useful. In other instances the block effect may have a well-defined effect. For example, suppose that a response surface study involves exposing painted samples in an accelerated weathering test machine. Suppose, however, that all of the samples corresponding to the design runs will not fit in a single level of the test machine.* In such a situation, an orthogonally blocked experimental design might be run with block I representing the samples on the upper level and block II those on the lower level. In this example, the block effect, B_1, would have a specific interpretation. It represents the difference in exposure conditions between lower and upper machine rows. The size and significance of B_1, in this case, is of importance to the experimenter. If it is small and nonsignificant, he has assurance that exposure conditions between lower and upper levels are the same. One may, therefore, choose to ignore it in a future experiment. If it is significantly large, on the other hand, it may be useful to express the results of the response surface study with $B_1 = 0$, so that the results represent an average exposure between upper and lower rows. Such an assignment amounts to adjusting the results for exposure level effect and may be particularly useful if comparisons are to be made of runs made at different times, temperatures, and so on.

Blocking for Sequential Cost-Effective Experimentation

Perhaps the most useful feature of blocking is the resulting sequential nature of experimentation when the blocking variable(s) is time. For

---

*Accelerated weathering machines conventionally have upper and lower levels for sample placement.

example, consider an experiment in two variables where little can be assumed about the nature of the relationship between the Y and X variables and no prior error estimate is available. If the experiments are inexpensive, we may simply run a complete second-order design such as the two-variable central composite design, Design C. This will enable us to estimate all linear and quadratic regression coefficients. In addition, an estimate of experimental error with which to judge the adequacy of the fitted equation will be obtained. If experiments are rather expensive, however, we may want to run smaller groups of experiments in a sequential manner with the option of discontinuing the study when sufficient information is obtained. For example, it frequently happens that after a large designed experiment is completed, the data indicate absolutely nothing of interest. Usually such a situation can be diagnosed after only a few well-chosen experiments. For two variables the following sequence of experiments is extremely efficient.

X_1	X_2	
-1	-1	
$+1$	-1	
-1	$+1$	
$+1$	$+1$	
0	0	Block 1
0	0	
0	0	
-1.414	0	
1.414	0	
0	-1.414	
0	1.414	Block 2
0	0	
0	0	
0	0	
0.667	0.836	
-0.238	1.042	
-0.963	0.464	
-0.963	-0.464	Block 3
-0.238	-1.042	
0.667	-0.836	
1.069	0	

For example, after data from the first block are collected a linear equation can be fitted and diagnosed for adequacy. If inadequate, the second block can be run and a quadratic equation fitted to the combined data and evaluated for adequacy. Finally, if a quadratic is unsatisfactory, block III can be run and a cubic model fitted to the combined data. This design and sequential decision scheme is highly efficient because experimentation can be stopped as soon as an adequate fit to the data is obtained. An example of this sequential methodology was given by Derringer (1).

Regression Analysis and Model Diagnosis

Once the data are obtained, regression analysis is employed to obtain the coefficients of the fitted equation as well as several diagnostic statistics. Regression analysis was the subject of Chapter 10 and is also given an excellent treatment by Draper and Smith (2) at a level understandable with nothing more than basic algebra as a prerequisite. Since computer programs for regression analysis are readily available the thrust of this section will be on interpretation of the output rather than on mathematical derivations. The most important diagnostic statistics on the computer output are the F ratio for the fitted equation (that is, regression F ratio) and the F ratio for lack of fit. The former should be statistically significant and the latter should not, both at some preselected level of significance, usually 5%. If this is the situation, then the equation can be considered useful for prediction and/or trend illustration. Given these two statistics, each with two outcomes, four possible situations are possible.

Case	F Regression	F Lack of Fit
A	S (significant)	S
B	S	NS
C	NS (not significant)	S
D	NS	NS

The graphs in Figure 14.2 depict an example of data structures that would give such outcomes for the fit of the equation

$$Y = b_0 + b_1 X$$

to the data points. Note that the problem for cases A and C is an inadequate model. In both examples the form of the equation can be changed to rectify the situation. More specifically, a quadratic term X^2 can be added. Case B is an example of a useful equation while case D indicates no relationship between Y and X. The same interpretation applies regardless of the number of X variables and the form of the equation and so is

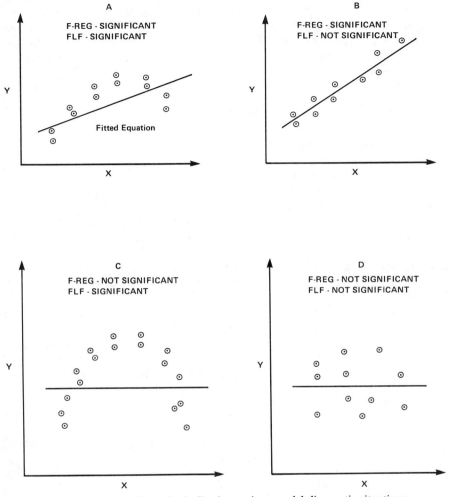

FIGURE 14.2. Examples indicating various model diagnostic situations.

useful in visualizing the adequacy or inadequacy of a general response surface. Thus, if outcomes A or C are obtained, model modification is needed before the regression equation can meet both F criteria. For outcome D, no amount of model manipulation will help and, of course, for outcome B a satisfactory equation has been found and no further action is necessary.

Other pertinent regression statistics are r^2 and the standard error of regression. r^2 or the coefficient of determination is a useful statistic as it indicates the fraction of the total variation in the data which has been accounted for by the fitted equation. r^2, however, is very sensitive to the number of residual degrees of freedom and can give false impressions of adequacy of fit when there are only a few such degrees of freedom. The standard error represents the spread of the data from the equation and should be approximately the same magnitude as the standard deviation of replicates for replicated data.

Transformation of Scale to Achieve Adequate Fit

When the fitted response surface equation exhibits significant lack of fit, two courses of action are most commonly taken. Additional terms could be added to the equation to make it more flexible. For example, in the previous section it was indicated that addition of an X^2 term would rectify the situation for cases A and C. Thus, if a linear response surface equation exhibits significant lack of fit, quadratic terms could be added and if a quadratic equation does not fit, cubic terms could be added. These terms would be: X_1^2, X_2^2, $X_1 X_2$ and X_1^3, X_2^3, $X_1^2 X_2$, $X_1 X_2^3$, respectively, for the case of two independent variables. Expanding the fitted model in this way, however, can be done only if the design has sufficient degrees of freedom for lack of fit and this is usually not the case.

Instead of expanding the model, adequate fit can often be achieved by transforming the scale of the dependent variable. For example, if the original response, Y, leads to poor fit, $\log Y$, $\sqrt{Y}$, $1/Y$, and so on, will often give adequate fit. Box and Cox (3) made the transformation process more systematic by proposing the following family of transformations

$$T = Y^\lambda \quad \lambda \neq 0 \tag{14.7}$$
$$T = \ln Y \quad \lambda = 0 \tag{14.8}$$

where Y is the dependent, or response variable, T is the transformed variable, and λ is a variable parameter generally in the interval $-3 \leq \lambda \leq +3$. In practice, several λ values in the above interval are selected. For each λ all of the Y values are transformed according to the transformation equations above and the regression analysis is run on the transformed data. For example, for $\lambda = 0.5$, the square root would be taken for all original data values and the resulting square root values subjected to regression analysis. The same procedure would be repeated for several λ values. After all of the regression analyses have been made, a plot would be made of F regression and F lack of fit as a function of λ. Such a plot will indicate what λ value will result in the desired result of maximum F regression and minimum F lack of fit.

This procedure is not as tedious as it appears because most modern regression analysis computer programs have built in options for data transformation and multiple regression runs. A good example of the transformation methodology is given by Derringer (4) and the λ plot from that article is shown in Figure 14.3. Here it will be seen that the F statistics

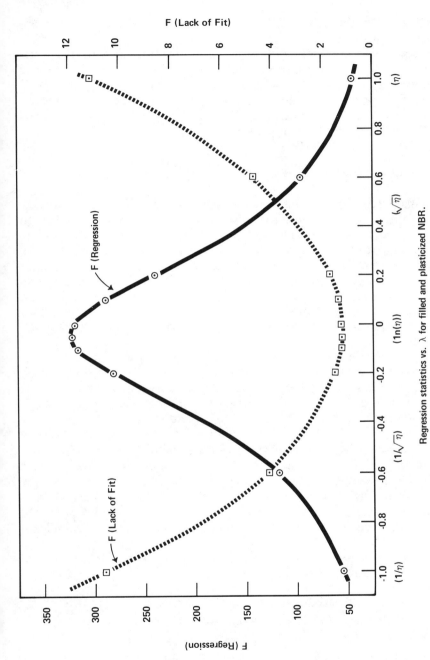

FIGURE 14.3. Plot of λ versus regression statistics for reference and example. η is the response variable.

are optimized in the vicinity of $\lambda = 0$ indicating that $\ln(Y)$ is the appropriate transformation.

This transformation methodology is useful not only in rectifying situations where quadratic response surface models fail to fit the data but also to simplify adequate fitting quadratic models. For example, in the above work by Derringer the methodology was used to change the scaling on viscosity so that a linear model in filler and oil variables could be used instead of a more complicated quadratic model.

Selection of Appropriate Design

Having discussed how the data from a response surface design are analyzed, it is helpful to take a more rigorous look at design selection. Design selection is an extremely important step because the most clever data analyst cannot answer a question which the design did not ask. Therefore, considerable thought should be given to choice of design. Many things must be considered before selecting a design, including available resources (i.e., money, equipment, personnel, etc.), time frame for carrying out experiments, necessity of blocking, amount of random error in important responses, consequences of decisions to be made based upon experimental results, and so on.

The primary condition that must be fulfilled is that the design be large enough (contain enough distinct experiments) to fit the intended regression equation. For example, if a full second-degree polynomial equation [i.e., equation (14.2)] is to be fitted to k independent variables, then the design must contain at least $N_{\min} = [(k + 1)(k + 2)]/2$ distinct points since this is the number of coefficients in the equation to be fitted. This is the only absolute rule in design selection, since violation will lead to inability to fit the equation and illustrate the response surface. All design points in excess of $N_{\min}$ increase the ability to detect lack of fit and/or increase the precision of the estimate of pure experimental error depending on whether or not they duplicate existing points. In both cases, adding data will increase the precision of the fitted equation in the absence of significant lack of fit.

The designs in the chapter appendix are indexed by four numbers in parentheses. These represent, in order,

1. Number of X variables.
2. Total number of experiments (i.e., total degrees of freedom).
3. Degrees of freedom available for estimating lack of fit.
4. Degrees of freedom available for estimating pure error.

Of these, items 1 and 2 are self-explanatory. Items 3 and 4, however, require further discussion.

The degrees of freedom available for estimation of lack of fit can also be used to accommodate additional terms in the model if necessary. For example, if lack of fit is found significant it can often be corrected by increasing the number of parameters in the regression model, usually by adding terms of the next higher order, such as cubic terms in a quadratic equation. Thus, if it is suspected that a second-order equation may prove inadequate in representing the behavior of the response, it might be advisable to select a design with excess lack of fit degrees of freedom so that the model may be expanded.

Increasing the degrees of freedom for pure error (DFPE), on the other hand, results solely in a more precise estimate of pure error and does not increase the number of terms which may be added to the model. If the response of interest is a familiar one for which a reliable estimate of pure error already exists, it is not absolutely necessary for the design to include replicate points since the prior estimate for pure error can be used for the lack of fit test. In such cases it is perfectly reasonable to choose a design for which DFPE = 0, such as Design A for two X variables. A few examples of design selection problems will illustrate the various trade-offs.

A commonly encountered situation is one where (1) resources are scarce, (2) the behavior of the response is expected to be relatively smooth so that a second-order polynomial should be sufficient for adequate representation, and (3) an estimate of pure error is available from previous studies. In such a situation Design E (3-11-1-0) would be a wise selection since it contains only one experiment more than the minimum of 10 required to fit the equation. If lack of fit turns out to be pronounced, the available degree of freedom should be sufficient to detect it. Most important, however, is the fact that should this design prove to be too small, additional points can be added in a future set of experiments to correct the situation. Thus nothing but time is lost.

In another situation, although the cost of the experiment is of secondary importance, the testing machine can accommodate no more than eight specimens at a time, the test takes several days and is known to drift over time. Furthermore, no estimate of experimental error is available and complexity of the desired response surface is a possibility. This situation requires a design which can be blocked into groups consisting of no more than eight experiments each. Also, degrees of freedom for both lack of fit and pure error must be available. For three X variables, Design G (3-20-5-3) would be a reasonable choice. In addition, since cost is not an important factor, two additional center points could be added to blocks 1 and 2 to give more precise estimates of pure error.

These are simple, but typical examples of the considerations involved in design selection. Many experimental situations arise, however, where for any of a number of reasons, an existing design is not completely suitable. When this happens it is not uncommon for the experimenter or consulting statistician to sacrifice some desired information so that the program will "fit" an existing design. This approach is generally to be discouraged unless no viable alternatives exist. In most cases, however, an existing design can be modified to fit the problem or alternatively a design may be custom made for the particular situation. It must be remembered that response surface methodology is a tool for problem solving. Modifying the problem to match the tool is generally a mistake. For additional response surface designs, some covering specific situations, the reader is referred to Meyers (5), Davies (6), Cox (7), and John (8), as well as numerous volumes of *Technometrics* and *Biometrics*.

14.3. PRIOR KNOWLEDGE AND DESIGN SELECTION

It should not be surprising that prior information about the relationship between Y and X variables will influence the choice of design. For example, suppose it is known that the response variable in question is a linear function of the three selected independent variables, X_1, X_2, and X_3. Furthermore, suppose that the range of response variables is large in comparison with the experimental error, σ. Then it can be shown that a 2^{3-1} factorial design* will be adequate for fitting the linear response surface.

Example 14.1. The design and collected data are as follows:

Run	Independent Variables in Design Units			Response Variable
	X_1	X_2	X_3	Y
1	-1	-1	$+1$	73.1
2	-1	$+1$	-1	71.9
3	$+1$	-1	-1	74.2
4	$+1$	$+1$	$+1$	82.5

The fitted regression equation is

$$Y = 75.425 + 2.925 X_1 + 1.775 X_2 + 2.375 X_3$$

A check on this equation could be made by running the center point $X_1 = X_2 = X_3 = 0$. The result should be close (in comparison to σ) to the predicted value of 75.425. The fitted equation can be used for constructing $X_1 - X_2$, $X_1 - X_3$, or $X_2 - X_3$ contour plots.

*See Chapters 12 and 13.

This example illustrates how the intelligent selection of an experimental design which takes prior information into account can result in a considerable savings in the cost of experimentation. For example, without an estimate of the experimental error, replicates would have to be included in the design. Furthermore, without the knowledge of a linear relationship between the response and independent variables, a 10-term quadratic relationship would generally have been assumed. Design G with 20 experiments, five times more than were required in the prior information case, would have been a reasonable choice for the latter case.

Mixture Designs

So far in this chapter the independent (X) variables were assumed to be such that they could be varied independently of the other X variables under study. In formulation problems where more than one independent variable is a component of the formulation, this assumption is invalid. Since the sum of all components of a formulation is constant, changing one component variable automatically results in a change in one or more of the other component variables. For example, consider a three-component solvent mixture with components X_1, X_2, and X_3. Any possible combination of these three components in units of fractional volume or weight can be represented as a point in a simplex plot such as shown in Figure 14.4. For example, equal amounts of all three components is represented as point A. Point B indicates 50% of X_1, 50% of X_2, and 0% of X_3, and point C represents pure X_3. Consideration of this graph will reveal that no X variable can be varied without changing the value of one or both of the others. A consequence of this phenomenon is that a separate class of response surface designs known as *mixture* designs must be employed for such problems. Furthermore the fitted regression equation cannot contain a constant or X^2 terms so that models of the following forms are required for fitting the data. For the special case of three variables we have

(a) First-order polynomial

$$Y = b_1 X_1 + b_2 X_2 + b_3 X_3 \qquad (14.9)$$

(b) Second-order polynomial

$$Y = b_1 X_1 + b_2 X_2 + b_3 X_3 + b_{12} X_1 X_2$$
$$+ b_{13} X_1 X_3 + b_{23} X_2 X_3 \qquad (14.10)$$

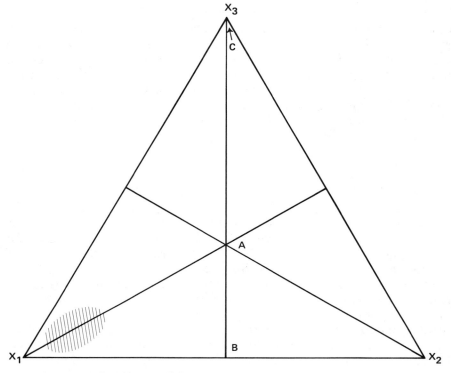

FIGURE 14.4.   Simplex for three-variable mixture design.

(c)   Third-order polynomial

$$Y = b_1 X_1 + b_2 X_2 + b_3 X_3 + b_{12} X_1 X_2 + b_{13} X_1 X_3$$
$$+ b_{23} X_2 X_3 + \gamma_{12} X_1 X_2 (X_1 - X_2)$$
$$+ \gamma_{13} X_1 X_3 (X_1 - X_3) + \gamma_{23} X_2 X_3 (X_2 - X_3)$$
$$+ b_{123} X_1 X_2 X_3 \tag{14.11}$$

where the b's and γ's are fitted coefficients. For more than three variables the equations take on analogous forms.

In many mixture problems the region of interest is located close to one vertex of the simplex so that one ingredient predominates. In addition, it is rare that the study would encompass all ingredient variables. In most cases

TABLE 14.1. Box–Behnken Design and Data for Example

	X_1	X_2	X_3	X_4	
1	-1	0	-1	0	2.04
2	-1	0	0	-1	1.81
3	-1	-1	0	0	2.69
4	-1	$+1$	0	0	2.01
5	-1	0	0	$+1$	3.71
6	-1	0	$+1$	0	4.20
7	0	0	-1	-1	0.00
8	0	-1	-1	0	1.66
9	0	$+1$	-1	0	2.22
10	0	0	-1	$+1$	2.53
11	0	-1	0	-1	1.13
12	0	$+1$	0	-1	0.52
13	0	0	0	0	3.54
14	0	0	0	0	2.86
15	0	0	0	0	2.65
16	0	-1	0	$+1$	3.69
17	0	$+1$	0	$+1$	3.52
18	0	0	$+1$	-1	2.83
19	0	-1	$+1$	0	3.93
20	0	$+1$	$+1$	0	4.40
21	0	0	$+1$	$+1$	5.00
22	$+1$	0	-1	0	1.60
23	$+1$	0	0	-1	0.61
24	$+1$	-1	0	0	2.87
25	$+1$	$+1$	0	0	2.68
26	$+1$	0	0	$+1$	3.33
27	$+1$	0	$+1$	0	4.82

X_1 = cure temperature of first coat
$-1 = 360°F$
$0 = 375°F$
$+1 = 390°F$

X_2 = degree of sanding prior to second coat
$-1 =$ no sanding
$0 =$ light sanding
$+1 =$ heavy sanding

X_3 = cure temperature of second coat
$-1 = 360°F$
$0 = 375°F$
$+ = 390°F$

X_4 = plasticizer contamination
$-1 = 0$ relative amount
$0 = 1$ of plasticizer mist
$+1 = 2$ applied to panel

the levels of a few key ingredients are varied independently of *each other*, the level of the predominant component being changed to "take up the slack" caused by changes in the variables of interest. For example, in Figure 14.4, if the region of interest is the shaded area where X_1 constitutes about 90% of the mixture, X_2 and X_3 can be varied independently of each other by adjusting X_1 to maintain the relationship $X_1 + X_2 + X_3 = 100\%$.

Gorman and Hinman (9) provide an excellent introduction to mixture designs. Additional references are voluminous and can be found dispersed throughout volumes of *Technometrics* and *Biometrics*. Typical mixture designs are included in the chapter appendix. For these designs, table entries indicate fractional composition.

Example of Typical Response Surface Study

An example of a typical response surface study is a study done by Mueller and Olsson (10) and reproduced here with permission of the authors. They studied the effect of four variables; cure temperature of first coat X_1, degree of sanding X_2, cure temperature of second coat X_3, and plasticizer contamination, X_4 on the yellowing of appliance coatings. They employed a Box–Behnken design (Design J in chapter appendix) which is given in Table 14.1 along with the resultant data. A quadratic equation in four

TABLE 14.2. Regression Output for Example

Regression Coefficients			
b_0	3.0167	b_{11}	-0.1204^a
b_1	-0.0458^a	b_{22}	-0.2517^a
b_2	-0.0517^a	b_{33}	0.2371^a
b_3	1.2608	b_{44}	-0.5817
b_4	1.24		
b_{12}	0.1225^a		
b_{13}	0.265^a		
b_{14}	0.205^a		
b_{23}	-0.022^a		
b_{24}	0.110^a		
b_{34}	-0.090^a		
F regression	16.5		
R^2	0.95		
Standard error	0.423		
F lack of fit	0.8^a		

^{a}Not statistically significant at 5% level of significance.

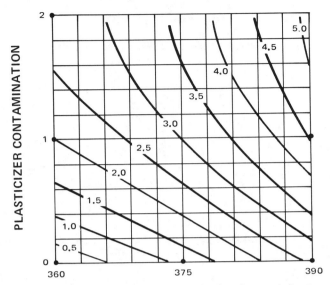

FIGURE 14.5. X_3–X_4 response surface for example.

variables was fitted using regression analysis. The equation coefficients and regression statistics are given in Table 14.2. As can be seen the equation is acceptable since F regression is significant and F lack of fit is not significant. It will be noted, however, that no terms involving either X_1 or X_2 exhibited significance, indicating that a two-variable design in X_3 and X_4 would have been sufficient. The response surface for the significant variables, X_3 and X_4, is shown in Figure 14.5.

REFERENCES

1. Derringer, G. C., *Industrial and Engineering Chemistry*, Vol. 61, No. 12, 1969, p. 6.
2. Draper, N. R., and Smith, H., *Applied Regression Analysis*, Wiley, New York, 1981.
3. Box, G. E. P., and Cox, D. R., *J. Royal Stat. Soc.*, Vol. B-26, 1964, p. 211.
4. Derringer, G. C., *Journal of Applied Polymer Science*, Vol. 18, 1974, p. 1083.
5. Myers, R. H., *Response Surface Methodology*, Allyn and Bacon, Boston, 1971.
6. Davies, O. L., *Statistical Methods in Research and Production*, Oliver and Boyd, London, 1957.
7. Cochran, W. G., and Cox, G. M., *Experimental Design*, 2nd ed., Wiley, London, 1957.
8. John, P. W. M., *Statistical Design and Analysis of Experiments*, Macmillan, New York, 1971.
9. Gorman, J. W., and Hinman, J. E., Technometrics, Vol. 4, No. 4, 1962, p. 463.
10. Mueller, F. X., and Olsson, D. M., *Journal of Paint Technology*, Vol. 43, No. 556, 1971, p. 54.

APPENDIX.   RESPONSE SURFACE DESIGNS

Design A (2-9-3-0): 3^2 Factorial Design

No.	X_1	X_2
1	-1	-1
2	-1	0
3	-1	$+1$
4	0	-1
5	0	0
6	0	$+1$
7	$+1$	-1
8	$+1$	0
9	$+1$	$+1$

Design B (3-27-17-0) 3^3 Factorial Design

No.	X_1	X_2	X_3
1	-1	-1	-1
2	-1	-1	0
3	-1	-1	$+1$
4	-1	0	-1
5	-1	0	0
6	-1	0	$+1$
7	-1	$+1$	-1
8	-1	$+1$	0
9	-1	$+1$	$+1$
10	0	-1	-1
11	0	-1	0
12	0	-1	$+1$
13	0	0	-1
14	0	0	0
15	0	0	$+1$
16	0	$+1$	-1
17	0	$+1$	0
18	0	$+1$	$+1$
19	$+1$	-1	-1
20	$+1$	-1	0
21	$+1$	-1	$+1$
22	$+1$	0	-1
23	$+1$	0	0
24	$+1$	0	$+1$
25	$+1$	$+1$	-1
26	$+1$	$+1$	0
27	$+1$	$+1$	$+1$

Design C (2-13-3-4) Rotatable[a] Central Composite Design

No.	X_1	X_2
1	−1	−1
2	−1	+1
3	+1	−1
4	+1	+1
5	−1.414	0
6	+1.414	0
7	0	−1.414
8	0	1.414
9	0	0
10	0	0
11	0	0
12	0	0
13	0	0

[a]A rotatable design results in a response surface for which predictions at equal distances from the design center have equal variance.

Design C1 (2-14-3-4) Central Composite Design In Two Blocks

No.	X_1	X_2	
1	−1	−1	
2	−1	+1	
3	+1	−1	
4	+1	+1	Block 1
5	0	0	
6	0	0	
7	0	0	
8	+1.414	0	
9	−1.414	0	
10	0	+1.414	Block 2
11	0	−1.414	
12	0	0	
13	0	0	
14	0	0	

Design D (3-20-5-5) Rotatable Central Composite Design

No.	X_1	X_2	X_3
1	-1	-1	-1
2	$+1$	-1	-1
3	-1	$+1$	-1
4	$+1$	$+1$	-1
5	-1	-1	$+1$
6	$+1$	-1	$+1$
7	-1	$+1$	$+1$
8	$+1$	$+1$	$+1$
9	-1.682	0	0
10	1.682	0	0
11	0	-1.682	0
12	0	1.682	0
13	0	0	-1.682
14	0	0	1.682
15	0	0	0
16	0	0	0
17	0	0	0
18	0	0	0
19	0	0	0
20	0	0	0

Design E (3-11-1-0) Hartley Central Composite

No.	X_1	X_2	X_3
1	-1	-1	$+1$
2	-1	$+1$	-1
3	$+1$	-1	-1
4	$+1$	$+1$	$+1$
5	$+1.682$	0	0
6	-1.682	0	0
7	0	$+1.682$	0
8	0	-1.682	0
9	0	0	$+1.682$
10	0	0	-1.682
11	0	0	0

Design F (3-15-3-2) Box–Behnken Design

No.	X_1	X_2	X_3
1	+1	+1	0
2	+1	−1	0
3	−1	+1	0
4	−1	−1	0
5	+1	0	+1
6	+1	0	−1
7	−1	0	+1
8	−1	0	−1
9	0	+1	+1
10	0	+1	−1
11	0	−1	+1
12	0	−1	−1
13	0	0	0
14	0	0	0
15	0	0	0

Design G (3-20-5-3) Central Composite Design in Three Blocks

No.	X_1	X_2	X_3	Block
1	−1	−1	1	1
2	1	−1	−1	1
3	−1	1	−1	1
4	1	1	1	1
5	0	0	0	1
6	0	0	0	1
7	−1	−1	−1	2
8	1	−1	1	2
9	−1	1	1	2
10	1	1	−1	2
11	0	0	0	2
12	0	0	0	2
13	a	0	0	3
14	$-a$	0	0	3
15	0	a	0	3
16	0	$-a$	0	3
17	0	0	a	3
18	0	0	$-a$	3
19	0	0	0	3
20	0	0	0	3

$a = 1.633$ for orthogonal blocking.

Design H (4-31-10-6) Rotatable Central Composite Design

No.	X_1	X_2	X_3	X_4
1	−1	−1	−1	−1
2	−1	−1	−1	1
3	−1	−1	+1	−1
4	−1	−1	+1	1
5	−1	1	−1	−1
6	−1	1	−1	1
7	−1	1	+1	−1
8	−1	1	+1	1
9	1	−1	−1	−1
10	1	−1	−1	1
11	1	−1	+1	−1
12	1	−1	+1	1
13	1	1	−1	−1
14	1	1	−1	1
15	1	1	+1	−1
16	1	1	+1	1
17	−2	0	0	0
18	2	0	0	0
19	0	−2	0	0
20	0	2	0	0
21	0	0	−2	0
22	0	0	2	0
23	0	0	0	−2
24	0	0	0	2
25	0	0	0	0
26	0	0	0	0
27	0	0	0	0
28	0	0	0	0
29	0	0	0	0
30	0	0	0	0
31	0	0	0	0

Design I (4-17-2-0) Hartley Design

No.	X_1	X_2	X_3	X_4
1	−1	−1	1	−1
2	−1	−1	1	1
3	−1	1	−1	−1
4	−1	1	−1	1
5	1	−1	−1	−1
6	1	−1	−1	1
7	1	1	1	−1
8	1	1	1	1
9	2	0	0	0

Design I Hartley Design (*continued*)

No.	X_1	X_2	X_3	X_4
10	-2	0	0	0
11	0	2	0	0
12	0	-2	0	0
13	0	0	2	0
14	0	0	-2	0
15	0	0	0	2
16	0	0	0	-2
17	0	0	0	0

Design J (4-27-10-0) Box–Behnken Design in Three Blocks

No.	X_1	X_2	X_3	X_4	
1	$+1$	$+1$	0	0	
2	$+1$	-1	0	0	
3	-1	$+1$	0	0	
4	-1	-1	0	0	
5	0	0	$+1$	$+1$	Block 1
6	0	0	$+1$	-1	
7	0	0	-1	$+1$	
8	0	0	-1	-1	
9	0	0	0	0	
10	$+1$	0	0	$+1$	
11	$+1$	0	0	-1	
12	-1	0	0	$+1$	
13	-1	0	0	-1	
14	0	$+1$	$+1$	0	Block 2
15	0	$+1$	-1	0	
16	0	-1	$+1$	0	
17	0	-1	-1	0	
18	0	0	0	0	
19	$+1$	0	$+1$	0	
20	$+1$	0	-1	0	
21	-1	0	$+1$	0	
22	-1	0	-1	0	
23	0	$+1$	0	$+1$	Block 3
24	0	$+1$	0	-1	
25	0	-1	0	$+1$	
26	0	-1	0	-1	
27	0	0	0	0	

Design K (5-32-6-5) Rotatable Central Composite Design

Run	X_1	X_2	X_3	X_4	X_5
1	-1	-1	-1	-1	$+1$
2	$+1$	-1	-1	-1	-1
3	-1	1	-1	-1	-1
4	$+1$	1	-1	-1	1
5	-1	-1	1	-1	-1
6	$+1$	-1	1	-1	1
7	-1	1	1	-1	1
8	$+1$	1	1	-1	-1
9	-1	-1	-1	1	-1
10	$+1$	-1	-1	1	1
11	-1	1	-1	1	1
12	$+1$	1	-1	1	-1
13	-1	-1	1	1	1
14	$+1$	-1	1	1	-1
15	-1	1	1	1	-1
16	$+1$	1	1	1	1
17	-2	0	0	0	0
18	2	0	0	0	0
19	0	-2	0	0	0
20	0	2	0	0	0
21	0	0	-2	0	0
22	0	0	2	0	0
23	0	0	0	-2	0
24	0	0	0	2	0
25	0	0	0	0	-2
26	0	0	0	0	2
27	0	0	0	0	0
28	0	0	0	0	0
29	0	0	0	0	0
30	0	0	0	0	0
31	0	0	0	0	0
32	0	0	0	0	0

Design L (5-23-2-0) Westlake Design

No.	X_1	X_2	X_3	X_4	X_5
1	1	1	1	1	-1
2	1	1	-1	-1	-1
3	-1	1	1	-1	-1
4	-1	1	-1	1	-1
5	1	-1	1	1	1
6	1	-1	-1	-1	1
7	-1	-1	1	-1	1
8	-1	-1	-1	1	1

Design L Westlake Design (*continued*)

No.	X_1	X_2	X_3	X_4	X_5
9	−1	1	−1	−1	1
10	−1	1	1	1	1
11	1	1	−1	1	1
12	1	1	1	−1	1
13	2	0	0	0	0
14	−2	0	0	0	0
15	0	2	0	0	0
16	0	−2	0	0	0
17	0	0	2	0	0
18	0	0	−2	0	0
19	0	0	0	2	0
20	0	0	0	−2	0
21	0	0	0	0	2
22	0	0	0	0	−2
23	0	0	0	0	0

Design M (5-46-20-4) Box–Behnken Design in Two Blocks

No.	X_1	X_2	X_3	X_4	X_5	
1	+1	+1	0	0	0	
2	+1	−1	0	0	0	
3	−1	+1	0	0	0	
4	−1	−1	0	0	0	
5	0	0	+1	+1	0	
6	0	0	+1	−1	0	
7	0	0	−1	+1	0	
8	0	0	−1	−1	0	
9	0	+1	0	0	+1	
10	0	+1	0	0	−1	
11	0	−1	0	0	+1	
12	0	−1	0	0	−1	
13	+1	0	+1	0	0	Block 1
14	+1	0	−1	0	0	
15	−1	0	+1	0	0	
16	−1	0	−1	0	0	
17	0	0	0	+1	+1	
18	0	0	0	+1	−1	
19	0	0	0	−1	+1	
20	0	0	0	−1	−1	
21	0	0	0	0	0	
22	0	0	0	0	0	
23	0	0	0	0	0	
24	0	+1	+1	0	0	
25	0	+1	−1	0	0	

Design M Box-Behnken Design in Two Blocks (*continued*)

No.	X_1	X_2	X_3	X_4	X_5	
26	0	−1	+1	0	0	
27	0	−1	−1	0	0	
28	+1	0	0	+1	0	
29	+1	0	0	−1	0	
30	−1	0	0	+1	0	
31	−1	0	0	−1	0	
32	0	0	+1	0	+1	
33	0	0	+1	0	−1	
34	0	0	−1	0	+1	Block 2
35	0	0	−1	0	−1	
36	+1	0	0	0	+1	
37	+1	0	0	0	−1	
38	−1	0	0	0	+1	
39	−1	0	0	0	−1	
40	0	+1	0	+1	0	
41	0	+1	0	−1	0	
42	0	−1	0	+1	0	
43	0	−1	0	−1	0	
44	0	0	0	0	0	
45	0	0	0	0	0	
46	0	0	0	0	0	

Design N (5-33-6-5) Central Composite Design in Two Blocks

No.	X_1	X_2	X_3	X_4	X_5	Block
1	−1	−1	−1	−1	1	1
2	1	−1	−1	−1	−1	1
3	−1	1	−1	−1	−1	1
4	1	1	−1	−1	1	1
5	−1	−1	1	−1	−1	1
6	1	−1	1	−1	1	1
7	−1	1	1	−1	1	1
8	1	1	1	−1	−1	1
9	−1	−1	−1	1	−1	1
10	1	−1	−1	1	1	1
11	−1	1	−1	1	1	1
12	1	1	−1	1	−1	1
13	−1	−1	1	1	1	1
14	1	−1	1	1	−1	1
15	−1	1	1	1	−1	1
16	1	1	1	1	1	1
17	0	0	0	0	0	1
18	0	0	0	0	0	1
19	0	0	0	0	0	1
20	0	0	0	0	0	1

Design N Central Composite Design in Two Blocks (*continued*)

No.	X_1	X_2	X_3	X_4	X_5	Block
21	0	0	0	0	0	1
22	0	0	0	0	0	1
23	−2	0	0	0	0	2
24	2	0	0	0	0	2
25	0	−2	0	0	0	2
26	0	2	0	0	0	2
27	0	0	−2	0	0	2
28	0	0	2	0	0	2
29	0	0	0	−2	0	2
30	0	0	0	2	0	2
31	0	0	0	0	−2	2
32	0	0	0	0	2	2
33	0	0	0	0	0	2

Design O (6-53-17-8) Central Composite Design

No.	X_1	X_2	X_3	X_4	X_5	X_6
1	−1	−1	−1	−1	−1	−1
2	+1	−1	−1	−1	−1	1
3	−1	1	−1	−1	−1	1
4	+1	1	−1	−1	−1	−1
5	−1	−1	1	−1	−1	1
6	+1	−1	1	−1	−1	−1
7	−1	1	1	−1	−1	−1
8	1	1	1	−1	−1	1
9	−1	−1	−1	1	−1	1
10	1	−1	−1	1	−1	−1
11	−1	1	−1	1	−1	−1
12	1	1	−1	1	−1	1
13	−1	−1	1	1	−1	−1
14	1	−1	1	1	−1	1
15	−1	1	1	1	−1	1
16	1	1	1	1	−1	−1
17	−1	−1	−1	−1	1	1
18	1	−1	−1	−1	1	−1
19	−1	1	−1	−1	1	−1
20	1	1	−1	−1	1	1
21	−1	−1	1	−1	1	−1
22	1	−1	1	−1	1	1
23	−1	1	1	−1	1	1
24	1	1	1	−1	1	−1
25	−1	−1	−1	1	1	−1
26	1	−1	−1	1	1	1
27	−1	1	−1	1	1	1
28	1	1	−1	1	1	−1
29	−1	−1	1	1	1	1

Design O Central Composite Design (*continued*)

No.	X_1	X_2	X_3	X_4	X_5	X_6
30	1	-1	1	1	1	-1
31	-1	1	1	1	1	-1
32	1	1	1	1	1	1
33	$-a$	0	0	0	0	0
34	a	0	0	0	0	0
35	0	$-a$	0	0	0	0
36	0	a	0	0	0	0
37	0	0	$-a$	0	0	0
38	0	0	a	0	0	0
39	0	0	0	$-a$	0	0
40	0	0	0	a	0	0
41	0	0	0	0	$-a$	0
42	0	0	0	0	a	0
43	0	0	0	0	0	$-a$
44	0	0	0	0	0	a
45	0	0	0	0	0	0
46	0	0	0	0	0	0
47	0	0	0	0	0	0
48	0	0	0	0	0	0
49	0	0	0	0	0	0
50	0	0	0	0	0	0
51	0	0	0	0	0	0
52	0	0	0	0	0	0
53	0	0	0	0	0	0

$a = 2.378$ for rotatability.

Design P (6-40-17-7) Central Composite Design in Three Blocks

No.	X_1	X_2	X_3	X_4	X_5	X_6	Block
1	-1	-1	-1	-1	-1	-1	1
2	1	-1	-1	-1	1	-1	1
3	-1	1	-1	-1	1	-1	1
4	1	1	-1	-1	-1	-1	1
5	-1	-1	1	-1	-1	1	1
6	1	-1	1	-1	1	1	1
7	-1	1	1	-1	1	1	1
8	1	1	1	-1	-1	1	1
9	-1	-1	-1	1	-1	1	1
10	1	-1	-1	1	1	1	1
11	-1	1	-1	1	1	1	1
12	1	1	-1	1	-1	1	1
13	-1	-1	1	1	-1	-1	1

Design P Central Composite Design in Three Blocks (*continued*)

No.	X_1	X_2	X_3	X_4	X_5	X_6	Block
14	1	-1	1	1	1	-1	1
15	-1	1	1	1	1	-1	1
16	1	1	1	1	-1	-1	1
17	0	0	0	0	0	0	1
18	0	0	0	0	0	0	1
19	0	0	0	0	0	0	1
20	0	0	0	0	0	0	1
21	-1	-1	-1	-1	1	1	2
22	1	-1	-1	-1	-1	1	2
23	-1	1	-1	-1	-1	1	2
24	1	1	-1	-1	1	1	2
25	-1	-1	1	-1	1	-1	2
26	1	-1	1	-1	-1	-1	2
27	-1	1	1	-1	-1	-1	2
28	1	1	1	-1	1	-1	2
29	-1	-1	-1	1	1	-1	2
30	1	-1	-1	1	-1	-1	2
31	-1	1	-1	1	-1	-1	2
32	1	1	-1	1	1	-1	2
33	-1	-1	1	1	1	1	2
34	1	-1	1	1	-1	1	2
35	-1	1	1	1	-1	1	2
36	1	1	1	1	1	1	2
37	0	0	0	0	0	0	2
38	0	0	0	0	0	0	2
39	0	0	0	0	0	0	2
40	0	0	0	0	0	0	2
41	$-a$	0	0	0	0	0	3
42	a	0	0	0	0	0	3
43	0	$-a$	0	0	0	0	3
44	0	a	0	0	0	0	3
45	0	0	$-a$	0	0	0	3
46	0	0	a	0	0	0	3
47	0	0	0	$-a$	0	0	3
48	0	0	0	a	0	0	3
49	0	0	0	0	$-a$	0	3
50	0	0	0	0	a	0	3
51	0	0	0	0	0	$-a$	3
52	0	0	0	0	0	a	3
53	0	0	0	0	0	0	3
54	0	0	0	0	0	0	3

$a = 2$.

Design Q (3-6-0-0) Quadratic Mixture Design

No.	X_1	X_2	X_3
1	1	0	0
2	0	1	0
3	0	0	1
4	0	0.5	0.5
5	0.5	0	0.5
6	0.5	0.5	0

Design R (3-10-0-0) Cubic Mixture Designs

No.	X_1	X_2	X_3
1	1	0	0
2	0	1	0
3	0	0	1
4	1/3	2/3	0
5	2/3	1/3	0
6	0	1/3	2/3
7	0	2/3	1/3
8	2/3	0	1/3
9	1/3	0	2/3
10	1/3	1/3	1/3

Design S (3-15-0-0) Quartic Mixture Design

No.	X_1	X_2	X_3
1	1	0	0
2	0	1	0
3	0	0	0
4	0.25	0.75	0
5	0.5	0.5	0
6	0.75	0.25	0
7	0.25	0	0.75
8	0.50	0	0.50
9	0.75	0	0.25
10	0	0.25	0.75
11	0	0.50	0.50
12	0	0.75	0.25
13	0.25	0.25	0.5
14	0.25	0.5	0.25
15	0.50	0.25	0.25

Design T (4-10-0-0) Quadratic Mixture Design

No.	X_1	X_2	X_3	X_4
1	1	0	0	0
2	0	1	0	0
3	0	0	1	0
4	0	0	0	1
5	0.5	0.5	0	0
6	0.5	0	0.5	0
7	0.5	0	0	0.5
8	0	0	0.5	0.5
9	0	0.5	0	0.5
10	0	0.5	0.5	0

Design U (4-20-0-0) Cubic Mixture Design

No.	X_1	X_2	X_3	X_4
1	1	0	0	0
2	0	1	0	0
3	0	0	1	0
4	0	0	0	1
5	2/3	1/3	0	0
6	1/3	2/3	0	0
7	0	1/3	2/3	0
8	0	2/3	1/3	0
9	1/3	0	2/3	0
10	2/3	0	1/3	0
11	1/3	1/3	1/3	0
12	0	0	2/3	1/3
13	0	2/3	0	1/3
14	2/3	0	0	1/3
15	0	1/3	1/3	1/3
16	1/3	1/3	0	1/3
17	1/3	0	1/3	1/3
18	0	0	1/3	2/3
19	0	1/3	0	2/3
20	1/3	0	0	2/3

SIMPLEX EVOP

JOHN BAX

Scott Bader (USA), Inc.
Richmond, California

In much of the experimental work required in formulating a paint, or in evaluating raw materials to be used in a paint, the paint chemist is faced with problems of evaluating several variables in obtaining an optimum formulation. These variables can range from different grades of a particular pigment, various levels of solvent, the use of different thickeners in latex paints, and a host of other variables that could all have a significant effect on either the cost or the performance of the final paint. These calculations are often achieved by intuitive experimentation using the formulator's past experience.

In order to completely assess the effect of variables, it is necessary to take a logical and formal approach to such experimental work. Traditionally, variations have been examined by changing one factor at a time. This is useful, but can be misleading. Simplex EVOP offers a means of controlling the experimentation in a logical way, so that a progression is made hopefully away from a "worst" area and toward an optimum.

Simplex EVOP begins with a designed set of experiments using all the pertinent variables. The design is an equilateral triangle in two variables, an equal-sided tetrahedron in three variables, or a simplex (a multidimensional triangle) in four or more variables. This method of optimization is also called the rotating simplex method of optimization and has also been given the popular names of self-directing optimization or simplex evolutionary operation (EVOP).

15.1. BACKGROUND OF SIMPLEX EVOP

EVOP is an abbreviation for evolutionary operation, for the progress toward an optimum by logical steps. The simplex method was first suggested by Spendley (1); in this an optimum is reached by using simple mathematics and applying certain specific rules. The technique was first used, and has been more widely used in chemical engineering applications where variations in factors such as temperature, reaction, feed rates of components, and so on, are varied and the effect of the variation is assessed on the final product. Much of the published work is relevant to chemical engineering and optimizing analytical conditions, but the technique can be applied to other areas. It has the advantage that the mathematics required are relatively simple and well within the scope of any chemist or engineer, and that unexpected results are often found, that is, results that the experimenter would not normally have predicted. Additionally, it offers a planned approach of searching for an optimum response rather than merely groping for a trend or optimum. Optimization will be understood to mean maximization of response, although it could also apply to the process of finding a minimum. The technique is also a logical extension of graphical techniques that have been used widely by all chemists.

The simplex is a geometric figure having a number of vertices, one more than the number of dimensions being considered. Each "dimension" is a function of an experimental variable such as solids content, mixing speed, and so on. Single experiments are run to determine the principal response values representing the vertices of the simplex. In some cases, the variable need not be a physical quantity, but can be a response such as closeness to a visual color match. Thus two variables give a triangle, three variables a tetrahedron, and so on. Greater numbers of variables are possible, but are difficult to visualize. A simplex with more than three variables is handled by calculation. The first step in the method is to define the response to be optimized. The second step is to identify the variables affecting it and the limits of each variable.

The technique in a two-factor simplex, or triangle, is similar to the triangulation methods used in map-making. Figure 15.1 shows a contour map with a triangle ABC. To find the peak, or optimum, a series of triangles is constructed, as in Figure 15.2. Knowing the contours, based on prior experience for instance, the progress of the simplex toward the peak would be obvious. If the contours are not known, as is generally the case, then simple rules are required to ensure a logical progression.

Continuing the analogy of the map and the peak, the heights of points ABC are determined. The lowest point (C) is discarded, and a second triangle constructed using the line AB as the base and an added new point which is positioned at the mirror image of the eliminated point. This lowest point in an experiment is commonly the worst result. This process is repeated during progress up the hill to the top. This then is the basis of the graphical technique.

When more than two variables are involved, an alternative, nongraphical technique is required. Having reached the top of our imaginary hill it is possible to move on to consider the type of problems faced by the paint chemist.

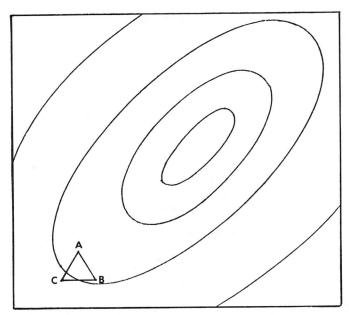

FIGURE 15.1. Contour plot of a two-variable system.

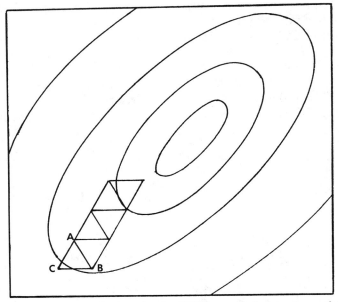

FIGURE 15.2. Illustrating simplex EVOP on a static response surface.

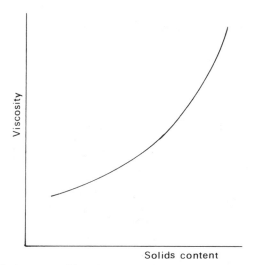

FIGURE 15.3.   Viscosity as a function of solids content.

15.2.   THE GRAPHICAL TECHNIQUE

Example 15.1.   Figure 15.3 shows the effect of solids content of a resin on viscosity. As the solution increases in solids content, so the viscosity increases. To obtain this data one makes a series of solutions at varying solids contents, measures the viscosity, and plots a graph. The desired points can then be read off the two axes— a basic technique used in almost all laboratories and familiar to most chemists. This is a one-factor (solids content) experiment, and the results can be easily plotted on a graph. A two-factor experiment needs to be plotted in three dimensions.

Consider the effect of two factors: the ratio of two reactive resins on film hardness. This could be plotted in three dimensions, but would involve a large number of experiments and the construction of solid models as in Figure 15.4.

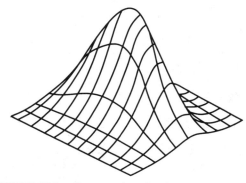

FIGURE 15.4.   Contour plot of a three-variable system.

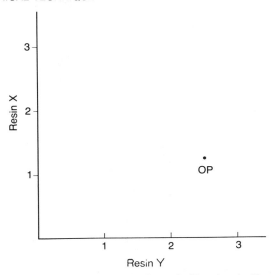

FIGURE 15.5. Unknown optimum point for resin X and resin Y combinations.

Consider Figure 15.5. This shows a point OP which represents an unknown optimum of film hardness at a particular level of resin X and resin Y. Using simplex EVOP, one would construct an equilateral triangle ABC (Figure 15.6) and carry out the experiments at the points of the triangle, using the levels indicated and giving the results shown in Table 15.1. Experiment A gives the lowest hardness, and so this point is eliminated and a second triangle constructed, BCD, which is a mirror image of the first, Figure 15.7. Experiment D is carried out giving a hardness of 7, Table 15.2. Experiment B is lowest, and is eliminated, and a further triangle DCE constructed. Experiment E is carried out and the lowest eliminated. This process is repeated until the optimum combination of the two resins is reached (Figure 15.8, Table 15.3).

The time-saving aspect of this system is apparent in that only one experiment is required to construct each new triangle.

When the optimum is reached the triangles begin to circle this point, G in Figure 15.8. To confirm that the result is genuine and not caused by experimental error one can either repeat experiment G, or construct a triangle assuming G is low. If the optimum is correct one should be automatically returned by the triangle mirror image or a repeat should confirm the result.

This introduces the first rule: where a result has appeared $f + 1$ times it should be regarded as suspicious and eliminated in the next design. F represents the number of factors, in this case, 2, which would result in its elimination if result G had appeared in three triangles. A genuine high result would automatically appear in the next design.

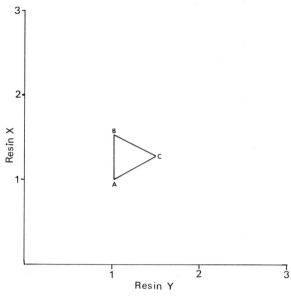

FIGURE 15.6. Data of Table 15.1.

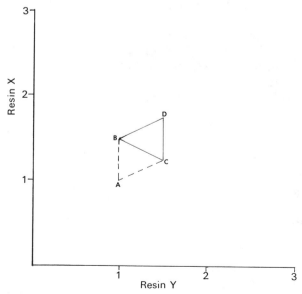

FIGURE 15.7. The "reflection" of experiments A, B, C.

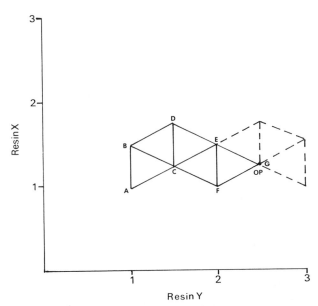

FIGURE 15.8. Optimum seeking movement of the simplex.

TABLE 15.1. Data for Initial Vertex Selection—Hardness Optimization Example

Experiment	Level Resin X	Level Resin Y	Hardness
A	1.0	1.0	5
B	1.5	1.0	6
C	1.25	1.5	8

TABLE 15.2. Hardness Optimization Example—Further Data

Experiment	Level Resin X	Level Resin Y	Hardness
A	1.0	1.0	5
B	1.5	1.0	6
C	1.25	1.5	8
D	1.75	1.5	7

TABLE 15.3.   Hardness Optimization—Final Vertex Selection

Experiment	Level Resin X	Level Resin Y	Hardness
A	1.0	1.0	5
B	1.5	1.0	6
C	1.25	1.5	8
D	1.75	1.5	7
E	1.5	2.0	10
F	1.0	2.0	9
G	1.25	2.5	12

This then describes a two-factor design, which can be shown graphically. Before moving on to three or more factor designs, one can introduce some formality into the design which will become more relevant when it is no longer possible to draw the figures on a plane surface. Referring back to Table 15.1 the data can be expressed as shown in Table 15.4.

In any design one will choose a suitable S, or start point, and p, or variation, for each factor. In this case the same value was chosen for each for ease of working; but one might be considering two factors, one of which could be varied by a level of 5% while the other could be varied by only 0.1%.

In the case of a two-factor experiment it is easy to construct the mirror image of the triangle graphically, but when one moves on, a more formal approach is required to determine the next point.

Considering Table 15.2 again, the levels for experiment D are determined as follows:

$$(2 \times \text{average of retained points}) - (\text{discarded point})$$

$$\text{resin X} = 2 \times \frac{(1.5 + 1.25)}{(2)} - 1.0 = 1.75$$

$$\text{resin Y} = 2 \times \frac{(1.0 + 1.5)}{(2)} - 1.0 = 1.5$$

This gives the levels of the two resins for experiment D.

Example 15.2.   In some cases with a two-variable experiment, it may happen that the outcomes of the three experiments may give two results which are both equally bad (i.e., nonoptimum). This is illustrated in Table 15.5 and Figure 15.9. In this case, as a general rule, it is advisable to discard both A and B, as shown in Table 15.5 and add experiments D and E. We would then drop point C on the next move as it is now the single "worst" experiment in the design. Using the same calculation rule as in

TABLE 15.4. Sample Format

A	Sx	Sy
B	$Sx + px$	Sy
C	$Sx + \dfrac{px}{2}$	$Sy + py$

Where Sx is start point resin $X = 1.0$
Sy is start point resin $Y = 1.0$
px is variation of level
resin $X = 0.5$
py is variation of level
resin $Y = 0.5$

TABLE 15.5. Data for Example 15.2[a]

Experiment	Level Resin X	Level Resin Y	Hardness
(A)	(1.0)	(1.0)	(5)
(B)	(1.0)	(3.0)	(5)
C	2.0	2.0	8
D	3.0	3.0	11
E	3.0	1.0	10

[a]Parentheses indicate that data were examined and then further runs were made.

TABLE 15.6. Optimization Process for Data of Example 15.2[a]

Experiment	X	Y	Hardness
(A)	(1.0)	(1.0)	(5)
(B)	(1.0)	(3.0)	(5)
C	2.0	2.0	8
	—	—	
Sum of best points	2.0	2.0	
Average of best points	2.0	2.0	
2 × average of best points	4.0	4.0	
Minus (A)	−1.0	−1.0	
New experiment D	3.0	3.0	
2 × average of best points	4.0	4.0	
Minus (B)	−1.0	−3.0	
New experiment E	3.0	1.0	

[a]Parentheses indicate that data were examined and then further runs were made.

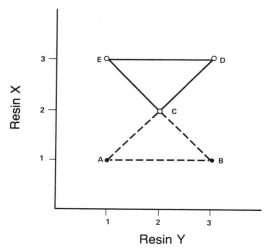

FIGURE 15.9.   The case of two "worst" results.

Example 15.1 we arrive at the points shown in Table 15.6 for the further experiments.

Example 15.3.   In other cases, the simplex may be applied to subjective data such as closeness to a visual color match. As an illustration, suppose we are preparing a transparent colored lacquer with two dyes and our goal is to match a competitive color, perhaps even in a different vehicle. In this case, experimental amounts of the two dyes would be the variable and the resultant three compositions the vertices of the simplex. Closeness to the competitive color could be ranked, the worst color match rejected, and a new point constructed as before. As described in the following sections, additional variables (other dyes) could be added to the experimental plan and analyzed for their optimum combination.

15.3.   NONGRAPHICAL TECHNIQUES

Using the same basic principles, one can move on to a three-factor experiment which has to be charted mathematically rather than by geometry. Consider, for example, a problem of dispersion, which could be evaluated either on a laboratory scale or in production. We have a high-speed disperser and we wish to produce a dispersion in the minimum time. Three factors we are considering are (1) solids content of the dispersion = x, (2) volume of the mix = y, and (3) rotor speed = z.

TABLE 15.7. Data for Initial Vertex Selection—Dispersion Example

Experiment	Solids Content X	Volume of Mix Y	Rotor Speed Z	Dispersion Time, min
A	54%	90 gal	3600 rpm	15.5
B	57%	90 gal	3600 rpm	12.0
C	56.5%	99 gal	3600 rpm	16.0
D	56.5%	94.5 gal	3780 rpm	14.0

Referring back to Table 15.4 we set the following figures:

$$Sx = 54\%$$
$$Sy = 90 \text{ gal}$$
$$Sz = 3600 \text{ rpm}$$
$$px = 3\%$$
$$py = 9 \text{ gal}$$
$$pz = 180 \text{ rpm}$$

As there are three factors, the first set will consist of four experiments, A, B, C, D, as in Table 15.7. Each of these tests are made, and the time required to obtain a good dispersion noted and entered in the last column. Experiment C gives the longest time, 16.0 min, and so experiment E is set up as in as in Figure 15.10.

$$2 \times (\text{average of retained points}) - \text{discarded point.}$$

Factor 1 S = Factor 2 S = Factor 3 S =		p = p = p =	
Experiment A Experiment B Experiment C Experiment D			
a. sum of retained levels b. Average x 2. $\dfrac{2 \times a}{3}$ c. Rejected level d. New level b - c			

FIGURE 15.10. Three-factor simplex design.

For solids content this is:

$$2 \times \frac{(54 + 57 + 56.5)}{(3)} - 56.5 = 55.16\%$$

For volume of mix:

$$2 \times \frac{(90 + 90 + 94.5)}{(3)} - 99 = 84 \text{ gal}$$

For rotor speed:

$$2 \times \frac{(3600 + 3600 + 3780)}{(3)} - 3600 = 3720 \text{ rpm}$$

Experiment E is carried out and entered against A, B, and D in Table 15.8.

In this case experiment A gives the highest dispersion time, and so the levels for experiment F are calculated and give the following data:

Solids content	58.4%
Volume of mix	89 gal
Rotor speed	3800 rpm
Dispersion time	10.5 mins

This makes experiment D the highest and so figures for experiment G can be calculated. Table 15.9 shows these, and further calculations up to experiment J. At this point experiment F has been in four series, that is $f + 1$ times, and so should be viewed with suspicion. If experiment K is calculated, this will be a repeat of experiment B and there has been rotation around point F, which means that it is either the optimum or a false result. This can easily be proved by a repeat of F.

This example illustrates a three-factor experiment, but also illustrates how this technique can be used in production. Slight variations from batch

TABLE 15.8. Dispersion Example—Further Data

Experiment	Solids Content X	Volume of Mix Y	Rotor Speed Z	Dispersion Time, min
A	54%	90 gal	3600 rpm	15.5
B	57%	90 gal	3600 rpm	12.0
D	56.5%	94.5 gal	3780 rpm	14.0
E	55.16%	84 gal	3720 rpm	13.5

TABLE 15.9. Dispersion Optimization—Final Vertex Selection

Experiment	Solids Content	Volume of Mix	Rotor Speed	Dispersion Time
B	57%	90 gal	3600 rpm	12.0
D	56.5%	94.5 gal	3780 rpm	14.0
E	55.16%	84.0 gal	3720 rpm	13.5
F	58.4%	89.0 gal	3800 rpm	10.5
G	57.21%	80.83 gal	3633.3 rpm	11.0
H	59.9%	89.2 gal	3635.5 rpm	11.5
J	60.0%	82.7 gal	3779.2 rpm	12.5

to batch, if carefully organized and monitored, can show process improvements without too much disruption and give a result quite quickly. In some cases, two of the factors in a three-factor experiment may be combined (such as a ratio of alkyd resin to melamine–formaldehyde resin), thus allowing the use of the simplex triangular graphical technique.

The simplex technique may be applied to any number of variables in the same manner. The basic rule remains the same:

$$\text{New Point Coordinate} = 2 \times (\text{average of old good}) - \text{old bad} \quad (15.1)$$

Several additional rules, discussed in detail in Ref. 3, that aid experimentation are:

1. If a vertex has been retained in $k + 1$ simplexes, where $k + 1$ is the number of variables under consideration, redetermine the experimental result at that vertex before continuing. This protects against obtaining good results due to chance alone.

2. If the newest point in the simplex has the poorest response, then instead of replacing this point, replace the next poorest point in the simplex. This protects against bouncing back and forth on the same points.

3. If the calculated coordinates of a new point fall outside the boundaries of one or more of the variables, making it impossible to do an experiment under the predicted conditions, assign an undesirable response to that point and continue.

15.4. USE OF SUBJECTIVE RESULTS

Statistical techniques often require results which can be expressed in absolute numbers, for example, temperature or viscosity. But many characteristics of paint are measured in a mathematically vague way, for example, ease of application, or flow (from a brushed film). In a simplex, as we are rejecting the "worst" paint, it is often possible to use subjective assessments. As an example, consider the effect of variations in a latex paint where we are concerned with obtaining optimum application properties, brush feel, and flow, while maintaining a maximum viscosity. In this we have four factors: solids content; level of fine clay; level of two polyacrylate thickeners, one a low viscosity and one a high viscosity. The solids content has an effect on viscosity and flow; the fine clay, used as a replacement for calcium carbonate to maintain a constant pigment volume concentration will affect the rheology and thus application properties. The two polyacrylate thickeners will affect viscosity and flow.

Using the plan previously outlined we establish the following levels for S and P.

	S	p
Solid content	50%	2%
Clay	50 lb/100 gal	5 lb/100 gal
Low-viscosity polyacrylate	20 lb/100 gal	5 lb/100 gal
High-viscosity polyacrylate	10 lb/100 gal	1 lb/100 gal

Our first series of five paints is as shown below.

Paint	Solids Content	Clay	Low-Viscosity Polyacrylate	High-Viscosity Polyacrylate	Application Properties	Viscosity ku
A	50	50	20	10		84
B	52	50	20	10		90
C	51	55	20	10	X	88
D	51	52.5	25	10		84
E	51	52.5	22.5	11		92

X indicates worst.

Rejecting paint C which has the worst flow we can then calculate new levels for our four factors using equation (15.1) as described previously.

$$(2 \times \text{average of retained points}) - (\text{discarded point})$$

For solids content this is:

$$2\frac{(50 + 52 + 51 + 51)}{4} - 51 = 51$$

For clay level this is:

$$2\frac{(50 + 50 + 52.5 + 52.5)}{4} - 55 = 47.5$$

For low-viscosity polyacrylate:

$$2\frac{(20 + 20 + 25 + 22.5)}{4} - 20 = 23.75$$

For high-viscosity polyacrylate:

$$2\frac{(10 + 10 + 10 + 11)}{4} - 10 = 10.5$$

This then gives us paint F and a second series:

Paint	Solids Content	Clay	Low-Viscosity Polyacrylate	High-Viscosity Polyacrylate	Application Properties	Viscosity ku
A	50	50	20	10		84
B	52	50	20	10		90
D	51	52.5	25	10	X	84
E	51	52.5	22.5	11	X	92
F	51	47.5	23.75	10.5		90

In this case both D and E were rated similar as being the worst. As we are looking for a high viscosity, D, with the lowest viscosity was rejected, giving us a new paint G with the following results:

A	50	50	20	10	3	84
B	52	50	20	10	4	90
E	51	52.5	22.5	11	$5X$	92
F	51	47.5	23.75	10.5	1	90
G	51	47.5	18.125	10.75	1	88

E showed as the worst, but we also began to rank the other paints in order to try to establish a trend.

A further paint, H, was made giving us the following table.

A	50	50	20	10	4	84
B	52	50	20	10	$5X$	90
F	51	47.5	23.75	10.5	1	90
G	51	47.5	18.125	10.75	1	88
H	51	45.0	18.45*	9.625	1	83

At this stage we would normally have rejected B, but the paints F, G, and H all gave similar application properties. However H, our latest paint, had an unacceptable viscosity and any further move could push this lower. Paints F and G were both very acceptable products commercially in terms of application properties and viscosity and so a comparison of other properties, scrub resistance, storage stability, color acceptance, and so on, was made, together with cost comparisons and paint G was judged to be the better of the two in overall appearance.

15.5. SIMPLEX SIZE VARIATION

The size of the simplex used can be important in obtaining an optimum. Ideally, the simplex should be small so that the optimum is accurately pinpointed. However, this can be time consuming, particularly if a large area is considered. A larger simplex will achieve an optimum more quickly but with less accuracy. For simplicity, consider a two-factor experiment which can be plotted graphically and then considered by nongraphical techniques.

As an example, consider the effect of solvent blends on the viscosity of a 40% resin solution, using three solvents, A, B, and C.

Since any combination is limited to a level of 60%, there is a natural restriction which allows consideration of only a two-factor experiment, using two solvents A and B, the levels of which will automatically set the level of solvent C. For example, if A = 10% and B = 20%, C must be 30%.

Figure 15.11 shows the situation graphically. Any point within the triangle represents possible mixes of three solvents, with the level of C being $60 - $ (A and B).

A and B are given large values of p so that the simplex moves quickly. Table 15.10 shows experiments G $-$ K, where I has been included $f + 1$ times and is rechecked. Note that the levels for solvent C are determined by the levels of A and B.

Since I is the optimum, this is used as the start for A and B and a lower value for p is introduced, Table 15.11, Figure 15.12. This gives 0 as the minimum, confirmed by a recheck. The point to note about this design is that the limitation of a total of 60% allows us to use a two-factor design in evaluating three factors.

The simplex size variation can be taken a stage further by evaluating the size of the variation from point to point in deciding the size of the next simplex. However, this involves rather more involved mathematics and begins to move away from a simple, easily applied technique. Further information on this has been published by Nelder and Mead (2).

The size of the original triangle will determine how accurately the optimum vertex is located. If the increments in X_1 and X_2, that is, the step size, approximate the experimental error, the simplex is obviously too

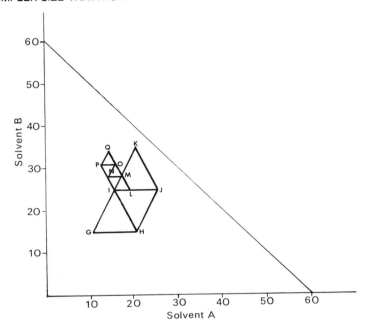

FIGURE 15.11. Solvent compositions.

small. In this case, the simplex will tend to wander erratically both before and after reaching the optimum. In fact, the optimum may never be located. With a simplex that is too large, obviously the point of maximum response may probably be missed and may never be found. For best results, the step size and therefore, the size of the simplex should be at least as large as two to three times the error of each measurement (14). To summarize, the choice of step size, while arbitrary, nevertheless should result in a comparable change in response. Generally speaking, it is advisable to select a large step size and perhaps repeat the experiment with smaller step sizes near the optimum.

TABLE 15.10. $SA = 10$, $SB = 15$, $pA = 10$, $pB = 10$

Experiment	Solvent A	Solvent B	Solvent C	Viscosity
G	10	15	35	17.6
H	20	15	25	16.8
I	15	25	20	14.2
J	25	25	10	15.3
K	20	35	5	14.8
Recheck I				14.1

TABLE 15.11. *S*A = 15, *p*A = 3, *S*b = 25, *p*B = 3

Experiment	Solvent A	Solvent B	Solvent C	Viscosity
I	15.0	25.0	20.0	14.1
L	18.0	25.0	17.0	14.3
M	16.5	28.0	15.5	13.8
N	13.5	28.0	18.5	13.6
O	15.0	31.0	14.0	12.9
P	12.0	31.0	17.0	13.3
Q	13.5	34.0	12.5	13.2
		Recheck O		
	15.0	31.0	14.0	12.8

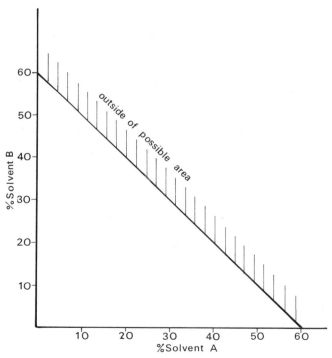

FIGURE 15.12. Solvent compositions—optimization.

478

15.6. SOME ADDITIONAL CONSIDERATIONS

As described by Hendrix (9), it is possible to systemitize the initial variable levels (i.e., X_1, X_2, X_3) in a three-factor experiment as in the following:

	X_1	X_2	X_3
A	−	−	+
B	+	−	−
C	−	+	−
D	+	+	+

where + indicates the high level of a variable and − indicates a low level. Additional designs are given by Hendrix (9) for as many as seven variables.

A more quantitative scheme for the initial location of the vertices has been given by Long (12). In this, fractions of step sizes are taken from the experimental origin defined as having coordinates of zero. The actual beginning level of the factor(s) would be based on considerations such as prior knowledge of the system before the current optimization attempt or perhaps just an experimental probe which was prepared based on the experimenter's "best guess." For the remaining vertices, the step size for each factor is multiplied by an appropriate fraction, and this in turn is added to the experimental origin. For example, a three-factor simplex would be based on the following:

	X_1	X_2	X_3
A	0.0	0.0	0.0
B	1.0	0.0	0.0
C	0.5	0.866	0.0
D	0.5	0.289	0.817

If at any stage it is desired to study the variation of a factor previously held constant, this is readily done. The addition of one single point in another dimension thus converts an equilateral triangle into a tetrahedron. Another factor variation, for example *replication* of individual experiments, is not generally recommended (12). However, in some cases of semi-quantitative responses, such as performance in salt spray, duplicate or triplicate panels of the individual formulations may be tested and the performance averaged (8).

In most coating problems of optimization, only one principal response is generally selected. The choice of this response will usually be dictated by the actual laboratory situation such as the need for improved abrasion resistance. Sometimes laboratory experiments will provide secondary responses. In these cases, these should be retained and, depending on the outcome of the simplex experimentation, perhaps plotted (via simplex optimization) and compared to the most important response. It may happen that the found optimum may be regarded with suspicion. A helpful course to doublecheck the optimum is to repeat the experiment from a widely different region of the response space.

15.7. CONCLUSION

This then is the basis of simplex EVOP— a technique which, correctly used, can assist the paint chemist in obtaining an optimum with some speed in a controlled fashion.

The method described has been deliberately kept to the simplest form. However, there are modifications and more sophisticated techniques and readers who are interested in pursuing these should consult the references.

REFERENCES

1. Spendley, W., Hext, G. R., and Himsworth, F. R., *Technometrics*, Vol. 4, 1962, p.441.
2. Nelder, S. A., and Mead, R., *Computer J.*, Vol. 7, 1965, p. 308.
3. Deming, S. N., and Morgan, S. L., *Anal. Chem.*, Vol. 45, 1973, p. 278A.
4. Deming, S. N., and Morgan, S. L., *Anal. Chem.*, Vol. 45, No. 3, 1973, p. 279A.
5. Olsson, D. M., *J. of Quality Technology*, Vol. 6, No. 1, 1974, p. 53.
6. Carpenter, B. H., and Sweeny, H. C., *Chemical Engineering*, Vol. 72, No. 14, 1965, p. 117.
7. Dean, W. K., Heald, K. N., and Deming, S. N., *Screner*, Vol. 189, 1975, p. 805.
8. Brooker, D. W., et al, *J.O.C.C.A.*, Vol. 52, 1969, pp. 989–1034.
9. Hendrix, C., *Chemtech*, August 1980, pp. 488–497.
10. Lowe, W. C., *Trans. Inst. Chem. Engr.*, Vol. 42, 1964, pp. T334–T344.
11. Kenworthy, I. C., *Appl. Statistics*, Vol. 16, 1967, pp. 211–224.
12. Long, D. E., *Analytica Chimica Acta*, Vol. 46, 1969, pp. 193–208.
13. Lowe, W. C., *Appl. Statistics*, Vol. 23, 1974, p. 218.
14. Krause, R. D., and Lott, J. A., *Clinical Chem.*, Vol. 20, No. 7, 1974, p. 775.

16

CONTROL CHARTS
FOR VARIABLES

JOHN COMPTON

College of Graphic Arts & Photography
Rochester Institute of Technology
Rochester, New York

16.1. INTRODUCTION TO CONTROL CHARTS

Central to the study of any production operation is the question "Is the process behaving normally?" The answer to this question will often dictate whether or not action must be taken to make changes in the operation. The term "process" is used here to describe any set of conditions or sequence of events which work together to produce a given result. In an industrial plant we can think of the manufacturing process as including all steps in the process from the original raw materials to the finished product. In a narrower sense, a process could be as basic as the function of a single machine or even a single element of a machine. The process to be studied is dictated by the concerns that are held. If, for example, in the manufacture of paint, we are worried about the tinting strength of titanium dioxide, there is no need to study the entire manufacturing process. It is imperative that the process to be studied is clearly defined before any information is gathered.

Equally as important as the proper definition of the process is the determination of what to measure. This item (the response variable) should be chosen by: (1) its relationship to that property of interest and (2) its sensitivity to changes in the process. If the response variable is not directly related to the characteristic of interest then little, if any, useful information will be obtained. The second factor, dealing with sensitivity, is also critical from the standpoint of detecting changes in the process when they occur.

The control chart is the statistical method primarily used for the evaluation and control of any repetitive process. This graphical method of displaying the process performance provides an easily interpreted pictorial representation of the process behavior. The use of a control chart produces a sound basis for answering the critical question "Is the process behaving normally?" Control chart techniques are used in the coatings industry in applications such as fineness of grind, weight per gallon, viscosity, color, solids content, acid value, and so on.

16.2. SOURCES OF VARIABILITY

Whenever we study the results of a repetitive process, we will always find differences. This is suggested by one of the fundamental laws of nature which states that no two things are exactly alike and no single process can behave exactly the same way, time after time. If we examine the results closely enough, differences will always be evident. Stated more simply, variability is a constant phenomenon of our world. These naturally occurring variations in a process are attributed to *chance* causes. This "chance caused" variability is the sum of a complex system of chance effects, each of which is small and for all practical purposes, untraceable. In other words, even though a process is performing to the best of its ability, we must be content to live with these natural chance caused changes as normal for the operation. The set of data in Table 16.1 is from a process that is governed primarily by chance. When these data are plotted as a frequency histogram the distribution in Figure 16.1 results.

Figure 16.1 illustrates the pattern of variability associated with a process that is primarily chance controlled. This bell-shaped curve will be recognized as the normal distribution. When a large number of chance factors with equal influence are affecting the process, statistical theory indicates that a normal distribution is the model to be expected. Consequently, when distributions other than the normal occur, it is an indication that other than chance causes are affecting the system.

The other type of variability which can affect the outcome of a process is that which is traceable to its origin. These variations are either larger than or different in pattern from those which are chance caused. We refer to this as assignable cause variability. The occurrence of a non-normal distribution is an indication that one (or more) assignable causes are affecting the system. Since these causes can be identified, steps are usually taken to remove them.

Although the use of a frequency histogram is important in the detection of assignable cause variability, it fails to consider the sequence in which the data occurred. This element of time is necessary for insight into the *stability* of the process. For an ongoing manufacturing process, the concept of stability is equally important to the concept of chance effects. In other words, it is not sufficient for a process to be only governed by chance; it

TABLE 16.1. Typical Data from a Process That Is Chance Governed

	Date	Data		Date	Data
Feb.	5	12	Mar.	2	19
	6	18		3	20
	7	17		4	20
	8	18		5	19
	9	20		6	17
	10	19		7	16
	11	14		8	15
	12	22		9	15
	13	16		10	16
	14	15		11	17
	15	18		12	18
	16	17		13	18
	17	16		14	21
	18	13		15	23
	19	21		16	22
	20	18		17	19
	21	19		18	14
	22	20		19	14
	23	16		20	15
	24	15		21	16
	25	17		22	18
	26	17		23	22
	27	17		24	21
	28	18		25	19
				26	17
				27	17

must also be operating at a stable level. The control chart is much like a histogram, except that the order of production is retained and therefore, the response variable is plotted as a function of time. Figure 16.2 illustrates a time plot of the data from Table 16.1. If this process was stable and chance controlled, the fluctuations as a function of time would be random in nature. In other words, the points would have no predictable pattern. Any departure from this random "noise" pattern thus signals the presence of assignable cause variability. During the first 20 days of this operation the process was chance controlled at a stable level. This is indicated by the random fluctuations on the graph. However, the data from the last 20 days indicate a predictable cycling pattern with definite peaks and valleys. This is a nonrandom pattern and indicates that assignable cause variability is occurring. Simply stated, the illustrative process is presently *not* behaving normally. It is important to know that the use of a histogram, in this case, would not have allowed for the detection of this time-related assignable cause problem. A process is defined as being "in control" only when a stable system of chance-caused effects is working.

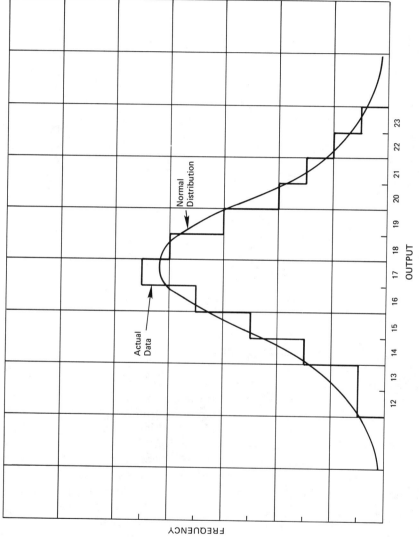

FIGURE 16.1. Frequency histogram from data in Table 16.1.

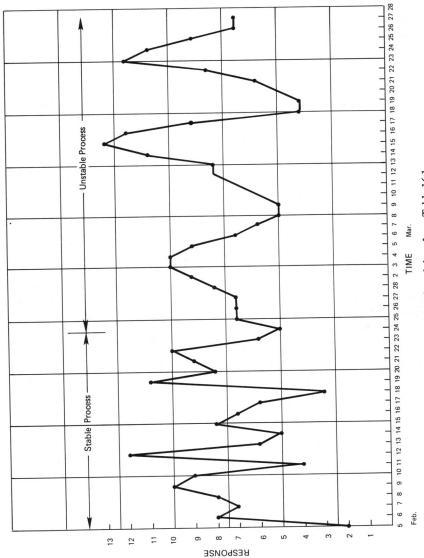

FIGURE 16.2. Time plot of data from Table 16.1.

16.3. CONTROL CHARTS FOR INDIVIDUAL OBSERVATIONS

The most basic method of control chart evaluation is that which is based on single measurements from the process. For this plan, the data are collected one by one and treated on the control chart as individual observations. If the process under observation is a new one or if the past performance is unknown, the first step is to accumulate sufficient data to determine the process capability. A sample size of at least 30, plotted as a frequency histogram, will give a good first indication as to whether the process is chance controlled. If it appears that the process is primarily governed by chance, the next step is to calculate the sample average ($\overline{X}$) and plot the data on a time scale. If the time plot exhibits a stable–random pattern the control limits may then be calculated. These control limits are derived from the sample standard deviation, (s), and represent the range of variations to be expected when only chance (random) causes are affecting the process. The limit lines are extended forward in time and data continue to be collected. The pattern indicated by the plotted points helps the operator to decide if the process should continue without changing it.

The upper and lower control limits are based on the belief that the sample data came from a normally distributed population. Therefore, these limits should include all data representing an unchanged process. However, since the normal curve never reaches zero frequency, it is impossible to place the limit lines in such a position as to include all possible members of the population. As an example, suppose that in a paint can filling operation a large sample size shows that the average gross weight (paint and can) is 10.00 lb., and the estimate of the standard deviation is 0.10 lb. If on a future test a gross weight of 10.40 lb is found, should the process be changed? The answer must be based on the probability involved in the relationship between the estimates of the population parameters and the new data. Since 10.40 is greater than three standard deviations from the average, there are less than 3 chances in 1000 that 10.40 belongs to the original population. This also means that there are less than 3 chances in 1000 that the process has *not* changed.

16.4. TYPES OF RISKS

The limit lines on a control chart indicate the levels above and below the average into which the process must drift before it can be assumed that the process has changed, and consequently when action must be taken. Where these lines are placed depends on the cost (financial and labor) of two risks.

1. *The alpha risk.* This is the risk of changing the process when it has not really changed, but you believe it has.

2. *The beta risk.* This is the risk of failing to change the process when it has really changed, but you think it has not.

Neither of these risks can be entirely eliminated because of the pattern of the normal curve. One risk, however, is usually more costly than the other and the limits are set accordingly. The limits are placed far from the mean when the alpha risk must be avoided as much as possible. The limits are placed close to the mean when the beta risk must be avoided as much as possible. The most frequently used control limits are set at $\pm 3s$ from the mean. Limits at $\pm 2s$ are occasionally used, and $\pm 4s$ rarely used. The alpha risk is approximately 3 in 1000 for limits of $\pm 3s$.

The determination of whether a process is in control or out of control arises in two basic ways. First, the test of whether an unknown process is in control is mainly a test for uniformity, or stability. The determination of whether an unknown process is in control involves taking a series of samples from the process, and testing these samples for significant differences from the aggregate of all the samples.

Second, the test of whether a process known to have been in control remains in control is mainly a test for conformance to a predetermined standard. The determination of whether a process remains in control involves taking continuous samples from the process, and testing these samples for significant differences from the expected performance of the process.

While this testing can be done by a series of significance tests, Dr. W. A. Shewhart devised, in 1924, a graphic method for doing this testing continuously. This method, known as the Shewhart control chart, or simply as the control chart, is thus a graphic perpetual test of significance. The control chart has been found to have an enormous range of application, not merely in quality control, but in all forms of continuing tests of significance and quality assurance. For a normal distribution the average $\overline{X}$ locates the center of the distribution, and the standard deviation σ describes the variability of the data. As previously discussed, for a normal distribution 99.7% of the values will lie within $X \pm 3\sigma$, 95% within $X \pm 2\sigma$, and 68% within $X \pm 1\sigma$. Limits of 1.96, 2.57, and 3.09 standard deviations on either side of the average are occasionally used since they correspond to normal probabilities of 1 in 20, 1 in 100, and 1 in 500, respectively.

Table 16.2 contains a set of seventy individual measurements of yield in percent from a chemical manufacturing process. Figure 16.3 presents the frequency histogram for this data. The histogram appears to be normal, and is therefore telling us that the process is primarily chance governed. The next step is to calculate the average of the subgroup ($\Sigma X/n$) and plot the data in order of production, as shown in Figure 16.4. An inspection of this graph indicates that the data came from a stable system. The next step is to calculate the sample standard deviation and place the limit lines on either side of the average as dictated by the alpha risk. Here we will choose

TABLE 16.2. Yield in Percent of Chemical Manufacturing Process ($n = 70$)

8:00 A.M.	0.70	3:15 P.M.	0.81	10:45 P.M.	0.77
15	0.75	30	0.80	11:00	0.79
30	0.77	45	0.71	15	0.76
45	0.75	4:00	0.76	30	0.71
9:00	0.73	15	0.76	45	0.77
15	0.78	30	0.75	12:00 A.M.	0.74
30	0.76	45	0.75	15	0.77
45	0.78	5:00	0.77	30	0.80
10:00	0.76	15	0.77	45	0.78
15	0.73	30	0.72	1:00	0.75
30	0.79	45	0.77	15	0.75
45	0.78	6:00	0.72		
11:00	0.71	15	0.76		
15	0.75	30	0.82		
30	0.76	45	0.76		
45	0.80	7:00	0.77		
12:00	0.78	15	0.77		
15	0.76	30	0.74		
30	0.76	45	0.75		
45	0.74	8:00	0.76		
1:00 P.M.	0.76	15	0.76		
15	0.70	30	0.77		
30	0.79	45	0.73		
45	0.77	9:00	0.77		
2:00	0.77	15	0.74		
15	0.76	30	0.72		
30	0.75	45	0.74		
45	0.76	10:00	0.75		
3:00	0.75	15	0.74		
		30	0.73		

$$\overline{X} = 0.757$$
$$s = 0.025$$

an alpha risk of 3 in 1000 and consequently place the limit lines at $\pm 3s$ as in Figure 16.5. The fact that the points *randomly fluctuate* between the limit lines indicates that the process is in a state of statistical control.

At this point in time the limit lines would be extended forward and samples would continue to be taken and plotted on the control chart. Action is taken on the process when nonrandom (or nonchance) patterns are exhibited. Figures 16.6a–d, illustrate a typical variety of nonchance patterns as follows:

(a) A point out of control, with the limit lines placed at $\pm 3s$ (alpha risk = .003). A point will fall outside of these boundaries only 3 times out of 1000 due to chance. Therefore, the odds overwhelm-

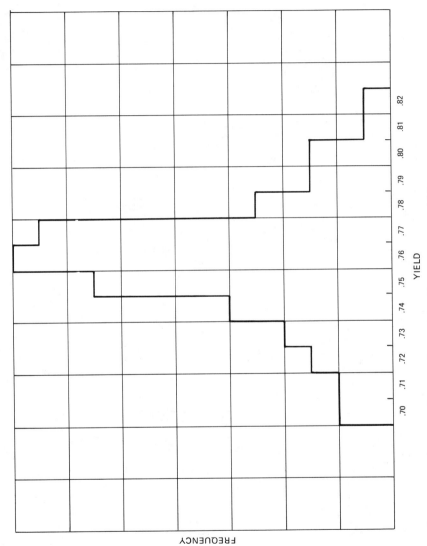

FIGURE 16.3. Frequency versus yield.

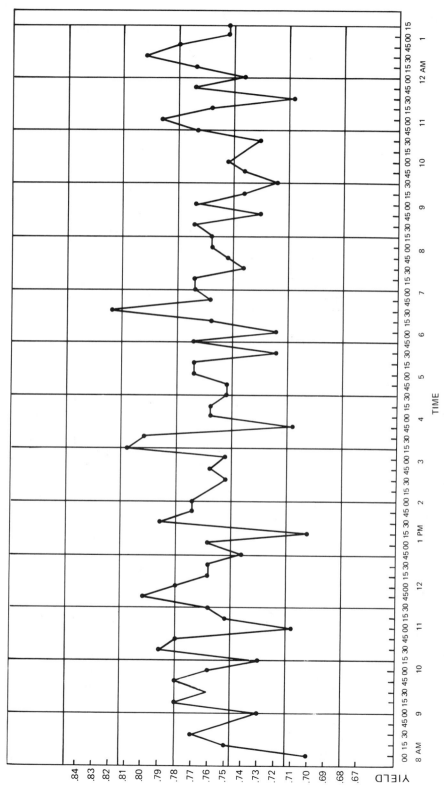

FIGURE 16.4. Time plot of percent yield data (Table 16.2).

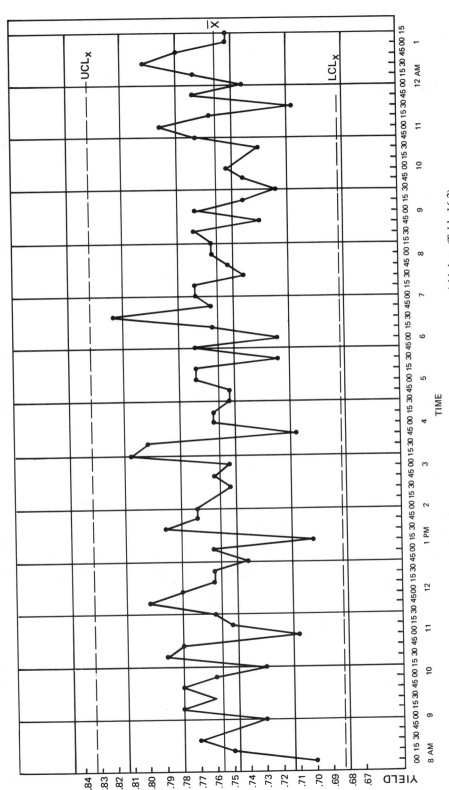

FIGURE 16.5. Control chart (time plot plus limit lines) of percent yield data (Table 16.2).

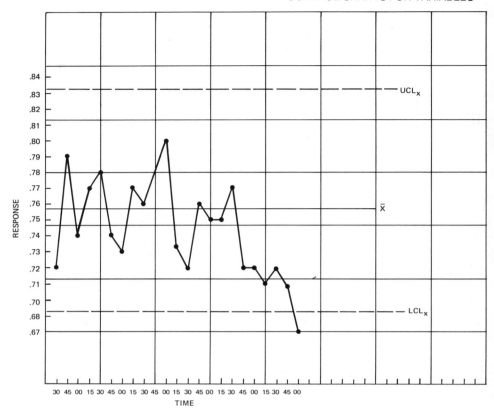

FIGURE 16.6.   (*a*) Nonchance pattern, "point out of control."

ingly point to an assignable cause and necessary action must be taken.

(b)   An upward trend. Frequently it is found that no points are out of control, but that there is a gradual climbing or falling tendency exhibited in the points. This is an indication that the process is changing and consequently, corrective measures should be taken. Statistical theory suggests that at least five points are needed to indicate such a trend.

(c)   A run. The points may be within the limit lines without upward or downward trends, but all falling above (or below) the average. This is an indication of a shift in the process level and is an indication that a change has occurred. Again, five of these points in a row indicate a nonrandom pattern requiring investigation.

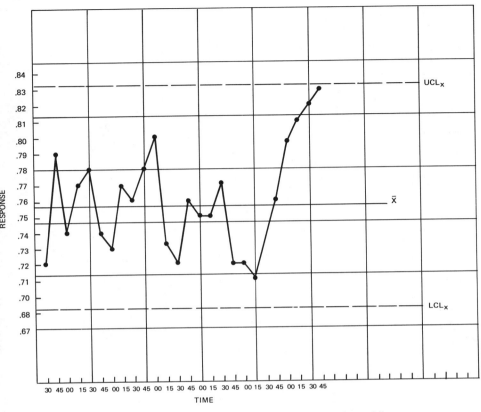

FIGURE 16.6. (*b*) Nonchance pattern, "an upward trend."

(d) Cycling. Without careful inspection this sawtooth pattern appears to be random. However, when this up–down pattern becomes predictable, it is an indication of assignable cause variability. Seven to ten such points are cause for action.

While the use of control charts for individual observations is a basic quality control method, it does contain a number of drawbacks. The most worrisome of these drawbacks is the fear that the underlying process distribution is not normal. Since the control limits are derived from the standard deviation of the assumed normal distribution, it can cause a critical error in the interpretation of the charts. Secondly, control charts based on individuals give little insight into short-term variations. Since data are taken only one at a time, there is no within sample estimate of variability. The following technique provides a method for dealing with these problems.

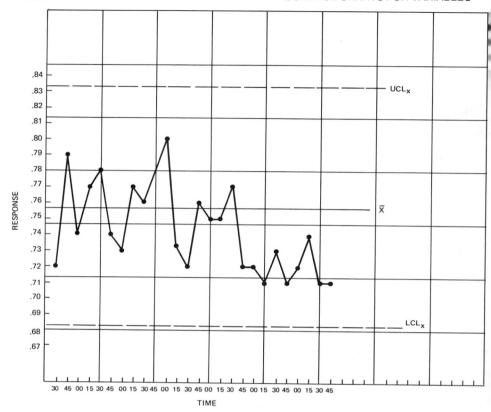

FIGURE 16.6.   (c) Nonchance pattern, "a run."

16.5.   CONTROL CHARTS FOR AVERAGES ($\overline{X}$) AND RANGES (R)

So far, we have considered control charts as consisting of individual pieces of data. In many industrial operations samples are often taken in groups of four, five, or more items. The size of these subgroups is referred to as n. The total number of samples is referred to as k. Averages of the subgroups are taken and the resulting values are plotted as control chart data.

Two important benefits are contained in this method. First, subgroup averages tend to produce a normal distribution, even though the individuals may be drawn from a population that is definitely non-normal. The benefit here is obvious. When the data are taken in subgroups and the average values used, we are always assured of an approximately normal distribution. Thus, the standard deviation can be used for control limits with confidence that it is justified. Second, the standard deviation of subgroup averages is smaller than the standard deviation of individuals

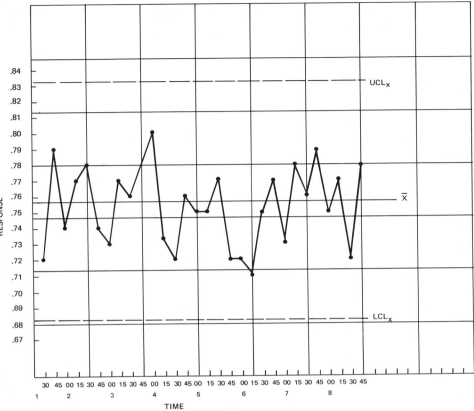

FIGURE 16.6. (*d*) Nonchance pattern, "cycling."

from the same population. The relationship is as follows:

$$\sigma_{\overline{x}} = \frac{\sigma_x}{\sqrt{n}}$$

This is useful because it improves our ability to make inferences from the samples. Since individual observations usually contain large variations, the resulting standard deviation is large. Consequently, many times it is hard to tell whether a piece of data that is different from the average represents a change in the process or merely a random fluctuation. Averages have less variability ($\sigma_{\overline{x}}$) and therefore make it easier to discriminate between an actual change and chance differences. The employment of these principles for control chart use was first proposed by Dr. Shewhart.

The data in Table 16.3 are typical of an industrial process. In this case they are color difference measurements made on successive batches of a

TABLE 16.3. Color Difference Measurements Made on Successive Batches of a Gray Gloss Coating

Time	X_1	X_2	X_3	X_4	ΣX	$\overline{X}$	R
Jan. 3	1.5	2.2	1.6	1.6	6.9	1.73	0.7
4	2.3	2.1	2.1	2.2	8.7	2.18	0.2
5	1.9	2.2	1.7	1.7	7.5	1.88	0.5
6	1.3	2.1	2.0	2.0	7.4	1.85	0.8
7	1.0	1.9	2.3	1.9	7.1	1.78	1.3
8	2.1	2.5	1.6	2.6	8.8	2.20	1.0
9	2.3	2.5	2.1	2.5	9.4	2.35	0.4
10	2.5	2.2	1.6	1.8	8.1	2.03	0.9
11	1.5	2.6	2.1	2.2	8.4	2.10	1.1
12	1.3	2.1	2.6	1.6	7.6	1.90	1.0
13	2.2	2.0	2.3	2.0	8.5	2.13	0.3
14	1.2	1.9	1.9	2.2	7.2	1.80	1.0
15	1.7	1.4	1.3	2.0	6.4	1.60	0.7
16	2.1	1.9	1.7	1.3	7.0	1.75	0.8
17	2.2	2.1	1.8	2.0	8.1	2.28	0.3
18	1.7	2.1	1.6	1.9	7.3	1.83	0.5
19	1.8	1.6	2.0	2.2	7.6	1.90	0.6
20	1.8	2.1	2.2	2.0	8.1	2.03	0.3
21	2.1	2.0	2.1	1.5	7.7	1.93	0.6
22	1.5	1.9	1.8	2.2	7.4	1.85	0.7
23	2.2	2.6	2.3	1.9	9.0	2.25	0.7

$\overline{\overline{X}} = 1.98$
$\overline{R} = 0.69$

gray gloss coating. The data were collected in subgroups with four values ($n = 4$) in each subgroup. In Figure 16.7 the data have been plotted and the upper and lower control limits added. Two graphs are employed to illustrate the variability of the data. The upper graph is an $\overline{X}$ chart and shows variations in the averages of the samples. The central line is drawn at $\overline{X}$, the average of all the averages, $\overline{\overline{X}} = (\Sigma \overline{X}/k)$. The lower graph is a plot of the ranges of the samples. The central line is drawn at $\overline{R}$, the average of the ranges ($\Sigma R/k$). Both graphs use the same time scale.

The control limits are derived from an estimate of the chance-caused variability in the process. This is most often obtained by organizing the data into *rational subgroups* which contain differences that can be attributed only to chance. Consequently, close attention must be paid to the selection of these subgroups. Three σ limits are most commonly used with Shewhart charts. The calculation of limit lines is made simpler through the use of coefficients used in conjunction with $\overline{R}$. The process depicted in Figure 16.7 can be said to be in a state of statistical control.

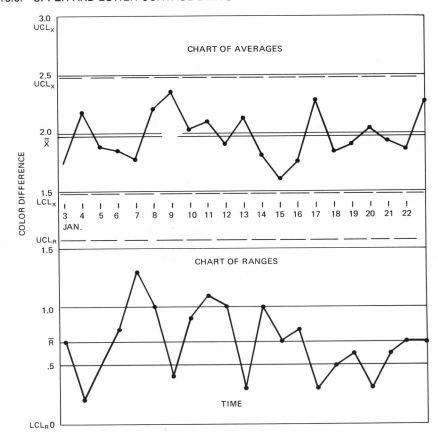

FIGURE 16.7.   Control chart for color-difference data (Table 16.3).

16.6.   UPPER AND LOWER CONTROL LIMITS

Consider a filling operation on a new machine in a paint factory, where we are filling gallons of white enamel with an average gross weight (paint and can) of 10.00 lb. The allowable limits are ±0.25 lb. We would like to construct a control chart to systematically keep our process under study.

Let us assume that although we do not yet know the sample standard deviation, s, of the process on the new machine, not more than 1 can in 1000 is to fall outside the limits of 9.75–10.25 lb. Therefore, the filling operation should be established so that the average weight will be larger than 9.75 (and smaller than 10.25) by an amount equal to $3s$, since the tail of a normal distribution curve beyond $3s$ contains about 1.3/1000 of the area.

Therefore, if

$$3\sigma' = 0.25 \text{ lb}$$
$$\sigma' = 0.083 \text{ lb}$$
$$\overline{X}' = 10.000 \text{ lb}$$

It is now possible to estimate the limits within which the means of samples of size n should fall if drawn from the universe of cans. These limits are

$$\overline{X}' \pm 3s_{\overline{x}} \quad \text{for} \quad 99.7\%$$

of the sample means.

The sample size n, the rational subgroup, in quality control is usually 4 or 5, within which variations may be considered due to nonassignable chance causes only (error group) and between which there may be differences due to assignable causes whose presence is considered possible.

Let

$$n = 4$$
$$s_{\overline{x}} = \frac{\sigma'}{\sqrt{n}} = \frac{0.083}{\sqrt{4}} = 0.0415$$

and

$$3s_{\overline{x}} = 0.1245; \qquad \overline{\overline{X}} \pm 3s_{\overline{x}} = 10.000 \pm 0.1245$$

Therefore, the control chart for paint can weight averages would be constructed with the following limits:

$$\text{central line} \quad = \overline{\overline{X}} \quad\quad\quad = 10.00 \text{ lb}$$
$$\text{upper control limit} = \overline{\overline{X}} + 0.1245 = 10.1245 \text{ lb}$$
$$\text{lower control limit} = \overline{\overline{X}} - 0.1245 = 9.8755 \text{ lb}$$

To further assure satisfactory uniformity, it may be desirable to also set up in analogous fashion control limits for the standard deviation (or range) of the samples and to record the results of successive samples.

For example,

$$s_s = \frac{\sigma'}{\sqrt{2n}} = \frac{0.083}{\sqrt{8}} = 0.0294$$
$$3s_s = 0.088$$

TABLE 16.4. Factors for Computing Control Chart Lines

Sample Size n	Chart for Averages				Chart for Individuals		Charts for Ranges					Factors for Central Line	Charts for Sample Standard Deviation (s)				Sample Size n
	Factors for 3σ Control Limits				Factors for 3σ Control Limits		Factors for 3σ Control Limits						Factors for 3σ Control Limits				
	A	A_1	A_2	A_3	E_2	E_3	D_1	D_2	D_3	D_4	d_2	c_3	B_5	B_6	B_7	B_8	
2.	2.121	3.760	1.880	2.659	2.660	3.760	0	3.686	0	3.267	1.128	0.7979	0	2.606	0	3.267	2.
3.	1.732	2.394	1.023	1.954	1.772	3.385	0	4.358	0	2.575	1.693	0.8862	0	2.276	0	2.568	3.
4.	1.500	1.880	0.729	1.628	1.457	3.256	0	4.698	0	2.282	2.059	0.9213	0	2.088	0	2.266	4.
5.	1.342	1.596	0.577	1.427	1.290	3.192	0	4.918	0	2.115	2.326	0.9400	0	1.964	0	2.089	5.
6.	1.225	1.410	0.483	1.287	1.184	3.153	0	5.087	0	2.004	2.534	0.9515	0.029	1.874	0.030	1.970	6.
7.	1.134	1.277	0.419	1.182	1.109	3.127	0.205	5.203	0.076	1.924	2.704	0.9594	0.113	1.806	0.118	1.882	7.
8.	1.061	1.175	0.373	1.099	1.054	3.109	0.387	5.307	0.136	1.864	2.847	0.9650	0.179	1.751	0.185	1.815	8.
9.	1.000	1.094	0.337	1.032	1.010	3.095	0.546	5.394	0.184	1.816	2.970	0.9693	0.232	1.707	0.239	1.761	9.
10.	0.949	1.028	0.308	0.975	0.975	3.084	0.687	5.469	0.223	1.777	3.078	0.9727	0.276	1.669	0.284	1.716	10.
11.	0.904	0.973	0.285	0.927	0.946	3.076	0.812	5.534	0.256	1.744	3.173	0.9754	0.313	1.637	0.321	1.679	11.
12.	0.866	0.925	0.266	0.886	0.921	3.069	0.924	5.592	0.284	1.716	3.258	0.9776	0.346	1.610	0.354	1.646	12.
13.	0.832	0.884	0.249	0.850	0.899	3.063	1.026	5.646	0.308	1.692	3.336	0.9794	0.374	1.585	0.382	1.618	13.
14.	0.802	0.848	0.235	0.817	0.881	3.058	1.121	5.693	0.329	1.671	3.407	0.9810	0.399	1.563	0.406	1.594	14.
15.	0.775	0.816	0.223	0.789	0.864	3.054	1.207	5.737	0.348	1.652	3.472	0.0923	0.421	1.544	0.428	1.572	15.

The limits of the standard deviation control chart then would be

$$\sigma' \pm 3s_s = 0.083 \pm 0.088 = 0 \text{ to } 0.171$$

$$\text{upper control limit} = \sigma' + 0.088 = 0.171$$

$$\text{lower control limit} = \sigma' - 0.088 = 0$$

This example (controlling performance to standard values) is presented as an introduction to the derivation of the conversion factors which are to be found in control limit tables such as published by the ASTM (see Table 16.4). For specified quality levels the central line will be drawn at $\overline{X}:\mu$ and the limits will be $\sigma \pm 3\sigma'/\sqrt{n}$. As an aid in computation, the quantity $3/\sqrt{n} = A$ has been calculated and is given for $n = 4$ in Table 16.4 as 1.5. Therefore, the control limits can be given as $\mu = \overline{X} = \pm 1.5(.083)$. In a similar manner, for the standard deviation, the UCL is given by $B_6\sigma'$ and the LCL is $B_5\sigma'$ which are 2.088 and 0.0, respectively. Formulas for the various control charts, standard values given or not known, can be found in Rickmers, (20, p. 473) or Natrella (18, p. 183). The basic selection criteria is that of known or unknown μ and σ, the size of the subgroups, and the number of the subgroups. Other factors are derived from estimates of the standard deviation based on the range corrected for bias due to the size of the subgroup.

16.7. SELECTION OF RATIONAL SUBGROUPS

The key to the successful use of the Shewhart charts is through the proper method of selecting rational subgroups. Detailed discussion of rational subgroups and subgroup selection may be found in Refs. 16 and 20. The points on Shewhart control charts should represent subgroups of data that are as similar as possible. In other words, the subgroups should be chosen in such a way that assignable cause variations are not likely to be present, but chance-caused variations are. If assignable causes are present, they should occur as differences *between* the subgroups rather than as differences *within* the subgroup.

The most common method for subgrouping is the order of production. For example, if it is believed that different machine settings have an effect on the characteristic being plotted, then all of the samples in the subgroup should come from the same batch. A series of subgroups will then show the effect of differences in machine settings, batches, and so on.

A small group (4 or 5) of consecutively produced items from a process is the most usual method for rational subgrouping. These subgroups are likely to be made up of a randomly produced set of samples which represent the state of the process at the time the subgroup was taken.

However, by selecting a rational subgroup that is so homogeneous that only a few of the sources of chance-caused variability can affect the process, the differences within the subgroup will be unusually small. Since the control limits are based on these differences (as given by $\overline{R}$), this will lead to limits which are unrealistically tight and therefore lead to *overcontrol*. If the rational subgroup is too large and includes some assignable causes in addition to the chance variations, the differences within the subgroup will be unusually wide and therefore lead to *undercontrol*. The effects of some assignable causes may be trivial and therefore not require changes to be made in the process. The effects can be incorporated within the subgroup and therefore treated as being natural to the process. It is important to remember that the proper selection of subgroups is basic to the success of the Shewhart charts.

This method of sampling provides two different types of information about the process. The means of the subgroups give insight into changes in the process level. The ranges of the subgroups give insight into the changes of the within-subgroup variability. This may be interpreted as *long-term variability* (as shown by the subgroup means) and *short-term variability* (as shown by the subgroup ranges).

16.8. CONSTRUCTION OF $\overline{X}$ AND R CHARTS (μ AND σ UNKNOWN)

Step One. The data are collected in subgroups, as in Table 16.3. The minimum subgroup size is 2, with 4 or 5 being used most commonly. At least 10 subgroups are needed initially. This means that the Shewhart method requires fewer data to begin with than charts for individuals.

Step Two. Calculate the mean and range of each subgroup.

Step Three. Calculate the grand average ($\overline{\overline{X}}$) and average range ($\overline{R}$) from all subgroups. These values represent the centerlines for the chart of averages and chart of ranges, respectively. For the data in Table 16.3, the $\overline{\overline{X}}$ equals 1.98 and the $\overline{R}$ equals 0.69.

Step Four. Plot a chart of subgroup averages using the grand average as the centerline and a chart of subgroup ranges using the average range ($\overline{R}$) as the centerline. As shown in Figure 16.7, both graphs should have the same time scale.

Step Five. Examine these charts for stability. If the process is behaving naturally, the fluctuations should have no recognizable pattern. Should the points take any of the patterns illustrated in Figure 16.6, the process is not ready for a control chart. The reasons for the nonrandom patterns must be found and corrected. When the process exhibits control, the limit lines may be determined.

Step Six. Determine the tentative upper and lower control limits for both charts. In the Shewhart method this is achieved through the use of coefficients which convert the average range to give limit lines at ± 3

standard deviations of averages from the mean. The formulas for the chart of averages are as follows:

$$\overline{\overline{X}} + A_2\overline{R} = \text{upper control limit}$$

$$\overline{\overline{X}} = \text{centerline}$$

$$\overline{\overline{X}} - A_2\overline{R} = \text{lower control limit}$$

The formula for the chart of ranges is as follows:

$$D_4\overline{R} = \text{upper control limit}$$

$$\overline{R} = \text{centerline}$$

$$D_3\overline{R} = \text{lower control limit}$$

The coefficients are found in Table 16.4. For example, the average range for the data in Table 16.3 is 0.69. The table value for A_2 ($n = 4$) is 0.729. These two values are multiplied together to give the value 0.50. This last value is added to and subtracted from $\overline{X}$ to obtain the upper and lower control limits for the $\overline{X}$ chart, that is, 1.98 + 0.50 = 2.48 and 1.98 − 0.50 = 1.48.

The table value for D_4 ($n = 4$) is 2.282. This number multiplied by the $\overline{R}$, which is 0.69 gives 1.57, which is the upper control limit for the range chart. Since the table value for D_3 ($n = 4$) is zero, the lower control limit for the ranges is placed at zero.

Step Seven. The limit lines are extended forward in time. The data continue to be collected in subgroups with the averages and ranges plotted on the charts. Action is taken when a lack of control occurs. This may be shown as assignable changes in the process average, which would be reflected as nonrandom patterns in the $\overline{X}$ chart. It also may be shown as assignable changes in the within-subgroup variability, which would be reflected as nonrandom patterns in the R chart. Third, it is conceivable that there could be assignable changes both between and within subgroups.

Step Eight. The limit lines are evaluated for possible revision. As additional data become available, a more accurate estimate of $\overline{R}$ may be obtained. Since in this method the limits are derived from the estimate of the average range, it may be necessary to revise the control limits if the newer estimate of $\overline{R}$ is much different from the initial value.

For example, the data in Table 16.5 represent additional samples from the process for Table 16.3. It is obvious with this new data that the process average has remained stable but that the estimate of $\overline{R}$ has been reduced to 0.39. This indicates that the variability has been reduced and therefore the process has been improved. The old and revised limit lines are shown in Figure 16.8.

TABLE 16.5. Additional Color Difference Measurements

Time	X_1	X_2	X_3	X_4	ΣX	$\overline{X}$	R
Jan. 25	2.2	2.1	2.0	1.8	8.1	2.03	0.3
26	2.1	2.5	2.3	2.5	9.4	2.35	0.4
27	1.6	2.1	1.7	1.9	7.3	1.83	0.5
28	2.3	2.0	2.2	2.0	8.5	2.13	0.3
29	2.2	1.9	1.7	1.9	7.5	1.88	0.5
30	1.6	1.8	1.8	1.9	7.1	1.78	0.3
31	1.9	1.8	1.9	1.7	7.3	1.82	0.2
Feb. 1	2.2	2.0	1.8	2.0	8.0	2.00	0.4
2	2.5	2.0	2.3	2.2	9.0	2.25	0.5
3	1.6	2.2	1.8	1.6	7.2	1.80	0.6
4	1.8	1.9	2.1	1.9	7.7	1.93	0.3

$\overline{\overline{X}} = 1.97$

$\overline{R} = 0.39$

16.9. CONTROL CHARTS USING A MOVING AVERAGE

For many industrial processes it is either unwise or impossible to collect the samples in rational subgroups. Those processes that have small within-batch (or short-time) variability will give unrealistically small estimates of the process variability when sampled via conventional subgroups. Still other processes require long periods of time to obtain a single sample. This makes the collection of multiple samples impractical, if not impossible. Processes such as these may be more intelligently monitored through the use of a control chart employing a *moving* average.

In this method individual samples are taken from the process and ordered in the sequence of their collection: X_1, X_2, X_3, ..., X_i. The averages of successive sets of two (or more) $\overline{X}_n$ values are calculated. After a sufficient number of successive averages have been calculated (at least 15) the $\overline{\overline{X}}$ of the values is determined. The $\overline{\overline{X}}$ is used as the central line of the control chart. An estimate of the standard deviation of the average values ($s_{\overline{x}}$) is obtained. The control limits are placed at $+3s_{\overline{x}}$. Successive $\overline{X}$ values are plotted on the chart.

The data in Table 16.6 represent individual samples taken on a chemical manufacturing process. Moving averages of each successive set of values have been determined. The $\overline{X}$ for these values is 0.17. To determine the control limit lines we must obtain an estimate of $s_{\overline{x}}$. This is found first by dividing the average range by the coefficient d_2 as determined by the information in Table 16.4. Therefore,

$$s_x \approx \frac{\overline{R}}{d_2}$$

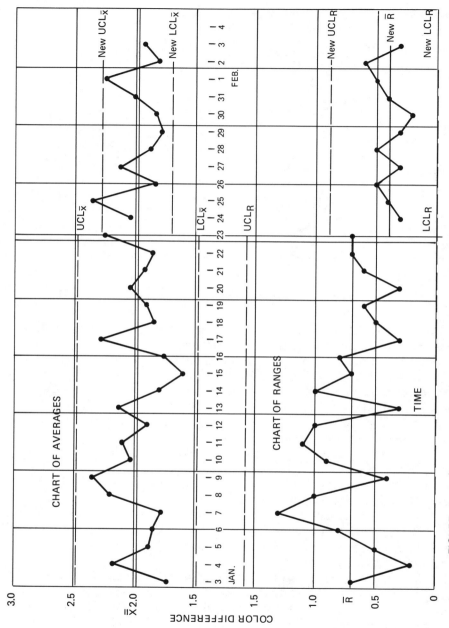

FIGURE 16.8. Control chart for color-difference data (Tables 16.3 and 16.4) with old and revised limits.

TABLE 16.6. Percent Impurities in Daily Batches of Organic Solvents

Date	Daily Values	Three-Day Moving Average	Three-Day Moving Range
June 3	0.10		
4	0.19		
5	0.21	0.17	0.11
6	0.17	0.19	0.04
7	0.12	0.17	0.09
8	0.14	0.14	0.05
9	0.16	0.14	0.04
10	0.14	0.15	0.02
11	0.17	0.16	0.03
12	0.20	0.17	0.06
13	0.27	0.21	0.10
14	0.16	0.21	0.11
15	0.08	0.17	0.19
16	0.17	0.14	0.09
17	0.20	0.15	0.12
18	0.12	0.16	0.08
19	0.15	0.16	0.08
20	0.22	0.16	0.10
21	0.16	0.18	0.07
22	0.10	0.16	0.12

$\overline{R}$ is the average of the ranges from each successive set of three observations. The resulting value represents an estimate of the standard deviation for individuals (s_x). Since $s_{\bar{x}}$ is desired the resulting value must then be divided by $\sqrt{n}$. In the example $\overline{R} = 0.08$, the table value for d_2 ($n = 3$) is 1.693 and $\sqrt{3} = 1.73$, consequently,

$$s_{\bar{x}} = (0.08 \div 1.693) \div 1.73$$
$$s_{\bar{x}} = (0.05) \div 1.73$$
$$s_{\bar{x}} = 0.03$$

Therefore,

$$\text{UCL}_{\bar{x}} = 0.17 + 3(0.03) = 0.26$$
$$\text{centerline} \ (\overline{\overline{X}}) = 0.17$$
$$\text{LCL}_{\bar{x}} = 0.17 - 3(0.03) = 0.08$$

In Figure 16.9 are plotted the daily values (a) and the three-day moving averages with control limits (b). When the two graphs are compared it is

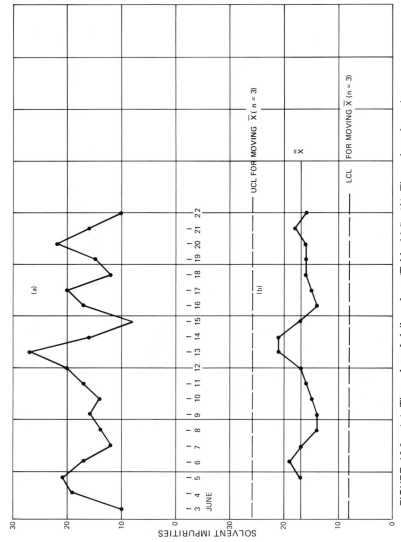

FIGURE 16.9. (*a*) Time plot of daily values (Table 16.6). (*b*) Time plot of moving averages ($n = 3$) with control limits (Table 16.6).

obvious that the use of the moving average reduces the magnitude of the random fluctuations in the process. However, the basic pattern of the chart of averages is similar to that on the chart of individuals. This effect provides more sensitivity for the detection of real changes in the process.

When utilizing a control chart based on a moving average, it is important to remember that the points on the chart are *not* independent of each other. Each point is influenced by the adjacent observations and as the number of values in each set increases, this influence will cover a greater distance.

The interpretation of a point outside control limits on a moving average chart is the same as for a point outside control limits on a conventional chart. However, this is not the case when several points in a row exceed the control limits because of the interobservation effect. Likewise, runs occurring about or below the central line do not have the same significance on the moving average chart as they do on the conventional $\overline{X}$ chart.

16.10. CONTROL CHARTS USING A MOVING RANGE

An alternative of employing control charts for individuals involves the use of the moving range. The data are arranged in order of production. The absolute differences between successive observations are determined and are given as the moving ranges. The average range $\overline{R}$ is calculated and multiplied by the coefficient E_2 to give $3s$ limits for the chart of X's. The central line for this chart is $\overline{X}$.

The data in Table 16.7 come from a process that lends itself well to the use of a moving range. The samples are presented in the order of production and the ranges of adjacent values determined. The $\overline{X}$ is 5.87 and the $\overline{R}$ is 0.02. The value of the coefficient E_2 ($n = 2$) is 2.66, as seen in Table 16.4. The estimate of the $3s$ interval is found by multiplying $0.20 \times 2.66 = 0.53$. Therefore, for the chart of X's

$$\text{UCL}_x = 5.87 + 0.53 = 6.40$$
$$\text{centerline} = 5.87$$
$$\text{LCL}_x = 5.87 - 0.53 = 5.34$$

The chart of individuals with limit lines is seen in Figure 16.10*a*.

The interpretation of this graph is identical to the interpretation of the chart for individuals previously discussed in this chapter. The fear that the underlying distribution of the process is non-normal is likewise justified here. The major benefit is that this method requires less data and fewer calculations to obtain the working chart.

With the data in this form it is also possible to construct a chart of moving ranges. This second chart will have its central line at $\overline{R}$ with the

TABLE 16.7. Impurities in Grams per Liter for a Chemical Manufacturing Process

Batch Number	Individual X Values	Range of Each Two Successive Values
1	5.8	
2	5.7	0.1
3	6.0	0.3
4	5.9	0.1
5	5.7	0.2
6	5.9	0.2
7	5.8	0.1
8	6.0	0.2
9	5.7	0.3
10	5.8	0.1
11	5.8	0.0
12	5.9	0.1
13	6.1	0.2
14	6.0	0.1
15	6.2	0.2
16	5.8	0.4
17	5.7	0.1
18	5.9	0.2

$\overline{X} = 5.87$

$\overline{R} = 0.20$

limit lines placed at $D_4\overline{R}$ and $D_3\overline{R}$. The coefficients D_4 and D_3 may be found in Table 16.4.

For the example, $D_4(n = 2)$ is 3.267 and D_3 ($n = 2$) is zero. Therefore, for the chart of moving ranges;

$$UCL_R = 0.20 \times 3.267 = 0.65$$

$$\text{centerline} = 0.20$$

$$LCL_R = 0.20 \times 0 = 0.0$$

The control chart for moving ranges is shown in Figure 16.10b.

The interpretation for the chart of moving ranges is very similar to that for the chart of moving averages. In other words, since the points are not independent, significant changes in the process will greatly affect succeeding points. When used together, the chart of individuals gives insight into long-term variations while the chart of moving ranges shows short-term variability.

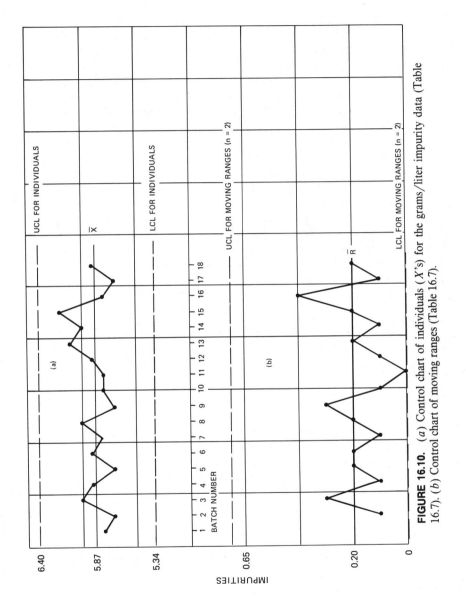

FIGURE 16.10. (*a*) Control chart of individuals (*X*'s) for the grams/liter impurity data (Table 16.7). (*b*) Control chart of moving ranges (Table 16.7).

16.11. CONTROL CHARTS BASED ON A GEOMETRIC MOVING AVERAGE

In the two previous methods it is obvious that the employment of arithmetic moving averages provided an increased ability to identify smaller changes in the process. This was achieved by combining successive observations so that the cumulative effect of the process shift would quickly overcome the random noise level. The use of this technique assumes that past and present data are equally important. However, for many processes it is more sensible to make use of a weighting method which will give the greatest importance to the more recent data since, for example, process changes may have been made in an attempt to reduce the variability of the data obtained early in the process. This weighting method involves the use of a geometric moving average. Here, the most recent observation is given the greatest weight, with all previous observations given weights which are decreasing in geometric progression.

The value plotted on the control chart is a weighted average from a given point in time identified as Z_t. The point Z_t is found through a weighting factor which is a fraction r, having a value between zero and 1. Therefore, the weighted value at time t is defined by

$$Z_t = rX_t + (1 - r)Z_t - 1 \tag{16.1}$$

where X_t is the most recent observation and Z_{t-1} is the weighted value at the immediately preceding time. The standard deviation of the weighted values (Z_t) is given as

$$\sigma Z_t = \frac{r}{2 - r} \times \sigma \tag{16.2}$$

where σ is the standard deviation of the X values from which X_t is determined.

The data in Table 16.8 provide an example. The X values were taken from a process with a mean of 10.0 and a standard deviation of 1.0. The Z_t values were found by using a weighting factor of $r = \frac{1}{4}$ so that

$$Z_t = \tfrac{1}{4}X_t + \tfrac{3}{4}Z_{t-1}$$

First we plot a conventional Shewhart control chart for the subgroup averages ($\overline{X}_t$) using three σ limits as seen in Figure 16.11. There are no points exceeding the control limits and no substantial evidence of nonrandom patterns which would indicate a process shift. Figure 16.12 shows a control chart based on the geometric moving average, Z_t, from the same data. The centerline of the chart is placed at the mean value of 10.0. The standard deviation of the Z_t values was found from the previous formula to

TABLE 16.8. Viscosity Measurements

Subgroup	$\overline{X}_t$	$\frac{1}{4}\overline{X}_t$	$\frac{3}{4}Z_{t-1}$	Z_t
1	11.6	2.9	7.5	10.4
2	10.4	2.6	7.8	10.4
3	8.9	2.2	7.8	10.0
4	9.3	2.3	7.5	9.8
5	11.6	2.9	7.4	10.3
6	10.1	2.5	7.7	10.2
7	8.6	2.2	7.7	9.9
8	10.8	2.7	7.4	10.1
9	9.8	2.5	7.6	10.1
10	8.6	2.2	7.6	9.8
11	10.2	2.6	7.4	10.0
12	9.8	2.5	7.5	10.0
13	10.4	2.6	7.5	10.1
14	12.1	3.0	7.6	10.6
15	9.8	2.5	8.0	10.5
16	12.4	3.1	7.9	11.0
17	11.9	3.0	8.3	11.3
18	11.1	2.8	8.5	11.3
19	9.9	2.5	8.5	11.0
20	12.4	3.1	8.3	11.4
21	12.9	3.3	8.6	11.9
22	11.4	2.9	8.9	11.8
23	11.7	2.9	8.9	11.8
24	11.5	2.9	8.9	11.8

be 0.40. With $\overline{X} + 3\sigma$ limits on the chart the upper control limit is placed at 11.20 and the lower control limit at 8.80. In evaluating this chart it is obvious that there is a gradual rise in the process beginning at about the thirteenth point. The seventeenth point falls outside of the limit and indicates a shift in the process level. The cumulative effect of the rising points allows for a detection of this small shift in the process average to be more quickly identified. (NOTE: The average of the first 12 subgroups is 10.0 with $\sigma = 1$ and the average of the second 12 subgroups is 11.5, with $s_{\overline{x}} = 1.0$.)

The geometric moving average method is far more sensitive to small shifts in the process level than the Shewhart method. Conversely, when the process level involves a large change, the Shewhart method more promptly signals the change. For the geometric moving average method to provide the most benefits the weighting value r should be selected with care. The weighting value r is related to the average length of time that the process must run before the chart will indicate a shift in the process level when one has actually occurred. This average run length (ARL) data for different r

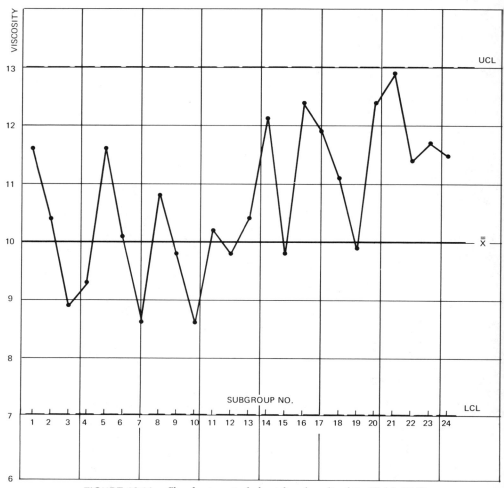

FIGURE 16.11. Shewhart control chart for viscosity data (Table 16.8).

values may be found in the article "Control Chart Tests Based on Moving Geometric Averages," by S. W. Roberts, to be found in the August 1959 edition of *Technometrics*.

16.12. CUMULATIVE-SUM CONTROL CHARTS

The regular control chart is a statistical control based on the underlying assumption that successive samples are unrelated and represent one process level. However, in many chemical processes, particularly those based on natural raw materials (mined minerals, vegetable products, etc.), this is not the case. In these processes, systematic differences show up over a period of time in such things as purity and chemical properties of the product. Cumulative sum charts, as developed by Dr. George Barnard using a basic concept for the cumulative sum chart as proposed by an English mathematician Dr. E. S. Paige, are used to control systems where sudden but slight changes in process level occur.

In the use of cumulative sum control charts, as in those using the geometric moving average technique, past data are combined with current observations. For this method, however, the current observations are simply added to the previous sum. Consequently, the plots represent the sum of the deviations of the observations from the process aim point. To reduce the contribution of the older data, a weighting factor is incorporated in the control limits. The cumulative sum chart resolves the "shotgun" or scattered appearance of the regular control chart into a true visual

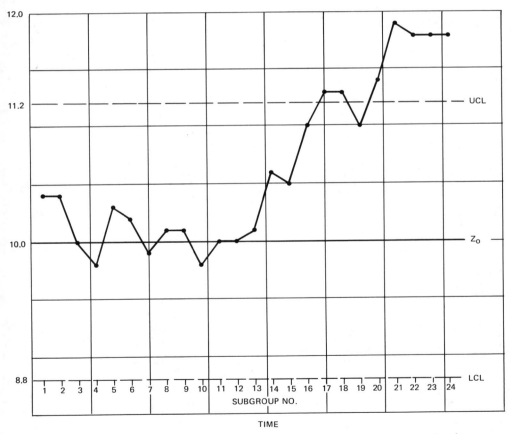

FIGURE 16.12. Control chart based on a geometric moving average of the viscosity data (Table 16.8).

tracking of actual process changes. It graphically cuts out the random "noise" (inherent variability such as in sampling, measurement, and the like) in the process. Simultaneously, it magnifies the actual systematic changes.

The process aim point (or control) must first be determined. Since the chart is essentially a running summation of process deviation from aim, the chart will display a horizontal pattern when the observations are closely grouped around the aim value. Therefore, the level of the pattern is of no interest. When the process is operating at a level below the aim, the points will continually drop. If the process is operating above the aim the points will continually rise. The interpretation of these deviations is associated with the slope of the line connecting the points. If the slopes are great then the indication is that a large change has occurred. Moderate slopes indicate slight shifts from the process aim. In practice, the scale brings the points close together. Thus, the operator can see the slopes without drawing a line though the points.

Although a change in slope on a cumulative sum chart will indicate that a change has occurred in the process, it is desirable to use a more objective method for determining when action is necessary. The method commonly employed involves the use of a V-shaped mask, as seen in Figure 16.13. The edges of this mask represent control limits that are similar in nature to those on a Shewhart chart. The mask is centered at a distance d units in front of the most recent data point. The tangent of θ, the half-angle of the V, is the second parameter needed to define the mask. With the mask in place, all of the preceding points will be within the limits of the V if the process is in control at that point. Therefore, if the points become obscured behind the edges of the V the control limits have been exceeded and thus indicate a change in the process level.

The dimensions of the V mask (d and θ) may be calculated from the following data:

1. The process standard deviation (θ_x).
2. The magnitude of the shift (D) in the process average, which should be detected.
3. The desired α risk.

The first step is to specify the shift in the process average that is to be detected in units of the standard deviation:

$$\delta = \frac{D}{\sigma_{\bar{x}}}$$

With this information

$$\theta = \tan^{-1}\frac{\delta}{2}$$

where the angle θ is the angle with a tangent of one-half the **allowable**

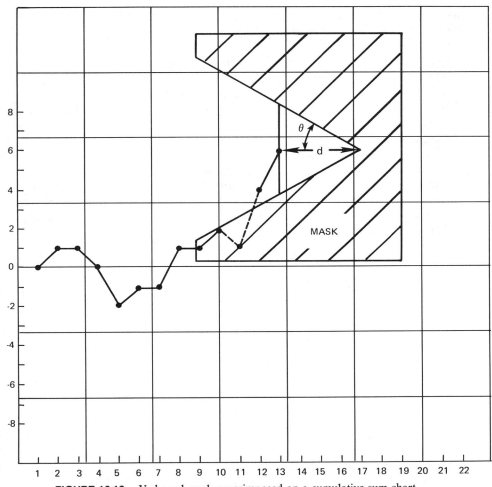

FIGURE 16.13. V-shaped mask superimposed on a cumulative sum chart.

deviation. The value of d may be found from

$$d = \frac{-2}{\delta^2}(\ln \alpha)$$

where ln represents the natural logarithm and α is the desired α risk.

These equations provide V mask dimensions such that the running sums of the deviations from the process aim are plotted in units of $\theta_{\bar{x}}$. These $\theta_{\bar{x}}$ units must occupy identical dimensions for both the vertical and horizontal axes of the chart.

Like the geometric moving average method the cumulative sum chart is very sensitive to relatively small shifts in process level. Also, through the use of the V shaped mask cumulative sum charts provide a simple, visual determination of these changes and serve to indicate when a process may be slipping off course.

The data in Table 16.9 are from a process with an aim value of 100 and a standard deviation of 2. The first column represents the subgroup number with the second column containing the subgroup averages. The fourth column contains the differences between the process aim and each of the subgroup averages. In column five is the running sum of the differences.

The plotted cumulative sums are shown in Figure 16.14. Notice that the intervals on both the vertical and horizontal axes are of equal magnitude. It is important to note that if the magnitude of the horizontal intervals is greater than those on the vertical axis, the slopes of the connecting lines will be changed. In the design of the mask this does not affect the value of d. However, the formula for θ becomes

$$\theta = \tan^{-1} \frac{\delta}{2K}$$

where K is the ratio of $\theta_{\bar{x}}$ to the desired plotting scale.

TABLE 16.9. Coating Process With an Aim Value of 100

Sample No.	X	Aim	D	Cumulative Sum
1	102	100	+2	2
2	101	100	+1	3
3	98	100	−2	1
4	99	100	−1	0
5	101	100	+1	1
6	102	100	+2	3
7	101	100	+1	4
8	100	100	0	4
9	99	100	−1	3
10	98	100	−2	1
11	99	100	−1	0
12	102	100	+2	+2
13	100	100	−0	+2
14	101	100	+1	+3
15	99	100	−1	+2
16	98	100	−2	0
17	99	100	−1	−1
18	99	100	−1	−2
19	97	100	−3	−5
20	101	100	+1	−4
21	100	100	0	−4
22	99	100	−1	−5

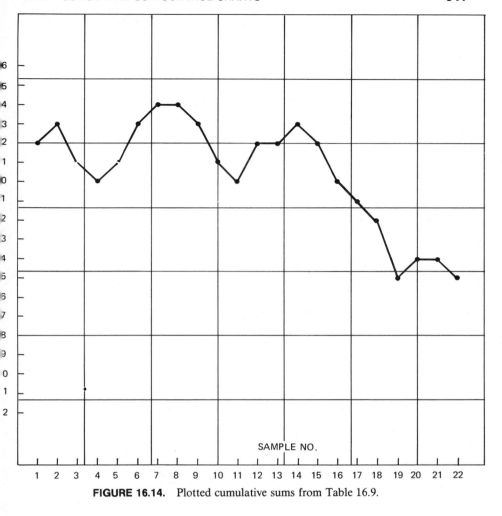

FIGURE 16.14. Plotted cumulative sums from Table 16.9.

To construct the decision making mask for the example it is necessary to determine the dimensions d and θ. Three values are required:

1. D, which is the magnitude of the shift in process average that is to be detected. Here, the desire is to detect a change equal to 1.5 standard deviations or more. Therefore,

$$D = 1.5\sigma_{\bar{x}} = 1.5(2) = 3.0$$

2. δ, the specification of the change to be detected in standard deviation units. Therefore,

$$\delta = \frac{D}{\sigma_{\bar{x}}} = \frac{3.0}{2.0} = 1.5$$

3. An assumed α risk of .003. The calculations for θ are as follows:

$$\theta = \tan^{-1}\frac{\delta}{2}$$

$$\theta = \tan^{-1}\frac{1.5}{2}$$

$$= 37°$$

The calculations for d are as follows:

$$d = \frac{-2}{\sigma^2}(\ln \alpha)$$

$$d = \frac{-2}{(1.5)^2}(\ln 0.003)$$

$$d = 5.17$$

The finished mask is shown in Figure 16.15. The mask is placed on the chart of the data as shown in Figure 16.16. It is positioned by

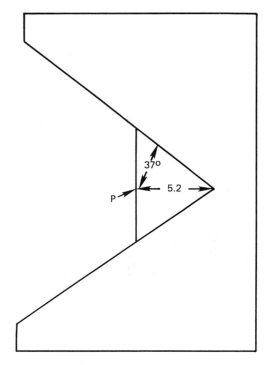

FIGURE 16.15. Finished mask for use with Figure 16.4.

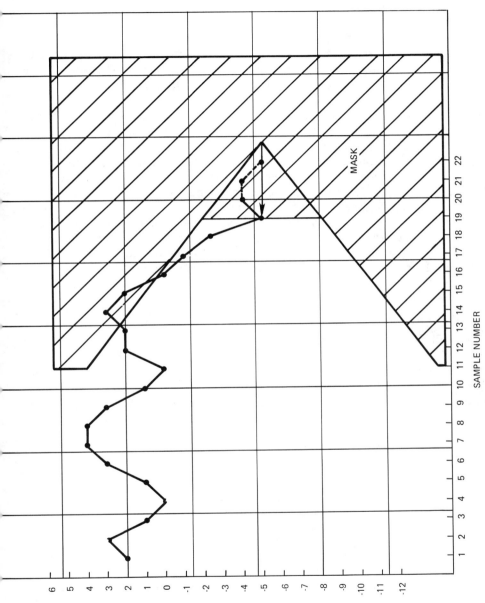

FIGURE 16.16. Mask in use with cumulative sum chart.

SAMPLE NUMBER

MASK

locating point P on the mask at the last plotted data point. This example indicates that the first sign of a change in the process average equal to 1.5 standard deviations or more is found at the plot of the nineteenth point. With the mask in this position it can be seen that the fourteenth and fifteenth points are obscured by the mask, thus indicating a shift in the process average at that point in time. Steps should now be taken to correct for this shift in the process average. In this example, the correction was apparently made as the mask indicates no problem with the next few points.

Although the shift occurred after the fourteenth sample, it is obvious that five *additional* samples had to be collected before the change was detected. The number of samples required between the time that the shift occurred and the time at which it was detected is identified as the run-length time. One of the benefits associated with the use of cumulative sum charts is that their average run length is generally less than that for the conventional Shewhart charts.

The use of this method is best tempered with experience. For example, if the points that indicate trouble in the process occurred far in the past they are probably not meaningful. In general, if the points indicating lack of control occur at a distance of more than $3d$ before the last point, they are doubtful in their value. However, experience in the use of this method would best dictate its use.

REFERENCES

Periodicals

1. Freund, R. A., A reconsideration of the variables control chart, *Ind. Qual. Control*, Vol. 16, No. 11, May 1960, pp. 35–41.

2. Hillier, F. S., X and R chart control limits based on a small number of subgroups, *Qual. Tech.*, No. 1, 1969, p. 17. *Ind. Quality Control*, Vol. 20, No. 8, February 1964, pp. 24–29.

3. Johnson, N. L., A simple theoretical approach to cumulative sum control charts, *J. Ame. Statistical Assn.*, December 1961.

4. Johnson, N. L., and Leone, F. C., Cumulative sum control charts, *Ind. Qual. Control*, Parts I, II, and III, June–August 1962.

5. Olds, E. G., Power characteristics of control charts, *Ind. Qual. Control*, Vol. 18, No. 1, July 1961, pp. 4–9.

6. Page, E. S., Cumulative sum charts, *Technometrics*, Vol. 3, No. 1, February 1961, pp. 1–9.

7. Prane, J. W., Statistical methods in the coatings industry, *Pt. & Varn. Prod.*, October 1956, 1–42.

8. Proschan, F., and Savage, I. R., Starting a control chart, *Ind. Qual. Control*, Vol. 17, No. 3, September 1960, pp. 12–13.

9. Roberts, S. W., Control charts based on geometric moving averages, *Technometrics*, Vol. 1, No. 3, August 1959, pp. 239–250.

10. Truax, H. M., Cumulative sum charts and their application to the chemical industry, *Ind. Qual. Control*, Vol. XVIII, No. 6, December 1961, pp. 18–25.

Books

11. Bauer, E. L., *A Statistical Manual for Chemists*, 2nd ed., Academic Press, New York, 1971, Chapter 6.
12. Bennett, C. A., and Franklin, N. L., *Statistical Analysis in Chemistry and the Chemical Industry*, Wiley, New York, 1963, Chapter 10.
13. Burr, E. W., *Elementary Statistical Quality Control*, Marcel Dekker, New York, 1978.
14. Burr, I. W., *Engineering Statistics and Quality Control*, McGraw-Hill, New York, 1953.
15. Cowden, D. J., *Statistical Methods in Quality Control*, Prentice-Hall, Englewood Cliffs, NJ, 1957.
16. Grant, E. L., *Statistical Quality Control*, 3rd ed., McGraw-Hill, New York, 1964.
17. Juran, J. M., *Quality Control Handbook*, 3rd ed., McGraw-Hill, New York, 1951.
18. Natrella, M. G., *Experimental Statistics*, National Bureau of Standards Handbook 91, U.S. Government Printing Office, Washington, D.C., 1963.
19. Shewart, W. A., *Economic Control of Quality of a Manufactured Product*, Van Nostrand, Princeton, NJ, 1931.
20. Rickmers, A., *Statistics: An Introduction*, McGraw-Hill, New York, 1967, Chapters 21 and 24.
21. Smith, E. S., *Control Charts*, McGraw-Hill, New York, 1947.

Miscellaneous

22. American Standard Association, Inc., American Standard Guide for Quality Control and American Standard Control Chart Method of Analyzing Data, 1959.
23. ASTM, *Manual on Quality Control of Materials*, American Society for Testing and Materials, Philadelphia, 1951.
24. Shewart, W. A., *Statistical Method from the Viewpoint of Quality Control*, edited by Deming, W. E., The Graduate School, Dept. of Agriculture, Washington, D.C. 1939.

17

SEQUENTIAL METHODS

ANTHONY A. SALVIA

Behrend College
Pennsylvania State University
Erie, Pennsylvania

17.1. INTRODUCTION

In this chapter we consider problems of hypothesis testing in which the sample size is not fixed in advance. Historically, many statisticians recognized the potential for reduction in experimental size in specific cases (see the first example below) but it remained for Abraham Wald (1946) to develop from first principles a general theory. The bulk of this chapter is based on that theory.

Example 17.1. Suppose an incoming inspection plan calls for selecting a sample of 10 items from a shipment; the shipment is acceptable if not more than one defective item is found. Clearly, it is sometimes possible to terminate inspection before all 10 items are examined. Table 17.1 illustrates this.

If we let N be the sample size actually observed, it is clear that N is a random variable with a density function, say $f(n)$; further, the expectation

$$E(N) = \sum_{2}^{10} nf(n)$$

is certainly finite, and is in fact less than 10.

It should also be clear that $f(n)$, and therefore $E(N)$, depend on the proportion of defective items in the shipment. If the shipment is perfect, then we would stop at the ninth item, and $E(N) = 9$; if all the items are bad, $E(N) = 2$. If the shipment is somewhere between these extremes, $2 < E(N) < 9$.

TABLE 17.1.   Possible Termination Points for Sampling Plan

Terminate at	Composition of Sample	Decision
2	First two defective	Reject
3	Third item = second defective	Reject
.		
.		
.		
8	Eighth item = second defective	Reject
9	Ninth item = second defective	Reject
9	No defectives found	Accept
10	Exactly one defective in the first nine items	Either

Let us consider now the test of a simple hypothesis

$$H_0: \theta = \theta_0$$

against a simple alternative

$$H_1: \theta = \theta_1$$

Let $f(x; \theta)$ be the density function of the random variable X, and define

$$r_n = \frac{f(X_1; \theta_1)f(X_2; \theta_1) \cdots f(X_n; \theta_1)}{f(X_1; \theta_0)f(X_2; \theta_0) \cdots f(X_n; \theta_0)}$$

The sequential probability ratio test (SPRT) of H_0 versus H_1 is defined as follows:

i.   If the decision to terminate experimentation was not reached at the $(n - 1)$st observation, observe X_n and compute r_n.

ii.   If $r_n \leq B$, accept H_0. If $r_n \geq A$, accept H_1.

iii.   If $B < r_n < A$, continue sampling.

A and B are constants, with $A > B > 0$.

The SPRT may be visualized as shown in Figure 17.1. Under conditions which are usually met in practice, it has been shown that:

i.   The SPRT terminates with probability 1; that is, the probability of indefinite continuation is zero.

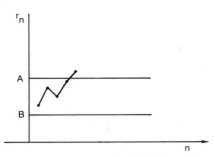

FIGURE 17.1. Graph of SPRT.

ii. The choice

$$A = \frac{1 - \beta}{\alpha} \qquad B = \frac{\beta}{1 - \alpha}$$

produces a test with type I and type II errors (see Chapter 6) approximately equal to α and β, respectively.

iii. The expected termination number $E(N)$ is smaller than the sample size required to attain the given values of α and β by any other procedure.

Example 17.2. SPRT, Binomial Distribution. We test $H_0: p = p_0$ versus $H_1: p = p_1 > p_0$. Here,

$$f(x_1; p)f(x_2; p) \cdots f(x_n; p) = p^{\Sigma x}(1 - p)^{n - \Sigma x}$$

and so

$$r_n = \left(\frac{p_1}{p_0}\right)^{S_n} \left(\frac{1 - p_1}{1 - p_0}\right)^{n - S_n}$$

where S_n is the number of "successes" in the first n trials. The points for which continuation is indicated satisfy the inequalities $B < r_n < A$; inserting the expression above for r_n and taking logarithms we obtain

$$\ln B < S_n \ln\left(\frac{p_1}{p_0}\right) + (n - S_n)\ln\left(\frac{1 - p_1}{1 - p_0}\right) < \ln A$$

or

$$\ln B - nK < S_n \ln\left(\frac{1 - p_0}{p_0} \cdot \frac{p_1}{1 - p_1}\right) < \ln A + nK$$

where

$$K = \ln \frac{1 - p_1}{1 - p_0}$$

We rewrite this as

$$h_0 + sn < S_n < h_1 + sn$$

with

$$h_0 = \frac{\ln B}{\ln \dfrac{1 - p_0}{p_0} \dfrac{p_1}{1 - p_2}}, \qquad h_1 = \frac{\ln A}{\ln \dfrac{1 - p_0}{p_0} \dfrac{p_1}{1 - p_1}}$$

and

$$s = \frac{-K}{\ln \dfrac{1 - p_0}{p_0} \dfrac{p_1}{1 - p_1}}$$

The boundaries of the "continuation" region are parallel straight lines, with slope s; the intercept of the lower line is h_0, that of the upper line is h_1. Graphing these lines we obtain Figure 17.2.

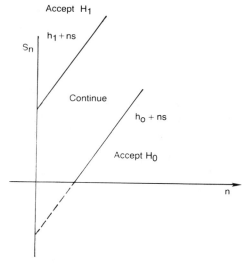

FIGURE 17.2. SPRT for the binomial case.

In practice, all of the quantities h_0, h_1, and s may be calculated before testing begins; we then simply graph the accumulated sum S_n, continuing to sample until the point (n, S_n) falls outside either of the parallel lines.

As a numerical example, suppose we wish to have

$$\alpha = \beta = .05$$

and

$$H_0: p = 0.2 \quad \text{versus} \quad H_1: p = 0.4$$

Then

$$B = \frac{.05}{.95} = \frac{1}{19}, \qquad A = 19$$

$$K = \ln \frac{0.6}{0.8} = \ln 3 - \ln 4$$

$$h_0 = \frac{-\ln 19}{\ln 8 - \ln 3}, \qquad h_1 = -h_0$$

and

$$S = \frac{\ln 4 - \ln 3}{\ln 8 - \ln 3}$$

Numerically these are

$$h_1 = \frac{2.945}{0.981} \simeq 3.002 = -h_0$$

$$S = \frac{0.288}{0.981} \simeq 0.29$$

the continuation region is thus

$$0.29n - 3 < S_n < 0.29n + 3$$

Note that the earliest possible termination is at the fifth observation, with rejection of H_0, if all five observations are successes; for then

$$S_5 = 5 \geq 0.29 \times 5 + 3 = 4.45$$

In adapting this SPRT to practice, it is advisable to convert the graphical procedure to a tabular one. The lower line is $.29n - 3$, the upper $.29n + 3$; substituting $n = 1, 2, 3, \ldots$ gives results as in Table 17.2; we have rounded up and down, since ΣX must be an integer. Such a table should extend to a value of n large enough so that the test will ordinarily terminate before we "run out of table." A high multiple (5 to 10) of the expected termination number (which we discuss below) will work in most cases.

TABLE 17.2. Tabular Version of SPRT

		Accept	
Column (1) Accumulated Sample Size	(2) Accumulated Number of Successes	H_0, if Column (2) Less Than or Equal to	H_1, if Column (2) Greater Than or Equal to
1		—	—
2		—	—
3		—	—
4		—	—
5		—	5
6		—	5
7		—	6
8		—	6
9		—	6
10		—	6
11		0	7
12		0	7
13		0	7
14		1	8
15		1	8
16		1	8
17		1	8
18		2	9
19		2	9
20		2	9

—Indicates decision not possible for that sample size.

Two quantities of interest in SPRT's are the operating characteristic L and the average termination number $E(N)$. Both of these are functions of the parameter θ; we now indicate a procedure by which they may be obtained.

Define

$$z = \ln \frac{f(x; \theta_1)}{f(x; \theta_0)}$$

$$z_i = \ln \frac{f(x_i; \theta_1)}{f(x_i; \theta_0)}, \qquad i = 1, 2, \ldots, n$$

$$Z_n = z_1 + z_2 + \cdots + z_n$$

(Note that $Z_z = \ln r_n$). Then

$$E(N) = \frac{E(Z_N)}{E(z)} \tag{17.1}$$

To determine the operating characteristic, we find first a function $h(\theta)$ which satisfies the equation

$$E(e^{zh(\theta)}) = 1 \tag{17.2}$$

Once $h(\theta)$ is determined, it can be shown that the operating characteristic $L(\theta)$ is approximately

$$L(\theta) \simeq \frac{A^{h(\theta)} - 1}{A^{h(\theta)} - B^{h(\theta)}} \tag{17.3}$$

and the numerator of equation (17.1) is approximately

$$E(Z_N) \simeq L(\theta)\ln B + [1 - L(\theta)]\ln A \tag{17.4}$$

We resume our example to illustrate the use of equations (17.1)–(17.4). For the binomial case under consideration, using p_0 and p_1 for θ_0 and θ_1

$$z = \ln \frac{f(x; p_1)}{f(x; p_0)} = \ln \frac{p_1^x(1 - p_1)^{1-x}}{p_0^x(1 - p_0)^{1-x}}$$

$$= x \ln \frac{p_1}{1 - p_1} \frac{1 - p_0}{p_0} + \ln \frac{1 - p_1}{1 - p_0}$$

If the parameter has a "true value" p,

$$E(z) = \ln \frac{1 - p_1}{1 - p_0} + \left(\ln \frac{p_1}{1 - p_1} \cdot \frac{1 - p_0}{p_0} \right) E(X)$$

$$= \ln \frac{1 - p_1}{1 - p_0} + \left(\ln \frac{p_1}{1 - p_1} \frac{1 - p_0}{p_0} \right) p$$

This expression will provide the denominator of equation (17.1); note that the logarithms will already have been calculated in determining the limits for the SPRT. Next, we determine $h(p)$. We have

$$e^{zh(p)} = \left[\frac{f(x; p_1)}{f(x; p_0)} \right]^{h(p)} = \left[\frac{p_1^x(1 - p_1)^{1-x}}{p_0^x(1 - p_0)^{1-x}} \right]^{h(p)}$$

We need to find $h(p)$ to satisfy $E(e^{zh(p)}) = 1$: The density function of X is $p^x(1-p)^{1-x}$, $x = 0, 1$, so that

$$E(e^{zh(p)}) = \left[\frac{p_1^0(1-p_1)^1}{p_0^0(1-p_0)^1}\right]^{h(p)}(1-p) + \left[\frac{p_1^1(1-p_1)^0}{p_0^1(1-p_0)^0}\right]^{h(p)} \cdot p$$

$$= \left(\frac{1-p_1}{1-p_0}\right)^{h(p)} \cdot (1-p) + \left(\frac{p_1}{p_0}\right)^{h(p)} \cdot p$$

Setting this expression equal to 1, we may solve the resulting equation for p:

$$p = \frac{1 - \left(\dfrac{1-p_1}{1-p_0}\right)^{h(p)}}{\left(\dfrac{p_1}{p_0}\right)^{h(p)} - \left(\dfrac{1-p_1}{1-p_0}\right)^{h(p)}} \tag{17.5}$$

Equation (17.5) gives p as a function of h, and, implicitly, h as a function of p. We may select values of h and determine the corresponding p. In this way, equations (17.3) and (17.4) may be employed to determine $L(p)$ and $E(N)$.

It is instructive to consider five particular values for h: ± 1, $\pm \infty$, and 0.

1. $h = 1$ gives

$$p = \frac{1 - \dfrac{1-p_1}{1-p_0}}{\dfrac{p_1}{p_0} - \dfrac{1-p_1}{1-p_0}} = p_0$$

2. $h = -1$ gives

$$p = \frac{1 - \dfrac{1-p_0}{1-p_1}}{\dfrac{p_0}{p_1} - \dfrac{1-p_0}{1-p_1}} = p_1$$

3. As $h \to +\infty$, since $p_1 > p_0$, we find $p \to 0$.
4. As $h \to -\infty$ we find likewise $p \to 1$.
5. For $h = 0$ we obtain the indeterminate form $0/0$.

Writing for convenience

$$a = \frac{1 - p_1}{1 - p_0}, \qquad b = \frac{p_1}{p_0}$$

we seek

$$\lim_{h \to 0} \frac{1 - a^h}{b^h - a^h}$$

Using l'Hôpital's rule, this is the same as

$$\lim_{h \to 0} \frac{-a^h \ln a}{b^h \ln b - a^h \ln a} = \frac{-\ln a}{\ln b - \ln a}$$

This is precisely s, the slope of the boundary lines for the SPRT.
Summarizing, we have found the points

p	0	p_0	s	p_1	1
h	$+\infty$	$+1$	0	-1	$-\infty$

These are sufficient to give a fair idea of the behavior of h as a function of p.

If necessary, additional points may be obtained in the same way. We may use equation (17.3) to obtain points on the operating characteristic curve as well. For example, corresponding to $h = 1$ we have $p = p_0$, and

$$L(p_0) = \frac{A^1 - 1}{A^1 - B^1} = \frac{A - 1}{A - B} = 1 - \alpha$$

When we recall that the operating characteristic gives the probability of accepting H_0, and that α is the probability of a type I error, this result is most reassuring. We find in the same way $L(0) = 1$, $L(p_1) = \beta$, and $L(1) = 0$. Graphing these, using additional points as necessary, we obtain Figure 17.3.

Following the same procedure we can determine points on the $E(N)$ curve; its general appearance is shown in Figure 17.4.

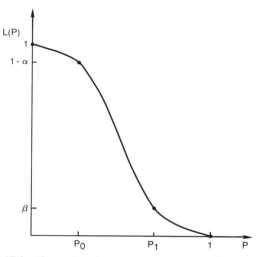

FIGURE 17.3. Operating characteristic curve for the binomial SPRT.

Taking the same numerical values as in our earlier example ($p_0 = 0.2$, $p_1 = 0.4$, $\alpha = \beta = 0.05$) and performing the required calculations we obtain the table below:

p	$E(N)$
0	11
.2	28.9
.4	25.4
1	5

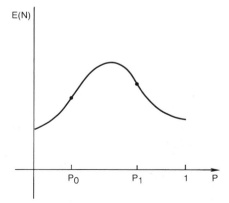

FIGURE 17.4. $E(N)$ for the binomial SPRT.

The maximum value of $E(N)$ occurs for a value of p between p_0 and p_1, usually quite near s, for example, at $h = 0.1$,

$$p = \frac{1 - \left(\frac{3}{4}\right)^{.1}}{(2)^{.1} - \left(\frac{3}{4}\right)^{.1}} = \frac{.0284}{.100} = .284$$

$$L(p) = \frac{(19)^{.1} - 1}{(19)^{.1} - \left(\frac{1}{19}\right)^{.1}} = \frac{.342}{.597} = .573$$

$$E(Z_N) = L(p)\ln B + [1 - L(p)]\ln A$$

$$= (0.573)(-2.944) + (0.427)(2.944)$$

$$= -0.4298$$

$$E(Z) = \ln \frac{3}{4} + \ln \frac{2}{3.4} \cdot p$$

$$= -0.288 + (.981)(.284)$$

$$= -0.0094$$

and

$$E(N) = \frac{E(Z_N)}{E(Z)} = 45.7$$

We have discussed so far only one SPRT, namely, the test for a binomial parameter. We could discuss in the same detail a variety of others; however, Table 17.3 provides the results for most cases of interest.

The use of the SPRT in the test of a simple hypothesis versus a simple alternative is not so restrictive as it seems. As we shall presently see, we can incorporate two-sided alternatives in the SPRT. Furthermore, in actual practice the choices of θ_0 and θ_1 can often be set to provide high confidence in the SPRT's conclusion.

Consider a typical situation in which one is interested in, say, the mean of a normal distribution. If $\mu = \mu_0$ represents current state-of-the-art output, we might consider the expense of modifying our manufacturing process if it was quite likely that μ would be improved to a value $\mu = \mu_1$. Figure 17.5 expresses the situation.

The middle region of Figure 17.5 represents what Wald termed a "zone of indifference"; μ may have improved, but not sufficiently to justify the additional expense of adopting the new process. The researcher may, then, by a judicious selection of μ_0, μ_1, α, and β, obtain an SPRT whose operating characteristic curve suits the need.

TABLE 17.3. Characteristics of Various SPRTs

Distribution	H_0	H_1	S	h_0	h_1	s	$E(Z)$	Relationship Between $h(\theta)$ and θ
Poisson	$\lambda = \lambda_0$	$\lambda = \lambda_1 > \lambda_0$	$\sum x_i$	$\dfrac{\ln B}{\ln \lambda_1/\lambda_0}$	$\dfrac{\ln A}{\ln \lambda_1/\lambda_0}$	$\lambda_1 - \lambda_0$	$\lambda \ln\left(\dfrac{\lambda_1}{\lambda_0}\right) - (\lambda_1 - \lambda_0)$	$\lambda = \dfrac{(\lambda_1 - \lambda_0)h(\lambda)}{(\lambda_1/\lambda_0)^{h(\lambda)} - 1}$
Normal $(\sigma^2 = 1)$	$\mu = \mu_0$	$\mu = \mu_1 > \mu_0$	$\sum x_i$	$\dfrac{\ln B}{\mu_1 - \mu_0}$	$\dfrac{\ln A}{\mu_1 - \mu_0}$	$\dfrac{\mu_1 + \mu_0}{2}$	$(\mu_1 - \mu_0)\mu - \dfrac{\mu_1^2 - \mu_0^2}{2}$	$h(\mu) = \dfrac{\mu_1 + \mu_0 - 2\mu}{\mu_1 - \mu_0}$
Normal $(\mu = 0)$	$\sigma = \sigma_0^2$	$\sigma = \sigma_1^2 > \sigma_0^2$	$\sum x_i^2$	$2T\ln B^{(1)}$	$2T\ln A$	$\dfrac{T\ln \sigma_1^2}{\sigma_0^2}$	$\dfrac{\sigma^2}{2T} - \dfrac{\ln \sigma_1}{\sigma_0}$	$\sigma^2 = \dfrac{\left[(\sigma_1^2/\sigma_0^2)^{h(\sigma^2)} - 1\right]T}{h(\sigma^2)\left(\sigma_1^{2h(\sigma^2)}/\sigma_0^2\right)}$

$$(1)\ T = \frac{\sigma_0^2 - \sigma_1^2}{\sigma_1^2 - \sigma_0^2}$$

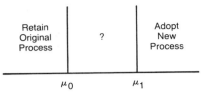

FIGURE 17.5. Decision problem based on the value of μ.

17.2. EXTENSION TO TWO-SIDED ALTERNATIVES

Frequently we encounter situations in which it is desirable to test a parameter value θ_0 against alternatives $\theta_0 \pm \Delta$; that is, the alternative hypothesis is really a pair of hypotheses,

$$H_1: \theta = \theta_0 + \Delta$$
$$H_{-1}: \theta = \theta_0 - \Delta$$

To conduct the SPRT we simply construct both pairs of parallel lines as shown in Figure 17.6. Care must be exercised in the choice of α and β, since they now apply to the entire graph, and not to just one of the pairs (H_0, H_1) or (H_0, H_{-1}).

For convenience α and β for the total experiment are usually halved for each "half-experiment."

These two-sided SPRT's have been adapted to quality control applications; in such applications they are known as cumulative sum control charts. Typically, the numbers $\theta_0 \pm \Delta$ represent upper and lower control limits for a process. If the cumulative sum remains within the "continue" region, the process is judged to be in control. A major contributor to this application is N. L. Johnson (see references).

17.3. OTHER EXTENSIONS

Despite the fact that Wald's original work in sequential methods was produced more than 35 years ago, extensions of the theory have been rather hard to come by. For example, it might appear that a natural problem to attack is the test of a normal mean, with the variance unknown.

One would quite naturally term this a sequential t test. Unfortunately, despite a number of attempts at constructing such tests in the past, no clearly best procedure has been discovered.

Sequential estimation has, likewise, had a very slow development. Some statisticians argue that there does not appear to be any valid way to "reuse" data to form good estimates.

The reason for this is easy to describe. In a fixed-sample-size experiment, suppose we are interested in estimating the mean of a normal

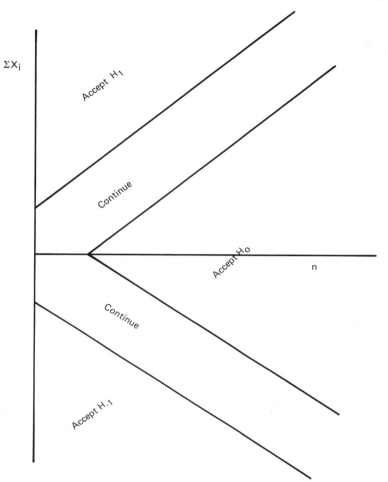

FIGURE 17.6. Two-sided SPRT illustrated for $H_0: \theta = \theta_0$ vs. $H_1: \theta = \theta_0 + \delta$ or $H_{-1}: \theta = \theta_0 - \delta$; θ is the mean of a normal random variable.

distribution. It is quite natural to use $\overline{X}$, the sample mean, and as we have seen earlier, the interval $\overline{X} \pm 2\sigma/\sqrt{n}$ is quite likely to contain the true population mean. Now $\overline{X} = \Sigma x/n$, and n is a fixed number. If, however, N is random, then $\overline{X}' = \Sigma x/N$ is the quotient of two random variables. We have no reason to believe that $\overline{X}$ and $\overline{X}'$ will be alike in their properties.

It is partly for this reason that sequential methods are not widely used in practice. The main reason, however, is that the researcher is conducting an essentially open-ended experiment with respect to time and money. While it is true that, on the average, observations will be fewer than with a fixed experiment, it is possible that, in any particular experiment, they will be much more.

At this point, one's natural inclination is to say something along the following lines: let us set up an SPRT, but agree that, if the test is still going when we reach some particular sample size, we shall arbitrarily choose to stop and determine a choice between H_0 and H_1 in some reasonable way. Then, even though N is random, certainly $N \le N_0$. Tests of this type are called *truncated* SPRT's.

Unfortunately, truncation affects the size of the type I and type II errors associated with the SPRT; if N_0 is too small relative to max $E(N)$, the effect may be very substantial.

Within the confines of a single chapter it is not possible to discuss all of these extensions and problems. For the reader with an interest in the subject, Wald's original text is understandable and informative, as is the text by Wetherill.

For the researcher with a serious interest in the subject, an excellent start is the reference text of Govindarajulu, which contains an extensive bibliography (335 references).

REFERENCES

1. Wald, A., *Sequential Analysis*, Wiley, New York, 1947.
2. Johnson, N. L., Cumulative sum control charts for the folded normal distribution, *Technometrics*, Vol. 5, 1963, pp. 451–458.
3. Johnson, N. L., Cumulative sum control charts and the Weibull distribution, *Technometrics*, Vol. 8, 1966, pp. 481–491.
4. Wetherill, G. B., *Sequential Methods in Statistics*, Metheum, London, 1966.
5. Govindarajulu, Z., *Sequential Statistical Procedures*, Academic Press, New York, 1975.
6. Whitehead, J., Large sample sequential methods with application to the analysis of 2×2 contingency tables, *Biometrika*, Vol. 65, 1978, pp. 351–356.
7. Whitehead, J., and Jones, D., The analysis of sequential clinical trials, *Biometrika*, Vol. 66, No. 3, 1979, pp. 443–452.
8. Anderson, T. W., A modification of the sequential probability ratio test to reduce sample size, *Ann. Math. Statistics*, Vol. 31, 1960, pp. 165–197.
9. Siegmund, D., Estimation following sequential tests, *Biometrika*, Vol. 65, 1978, pp. 341–349.
10. Sobel, M., and Wald, A., *Ann. Math. Statistics*, Vol. 20, 1949, pp. 502–522.

18

SYSTEMATIC ERROR IN STATISTICAL EXPERIMENTATION

JAMES R. KING

TEAM.
Tamworth, New Hampshire

John Tukey of Princeton University, in a challenging paper, "The Technical Tools of the Statistician" (1), stated that statisticians have done an admirable job of developing techniques for identifying and manipulating the problems caused by random variation, or error, but he deplored the lack of attention to methods for identifying and coping with systematic error, variation, or that which is therefore an unidentified variable of unknown magnitude. Studies by the author of this chapter over a period of 20 years have determined that unidentified error, which can be classed as systematic error, is responsible for unsuccessful experiments in one-third or more of the cases analyzed.

Random variation, or error, in its purest form is simply the chance variation which we observe in repeated trials, or experiments. We rarely, if ever, obtain identical results from repeated sampling but we do obtain results which are similar, or equivalent, for all practical purposes. The predictable variation among such repeated events is called sampling, or experimental error. The effects of experimental error on the various applied statistical methodologies are discussed in Chapters 6 and 7 with respect to hypothesis testing and estimation intervals; in Chapters 8 and 9 for acceptance sampling and the comparison of population behavior; in Chapter 16 for the control limits on control charts; in Chapter 10 for the estimation of correlation and regression coefficients; and in Chapters 13 and 14 which show how important the estimation of error is to the analysis of designed experiments. Successful use of all of these methodologies ultimately hinges on the assumptions which one can make about the validity or representativeness of data obtained for analysis. A key assumption which is required is that only random sampling error is occurring.

Nonrandom variation could also be called nonsampling variation because nonrandom variation can be highly erratic, at one extreme, to highly structured at the other extreme. Erratic behavior may occur as voids in web materials, interrupted flow of fluids, or a run of underfilled paint cans; as transient losses of electric power which can affect worker performance, proper machine operations, or the accuracy of measurement equipment; and from gross human error, better known as blunders. Such occurrences tend to have a one-shot nature and are therefore highly elusive and not good candidates for reliable discovery techniques. On the other hand, structured variations are errors that are systematic, repeated, and reproducible. Systematic errors are frequently associated with a particular instrument or experimental technique. They are often found to be more important than random errors and more difficult to treat. Once they begin to occur, they are essentially continuous and tend to become progressively more influential until, in the worst case, they produce catastrophic results. Systematic errors create undesirable biases in process and products which can result in poor yields, poor performance, and undesirable side effects on important characteristics. Systematic errors are not easily classifiable but it is useful to relate them to the mainstream of manufacturing and quality control by considering the ways in which they may occur in terms of materials, methods, machines, and measurement. The following case histories are, therefore, intended to be illustrative and provocative rather than exhaustive and definitive because of the numerous variations that may occur in the underlying problem sources that are revealed. These case histories are mainly illustrative of cases where the underlying cause was discovered. However, many times this is not the case and the problem, while recognized, takes considerable detective work to uncover the solution. Frequently, there is a time factor involved, and delay in seeking out a solution can result in failure to find an assignable cause due to reasons such as a change in the process. The coatings (polymer) industry is a fruitful field for the investigation, prevention, and correction of systematic errors.

18.1. MATERIALS SYSTEMATIC ERRORS

Example 18.1. A former colleague loved to tell the story of a process capability study conducted to determine the tolerance range of a random assembly of steel stampings assembled into rotors and stators for a fractional horsepower motor. This study indicated that the stackup error for 10 laminations of 0.100-in. sheet steel should not exceed 1.000 ± 0.003

in. However, when large-scale production began, it was found that there were three bands of results: 1.006 ± 0.003 in., 1.000 ± 0.003 in., and 0.994 ± 0.003 in. The results of a lengthy investigation led back to the steel rolling mill and ultimately disclosed that the rolling mill operators on each of the three shifts operated with three different concepts.

The commercial specification for the class of sheet steel involved allowed a $\pm 10\%$ variation in thicknesss and/or weight. As a consequence, one operator ran as close to the high end of the specification as possible because he was paid by tonnage and this practice maximized his earnings. Another operator tried to meet the nominal dimension as near as possible because he had been trained to "shoot for the target." The third operator ran toward the low side of the tolerance because he had been instructed that minimum materials usage meant more profit for the company.

Therefore, the result was that there was random error assignable to any one shift operator. This error was about the same for all operators. However, there was a systematic difference between the three shifts which was about twice as large as the extremes of the random error component. Such systematic errors can never be accounted for by any form of random error theory.

Example 18.2.   Another materials example occurred in a textile manufacturing company. In the process of dyeing light shades (pastels), there occurred a random light and dark point effect, called, descriptively, "salt and pepper." After weeks of investigation, the chief chemist was convinced that the problem was caused by the presence of tin in some phase of the process but he was unable to determine any source for tin contamination.

However, the chemistry laboratory had maintained for years a "library" of sample bottles of each shipment of the principal reagents used in formulating dye baths. The ultimate answer was found by the accidental observation that the chemical assay of a common constituent in dye baths, USP grade sodium sulfite, showed a shift, at a particular date, from "trace" to about 10–15 parts per million for tin. A telephone conversation with the supplier revealed that the source of raw material had changed at the time that the chemical assay changed. Shipment of sodium sulphite from the earlier source cured the problem. The specification for future procurement was modified to require no more than trace amounts of tin although the normal USP grade allows up to a maximum of 0.1% impurities.

Problems such as the above often occur in unspecified parameters which are assumed to be somehow covered in a general omnibus upper-limit specification.

Example 18.3.   An unusual double systematic error was experienced by the Land film operation of The Polaroid Corporation. Polaroid Land Film includes a small "pod" of special developer for each picture frame in a roll

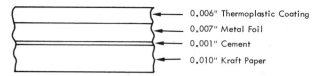

FIGURE 18.1.   Cross section of paper–metal foil–plastic laminate.

of film. This pod is made from a laminate of paper, metal foil, and a thermoplastic coating as shown in Figure 18.1.

The laminate is received in wide "master" rolls which are subsequently slit into narrow rolls. The slit rolls are fed through a folder; a heated pressure platen makes a back seal along the fold; another platen makes the end seals, forming a canoe shape; the developer, in gel form, is fed under pressure into the resulting cavity; and another platen makes the pressure seal which closes the pod. Subsequently, the end seals are cut to form the final pod, as shown in Figure 18.2.

During pod assembly, there were obvious failures to develop adequate pressure seals. Visual observation of the pod machine showed gross developer leaks from some pods. A hydraulic pressure seal test further indicated that some pods had burst strengths 10 times higher than the specification limit. Receiving inspection tests were reviewed but there was no evidence of nonconforming material.

Continued observation of the pod machines indicated a sometimes periodicity in the pattern of grossly leaking pods. Slit rolls causing unusual difficulty were taken off the machines and over a two-week period, these rolls were seen to have come from a series of consecutively numbered master rolls. Slits from several master rolls were measured for thickness at 1-in. intervals for 6 ft. The thermoplastic was then removed, another set of measurements was made at the same points, and the thickness of thermoplastic estimated by subtraction. The cement for the paper backing was then dissolved, the remaining metal foil was measured, and foil and paper thickness estimated by subtraction.

There was a significant longitudinal pattern in the metal foil which appeared to be sinusoidal with a period of about 38 in. The foil was supposed to be 0.007 in. $\pm$ 10%. The average thickness was almost exactly 0.007 in. but the peak variations were about ± 0.0025 in., resulting in

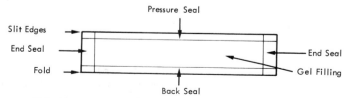

FIGURE 18.2.   Plan view and features of a developer pad.

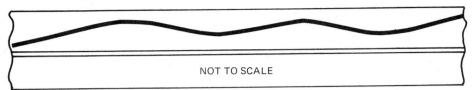

FIGURE 18.3.   Longitudinal cross section of laminate showing sinusoidal variation in metal foil.

thicknesses from 0.0045 to 0.0095 in. Since the thermoplastic layer was supposed to be 0.006 in., the hill and valley pattern in the foil caused the actual thermoplastic thicknesses to range from 0.0035 to 0.0085 in., which seemed a good explanation for the highly variable hydraulic burst pressures. The general result is illustrated in Figure 18.3 schematically, but not to scale.

A check of master roll serial numbers indicated that the sinusoidal variation was occurring in only about one-third of the material being used. A telephone call to the foil rolling mill disclosed that they rolled this foil on any of three mills. Further conversation determined that the main pressure roll had a 12-in. diameter giving a 37.7-in. circumference. A quick check at the rolling mill resulted in the finding that there was a bent shaft on one of the mills and that this mill had, in fact, rolled the material with the troublesome serial numbers.

Meanwhile, further analysis of the accumulating data revealed a secondary pattern in the thickness of the thermoplastic. As previously stated, this thickness was a nominal 0.006 in. However, there was a consistent side-to-side variation of 0.003 to 0.009 in., giving the requisite average of 0.006 in. This variation, added to the foil variation, resulted in actual thermoplastic thicknesses which varied from 0.0005 to 0.0115 in. This was now a much more plausible explanation of the visual observations of grossly leaking pods (no effective seal) and the test results 10 times higher than the specification limit for burst pressure (excessively thick seals). The (idea) of the side-to-side pattern is illustrated in Figure 18.4.

When the above information was given to the general manager of the supplier plant which applied the plastic coating, he found that the coating machine operator for this job was setting the coating machine improperly. Such machines should be adjusted by setting the thickness gap at each end

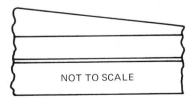

FIGURE 18.4.   Transverse cross section showing side-to-side variation of plastic coating.

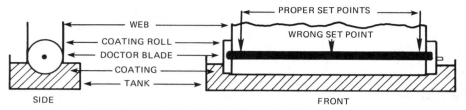

FIGURE 18.5.   Sketch of coating tank with doctor blade.

of the doctor blade. However, this operator was setting only in the middle and adjusting the setting on only one side of the doctor blade which caused the observed results. The general features of such a coating machine are shown in Figure 18.5.

To accentuate this problem, this supplier had recently had 250,000 lb of coated clear plastic returned by a large meat processing firm with the complaint that they could not obtain consistent heat seals on their packages. The general manager had been upset about this, naturally, but did not know what action to take to prevent a recurrence. However, after determining that a key operator was running a coating machine incorrectly, he checked out all of the plant's coating machines and found most of them incorrectly set. Procedures were changed quickly and he issued a credit of $85,000 for unusable plastic coated foil, almost cheerfully.

18.2.   METHODS SYSTEMATIC ERRORS

Example 18.4.   In the preceding example, one must have experienced some wonder as to how nonfunctional material was accepted into the user's system. The explanation is startlingly simple. All of the unusable material passed the incoming inspection tests, and quite easily, too. The incoming requirements were quite simple and were based on conventional ASTM and TAPPI standards for continuous web materials. That is, such materials are normally bought and sold on a weight per unit area basis. The normal sample for such materials is usually one or more pieces of 1 yard each taken from the running length of a roll of material. Most frequently, such samples are taken near the ends of rolls to eliminate splicing in the body of the roll.

Consider that the weight specification for the foil laminate was ±10% of a nominal specified weight. The sinusoidal foil error would contribute a maximum of 7% in the worse case and would average near to 3.5%, well within specification. The side-to-side pattern contributed nothing to variation since the plastic coating was nearly a perfect 0.006 in. Therefore, neither systematic error was disclosed by an accepted "standard" quality control test method.

The test method problem was resolved in two stages. The first stage was to institute two kinds of sampling: longitudinal and transverse. The longitudinal sample was measured at five points 10 in. apart for the thickness of each layer. Each layer average was required to be $\pm 10\%/\sqrt{5}$ and the range was required to be less than $1.49(\sigma_{\bar{x}}') \times 2.325(d_2) \times 2.11(d_4)$, or 7.3, from standard $\bar{X}, R$ control chart theory. The transverse sample was measured at five points 5 in. apart with the same requirements. This was a fairly costly and cumbersome method, however.

The second stage required considerable development and careful evaluation but it was finally successful. A dye was added to the plastic coating which showed significant graduations of color under black light from any of the causes of systematic variation. This was quicker, easier, and much cheaper than tedious indirect measurements.

Example 18.5. Another example of systematic error due to test methods occurred during life testing of electronic components. Measurements were scheduled to be made at specified time intervals within an oven at 85 °C, to minimize total test time. The parameter of interest was leakage current. After three readout points, the data were highly contradictory. Median values of the first and third readouts were much higher than the second readout. In addition, in each series of 24 readings, the first and third series showed readings increasing continuously from reading 1 through 24. However, the second series showed the readings to decrease consistently from reading 1 through 24. The controlling thermostat was set to control the oven temperature to a nominal $\pm 2°C$. However, the thermostat was located in the exit passage of the oven air flow. When a thermocouple was located in the center of the oven volume where the components were actually located, the thermocouple indicated that during the on–off cycles of the oven, the actual temperature varied by $\pm 6°C$. This was enough temperature variation to vary the absolute magnitude of the leakage current readings by about 2/1, which was statistically and technically significant.

This problem was easily resolved by modifying the procedure. The on–off time cycle was sufficiently long that consistent readings could be obtained, over the small time required for 24 readings, by timing the start of readings to occur 1 min after the heater "ON" light went out.

18.3. MACHINE SYSTEMATIC ERRORS

Example 18.6. One variety of machine systematic error which was observed some years ago involved an unattended automatic machine which would shut down unexpectedly. At first, it was assumed that the shutdown was some kind of random transient phenomenon. However, the use of a time recorder revealed that the shutdowns occurred within a 5-min interval of 5:30 P.M. Curiously, there had been an earlier period during which such shutdowns had not occurred.

It was finally determined that the shutdown occurred because a new electrician on the second shift switched power from an in-plant generating system to the local commercial power network when he went to supper. However, he was not properly skilled in the requirements for phasing in a relatively small power source to a large power network. Poor phasing-in caused a power surge which tripped current sensitive protective relays on the machinery. When the electrician was taught the proper method of phasing-in, the problem did not reoccur.

Example 18.7. Dr. Svente Wold of Umea University in Sweden has furnished the account of a continuous chemical reaction process programmed by a sophisticated electronic controller. The process shut down at 2:30 P.M. and at 5:00 P.M. each afternoon. It so happened that Dr. Wold had heard of a similar experience with an electronic controller in Copenhagen, Denmark, which had been traced to interference from radio frequencies used by local taxicabs. Dr. Wold investigated local sources of electromagnetic interference, such as taxis, police, and amateur radio, without success. Eventually, he determined that it was radio transmission from the daily flight to Stockholm. This flight arrived at 2:30 P.M. and departed at 5:00 P.M. A shielding system was installed around the controller which eliminated the problem.

18.4. MEASUREMENT SYSTEMATIC ERROR

Example 18.8. During a special test program at a semiconductor manufacturing plant, there was a highly significant increase in the base drive required to maintain a specified output at the collector of a germanium mesa transistor. This increase in base drive (decrease in current gain) was not believable because other transistor parameters were not showing any significant change.

The original test program contract called for the measurement of five standard transistors on each piece of test equipment before readings were taken for the record. The results of the readings of the standards were supposed to be evaluated by $\overline{X}$, R control charts using control limits based on 25 independent replicate readings taken before the start of the test program. As it occurred, the five standard samples were, in fact, measured before the beginning of each test sequence but the intended control charts were never established.

When the standard sample data were belatedly plotted on control charts, there was evidence of a very clear drift of the standard sample measurements outside of the preestablished control limits. Subsequently, the measurement equipment was carefully checked out and it was determined that an electrolytic capacitor supplying a required pulse voltage had become increasingly leaky. This caused a progressive drop in the actual pulse voltage output, thereby requiring the apparent base drive to be increased in order to maintain the required transistor output.

What was of particular interest from a statistical standpoint was that the variance of each successive set of data increased significantly along with the average value. These apparent changes were in no practical way related to the parts under test.

Example 18.9. Figure 18.6 is an $\overline{X}$, R control chart based on initial readings of a wire wound resistor before a life test (2). The points are the average readings of consecutive samples of five. A count of runs above and below the grand average line indicates that there is one significant run above the line for the first 11 points and another below the line for the last 10 points. This suggests an extremely significant trend in the measured values over the time required to take the readings.

As it happened, these data were taken on a digital bridge starting almost immediately after turning on the bridge. The observed trend was really caused by the drift in the internal reference standard as it warmed up.

Figure 18.7 is another control chart representing the remeasurement of the first 15 samples used in Figure 18.6. In this case, the digital bridge has been properly stabilized before readings were taken.

The amount of error reduction can best be illustrated by a simple analysis of variance table:

	Figure 18.6	Figure 18.7
Within samples	1850	780
Between samples	5020	125
Totals	6870	905

On an overall basis, 87% of the variation shown in Figure 18.6 is not present in Figure 18.7. The bulk of the excess variation was, of course, due to the warmup trend which is evidenced by the 50/1 contrast in the "between samples" variance components.

There is one more comment on this data which is of interest. The readings used in Figure 18.6 were rounded off to the nearest 100 ppm "to simplify manual recording." Since the bridge gave a readout to the nearest 1 ppm, the data of Figure 18.7 were recorded similarly. The "Within Samples" variance components compared to the "Between Samples" variance show that the use of smaller measurement increments results in an error reduction for "Between Samples" measurement error of about 60%. This comment could be readily added to the examples of methods systematic errors (Section 18.2).

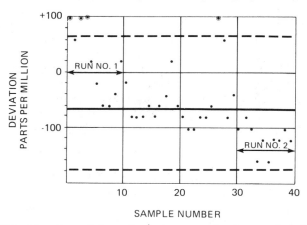

FIGURE 18.6.   Control chart of sample means of resistance deviation.

These kinds of systematic error in test data have been found repeatedly by the author in working with many different companies on many different products. If one has not positively demonstrated that such errors have been eliminated in a particular situation, then it is likely that one or more of these influences exist in a large majority of cases.

18.5. FURTHER COMMENT

This same kind of effect as described above has been observed in testing of semiconductors due to cyclical fluctuations in the ambient temperature of a test area. It has been observed in stability testing of three-axis stable platforms for airborne autonavigators and was caused by changes in power company line voltages due to varying power loads at different times of the

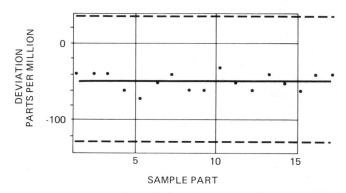

FIGURE 18.7.   Control chart for remeasured samples.

day. It has been observed in the acceleration testing of inertial switches for space vehicles due to amplifier warmup caused by insufficient heat sinking of an amplifier.

The time duration of such cycles of variation will influence the general appearance of the data. For example, if the cycle is quite long with respect to the time of making one measurement, or, approximately four or more times longer than the time required to complete one set of measurements, then a pattern similar to Figure 18.6, or an opposite pattern, can occur. Example 18.8 discussed the situation where both patterns occurred. If the cycle of variation is approximately as long as the time required to complete one set of measurements, then the data may include a periodic function such as a ramp, double-ramp or sawtooth, more usually a sinusoid, and, sometimes a wild erratic pattern.

18.6. THE PENALTIES OF SYSTEMATIC ERROR

When unsuspected systematic error is present, several problems occur. These include spurious trends, bogus significant differences, inflated variances, and flat contradictions. Each of these problems leads to typical difficulties. Spurious trends lead to determining correlations which do not exist in reality. Bogus significant differences lead to false conclusions. Inflated variances reduce the likelihood of establishing real differences of effects. Contradictions lead to mystery.

For the analysis of routine test data, new developments in probability plotting techniques identify systematic errors easily using simple visual tests for sample homogeneity, or consistency (3–5). In addition, when planning experiments, if one includes consideration of possible systematic errors, then some categorical questions concerning materials, methods, machines, and measurements will be introduced before the fact. Any such questions which are resolved will surely reduce the problems and uncertainty after the fact. The same comments can also apply to the planning function for routine control chart sampling.

REFERENCES

1. Tukey, J. W., The technical tools of statistics, *The American Statistician*, Vol. 19, No. 2, 1965.
2. King, J. R., Data analysis in physics of failure studies, *Transactions of the 19th Annual Convention*, ASQC, 1965.
3. King, J. R., Probability plotting techniques revisited, *Team Methods*, Vol. 6, No. 1, 1979, Box A25, Tamworth, NH, 03886.
4. King, J. R., Establishing homogeneous samples, *Team Methods*, Vol. 6, No. 4, 1979.
5. King, J. R., *Frugal sampling schemes*, TEAM, Inc., Box A25, Tamworth, NH, 03886.

DIMENSIONAL ANALYSIS

BHAGWATI P. GUPTA

Lord Corporation
Erie, Pennsylvania

Dimensional analysis is a tool that has many useful applications for the practicing chemist and engineer. Basically it is a shorthand system for dealing systematically with the relations between units of measurement of physical quantities. Dimensionless numbers are combinations of physical variables, in which all units of measurement cancel. The numbers are therefore purely quantitative, and values are independent of measurement systems or physical media, provided only that these are measured in consistent units.

The most common uses of dimensionless numbers are in analog and scale model studies of physical systems or processes. Dimensional analysis reduces the number of independently variable quantities that describe a problem by combining the individual variables into dimensionless groups. This reduces the number of constraints required to maintain similarity between prototypes and models, and thus decreases the number of separate quantities to be considered. Experimentation may be simplified due to the not uncommon reduction of a total set of variables by two-thirds. Dimensional analysis also permits factors governing physical situations to be put into a more convenient form for planning experiments or interpreting data. The resulting equation(s), furthermore, is fundamentally helpful in determining the form of the relationship among variables. In general, application of dimensionless numbers is limited by the validity and completeness of the assumptions made in forming the groups. Empirical verification is therefore generally necessary.

Dimensionless numbers relevant to a phenomenonon are traditionally derived using algebraic techniques such as those of Bridgman, the Rayleigh method, the Buckingham pi theorem, by normalizing terms in differential equations, from ratios governing system similarity parameters, or by matrix analysis.

In polymer science, dimensional analysis can be used for changing units, checking equations, deriving formulas, fitting formulas to experimental data, analyzing physical systems by the use of models, and in planning systematic experimentation. Among other uses, it has application in chemical reactions, heat transfer, viscosity, dispersion, pumping, filtration, defoaming, and so on. The use of dimensionless expressions is of particular value in dealing with phenomena too complicated for a complete treatment in terms of the fundamental equations of mass, energy, and momentum. Most of the physical problems facing, for example, the chemical engineer in the process industry are of this complicated nature. The following material will serve to introduce the use of dimensional analysis. Greater in-depth information can be obtained by consulting the references.

19.1. FUNDAMENTALS

Dimensions and Units

The properties of a body or a system that can be measured and can be used to describe its physical state are called the *dimensions* of the body or the system. For example, the physical state of a moving body could be described by its mass, size (length, area or volume), and velocity or acceleration. Time may also be involved in this description. Thus, all these quantities are dimensions. The dimension of an object is a concept which is easily recognized and does not depend on how it is measured. For example, measurements such as length = 3 miles, breadth = 3 m, and thickness = 0.2 in., all have one property in common: they are all "lengths" and are said to have the *dimension* of length.

The magnitude (but not the numerical value) of a dimension is called the *unit* of the dimension. For example, the measurement of a length may be represented by the "foot" or "meter," which are the *units* of length.

There are some quantities that are nondimensional. These quantities are ratios of two similar dimensional quantities such as strain, angle (radians), and specific gravity. Such quantities are unaffected by the system of units used.

Some Dimensional Notations

The dimensional notations are enclosed in parentheses. Some of these are illustrated here.

$$\text{dimension of length} = (L)$$
$$\text{dimension of mass} = (M)$$

$$\text{dimension of time} = (T)$$

$$\text{dimension of temperature} = (\theta)$$

$$\text{dimension of force} = (F)$$

Fundamental and Derived Dimensions

Basically, (M), (L), (T), and (θ) are considered the "fundamental dimensions." The dimensions of other properties related to mass, length, time, and temperature are termed as "derived." However, some of the fundamental dimensions can, on occasion, be treated as derived and vice versa. Before we proceed further, we should know that a dimensional equation, that is, a relation between the derived dimensions and the fundamental dimensions, must be *dimensionally homogeneous*. This means that the dimensions of the terms on one side of an equation must be equal to those on its other side.

Examples of Derived Dimensions

1. Area = length × length
   or (Area) = $(L)(L) = (L^2)$
   Similarly

2. (Volume) = $(L)(L)(L) = (L^3)$

3. (Velocity) = $\dfrac{\text{(distance)}}{\text{(time)}} = \dfrac{(L)}{(T)} = (LT^{-1})$

4. (Acceleration) = $\dfrac{\text{(velocity)}}{\text{(time)}} = (LT^{-2})$

5. (Angle) = $\dfrac{\text{(arc length)}}{\text{(radial length)}} = \dfrac{(L)}{(L)} = 1$
   = dimensionless

6. (Angular velocity) = $\dfrac{\text{(angle)}}{\text{(time)}} = (T^{-1})$

7. (Angular acceleration) = $\dfrac{(T^{-1})}{(T)} = (T^{-2})$

8. (Volume rate of discharge) = $\dfrac{\text{(volume)}}{\text{(time)}} = (L^3T^{-1})$

9. (Pressure) = (stress) = $\dfrac{\text{(force)}}{\text{(area)}} = \dfrac{(MLT^{-2})}{(L^2)}$
   $= (ML^{-1}T^{-2})$

10.   (Dynamic viscosity) $= \dfrac{\text{(shear stress)}}{\text{(velocity gradient)}}$

$$= \dfrac{(ML^{-1}T^{-2})}{(T^{-1})} = (ML^{-1}T^{-1})$$

11.   (Kinematic viscosity) $= \dfrac{\text{(dynamic viscosity)}}{\text{(mass density)}}$

$$= \dfrac{(ML^{-1}T^{-1})}{(M)/(L^{3})} = (L^{2}T^{-1})$$

12.   (Surface tension) $= \dfrac{\text{(energy)}}{\text{(area)}} = \dfrac{\text{(force)(distance)}}{\text{(area)}}$

$$= \dfrac{(MLT^{-2})(L)}{(L^{2})} = (MT^{-2})$$

The dimensions of all these properties, area, volume, velocity, and acceleration are "derived" from the "fundamental" dimensions, (L) and (T). As stated before, sometimes the derived dimensions can be treated as fundamental ones, for example:

$$\text{(force)} = \text{(mass)} \times \text{(acceleration)} \quad \text{or} \quad (F) = (M)(LT^{-2}) = (MLT^{-2})$$

Here, (F) is the derived dimension. But, also (M) can be treated as derived in terms of (F) as the fundamental:

$$(M) = (FL^{-1}T^{2})$$

The dimensions of any quantity can be obtained, as we have shown above, by its relationship with other quantities of known dimensions and by using the principle of homogeneity.

Types of Units

As is the case with dimensions, the units of a quantity are also derived from the fundamental units and by the use of the dimensional equations. For example,

$$\text{(force)} = (F) = (MLT^{-2})$$

or

$$\text{unit of force} = \dfrac{\text{(unit of mass} \times \text{unit of length)}}{\text{(unit of time)}^{2}}$$

The corresponding set of units define a system of units. For these units, there are two commonly used systems. The first one is the length, mass,

TABLE 19.1.   The Absolute and the Engineering System of Units

Fundamental Quantities	Absolute			Engineering		
	mks	cgs	fps	mks	cgs	fps
Mass	kilogram	gram	pound	9.81 kg	981 g	slug
Force	newton	dyne	poundal	kilogram force	gram force	pound force
Length	meter	centimeter	foot	meter	centimeter	foot
Time	second	second	second	second	second	second
Temperature	Celsius	Centigrade	Fahrenheit	Celsius	Centigrade	Fahrenheit

and time set, whose fundamental dimensions are L, M, and T. The second one is the length, force, and time set, with the dimensions L, F, and T as fundamental units. The first is called the dynamic or absolute system, in which the force is the derived unit and the second one is called the engineering or practical system, in which the mass is treated as the derived unit. Even with these two sets of fundamental dimensions, subdivision into many different systems of units is possible. For example, with the first set of dimensions, we could use centimeter-gram-second (cgs) or the meter-kilogram-second (mks) system, the latter being a part of the SI (System International). The second set of dimensions is illustrated by the foot-pound-second (fps) system which is being replaced by the SI.

Table 19.1 gives the fundamental units in these different systems.

It should be noted here that mass is derived from weight divided by acceleration due to gravity, and the unit for weight is that of force.

Conversion from One System of Units to Another

All the numerical values in a formula must belong to the same system of units. Therefore, if there are some values the units of which are in a different system than the units of other values, we will have to convert units of one set of values into those of the other set of values. To achieve this, we use a law which states: "The true physical value of any quantity must be independent of the system of units used to measure it."

Let θ_1 be the physical value of a quantity and n_1 be its numerical value measured in the units of size u_1 (size u is the physical value for a numerical value of 1). So,

$$\theta_1 = n_1 u_1 \tag{19.1}$$

Similarly, the physical value of the same quantity in another system of units of size u_2, when its numerical value in those units is n_2, will be

$$\theta_2 = n_2 u_2 \tag{19.2}$$

but θ_1 must be equal to θ_2. So

$$n_1 u_1 = n_2 u_2 = \text{a constant}$$

or

$$n \propto \frac{1}{u} \qquad (19.3)$$

Now, let us derive a formula for converting a numerical value given in one system of units into the other system of units.

Let the dimensional formula of a quantity θ be given by $(M^a L^b T^c)$.

Let the fundamental units of mass, length, and time in the two systems be m_1, l_1, t_1 and m_2, l_2, t_2.

Let the relations between these two systems be given by

$$m_1 = K_m m_2, \qquad l_1 = K_l l_2, \qquad t_1 = K_t t_2 \qquad (19.4)$$

where K are constants for conversion.

So, the derived units of θ in the two systems will be

$$u_1 = m_1^a l_1^b t_1^c = (K_m m_2)^a (K_l l_2)^b (K_t t_2)^c$$

and

$$u_2 = M_2^a l_2^b t_2^c$$

Then, if the numerical values of these quantities in the two systems are n_1 and n_2, we get

$$n_1 u_1 = n_2 u_2$$

or

$$n_1 K_m^a K_l^b K_t^c m_2^a l_2^b t_2^c = n_2 m_2^a l_2^b t_2^c$$

or

$$n_2 = n_1 K_m^a K_l^b K_t^c \qquad (19.5)$$

Example 19.1. Given the physical value of the velocity equal to 88 ft/sec. What is its value in miles per hour?

The dimensional formula for velocity is (LT^{-1}). So, $a = 0$, $b = 1$, $c = -1$. Given are: $n_1 = 88$, $u_1 = \text{ft/sec}$, $u_2 = \text{miles/hour}$.

We know that there are 5280 ft in 1 mile, so, by using $l_1 = K_l l_2$, we get

$$K_l = \frac{1}{5280}$$

Similarly $K_t = 1/3600$. So, by using equation (19.5), we get

$$n_2 = 88 \times (K_m)^0 \frac{(1)^1}{(5280)} \frac{(1)^{-1}}{(3600)}$$

$$= 60$$

So, 88 ft/sec = 60 miles/hour.

Example 19.2. Convert the mass density of water, which is given equal to 62.4 lb/ft^3 in fps system, into SI units. Dimensional formula for mass density is (ML^{-3}). So,

$$a = 1, \qquad b = -3, \qquad c = 0.$$

We know that 1 lb of mass is equal to 0.4536 kg mass. So the conversion constant for mass $K_m = 0.4536$. Similarly as 1 ft is equal to 0.3048 m, $K_l = 0.3048$. So,

$$n_2 = n_1 (0.4536)^1 (0.3048)^{-3} (K_t)^0$$

But $n_1 = 62.4$ as given, so, $n_2 = 1000$. Thus, the value of 62.4 lb/ft^3 is equal to 1000 kg/m^3.

19.2. CHECKING OF EQUATIONS BY USING DIMENSIONAL HOMOGENEITY

The principle of dimensional homogeneity states that the dimensions of each term of an equation must be the same. This principle is therefore a necessary check on the correctness of an equation, although the equation itself may not be necessarily correct.

Example 19.3. Pierce (10) gives the following example. The expression below represents the amount of heat necessary to raise a liquid to its boiling point and vaporize it. Is the equation dimensionally correct?

$$Q = MC_p (\theta_1 - \theta_2) + Mh_{fg}$$

Write the equation, substitute, and inspect the results.

$$Q = MC_p(\theta_1 - \theta_2) + Mh_{fg}$$
$$Q = (M)(QM^{-1}\theta^{-1})(\theta_1 - \theta_2) + (M)(QM^{-1})$$
$$= (Q\theta^{-1})(\theta) + Q$$
$$Q = Q + Q$$
$$\therefore \text{Heat} = \text{heat.}$$

Example 19.4. The energy equation for a flowing substance is of the form,

$$JE_1 + p_1v_1 + \frac{c_1^2}{2g} + Q + W = JE_2 + p_2v_2 + \frac{c_2^2}{2g}$$

where

E = energy stored in the substance
p = pressure
v = specific volume
c = flow velocity
Q = energy flowing into the substance in the form of heat per unit weight
W = energy flowing into the substance in the form of work per unit weight

Now,

$$\text{dimensions of } p = \frac{(MLT^{-2})}{(L^2)} = (ML^{-1}T^{-2})$$

$$\text{dimensions of } v = \frac{(\text{volume})}{(\text{weight})} = \frac{(L^3)}{(M)(LT^{-2})} = (M^{-1}L^2T^2)$$

So,

$$\text{dimensions of } pv = (ML^{-1}T^{-2})(M^{-1}L^2T^2) = (L)$$

$$\text{dimensions of } \frac{c^2}{2g} = \frac{(LT^{-1})^2}{(LT^{-2})} = (L)$$

and

$$\text{dimensions of mechanical work/weight} = \frac{(MLT^{-2})(L)}{(MLT^{-2})} = (L)$$

By substituting these dimensions in the given equation and comparing it with those of the mechanical work/weight, we see that the dimensions of EJ, Q, and W must be those of the mechanical work/weight, that is, they must be (L) if the equation has to be dimensionally correct. Also, because units of J are

$$(H^{-1}ML^2T^{-2})$$

where (H) is the dimension for heat, dimensions for E should be heat/weight

$$\frac{(H)}{(MLT^{-2})}$$

19.3. FORMATION OF EQUATIONS BY DIMENSIONAL ANALYSIS

Indicial Method

The dimensional homogeneity principle and the nature of the properties involved are used in forming a relation between these properties. This method is called the "indicial" method and may be carried out in the following steps:

1. Let A represent the dependent variable and $B, C, D, E, \ldots$ represent the independent variables, that is, a relation is to be found for A in terms of $B, C, D, E, \ldots$.
2. Write the relationship in the form

$$A = KB^bC^cD^dE^e \cdots \qquad (19.6)$$

where $K, b, c, d, e, \ldots$ are unknown.
3. For dimensional homogeneity, the fundamental dimensions of the left-hand side must be equal to those of the right-hand side. So,

$$(A) = (B)^b(C)^c(D)^d(E)^e \cdots \qquad (19.7)$$

4. Say,

$$(A) = (M^{ma}L^{la}T^{ta})$$
$$(B) = (M^{mb}L^{lb}T^{tb})$$
$$(C) = (M^{mc}L^{lc}T^{tc})$$

then, by substituting in equation (19.7), we get $(M^{m_a}L^{l_a}t^{t_a}) = (M^{m_b}L^{l_b}T^{t_b})^b(M^{m_c}L^{l_c}T^{t_c})^c\ldots.$ By equating the indices, we get

$$m_a = (bm_b)(cm_c)(dm_d)\cdots$$
$$l_a = (bl_b)(cl_c)(dl_d)\cdots \tag{19.8}$$
$$t_a = (bt_b)(ct_c)(dt_d)\cdots$$

5. By solving these simultaneous equations (19.8), we obtain values of $b, c, d, e, \ldots$ either absolutely or in terms of other indices depending on whether the number of simultaneous equations is equal to or less than the number of unknown indices.

6. By using these values, then, equation (19.6) gives the required relation which can be arranged in a desired form.

Example 19.5. The following example is adapted from Bridgman (3). An elastic pendulum is made by attaching to a weightless spring of elastic constant k a box of volume V which is filled with a liquid of density d. The mass of the liquid in the box is acted upon by gravity, and we are required to find an expression for the time of oscillation. We first make a list of the quantities and their dimensions taking volume as an independent unit of its own kind. Then we have:

Name of Quantity	Symbol	Dimensional Formula
Elastic constant	k	MT^{-2}
Time of oscillation	t	T
Volume of box	v	V
Density of liquid	d	MV^{-1}
Acceleration of gravity	g	LT^{-2}

We now have five variables, but four fundamental kinds of quantity, so that there is only one dimensionless product. We are particularly interested in t, so we choose the exponent of t equal to unity, and are required to find the other exponents so that $tk^\alpha v^\beta d^\gamma g^\delta$ is dimensionless.

We can solve for the unknowns by inspection, or if we prefer, write out the equations, which are

$$\alpha + \gamma = 0$$
$$\delta = 0$$
$$-2\alpha - 2\delta + 1 = 0$$
$$\beta - \gamma = 0$$

The solution of this set of equations is

$$\alpha = \tfrac{1}{2}, \qquad \beta = -\tfrac{1}{2}, \qquad \gamma = -\tfrac{1}{2}, \qquad \delta = 0$$

The dimensionless product is

$$tk^{1/2}v^{-1/2}d^{-1/2}$$

and the solution is

$$t = \text{const}\sqrt{\frac{vd}{k}}$$

Example 19.6. This example is also adapted from Bridgman (3). If a small sphere falls through a viscous liquid under gravity (such as in a falling-ball viscometer), it will experience a resistance from the fluid and will attain a constant or terminal viscosity. This is the problem of Stokes. The elements with which we have to deal are the velocity of fall, the density of the sphere, the diameter of the sphere, the density of the liquid, the viscosity of the liquid, and the intensity of gravity.

In solving this problem, it is an advantage to treat force as a mechanical unit of its own kind of quantity and not introduce, in addition, a compensating dimensional constant.

The following assumptions are also made: (1) the motion of the sphere is sufficiently slow; (2) the fluid is of infinite extent; (3) there is no slip between the fluid and the sphere; (4) the sphere is rigid.

Listing the quantities and their dimensions, we have:

Name of Quantity	Symbol	Dimensional Formula
Velocity of fall	v	LT^{-1}
Diameter of sphere	D	L
Density of sphere	d_1	ML^{-3}
Density of liquid	d_2	ML^{-3}
Viscosity of liquid	μ	$FL^{-2}T$
Intensity of gravity	g	FM^{-1}

The dimensional formula of viscosity is obtained directly from its definition as force per unit area per unit velocity gradient. The intensity of gravity is taken with the dimensions shown, because obviously the equations of motion will not mention the accelerational aspect of gravitational action, but only the intensity of the force exerted by gravity upon unit mass.

We now have six variables, and four fundamental kinds of unit. There are, therefore, two dimensionless products. One of them is evident on inspection and is d_2/d_1. Now of the remaining quantities, we are especially interested in v. We need combine this with only four other quantities to obtain a dimensionless product. We choose D, d_1, μ, and g, and seek a dimensionless product of the form

$$vD^\alpha d_1^\beta \mu^\gamma g^\delta$$

The exponents are found to be

$$\alpha = -2, \quad \beta = -1, \quad \delta = -1, \quad \gamma = 1$$

Hence, the dimensionless products are

$$vD^{-2}d_1^{-1}\mu g^{-1} \quad \text{and} \quad \frac{d_2}{d_1}$$

and the final solution is

$$v = \frac{D^2 d_1 g}{\mu} f\left(\frac{d_2}{d_1}\right)$$

The function f is arbitrary, so that we cannot tell how the result depends on the densities of the sphere and the liquid, but we do see that the velocity of fall varies as the square of the diameter of the sphere, and the intensity of gravity, and inversely as the viscosity of the fluid. The general solution to this is the well-known Stokes law:

$$\mu = \frac{8}{9} \frac{gD^2}{v}(d_1 - d_2)$$

Example 19.7. Velocity of a one-dimensional stress wave in a solid is to be related to the properties of the material.

Let us assume that this velocity, V, is a function of modulus of elasticity, E, and the mass density, ρ, of the material, and can be written by

$$V = f(E, \rho) \tag{19.9}$$

where f is an unknown function.

Let us assume that this relationship can be written as

$$V = AE^a\rho^b \tag{19.10}$$

where A is a numerical constant (dimensionless) and a and b are unknown powers which are to be determined by dimensional analysis. The dimensions of each quantity involved are

$$(V) = (LT^{-1})$$
$$(E) = (ML^{-1}T^{-2})$$
$$(\rho) = (ML^{-3})$$

By substituting these dimensions in equation (19.10), we get

$$(LT^{-1}) = (ML^{-1}T^{-2})^a (ML^{-3})^b$$
$$= (M^{(a+b)}L^{(-a-3b)}T^{(-2a)})$$

So, by equating the powers, we get

$$a + b = 0$$
$$-a - 3b = 1 \qquad (19.11)$$
$$-2a = -1$$

These are three simultaneous equations, but there are only two unknowns. One of these equations would be dependent. Solving, say the first and third equations, we get

$$a = \tfrac{1}{2} \quad \text{and} \quad b = -\tfrac{1}{2}$$

The second of these equations is identically satisfied. Equation (19.10) then becomes

$$V = AE^{1/2}\rho^{-1/2} = A\sqrt{\frac{E}{\rho}} \qquad (19.12)$$

Example 19.8. Let the volume rate of discharge, Q, of a fluid through an orifice be a function of p, d, D, μ, ρ, that is,

$$Q = f(p, d, D, \mu, \rho) \qquad (19.13)$$

where

$$p = \text{pressure difference}$$
$$d = \text{orifice diameter}$$
$$D = \text{diameter of the pipe}$$
$$\mu = \text{dynamic viscosity}$$
$$\rho = \text{mass density}$$

Let us write Q in the form

$$Q = Ap^a d^b D^c \mu^e \rho^f \qquad (19.14)$$

The dimensions of the quantities involved are

$$(Q) = (L^3 T^{-1})$$
$$(p) = (ML^{-1}T^{-2})$$
$$(d) = (L)$$
$$(D) = (L)$$
$$(\mu) = (ML^{-1}T^{-1})$$
$$(\rho) = (ML^{-3})$$

Substituting these in equation (19.14), we get

$$(L^3)(T^{-1}) = (ML^{-1}T^{-2})^a (L)^b (L)^c (ML^{-1}T^{-1})^e (ML^{-3})^f$$
$$= (M^{(a+e+f)})(L^{(-a+b+c-e-3f)})(T^{-2a-e})$$

So, by equating the powers, we get

$$a + e + f = 0$$
$$-a + b + c - e - 3f = 3 \qquad (19.15)$$
$$-2a - e = -1$$

There are three simultaneous equations but five unknowns. Therefore, we will have to determine values of three unknowns in terms of the remaining two. Let us choose, arbitrarily, a, b, and f to be solved in terms of c and e. We get $a = \frac{1}{2}(1 - e)$, $b = 2 - c - e$, and $f = -\frac{1}{2}(1 + e)$. Substituting these values in equation (19.14) and rearranging, we get

$$Q = Ad^2 \left(\sqrt{\frac{p}{\rho}} \right) \left(\frac{d}{D} \right)^{-c} \left(\frac{d\sqrt{p\rho}}{\mu} \right)^{-e}$$

It is seen here that (d/D), $(d\sqrt{p\rho}/\mu)$, and (A) are dimensionless quantities, and c and e are unknown. So, we can group these quantities in the form of an unknown function, as

$$Q = d^2 \sqrt{\frac{p}{\rho}} \cdot \Psi \left(\frac{d}{D}, \frac{d\sqrt{p\rho}}{\mu} \right) \qquad (19.16)$$

This equation can be modified further by using the Reynolds number. The pressure difference p can be written as

$$p = \text{constant} \times \rho V^2$$

so the quantity

$$\frac{d\sqrt{p\rho}}{\mu} = \frac{d\rho V}{\mu} \times \text{constant}$$

$$= \text{Re} \times \text{constant}$$

where Re, the Reynolds number, is equal to $(d\rho V/\mu)$. This changes equation (19.16) into

$$Q = d^2 \sqrt{\frac{p}{\rho}} \ \Psi\left(\frac{d}{D}, \text{Re}\right) \qquad (19.17)$$

19.4. DIMENSIONAL ANALYSIS BY THE GROUP METHOD — BUCKINGHAM'S PI METHOD

Buckingham's Pi Theorem

This method is based on Buckingham's pi theorem which can be stated as:

> If an equation is dimensionally homogeneous, it can be reduced to a relationship among a complete set of dimensionless products.

The proof of this theorem is not being given here but we will explain its meanings and use by the following:

Example 19.9. Equation (19.12) gives a dimensionally homogeneous equation

$$v = A\sqrt{E/\rho} \qquad (19.12)\text{rep.}$$

This can be rewritten as

$$\frac{(v)}{\left(\sqrt{E/\rho}\right)} = A \qquad (19.18)$$

We note here that $(v)/(\sqrt{E/\rho})$ is a nondimensional group (or product). By denoting it by π_1, we can write equation (19.18) as

$$\pi_1 = A$$

or

$$\Psi(\pi_1) = 0 \qquad (19.19)$$

By this example, we learn the following:

The number of physical quantities, say n, is equal to three (V, E, ρ). The number of independent indicial equations, say m, is two [equation (19.11)]. The number of fundamental dimensions, say K, is three (M, L, T). The number of nondimensional products (π_1) is one.

This gives an equality

$$\text{no. of dimensionless groups} = n - m$$

Example 19.10. Consider equation (19.16) derived by the indicial method:

$$Q = d^2 \sqrt{\frac{p}{\rho}} \cdot \Psi \left[\frac{(d)}{(D)}, \frac{(d\sqrt{p\rho})}{(\mu)} \right] \qquad (19.16)\text{rep.}$$

This can be rewritten as

$$\frac{(Q)}{(d^2\sqrt{p/\rho})} = \Psi \left[\frac{(d)}{(D)}, \frac{(d\sqrt{p\delta})}{(\mu)} \right] \qquad (19.20)$$

By denoting

$$\frac{(Q)}{(d^2\sqrt{p/\rho})} = \pi_1$$

$$\frac{(d)}{(D)} = \pi_2$$

and

$$\frac{(d\sqrt{p\rho})}{(\mu)} = \pi_3$$

We can rewrite equation (19.20) as

$$\pi_1 = \Psi(\pi_2, \pi_3)$$

or

$$\Psi(\pi_1, \pi_2, \pi_3) = 0 \qquad (19.21)$$

We note again in this example that π_1, π_2, and π_3 are dimensionless groups.

We also note the following observations:

Number of physical quantities (Q, d, p, ρ, D, μ), $n = 6$.
Number of independent indicial equations [equation (19.15)], $m = 3$.
Number of fundamental dimensions used, $k = 3$.
Number of dimensionless groups $= 3$.

We can say again that,

$$\text{no. of dimensionless groups} = 3 = 6 - 3$$
$$= n - m = n - k$$

Based on this trend, shown by these two examples, this theorem can be explained in a general form, as

Let us assume there are n physical quantities, $Q_1, Q_2, \ldots, Q_n$, to be related by a complete equation

$$F(Q_1, Q_2, \ldots, Q_n) = 0 \qquad (19.22)$$

Then, this equation can be reduced to

$$\Psi(\pi_1, \pi_2, \ldots, \pi_{n-m}) = 0 \qquad (19.23)$$

where m is the number of independent indicial equations.

Normally, m is equal to k, the number of fundamental dimensions. So, the minimum number of dimensionless groups may also be equal to $n - k$ and the equation can then be written as

$$\Psi(\pi_1, \pi_2, \ldots, \pi_{n-k}) = 0 \qquad (19.24)$$

Formation of Dimensionless Groups

The dimensionless groups, π_i, can be written, as noted in the previous examples, as products of the fundamental physical quantities and the derived physical quantities in the form

$$\pi_1 = Q_1^{a_1} \cdot Q_2^{b_1} \cdot \cdots \cdot Q_k^{k_1} \cdot P_1$$
$$\pi_2 = Q_1^{a_2} \cdot Q_2^{b_2} \cdot \cdots \cdot Q_k^{k_2} \cdot P_2 \qquad (19.25)$$
$$\pi_{n-k} = Q_1^{a}(n-k) \cdot \cdots \cdot Q_k^{k_{(n-k)}} \cdot P_{(n-k)}$$

where P_i represent the products $(Q_{k+1} \cdot \cdots \cdot Q_n)_i$ which depend on the fundamental quantities $Q_1, Q_2, \ldots, Q_k$.

The following problems will illustrate one of the methods of forming the dimensionless groups.

Problem 19.1.   For a lubricated journal bearing, determine a relationship between the resisting torque (t), the journal diameter (D), the length (l), the clearance (c), the viscosity (μ) of the oil, and the speed of rotation (N) of the shaft for a given load (W).

Solution.   The fundamental quantities $Q_1, Q_2, \ldots, Q_k$ are selected and cannot exceed three in accordance with the fundamental dimensions (M), (L) and (T). Let us select D, N, and W as Q_1, Q_2, and Q_3. The dimensions of these quantities are

$$(D) = (L), \quad (N) = (T^{-1}), \quad \text{and} \quad (W) = (MLT^{-2})$$

The fundamental dimensions, then, are written in terms of these three chosen quantities, called the fundamental quantities, as

$$(L) = (D), \quad (T) = (N^{-1}), \quad \text{and} \quad (M) = \frac{(W)}{(LT^{-2})} = \frac{(W)}{(DN^2)}$$

Now, the remaining quantities t, l, c, and μ can be derived to form the dimensionless groups.

Group π_1. Say t is Q_4, where

$$(t) = (ML^2 T^{-2}) = \left(\frac{W \cdot D^2 \cdot N^2}{DN^2} \right) = (WD)$$

So,

$$\pi_1 = \frac{t}{WD} = Q_4 Q_3^{-1} Q_1^{-1}$$

Group π_2. Say l is Q_5, where

$$(l) = (L) = (D)$$

So,

$$\pi_2 = \frac{l}{D} = Q_5 Q_1^{-1}$$

Group π_3. Say c is Q_6, where

$$(c) = (L) = (D)$$

So,

$$\pi_3 = \frac{c}{D} = Q_6 \cdot Q_1^{-1}$$

Group π_4. The remaining quantity, Q_7, is μ which can be written dimensionally as

$$(\mu) = (ML^{-1}T^{-1}) = \left(\frac{W \cdot D^{-1} \cdot N}{DN^2}\right) = (WD^{-2}N^{-1})$$

So

$$\pi_4 = \frac{\mu}{WD^{-2}N^{-1}} = Q_7 \cdot Q_3^{-1} \cdot Q_1^2 \cdot Q_2$$

Rearranging these groups we have

$$\pi_1 = Q_1^{-1}Q_2^0Q_3^{-1}(Q_4) = \frac{(t)}{(WD)}$$

$$\pi_2 = Q_1^{-1}Q_2^0Q_3^0(Q_5) = \frac{(l)}{(D)} \qquad (19.26)$$

$$\pi_3 = Q_1^{-1}Q_2^0Q_3^0(Q_6) = \frac{(c)}{(D)}$$

$$\pi_4 = Q_1^2Q_2^1Q_3^{-1}(Q_7) = \frac{(\mu D^2 N)}{(W)}$$

Comparing equation (19.26) with equation (19.25), we note that $P_1 = Q_4$, $P_2 = Q_5$, $P_3 = Q_6$, and $P_4 = Q_7$.

We also see that, since the number of physical quantities, n, is equal to 7 $(Q_1, Q_2, \ldots, Q_7)$ and k is equal to 3, we get the dimensionless groups equal to $n - k = 4$.

The relationship is, then,

$$\Psi(\pi_1, \pi_2, \pi_3, \pi_4) = 0$$

or

$$\pi_1 = \Psi(\pi_2, \pi_3, \pi_4)$$

or

$$\frac{(t)}{(WD)} = \Psi\left\{\left(\frac{l}{D}\right), \left(\frac{c}{D}\right), \left(\frac{\mu ND^2}{W}\right)\right\} \qquad (19.27)$$

Problem 19.2. Relate the propeller thrust (T) with the axial velocity (V), the diameter (D), the speed of rotation (N), the fluid density (ρ), and the fluid viscosity (μ).

Solution. In the functional form T can be written by

$$T = f(V, D, N, \rho, \mu)$$

The number of fundamental dimensions, k, is 3. Here, the number of physical quantities is 6. Therefore, the number of dimensionless groups will be $(6 - 3) = 3$. Also we will choose the fundamental quantities equal to the number of fundamental dimensions, that is, 3.

Let us select three quantities such that they involve all three fundamental dimensions. For example, consider D, N, and ρ, the fundamental quantities Q_1, Q_2, and Q_3 where dimensionally

$$(D) = (L), \quad (N) = (T^{-1}), \quad \text{and} \quad (\rho) = (ML^{-3})$$

Then we can write

$$(L) = (D), \quad (T) = (N^{-1}), \quad \text{and} \quad (M) = (\rho L^3) = (\rho D^3)$$

Now we form the groups.

Group π_1. Say T is Q_4, where

$$(T) = (MLT^{-2}) = (\rho D^3 \cdot D \cdot N^2) = (\rho D^4 N^2)$$

So,

$$\pi_1 = \frac{T}{\rho D^4 N^2} = Q_4 Q_3^{-1} Q_1^{-4} Q_2^{-2}$$

Group π_2. Say V is Q_5, where

$$(V) = (LT^{-1}) = (DN)$$

So

$$\pi_2 = \frac{V}{DN} = Q_5 Q_1^{-1} Q_2^{-1}$$

Group π_3. Say μ is Q_6, where

$$(\mu) = (ML^{-1}T^{-1}) = (\rho D^3 \cdot D^{-1} N) = (\rho D^2 N)$$

or

$$\pi_3 = \frac{\mu}{\rho N D^2} = Q_6 Q_3^{-1} Q_2^{-1} Q_1^{-2}$$

We can now write

$$\pi_1 = Q_1^{-4}Q_2^{-2}Q_3^{-1}(Q_4)$$
$$\pi_2 = Q_1^{-1}Q_2^{-1}Q_3^{0}(Q_5)$$
$$\pi_3 = Q_1^{-2}Q_2^{-1}Q_3^{-1}(Q_6)$$

This compares with equations (19.25). Now, since

$$\Psi(\pi_1, \pi_2, \pi_3) = 0$$

or

$$\pi_1 = \Psi(\pi_2, \pi_3)$$

We get

$$\left(\frac{T}{\rho N^2 D^4}\right) = \Psi\left[\left(\frac{V}{DN}\right), \left(\frac{\mu}{\rho N D^2}\right)\right] \qquad (19.28)$$

We can write this equation by considering different groups. For example, we change the following groups:

$$\pi_1 = \frac{T}{\rho D^4 N^2}$$

But $V = ND$(group π_2), or

$$N = \frac{V}{D}$$

So

$$\pi_1 = \frac{T}{\rho V^2 D^2}$$

Also by taking

$$\pi_3 = \frac{\rho N D^2}{\mu} \quad \text{instead of} \quad \frac{\mu}{\rho N D^2}$$

we get

$$\pi_3 = \frac{\rho N D^2}{\mu} = \frac{\rho \cdot (V/D) \cdot D^2}{\mu} = \frac{\rho V D}{\mu}$$

Thus, the relation becomes

$$\left(\frac{T}{\rho V^2 D^2}\right) = \Psi\left[\left(\frac{V}{DN}\right) , \left(\frac{\rho V D}{\mu}\right)\right] \tag{19.29}$$

This is the form of the relation given in (11), but has been derived by a different approach.

This example shows that the forming of nondimensional groups is arbitrary. However, we can form an equation in terms of some groups that may have some physical significance as will be shown next.

19.5. SIGNIFICANCE OF DIMENSIONLESS GROUPS AND REYNOLDS NUMBER

These groups, because they carry no dimensions, have to be ratios of two physical quantities of the same kind. For example, in Problem 19.1 of Section 19.4,

$$\pi_1 = \frac{t}{WD}$$

where

$$t = \text{resisting torque}$$
$$WD = \text{applied torque}$$

Consider equation (19.16) which is

$$Q = d^2\sqrt{\frac{p}{\rho}} \cdot \Psi\left[\left(\frac{d}{D}\right) , \left(\frac{d\sqrt{p\rho}}{\mu}\right)\right]$$

The third group

$$\pi_3 = \frac{d\sqrt{p\rho}}{\mu}$$

But $p \propto \rho V^2$ where $V = $ fluid velocity, so we can write

$$\pi_3 = \frac{d\rho V}{\mu}$$

This group, $d\rho V/\mu$, is the Reynolds number, Re, so

$$\text{Re} = \frac{\rho V d}{\mu} \quad \text{or} \quad \text{Re} = \frac{\rho V^2 d^2}{\mu V d}$$

Now mass $\propto \rho d^3$, and acceleration $\propto d/T^2$ where $T =$ time, so

$$\text{inertial force} \propto \rho d^3 \cdot \frac{d}{T^2} = \rho d^2 \left(\frac{d}{T}\right)^2$$

But $V \propto d/T$, so inertial force $\propto \rho d^2 V^2$.

Similarly, viscous force $\propto$ area $\times$ viscous shear stress where area $\propto d^2$, and viscous shear stress $\propto \mu \times$ velocity gradient $\propto \mu(V/d)$. Therefore,

$$\text{viscous force} \propto \mu V d$$

Thus,

$$\pi_3 = \text{Re} \propto \frac{\text{inertial force}}{\text{viscous force}} \tag{19.30}$$

Therefore, one of the groups in the equation of flow is the Reynolds number which represents the ratio of inertial force to viscous force. This proves that all the dimensionless groups must represent ratios of some physical quantities and have some physical significance.

19.6. REYNOLDS NUMBER AND ITS APPLICATION TO COATINGS

We have discussed in the previous section the Reynolds number as encountered in the derivation of the fluid flow equation. We now are going to discuss it in greater detail.

From the equation of fluid motion, Reynolds reasoned that the onset of turbulence should depend on a critical value for the dimensionless quantity, called the Reynolds number. For most flow conditions (Figure 19.1) let

$$\rho = \text{density of fluid}$$
$$V = \text{average fluid velocity}$$
$$\mu = \text{viscosity of fluid}$$
$$d = \text{diameter of the pipe}$$

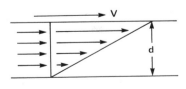
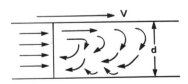

Laminar Regime　　　　　　　　　　Turbulent Regime

FIGURE 19.1. Laminar and turbulent flow patterns.

Then the Reynolds number is given by

$$\text{Re} = \frac{\rho V d}{\mu}$$

In pipes, the transition from laminar to turbulent flow occurs at a Reynolds number of 2000, see for example Refs. 12 and 13. This, however, may not apply to circular tanks. As noted by Patton (14), mixing is associated with turbulent flow and dispersion with laminar flow. The Reynolds number is composed of a kinetic component ($\rho V d$) that tends to induce turbulence and a viscous component (μ) that tends to induce laminar flow. Therefore, for the mixing operation, the kinetic component has to be greater than 2000 times the viscous component, while for the dispersion operation, it has to be less than 2000 times the viscous component. This gives the design criteria. The viscosity, velocity, and density of the fluid (vehicle) may be varied to get the proper design.

The following empirical equations have been established for turbulent flow of Newtonian fluids in pipes (15).

For smooth pipes,

$$\frac{1}{\sqrt{\phi}} = 2\log\left(\frac{\text{Re}}{\sqrt{\phi}}\right) - 0.8$$

For rough pipes,

$$\frac{1}{\sqrt{\phi}} = 2\log\left(\frac{d}{2\eta}\right) + 1.7$$

where

ϕ = friction factor

= 64/Re

η = an indicator of surface roughness

19.7. DIMENSIONAL ANALYSIS FOR AGITATION

Based on the Navier–Stokes equation for a mass and momentum balance in terms of local pressure and velocity, and using the dimensionless lengths and time, the velocity and pressure distributions can be expressed (16, 17) by

$$v^*(x^*, y^*, z^*, t^*) = f_1(\text{Re}, \text{Fr})$$

and

$$p^*(x^*, y^*, z^*, t^*) = f_2(\text{Re}, \text{Fr})$$

where

$$x^* = \frac{x}{d}, \quad y^* = \frac{y}{d}, \quad z^* = \frac{z}{d}, \quad t^* = tN$$

$$v^* = \frac{v}{Nd}, \quad p^* = \frac{(p - p_o)g_c}{\rho N^2 d^2}$$

d = diameter of the impeller

N = agitator rotational speed

ρ = liquid density

g_c = gravitational conversion factor

p = liquid pressure

p_o = reference pressure

v = velocity of liquid

Re = Reynolds number = $d^2 N\rho\mu = \dfrac{Vd\rho}{\mu}$

μ = liquid viscosity

Fr = Froude number = $\dfrac{dN^2}{g}$

These relations indicate how the diameter, rotational velocity, density, and viscosity are related to give the velocity and pressure of the fluid.

Also, the following relations can be established:

$$\theta^*(x^*, y^*, z^*, t^*) = f_3(Re, Pr)$$

and

$$X_A^*(x^*, y^*, z^*, t^*) = f_4(Re, Sc) \tag{19.31}$$

where

θ^* = dimensionless temperature

X_A^* = dimensionless concentration

Pr = Prandtl number = $\dfrac{c_p\mu}{k}$

Sc = Schmidt number = $\dfrac{\mu}{\rho D_{AB}}$

c_p = heat capacity

k = thermal conductivity

D_{AB} = diffusion coefficient

These relations describe the temperature and concentration variation in an agitated vessel.

For practical design aspects, the following relations are used:

$$\frac{(Pg_c)}{(\rho N^3 d^5)} = f_5(\mathrm{Re}, \mathrm{Fr})$$

$$v^* = \frac{v}{Nd} = f_6(\mathrm{Re})$$

$$\frac{(Q)}{(Nd^3)} = f_7(\mathrm{Re})$$

$$t_b^* = f_8(\mathrm{Re}, \mathrm{Sc}) \qquad (19.32)$$

where

$$P = \text{agitation power}$$
$$v = \text{average fluid velocity}$$
$$Q = \text{pumping capacity}$$
$$t_b^* = \text{dimensionless blend time}$$

19.8. APPLICATION OF DIMENSIONAL ANALYSIS TO FILTRATION

The process of filtration can be defined as a controlled separation of solid particles from a suspension in a fluid phase where all the particles collect on a filter septum of high porosity and form a cake while the fluid permeates through the combined interstices of the cake and the septum (18, 19, 20).

Mathematically, the average filtration rate can be represented by

$$\phi = f_1(\Delta p, \rho, \mu, \lambda, \beta)$$

ϕ = average filtration rate in cm^3/cm^2-sec

Δp = pressure drop in g/cm-sec^2

ρ = fluid density in g/cm^3

μ = fluid viscosity in g/cm-sec

λ = filtrate collected per unit area of filter in cm^3/cm^2

β = dimensionless resistance, a quantitative measure of solid–fluid interaction, in g water/g dry cake   (19.33)

Derivation of Equation (19.33) Using Buckingham's Pi Method

Let us select ρ, μ, and λ as the fundamental quantities which can be written dimensionally as

$$[\rho] = \left[\frac{M}{L^3}\right], \quad [\mu] = \left[\frac{M}{LT}\right], \quad \text{and} \quad [\lambda] = [L]$$

We then get

$$[L] = [\lambda]$$

so

$$[\rho] = \left[\frac{M}{\lambda^3}\right] \quad \text{or} \quad [M] = [\rho\lambda^3]$$

Then

$$[\mu] = \left[\frac{M}{LT}\right] \quad \text{or} \quad [T] = \left[\frac{M}{\mu L}\right] = \left[\frac{\rho\lambda^2}{\mu}\right]$$

Group π_1

We have

$$[\phi] = \left[\frac{L}{T}\right] = \left[\frac{\lambda\mu}{\rho\lambda^2}\right] = \left[\frac{\mu}{\rho\lambda}\right]$$

Let

$$\pi_1 = \frac{\phi}{\mu/\rho\lambda} = \left(\frac{\rho\phi\lambda}{\mu}\right)$$

Group π_2

We know

$$[\Delta p] = \left[\frac{M}{LT^2}\right] = \left[\frac{\rho\lambda^3 \cdot \mu^2}{\lambda \cdot \rho^2\lambda^4}\right] = \left[\frac{\mu^2}{\rho\lambda^2}\right]$$

Let

$$\pi_2 = \frac{\Delta p}{\mu^2/\rho\lambda} = \left(\frac{\Delta p \cdot \rho\lambda^2}{\mu^2}\right)$$

Group π_3

$$\pi_3 = \beta$$

We then get

$$\pi_1 = \psi(\pi_2, \pi_3)$$

or

$$\frac{\lambda\phi\rho}{\mu} = \psi\left(\frac{\Delta p\rho\lambda^2}{\mu^2}, \beta\right) \tag{19.34}$$

We can rewrite this equation by changing $(\Delta p \cdot \rho\lambda^2/\mu^2)$ as $\phi = \mu/\rho\lambda$ so $\mu = \rho\lambda\phi$. Therefore,

$$\left(\frac{\Delta p \cdot \rho\lambda^2}{\mu^2}\right) = \frac{\Delta p\rho\lambda^2}{\rho^2\lambda^2\phi^2} = \left(\frac{\Delta p}{\rho\phi^2}\right) \tag{19.35}$$

so

$$\left(\frac{\lambda\phi\rho}{\mu}\right) = \Psi\left\{\left(\frac{\Delta p}{\rho\phi^2}\right), (\beta)\right\} \tag{19.36}$$

The group $(\lambda\phi\rho/\mu)$ is called fluid flow rate, or the filtration number, or the pseudo Reynolds number, where ϕ is thought of as a superficial fluid velocity. The group $(\Delta p \cdot \rho\lambda^2/\mu^2)$ is called the driving force number and can also be written as

$$\frac{\Delta p\rho\lambda^2}{\mu^2} \equiv \left(\frac{\Delta p}{\rho\phi^2}\right)\left(\frac{\lambda\phi\rho}{\mu}\right)^2 \tag{19.37}$$

where $(\Delta p/\rho\phi^2)$ is called the pressure coefficient. The equation (19.34) can also be written in more generalized form as

$$\left(\frac{\lambda\phi\rho}{\mu}\right) = A\left(\frac{\Delta p \cdot \rho\lambda^2}{\mu^2}\right)^a (\beta)^b \tag{19.38}$$

where A, a, and b are the experimental constants.

The following assumptions are included in the above function.

1. The suspension is dilute and has constant consistency.
2. The resulting cake is uniform and incompressible.
3. The filtrate is a clear fluid and can be characterized by its density and viscosity which are functions of temperature only.

4. The cake thickness is directly proportional to the amount of filtrate collected and can therefore be controlled through controlling the amount of filtrate.

5. The mechanical resistance (to filtration) due to the filter medium and the cake thickness is mainly a function of the medium and the filtration pressure and can therefore be fixed by an appropriate choice of the medium and the filtration pressure.

19.9. OTHER APPLICATIONS

Some other applications of dimensional analysis in coatings are in the Federov number

$$d\left[\frac{4g\rho^2}{3\mu^2}\left(\frac{\lambda M}{\lambda g} - 1\right)\right]^{1/3}$$

where

$$d = \text{particle diameter, } l$$

$$g = \text{gravitational acceleration, } l/t^2$$

$$\rho = \text{mass density of fluid, } m/l^3$$

$$\mu = \text{absolute viscosity of fluid, } m/lt$$

$$\lambda M = \text{specific gravity of particles, } 1$$

$$\lambda g = \text{specific gravity of fluid, } 1$$

which is applicable to fluidized beds. The evaporation processes are as follows:

$$\frac{C_p}{H_v\beta}$$

where

$$C_p = \text{specific heat at constant pressure,}$$

$$l^2/t^2T$$

$$H_v = \text{heat of vaporization/mass, } l^2/t^2$$

$$\beta = \text{coefficient of volume expansion, } 1/T$$

and the Marangoni effect (21)

$$\frac{\Delta\sigma}{\Delta T}\frac{\Delta T}{\Delta l}\frac{d^2}{\mu D}$$

where

$$\frac{\Delta\sigma}{\Delta T} = \text{surface tension–temperature coefficient, } m/t^2 T$$

$$\frac{\Delta T}{\Delta l} = \text{temperature–length gradient, } T/l$$

$$d = \text{fluid depth, } l$$

$$\mu = \text{absolute viscosity, } m/lt$$

$$D = \text{thermal diffusivity, } l^2/t$$

which finds application in the theories of antifoam behavior or cellular convection due to surface tension gradients. The Marangoni effect is required for foam stability and must be overcome to destroy foam.

REFERENCES

General References

The first four sections dealing with topics on general dimensional analysis have been prepared using the following:

1. Douglas, J. F., *An Introduction to Dimensional Analysis for Engineers*, Sir Isaac Pitman & Sons Ltd., London, 1969.
2. Langhaar, H. L., *Dimensional Analysis and Theory of Models*, Wiley, New York, 1954.
3. Bridgman, P. W., *Dimensional Analysis*, Yale University Press, New Haven, 1943.
4. Taylor, E. S., *Dimensional Analysis for Engineers*, Clarendon Press, Oxford, 1974.
5. Wood, W. G., and Martin, D. G., *Experimental Method*, The Athlone Press, University of London, London, 1974.
6. Buckingham, E., *On Physically Similar Systems*, Vol. IV, No. 4, pp. 345–376, Second Series Bureau of Standards, June 18, 1914.
7. Buckingham, E., *Trans. Am. Soc. Mech. Eng.*, Vol. 37, 1918, pp. 263–296.
8. Streeter, V. L., *Fluid Mechanics*, 3d ed., McGraw-Hill, New York, 1962, pp. 162–164.
9. Land, N. S., *A Compilation of Nondimensional Numbers*, NASA SP-274, Supr. of Doc. U.S. Government Printing Office, Washington, D.C., No. 3300-0408.

Special References

10. Pierce, J. E., *Chemical Engineering*, pp. 185–190, April 1954.
11. Van Driest, E. R., On dimensional analysis and the presentation of data in fluid flow problems, *Journal of Applied Mechanics*, pp. A34–A40, March 1946.
12. Toms, B. A., Chapter 12, Fundamental techniques: fluids, in *Rheology-Theory and Applications*, Vol. 2, (Erich, F. R., ed.), Academic Press, New York, 1958.

581

13. Patton, T. C., *Paint Flow and Pigment Dispersion*, Wiley–Interscience, New York, 1964.
14. Patton, T. C., Theory of high speed disk impeller dispersion, *Journal of Paint Technology*, Vol. 42, No. 550, pp. 626–635, Nov. 1970.
15. Weltmann, R. N., Chapter 6, Rheology of pastes and paints, in *Rheology-Theory and Applications*, Vol. 3 (Erich, F. R., ed.), Academic Press, New York, 1960, p. 189.
16. Dickey, D. S., and Fenic, J. C., Dimensionless analysis for fluid agitation systems, *Chemical Engineering*, pp. 139–145, Jan. 5, 1976.
17. Fenic, J. G., and Fondy, P. L., *Application of Similarity Analysis to Blending of Miscible Liquids*, Technical Report, Chemineer, Inc., Dayton, Ohio.
18. Dickey, G. D., *Filteration*, Reinhold, New York, 1961.
19. Foust, A. S., Wenzel, L. A., Clump, C. W., Maus, L., and Anderson, L. B., *Principles of Unit Operations*, Wiley, New York, 1967.
20. McCabe, W. L., and Smith, J. C., *Unit Operations in Chemical Engineering*, McGraw-Hill, New York, 1956.
21. Bikerman, J. J., *Physical Surfaces*, Academic Press, 1970.

20

NOMOGRAPHY

ADAM ZANKER

Kiriat—Jam "G"
Israel

The word *nomography* is of Greek origin; it is composed from the word *nomos*, meaning "the law," and from the word *graphos*, meaning "the drawing." Hence, nomography in the widest sense represents graphical techniques of solving mathematical equations. In the wide variety of existing nomographical techniques, we find nomographs in the form of grids, nomographs that are solved by means of parallel movements of a straight edge, nomographs with transparent straight-angle or acute-angle crosses superimposed on the nomograph itself, nomographs that are solved by using a compass, and so on.

However, among the numerous nomographical possibilities, a special type exists. This is the so-called colinearic nomograph, which is almost an exclusive type appearing in recent publications. A colinearic nomograph means that in the basic nomograph given (which represents, as a rule, a mathematical equation with three variables), the values of all three variables solving the given equation lie on a common straight line. Hence, solving such a nomograph is performed by finding known values of the two independent variables on the appropriate scales, connecting these with a straight edge, and reading the third, dependent value (the result) on the intersecting point of this edge with the third scale.

The basic colinearic nomographs, representing a mathematical function with three variables, may occur in a wide variety of geometrical forms. The majority of nomographs designed and published in many technical journals and books consist of three straight scales. These scales may be either parallel, they may form a Z shape, or they may be arranged in the form of three lines intersecting in the same point, or finally, they may be built on three straight lines which form triangles of various shapes. Another useful application for nomographs of mainly parallel or Z scales is in the familiar

583

industrial slide rules showing conversions of units; costing; calculations, for example solids/volume conversions in paint coverage calculations; and similiar applications.

Furthermore there are nomographs with two straight and one curved scale, or with one straight and two curved scales, or with three independent curved scales. Sometimes they even occur in the form of a single curved line, usually a circle, ellipse, or parabola, bearing on itself two scales, the third scale placed on a straight line, which either intersects the curved line, or touches it in a single point, or lies outside this curved line. Finally, in rarely used cases, all three scales of the nomograph may be plotted on one single curved line, (e.g., the case of a Cartesian leaf).

Sometimes, mathematical expressions including more than three variables have to be put into nomographical form. In this case, a number of possibilities exist depending on the form of the expression given. In some cases, a nomograph may be built having a binary grid for two variables with two other variables being plotted on ordinary scales. The quantity of binary grids may even rise to three, thus allowing construction of a nomograph of up to six variables; this possibility is, however, rather rarely used.

In most cases, the mathematical expression containing more than three variables, is, if possible, divided into a number of subexpressions, consisting of three variables each; a number of intermediate variables is added, if necessary, and finally a series of separate, three-scaled nomographs is drawn. Afterwards, these nomographs may be graphically added forming one compound nomograph for solving the total mathematical expression.

20.1. INTRODUCTORY THEORETICAL CONSIDERATIONS OF NOMOGRAPHY

Each function of three variables, u, v, and w, may be expressed in the general algebraic form as follows:

$$F(u, v, w) = 0 \qquad (20.1)$$

This form is valid for a simple expression such as

$$u + v + w = 0 \qquad (20.2)$$

as well as for such a sophisticated one, as, for example

$$\left\{ \left[(\ln u)^2 \right]^{-\tan V} + \arccos\left(\log(u^2)\right) \right\}^{-\sin\sqrt{W}} = 0 \qquad (20.3)$$

Figure 20.1 is a geometrical expression of equation (20.1).

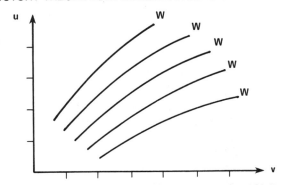

FIGURE 20.1. Geometrical expression of equation (20.1).

In Cartesian coordinates, the variables u and v may be plotted on the appropriate perpendicular axes, x and y, the variable w being represented by a family of curves. However, not every one of these algebraic expressions corresponding to the equation (20.1) may be put into nomographical form, or, in other words, not every one of these expressions is nomographable.

Fortunately, a single and sufficient criterion exists which allows us to determine whether a given equation is nomographable or not. It has been found that each mathematical expression that may be rearranged in the following form of a determinant:

$$\begin{bmatrix} f(u), & \phi(u), & 1 \\ f(v), & \phi(v), & 1 \\ f(w), & \phi(w), & 1 \end{bmatrix} = 0 \qquad (20.4)$$

will be nomographable. The converse is true as well: if it is impossible to construct a determinant of the above mentioned type, the mathematical expression is not nomographable.

In this determinant, the functions $[f(u), \phi(u)]$, $[f(v), \phi(v)]$, and $[f(w), \phi(w)]$ represent the equations of the separate scales for the variables u, v, and w in Cartesian coordinates. However, it is a very complicated task to try to put an algebraic function, given in the form of equation (20.1), into the form of a suitable determinant [equation (20.4)]. A simple and standard method for performing this operation does not exist. Therefore, resigning from the universality of transformation of an equation into the determinant form, we may find other, less universal, but extremely useful criterion which allow us to make a proper judgment concerning the possible nomographability of a given equation.

There exist a number of so-called canonical forms of mathematically nomographable equations. By comparison of an equation in a given

The possible geometrical form of a Nomograph

No.	The Criterion	Mathematical Expression	The simplest form:	Some of other possibilities:
1	First Canonical Form	$f(w) = f(u) + f(v)$ or $f(w) = f(u) - f(v)$	Three parallel lines 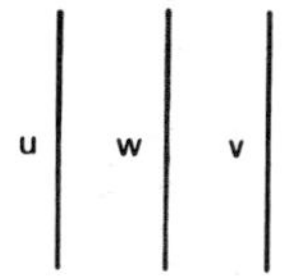	Three lines intersecting in one point 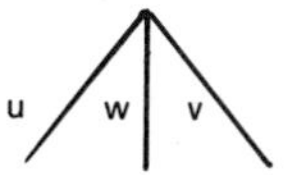 / Circle and Tangential 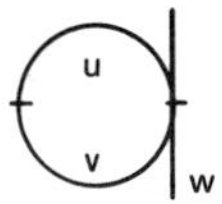
2	Second Canonical Form	$f(w) = f(u)/f(v)$ or $f(w) = f(u) \times f(v)$	"Z" Nomograph 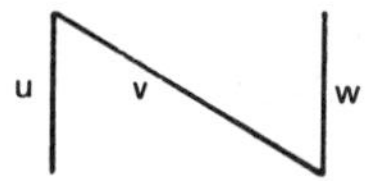	Circle and Diameter 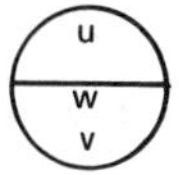
3	Third Canonical Form	$f(w) = \dfrac{f(u) + f(v)}{f(u) \times f(v) - 1}$	Circle and a straight line outside 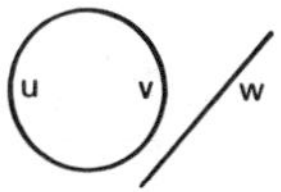	Parabola and a straight line outside 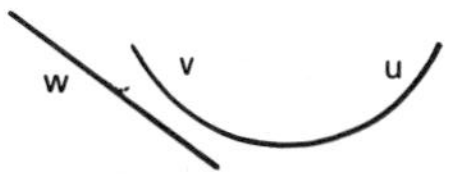
4	Cauchy Expression	$f(u) \times f(v) + f(u) \times f(v) \times \varphi(w) + \psi(w) = 0$	Two parallel lines and one curved line 	Two intersecting lines and one curved line 
5	Clark Expression	$f(u) \times f(v) \times f(w) + [f(u) + f(v)] \times \psi(w) + \psi(w) = 0$	Circle and an intersecting curved line 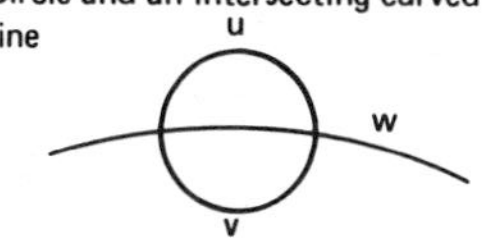	Hyperbola and an intersecting curved line 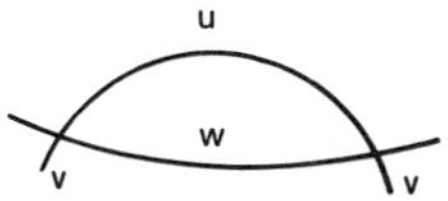

FIGURE 20.2. The possibilities of nomographs.

canonical form, we may find it relatively easy to determine whether a given equation may be transformed into a nomograph, and which form this nomograph will take. Figure 20.2 summarizes these various nomographable possibilities.

In this chapter we shall discuss only such nomographs which correspond to the first and the second canonical form, as well as to the Cauchy criterion. The construction of nomographs according to the third canonical form, and to the Clark criterion, is a little more complicated, rarely used, and therefore will not be described here. Further details may be found in the Bibliography.

20.2. PRACTICAL ELEMENTS OF NOMOGRAPHY

Scales

Each nomograph consists of scales; hence familiarity with the problem of scales has to be the first step in practical nomography. A scale may be defined as a sequence of numerical values, plotted on a linear scale support. A scale support may be either a straight line, or a curved one (see, for example, Figure 20.3). The numerical values may be plotted in any arbitrarily chosen way, provided that they function in an acceptable manner.

In nomographical practice, the scales usually represent a defined mathematical expression. For example, scales representing the following expressions

$$f(u) = ku; \quad f(u) = 10u; \quad f(u) = 0.67u$$

are called regular or uniform scales. All other types of scales are nonuniform or nonregular scales.

The scales representing the expressions

$$f(u) = \log u; \quad f(u) = \ln u; \quad f(u) = 25 \log u$$

are called the logarithmic scales.

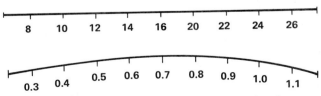

FIGURE 20.3. Examples of straight and curved scales.

In the same manner we may speak about a quadratic scale, $f(u) = u^2$; a square-root scale $f(u) = \sqrt{u}$; a sinusoidal scale $f(u) = \sin u$; and so on.

How To Construct a Scale

To construct a scale on a straight-line support, the following sequence of steps has to be performed:

(a) The length of a scale, L, has to be assumed (the units of length may be arbitrarily chosen, say, centimeter or inch). The beginning and the end of the scale are marked.

(b) The range of a variable, Δu, and the range of a function $\Delta f(u)$, have to be established.

(c) The modulus m of the scale has to be calculated according to the formula

$$m = \frac{L}{\Delta f(u)} \tag{20.5}$$

(d) The final form of the scale function has to be calculated according to the formula:

$$f'(u) = m \cdot f(u) \tag{20.6}$$

(e) A table of numerical values has to be prepared.

(f) The numerical values are plotted on the scale.

Example 20.1. A function is given: $f(u) = 10u$, which has to be put on a scale. The u values vary from $u_1 = 0$ to $u_2 = 8$.

Solution

(a) The length of the scale is assumed to be $L = 12$ cm.

(b) The range of variable u equals $u_2 - u_1$; $8 - 0$; $\Delta u = 8$.

(c) The range of the function is $f(u_2) - f(u_1)$; $80 - 0$; $\Delta f(u) = 80$.

(d) The modulus of the scale equals

$$m = L/\Delta f(u); \; m = 12/80; \; \underline{m = 0.15}.$$

(e) The final equation of the scale is

$$f'(u) = m \cdot f(u); f'(u) = 0.15 \cdot 10u; \underline{f'(u) = 1.5u}.$$

FIGURE 20.4.   Regular scale for the equation $f(u) = 1.5 \times u$.

(f)   Calculating the separate points on the scale according to the final equation: $f'(u) = 1.5u$, we get the following table of numerical values:

u	0	1	2	3	4	5	6	7	8
$f'(u)$	0	1.5	3	4.5	6	7.5	9	10.5	12

(g)   After plotting these numerical values on the scale, we get the complete scale shown in Figure 20.4. This scale is an example of a regular scale.

Example 20.2.   A function is given: $f(u) = u^2$. A scale is to be prepared for this function, where the ranges of u vary from $u_1 = 0$ to $u_2 = 5$.

Solution

(a)   The length of scale L is assumed as $L = 10$ cm.
(b)   The range of variable u is $u_2 - u_1 = 5 - 0 = 5$, and the range of the function $f(u)$ is $u_2^2 - u_1^2$; $5^2 - 0^2$; $\underline{\Delta f(u) = 25}$.
(c)   The modulus equals $m = L/\Delta f(u)$; $m = 10/25$; $\underline{m = 0.4}$.
(d)   The final form of the scale equation is then $f'(u) = m \cdot f(u)$; $\underline{f'(u) = 0.4u^2}$.
(e)   The table of calculated numerical values is

u	0	1	2	3	4	5
$f'(u)$	0	0.4	1.6	3.6	6.4	10

(f)   The final scale is shown in Figure 20.5.

This function is called a quadratic, or square function. It is obvious that this scale is not regular; the distances between $f'(u)$ values increase as the value of u grows. We say that this scale is incomplete, since it requires more points, especially in the higher value range. Therefore, the table of

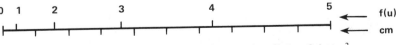

FIGURE 20.5.   Quadratic scale for the equation $f(u) = 0.4 \times u^2$.

numerical values has to be widened, and has to include, for example, the
following additional values:

u	1.5	2.5	3.5	4.2	4.4	4.6	4.8
$f'(u)$	0.9	2.5	4.9	7.056	7.744	8.464	9.216

The scale with these additional values added is shown in Figure 20.6.

When still more points on the scale are desired, this may be achieved in
two ways:

(a) By calculating some additional values of $f'(u)$.
(b) By plotting some points by means of "intuitive" interpolation,
    between the already existing points.

The "dense" scale, after this operation, is shown in Figure 20.7.

Example 20.3. A function is given: $f(u) = 1/u$. The scale for this function is called a "reciprocal" scale.

We assume the length of this scale as $L = 10$ cm, and the range of
variable u as $1 < u < 10$. Hence $1 > 1/u > 0.1$, $\Delta f(u) = 0.9$, and the
modulus is $m = 10/0.9 = 11.1111$. We may round the modulus, and fix it
as $\underline{m = 11}$.

The final equation is then $f'(u) = 11/u$, and the table of numerical
values looks as follows:

u	1	2	3	4	5	6	7	8	9	10
$f'(u)$	11	5.5	3.66	2.75	2.2	1.83	1.57	1.38	1.22	1.1

The final scale is shown in Figure 20.8.

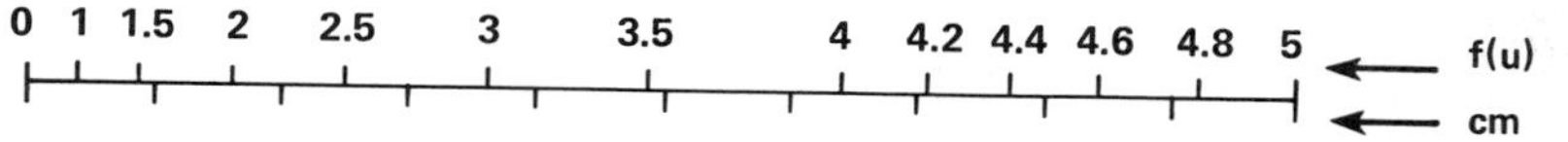

FIGURE 20.6. Quadratic scale with halves of values included $f(u) = 0.4 \times u^2$.

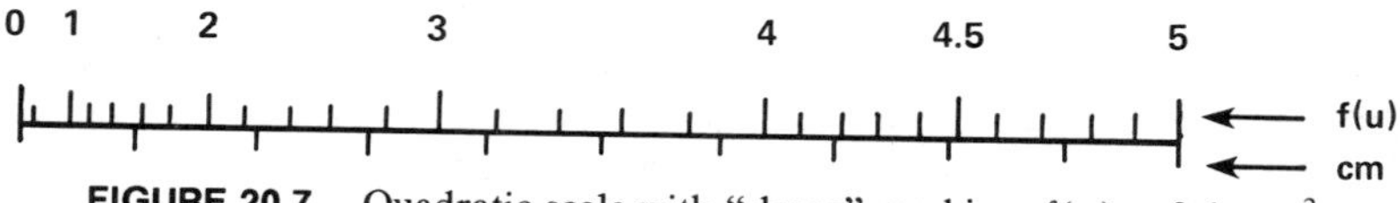

FIGURE 20.7. Quadratic scale with "dense" marking, $f(u) = 0.4 \times u^2$.

FIGURE 20.8. Reciprocal scale for the equation $f(u) = 11 \times 1/u$.

FIGURE 20.9. Logarithmic scale for the equation $f(u) = 15 \times \log u$.

This scale is not particularly convenient but nevertheless is sometimes used in nomographic practice.

Example 20.4. A function is given: $f(u) = \log(u)$. The range of u varies from $u = 1$ to $u = 10$. Obviously no zero value may ever be included in the logarithmic scale.

Solution

(a) The length of scale is assumed as $L = 15$ cm.
(b) The range of variables u is $u_2 - u_1 = 10 - 1 = 9$; the range of function $f(u) = \log u_2 - \log u_1 = 1 - 0 = 1$; $\Delta f(u) = 1$.
(c) The modulus m equals

$$m = L/\Delta f(u); \qquad m = 15/1; \qquad \underline{m = 15}$$

(d) The equation of scale equals

$$f'(u) = m \cdot f(u); \qquad \underline{f'(u) = 15 \log u}$$

(e) The table of numerical values equals

u	1	2	3	4	5	6	7	8	9	10
$f'(u)$	0	4.51	7.16	9.03	10.48	11.67	12.68	13.55	14.31	15

(f) The final scale is shown in Figure 20.9.

This is a so-called "logarithmic scale" and it has the utmost importance in nomographical practice.

Scales on a Cartesian Plane

The scales that have been discussed above may of course be put on a Cartesian plane. For greater convenience it is assumed that the straight-line scales shall be placed vertically, that is, on the y axis, or parallel to it (see Figure 20.10).

On the Cartesian plane, each scale has to be defined by its coordinates x and y. The vertical scale, overlapping the y axis, has the following equa-

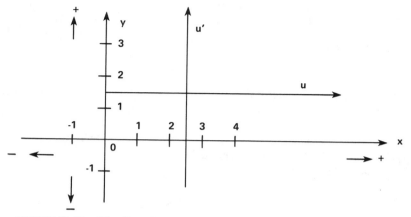

FIGURE 20.10.   The Cartesian plane with possible situation of scales for u, u'.

tions: $x = 0$, $y = f'(u)$. The vertical scale, parallel to the y axis, and intersecting the x axis at the point, say at $x = 4$, has the set of equations as follows: $x = 4$, $y = f'(u)$. If such a scale has to be put in the horizontal position, and say, it has to be superimposed on the x axis, the set of equations will be, of course, $y = 0$, $x = f'(u)$.

The case is more complicated when a scale is curved relative to the x–y coordinates. In this case, the equation of a scale has to be described by a set of two functions, when neither of these is zero, or constant (Figure 20.11).

The set of equations then looks generally as follows:

$$x = f'(u), \qquad y = f''(u) \tag{20.7}$$

The methods of formulating such a set of equations shall be discussed in detail further on in this chapter.

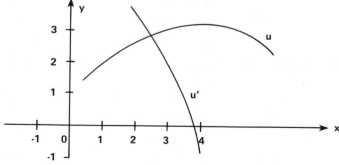

FIGURE 20.11.   The Cartesian plane with possible situation of curved scales for u, u'.

Double-Graduated Scales

It is common practice to put certain relationships either in a form of a graph or in a form of a table of numerical values. For example, tables exist for mutual conversion from degrees Fahrenheit to Celsius, from oil densities to API gravities, from efflux cup times to viscosity in centipoise, and so on. There are also graphs representing, say, the relationship between vehicle velocity and braking distance, between viscosity and temperature, and so on. All of these cases may be conveniently put on the so-called double-graduated scales, which are extremely useful for nomographic techniques.

We take as an example a typical viscosity/temperature relationship. One graph, illustrating this relationship, looks similar to Figure 20.12. (See John H. Perry, *Chemical Engineer's Handbook*, Ref. 11, pp. 3–228). The use of this graph is obvious: one draws a vertical line from the temperature axis up to the intersection with the curve and then draws a horizontal line to the intersection with the viscosity axis, where the desired result is read.

If we repeat this operation frequently, performing it each time for another temperature value, we shall obtain a dense set of temperature points, *projected* on the existing viscosity scale. Hence, the double-graduated scale may be drawn as in Figure 20.13.

In a somewhat similar manner one may obtain the double-graduated scale for, say, Fahrenheit/Celsius conversion. Here we take a desired range of Fahrenheit degrees, say between $t_1 = 0°F$ to $t_2 = 300°F$, and plot it in a regular way on a scale with the assumed length. When this scale is ready, one takes the conversion Fahrenheit/Celsius table, and plots the corre-

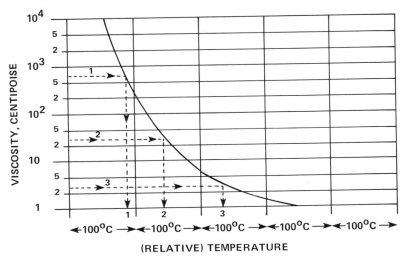

FIGURE 20.12. Projection of viscosity values on the temperature scale.

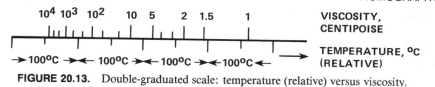

FIGURE 20.13.   Double-graduated scale: temperature (relative) versus viscosity.

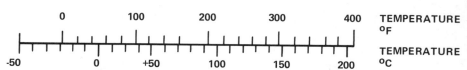

FIGURE 20.14.   Double-graduated scale for temperature conversion.

sponding values of degrees Celsius on the *opposite* side of the just prepared Fahrenheit scale. After this operation, the double-graduated scale looks as shown in Fig. 20.14.

Practical Hints

It is good practice to plot the scales in such a way that "round" values are provided with numbers, and the "fractional" values with marking lines. It is advisable to use the marking lines with two different lengths: the longer ones for the "round" values and the shorter ones for the intermediate values.

When using regular scales, "round" values are preferable to the even numbers, or numbers divided by 5 or by 10, or by 100, depending on the length of the scale.

For nonuniform scales (such as logarithmic, square, square root, etc.) no exact rules can be established. Frequently, working logic will indicate the proper method of scale marking, as well as the observation of the marking techniques as employed in numerous published nomographs.

20.3.   NOMOGRAPHS WITH STRAIGHT SCALES

Nomographs with Three Parallel Straight Scales

This type of nomograph (Figure 20.15) is the simplest and the most frequently used. It solves the first canonical form of the equations listed in Figure 20.2.

This equation appears as follows:

$$f(w) = f(u) + f(v) \tag{20.8}$$

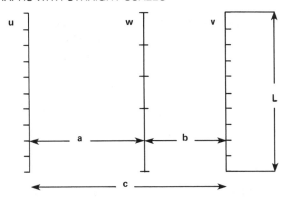

FIGURE 20.15. The dimensions of parallel-scale nomographs.

where $f(u)$ and $f(v)$ are the independent variables, and $f(w)$ is the *dependent* variable, that is, the value to be found.

It is good practice (although one not always possible to fulfill) to place the scale for the dependent variable, $f(w)$, in the *middle* of the nomograph, between the scales of $f(u)$ and $f(v)$.

We shall construct such a nomograph in its simplest possible form, assuming that

$$f(w) = w; \qquad f(u) = u; \qquad f(v) = v$$

Hence, the equation to be converted to a nomograph is as follows:

$$w = u + v$$

The order of performing this task consists of the following sequence of steps:

(a)   The length (L) of scales for u and v and the distance between these scales, c, have to be assumed. It is assumed as well that the u and v scales shall be the side ones, the resulting w scale becoming the central one.

(b)   The range of variables u and v has to be assumed.

(c)   The moduli, m_u and m_v have to be calculated. (Note: These moduli, m_u and m_v may be rounded for greater convenience.)

(d)   The modulus of the central, w, scale has to be calculated. (Note: This modulus may not be rounded; it has to be used exactly as calculated.)

(e)   The final functions for u, v, and w are chosen [$f'(u)$, $f'(v)$, and $f'(w)$].

(f)   The table of numerical values of $f'(u)$, $f'(v)$, and $f'(w)$ has to be computed.

(g) The distances, a and b, between the scales have to be found.
(h) The values for u and v have to be plotted on both side scales.
(i) One point on the w scale is now found, and from this point other points plotted.
(j) The finished nomograph is checked for accuracy.

Example 20.5.   Draw a nomograph for the equation

$$w = u + v$$

Solution

(a) The height of the scales equals $L = 15$ cm. The distance between side scales $c = 10$ cm.
(b) The range of variables is $0 < u < 10$; $0 < v < 20$.
(c) The moduli are $m_u = 15/10$; $\underline{m_u = 1.5}$; $m_v = 15/20$; $\underline{m_v = 0.75}$.
(d) The modulus of the central, w scale is calculated according to the following equation:

$$m_w = \frac{m_u \times m_v}{m_u + m_v} \tag{20.9}$$

$$m_w = \frac{1.5 \times 0.75}{1.5 + 0.75} = \frac{1.125}{2.25}; \qquad m_w = 0.5$$

(e) The final forms of functions u, v, w are

$$f'(u) = 1.5u; \qquad f'(v) = 0.75v; \qquad f'(w) = 0.5w$$

(f) The numerical values of the scales are

u	0	2	4	6	8	10
$f'(u)$	0	3	6	9	12	15

v	0	2	4	6	8	10	12	14	16	18	20
$f'(v)$	0	1.5	3	4.5	6	7.5	9	10.5	12	13.5	15

w	0	5	10	15	20	25	30
$f'(w)$	0	2.5	5	7.5	10	12.5	15

(g) The distances a and b are calculated as follows:

$$\frac{m_u}{a} = \frac{(m_u + m_v)}{a + b} \tag{20.10}$$

After rearranging this equation, and substituting $a + b = c$, we get the following expression:

$$a = \frac{m_u \times c}{m_u + m_v} \tag{20.11}$$

In our case, substituting the calculated numerical values we get

$$a = \frac{1.5 \times 10}{1.5 + 0.75}; \qquad a = 6.6667$$

and

$$b = c - a$$

The scale for w is then drawn at a distance of 6.6667 cm from the left, u scale.

(h) The values of $f'(u)$ and $f'(v)$ have to be plotted on both side scales for u and v. The direction for the ascending numbers has to be *identical* in both side scales (i.e., both scales pointing upward, or both scales pointing downward).

(i) Assuming $u = 5$ and $v = 10$, we find $w = u + v$; $w = 15$. With a ruler, we connect the values of $u = 5$ and $v = 10$ on both side scales and we find on the intersection point of the central, w scale, the point where $w = 15$. Taking this point as a guide, we plot the values of $f'(w)$ on the central, w scale, maintaining the direction of ascending w numbers *identical* to those of the u and v scales.

(j) After checking a number of additional points for accuracy, the nomograph is ready for use.

Figure 20.16 illustrates another worked out solution.

Example 20.6. The following equation is to be put into nomographic form:

$$w = \frac{3u}{1 + v}$$

Taking logarithms,

$$\log w = \log(3u) - \log(1 + v)$$

In this form, this equation conforms to the first canonical form of the nomographic equations and may be nomographed similarly to Example 20.5.

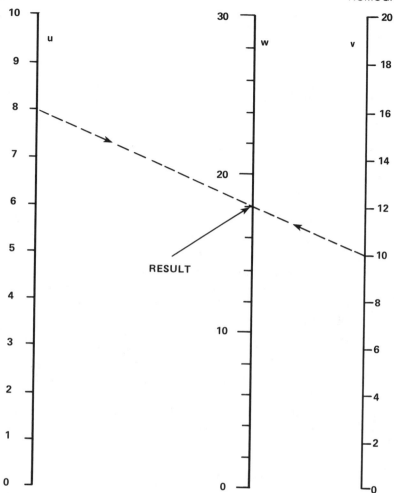

FIGURE 20.16. Nomograph from three parallel scales for the equation $w = u + v$.

The construction of the nomograph is as follows:

(a) The height of scales $L = 15$ cm; the distance between side scales $c = 10$ cm.
(b) The range of variables is $1 < u < 10$, $1 < v < 10$.
 The range of functions is

$$f(u): \quad \log(3) < \log(3u) < \log(30)$$
$$0.477 < \log(3u) < 1.477$$

hence, the difference, $\Delta f(u)$ equals $f(u_2) - f(u_1)$; and $\underline{\Delta f(u) = 1}$.

$$f(v): \qquad \log(2) < \log(1 + v) < \log(11)$$
$$0.301 < \log(1 + v) < 1.0414$$

and $\underline{\Delta f(v) = 0.7407}$.

(c) The moduli are

$$m_u = \frac{15}{1}; \qquad m_u = 15$$

$$m_v = \frac{15}{0.7407}; \qquad m_v = 20.25 \text{ (assumed: } m_v = 20)$$

(d) The modulus of the central scale is

$$m_w = \frac{m_v \times m_u}{m_v + m_u} = \frac{20 \times 15}{20 + 15} = \frac{300}{35} = 8.57$$

(e) The final forms of the functions are

$$f'(u) = 15 \log(3u); \qquad f'(v) = 20 \log(1 + v); \qquad f'(w) = 8.57 \log(w)$$

(f) The selected numerical values for separate scales are

u	1	2	4		6	8	10
$f(u)$	7.16	11.67	16.19		18.83	20.7	22.16

v	1	2	4		6	8	10
$f'(v)$	6.02	9.54	13.98		16.9	19.08	20.83

w	0.3	0.5	1	2	5	10	15
$f'(w)$	-4.48	-2.58	0	$+2.58$	5.99	8.57	10.08

(g) The distance a equals

$$a = \frac{m_u \times c}{m_u + m_v} = \frac{15 \times 10}{15 + 20} = \frac{150}{35}; \qquad a = 4.29$$

(h) The direction of scales for $f'(u)$ and $f'(v)$ have to be *reciprocal* since the functions for u and v are connected with the sign $(-)$ minus, and the direction of scale w is *identical* with that of scale u. The values for u and v have to be plotted on the appropriate scales.

(i)   Assuming $u = 5$ and $v = 5$, we compute the value of w for this case:

$$w = \frac{3u}{1 + v}; \qquad w = \frac{3 \times 5}{1 + 5} = \frac{15}{6} = 2.5$$

We calculate the dimension of this value, using the equation for $f'(w)$:

$$f'(w) = 8.57 \times \log w; \qquad f'(2.5) = 8.57 \times \log(2.5)$$
$$f'(2.5) = 3.41$$

(j)   After connecting the values of $u = 5$ and $v = 5$ on the appropriate scales (see Figure 20.17), we find the intersecting point of the ruler

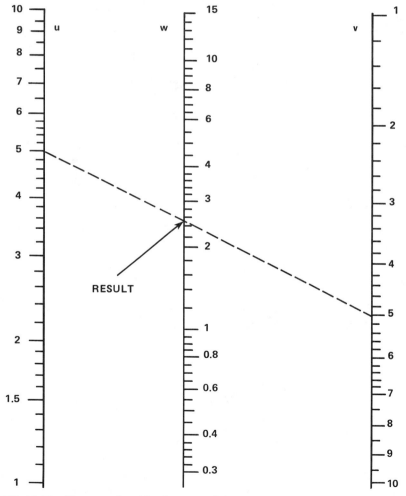

FIGURE 20.17. Nomograph with three parallel scales (logarithmic) for the equation $w = 3u/1 + v$.

with the w scale. The value of this point is $w = 2.5$, with a dimension of 3.41. After fixing this point and using the table of $f'(w)$ values, we plot the remaining values of the w scale. After plotting some intermediate values on the scales, the nomograph is ready for use.

Practical Hints

The first canonical form of the nomograph equation, in its simplest variant, may be written in two forms;

$$w = u + v \quad \text{and} \quad w = u - v$$

There are four possible configurations for the scales for each variant as shown in Figure 20.18.

As already mentioned the best practice is to maintain the scales in such a way that the resulting w scale is placed between the independent variable scales, u and v. Since, in numerous cases, logarithmic scales are in use, it is recommended practice that instead of repeated calculations of scale points the so-called "logarithmic harp" be used. This "harp" is shown in Figure 20.19. The vertical lines contain the number which represents the length of the scale (for the one decade range from 1 to 10) in centimeters. This measure represents as well the modulus of the logarithmic scale. By folding this scale along a chosen modulus number, we may directly copy its points on the constructed scale.

Nomographs with Scales Intersecting in One Point (The "Angle" Nomograph)

A special case, covered also by the first canonical form, is the equation for the sum of the reciprocals.

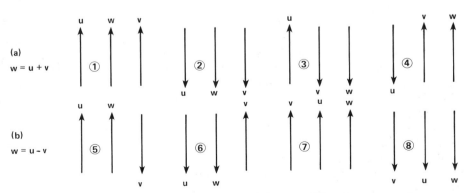

FIGURE 20.18. Possible configuration of three parallel scales nomograph.

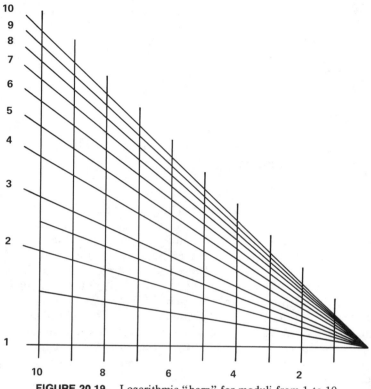

FIGURE 20.19. Logarithmic "harp" for moduli from 1 to 10.

The typical equation looks as follows:

$$\frac{1}{f'(w)} = \frac{1}{f'(u)} + \frac{1}{f'(v)}$$

(20.12)

and the typical nomograph is shown in Figure 20.20.

The important advantage of this form is that all three scales, u, v, and w, are the regular ones as previously defined. When the angles between scales α and β are chosen to be 60° each, and the moduli of scales u and v are equal, the modulus of the w scale is exactly double that of u and v. The zero point of all three scales is placed at the common intersecting point. If the angle $(\alpha + \beta)$ differs from 120°, and the moduli of the u and v scales are not the same, then the angles α and β are not identical; also the modulus of the central w scale does not equal the sum of the other moduli m_u and m_v.

An algebraical method exists for the calculation of the α, β, and m_w values. However, there is a graphical method to arrive at these values, without resorting to tedious calculations. This method follows.

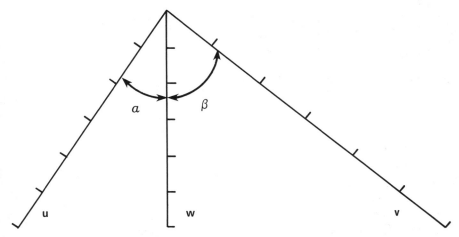

FIGURE 20.20. Schematic of an "angle" nomograph.

In Figure 20.21, u and v scales are drawn at an arbitrary angle and with arbitrary scale lengths. After plotting the values of u and v on both side scales (the zero mark is common for both scales), we choose two identical pairs of u and v values, for example;

$$\text{Pair 1:} \qquad u = 10 \qquad v = 6$$

$$\text{Pair 2:} \qquad u = 6 \qquad v = 10$$

We connect these pairs with a ruler. The two lines formed intersect in point (A).

Point (A) has to be connected with the common intersecting point with a straight edge, and this line forms the *scale support* for the variable w.

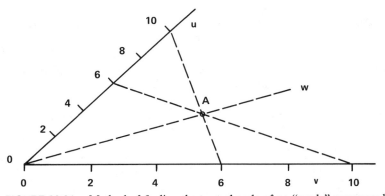

FIGURE 20.21. Method of finding the central scale of an "angle" nomograph.

Further, we calculate the w value for this particular point A. We measure the distance from this point to zero, we calculate the modulus m_w, and finally, we plot the w values on this scale.

Example 20.7. Construct a nomograph for the equation

$$\frac{1}{w} = \frac{1}{u} + \frac{1}{v}$$

when the range of variables is

$$0 < u < 20; \qquad 0 < v < 10$$

Solution. We arbitrarily choose the length of scales u and v as $L_u = L_v = 12$ and we choose the angle between the u and v scales as $90°$ (both scales perpendicular).

We calculate the moduli of scales for u and v:

$$m_u = \frac{12}{20} = 0.6; \qquad m_v = \frac{12}{10} = 1.2$$

We plot the values of u and v on both scales according to their moduli.

The chosen pair of values for the calculation of m_w has to be as far from the zero point as possible. We will choose the pairs $u = 10$, $v = 8$, and $u = 8$, $v = 10$.

We connect these two pairs of values with a ruler, and find the intersecting point. We connect this point with the zero point of the u and v scales, and we get the scale support for w.

We calculate the value of w at this intersecting point:

$$\frac{1}{w} = \frac{1}{8} + \frac{1}{10}; \qquad \frac{1}{w} = \frac{10}{80} + \frac{8}{80} = \frac{18}{80}; \qquad w = \frac{80}{18} = 4.444$$

The distance measured from point $w = 4.444$ to the common zero point of the nomograph, L_w, equals 5.93 cm, hence the modulus of the w scale equals $M_w = 5.93/4.444$, $m_w = 1.334$.

According to this modulus, we may now plot the w scale on its scale support. The final nomograph is presented in Figure 20.22.

Example 20.8. Owing to the fact that the scales in the nomograph of

$$\frac{1}{f(w)} = \frac{1}{f(u)} + \frac{1}{f(v)} \tag{20.12a}$$

type are regular, it is sometimes worthwhile to perform some algebraical

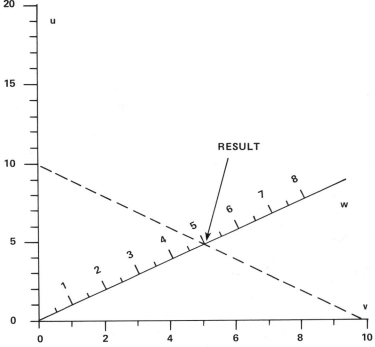

FIGURE 20.22.   The "angle" nomograph for the sum of reciprocals.

operations in order to bring an equation to be nomographed into this particular form.

We take the following equation:

$$2 \log R \left(\frac{Q}{4} + \frac{3}{4} P^2 - 1 \right) = 3P^2 (Q - 4) \qquad (20.13)$$

We may write it in this sequence of forms:

$$\frac{\log R}{2} \left[(Q - 4) + 3P^2 \right] = (Q - 4) \times 3P^2 \qquad (20.14)$$

$$\frac{\log R}{2} (Q - 4) + \frac{\log R}{2} \times 3P^2 = (Q - 4) \times 3P^2 \qquad (20.15)$$

$$\frac{\log R}{2} \times \frac{(Q - 4)}{3P^2} + \frac{\log R}{2} = (Q - 4) \qquad (20.16)$$

$$\frac{\log R}{2} \times \frac{1}{3P^2} + \frac{\log R}{2} \times \frac{1}{(Q - 4)} = 1 \qquad (20.17)$$

$$\frac{1}{3P^2} + \frac{1}{(Q - 4)} = \frac{1}{(\log R)/2} \qquad (20.18)$$

If we assume that $f(w) = (\log R)/2$, $f(u) = 3P^2$ and $f(v) = (Q - 4)$, we finally get the basic form of the desired equation:

$$\frac{1}{f(w)} = \frac{1}{f(u)} + \frac{1}{f(v)}$$

Further steps are exactly the same, as in Example 20.7. An additional step has to be added after fulfilling this task, namely, the regular scales for $f(u), f(v)$, and $f(w)$ have to be replaced by the proper values, P, Q, and R according to the previously prepared tables of numerical values found for each of the separate relationships between $f(w)$ and R, $f(u)$ and P, and $f(v)$ and Q.

Nomographs with Two Parallel Scales and a Third Slanting Line (Z or N Nomographs)

This type of nomograph is characteristic of the second canonical form of nomographable equations:

$$f(w) = \frac{f(u)}{f(v)} \tag{20.19}$$

or

$$f(w) = f(u) \cdot f(v) \tag{20.20}$$

As we have shown previously, a logarithmic transformation of this form converts it to the form of the sums of logarithms, and hence to the first canonical form. However, in many cases, it is preferable to stay with the original form and develop it without change.

The typical nomograph of this type consists of three scales for $f(u), f(v)$ and $f(w)$. The scales for $f(u)$ and $f(w)$ may be regular; and they run in opposite directions. The zero point of the $f(v)$ scale is common with the zero of the $f(w)$ scale and the $f(v)$ scale is *not* a regular one. It starts with the value of $f(v) = 0$ at one end and ends with the value of $f(v) = \infty$ at the opposite end (Figure 20.23).

The plotting of scales u and w on the appropriate scale supports is performed in exactly the same manner as the previously described cases, that is,

$$f'(u) = m_u \times f(u); \qquad f'(w) = m_w \times f(w)$$

In the case where the scale for $f(u)$ overlaps the y axis on the Cartesian plane, the equation for the coordinate x of the oblique $f(v)$ scale is as

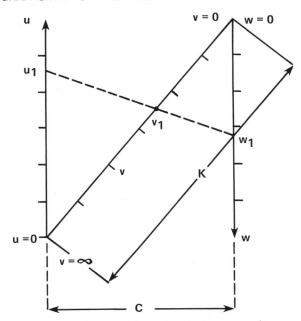

FIGURE 20.23. Schematic of a Z nomograph.

follows:

$$X_v = \frac{K \times m_u}{m_u + m_w \times f(v)} \tag{20.21}$$

where K is the total length of the $f(v)$ scale from the zero points of the u and w scales.

There are two possible cases with the Z nomograph. Case I exists when the two zero points of the u and w scales are of interest; in this case, the entire v scale is drawn in the body of the nomograph as in Figure 20.24.

Another, case II, exists when the zero values of either the u or w scale (or of both scales) are of no interest, and they lie outside the body of the nomograph. In this case only a part of the v scale is drawn on the nomograph, for example, $(v_2 - v_1), (u_2 - u_1), (w_1 - w_2)$ in Figure 20.25.

Example 20.9. A nomograph is to be designed for the equation

$$w = u \times v$$

where the range of variables is

$$0 < u < 10$$
$$0 < v < 5$$

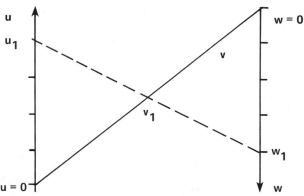

FIGURE 20.24. Z nomograph with valid values of $u = 0$ and $w = 0$ and with the whole v scale.

Solution. Since both zero values of the variables u and v are of interest, this example corresponds to case I.

We assume that the length of scales u and w shall be

$$L_u = L_w = 10 \text{ cm}$$

We calculate the range of the w variable:

$$w_{\min} = u_{\min} \times v_{\min} = 0 \times 0 = 0$$
$$w_{\max} = u_{\max} \times v_{\max} = 10 \times 5 = 50$$

The moduli of u and w scales are then

$$m_u = \tfrac{10}{10} = 1; \qquad m_w = \tfrac{10}{50} = 0.2$$

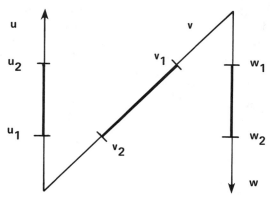

FIGURE 20.25. Z nomograph where only part of scales u, v, w are of interest.

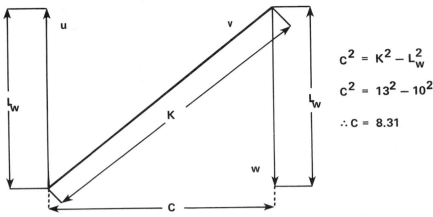

$$c^2 = K^2 - L_w^2$$

$$c^2 = 13^2 - 10^2$$

$$\therefore C = 8.31$$

FIGURE 20.26. Calculation of the distance C between parallel scales of Z nomograph.

We assume that the length K shall be $K = 13$ cm (see Figure 20.23). From the Pythagorean theorem, we find the horizontal distance, c, between the u and w scales (see Figure 20.26).

We then draw the nomograph according to the calculated dimensions, plotting the scales for u and w, and we start to calculate the X values for the oblique v scale, according to the equation

$$X_v = \frac{K \times m_u}{m_u + m_w \times f(v)} \qquad (20.22)$$

After substituting $K = 13$, $m_u = 1$, $m_w = 0.2$, we get the expression for X_v:

$$X_v = \frac{13 \times 1}{1 + 0.2 \times f(v)} = \frac{13}{1 + 0.2 \times v}$$

We prepare a table of variables for X_v according to the following expression:

v	0	0.5	1	1.5	2	2.5	3	3.5	4	4.5	5
X_v	13	11.82	10.93	10.0	9.28	8.67	8.13	7.64	7.22	6.84	6.5

The scale for w has to be graduated in centimeters (auxiliary graduation) and afterward the values from the table shown have to be plotted. The final nomograph is shown in Figure 20.27.

In many cases, especially when the functions u, v, and w are not very complicated, the step of calculating the v values may be omitted, and the appropriate scale points may be found graphically.

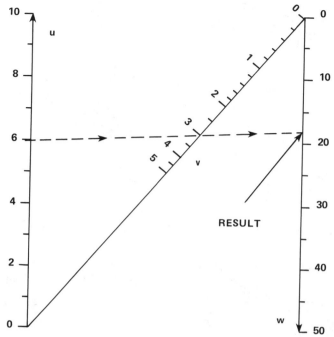

FIGURE 20.27. Z nomograph for the equation $w = u \times v$.

For example, in the calculation shown above, when the u and w scales are already plotted, we can perform the appropriate calculations mentally. We then connect the suitable points on the u and w scales with a ruler and, by simple intersections, we mark them on the oblique, v scale.

For example, we would like to find graphically the location of $v = 1$. We know that $w = u \times v$. We assume that $u = 10$ and $w = 10$, we connect these values on the u and w scales, and the ruler intersects the v scale exactly at the point $v = 1$.

Similarly, we assume that $u = 10$ and $w = 20$, and in this way we mark the point of $v = 2$, and so on.

Example 20.10. We want to construct a Z nomograph for the equation $w = u \times v$, when the range of variables is as follows:

$$100 < u < 150 \qquad \Delta u = 50$$
$$2 < v < 3 \qquad \Delta v = 1$$

We calculate the range of variable w:

$$w_{min} = u_{min} \times v_{min} = 2 \times 100 = 200$$
$$w_{max} = u_{max} \times v_{max} = 3 \times 150 = 450$$

Hence $200 < w < 450$; $\Delta w = 250$.

Since both zero marks for the u and w scales are outside the nomograph, case II is now under consideration.

Solution. First we calculate the moduli of the u and w scales, assuming that the length of those scales shall be $L_u = L_w = 10$ cm.

The moduli are

$$m_u = \tfrac{10}{50} = 0.2$$
$$m_v = \tfrac{10}{250} = 0.04$$

The dimensions of the extreme points of the v and w scales may be calculated according to the formulas

$$f(u) = 0.2 \times u; \qquad f(w) = 0.04 \times w$$

Hence,

$$f(u = 100) = 0.2 \times 100 = 20; \qquad f(w = 200) = 0.04 \times 200 = 8$$
$$f(u = 150) = 0.2 \times 150 = 30; \qquad f(w = 450) = 0.04 \times 450 = 18$$
$$f(u = 0) = 0.2 \times 0 = 0; \qquad f(w = 0) = 0.04 \times 0 = 0$$

We assume a distance between the vertical u and w scales of $c = 8$ cm, and that the lowest value of v shall lay exactly opposite the highest w value, and vice versa.

Figure 20.28 is a plot of these values; the parts of the u and w scales which are of interest are marked with brackets.

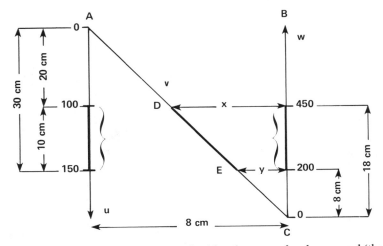

FIGURE 20.28. Construction of Z nomograph with only a part of scales engaged (the sketch is not to scale).

We take into account the triangle ABC, and we calculate its dimensions. The side AB equals 8 cm (from definition). The side BC is composed of the section of 18 cm (from the w scale) and the section of 20 cm (from the u scale); therefore the total length of BC equals 38 cm.

According to the theory of right triangles, we may build the following proportions on the triangle ABC:

$$\frac{38}{8} = \frac{18}{x} \quad \text{and} \quad \frac{38}{8} = \frac{8}{y}$$

which give

$$x = 3.79 \text{ cm}, \qquad y = 1.68 \text{ cm}$$

Now, when all the necessary geometrical data are collected, we can build a rectangle with the dimensions 8×10 cm. On the upper edge we put the point x at the distance 3.79 cm from the w scale; on the lower edge point y is set at the distance 1.68 cm from the w scale. We draw an oblique line, connecting these points x and y, and we get the part of interest of the scale support for the variable v.

Afterward, we plot the scales for u and w, and we find the suitable points of v values on the central, oblique v scale, using the geometrical method mentioned previously in Example 20.9. The final nomograph is presented in Figure 20.29.

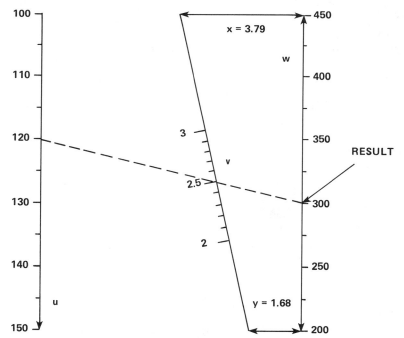

FIGURE 20.29. The finished Z nomograph for the equation $w = u \times v$ with limited range of variables and only part of the scales engaged.

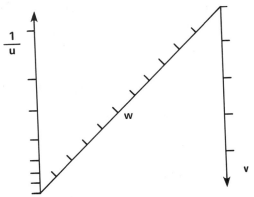

FIGURE 20.30. The Z nomograph with the resulting w scale as a central one, and the u scale as a reciprocal.

As stated previously, the recommended practice is to maintain the resulting scale *between* the scales of independent variables. In the cases just shown, this condition has not been fulfilled, since the resulting w scale has been placed as a side scale, and not as a central one.

This disadvantage may be overcome as follows. The equation $w = u \times v$ may be rearranged

$$v = w \times \frac{1}{u} \qquad (20.23)$$

and the nomograph looks as in Figure 20.30.

In this case, we have succeeded in placing the resulting w scale as a central one, which is a desired step. However, to do this, one of the side scales, namely the u scale, has to be plotted according to its reciprocal values. This is often undesirable, since this type of scale is extremely dense for higher u values, with superfluous accuracy in the low range of u values (e.g., Example 20.3).

20.4. NOMOGRAPHS WITH TWO STRAIGHT AND ONE CURVED SCALE

Cauchy Expression

The Cauchy expression for the fourth nomographable form of equation looks as follows:

$$f(u) \times f(w) + f(v) \times \psi(w) + \phi(w) = 0 \qquad (20.24)$$

Dividing both sides by the $\phi(w)$ function, we get the following form:

$$f(u) \times \frac{f(w)}{\phi(w)} + f(v) + \frac{\psi(w)}{\phi(w)} = 0 \qquad (20.25)$$

Substituting

$$f_1(w) = \frac{f(w)}{\phi(w)} \quad \text{and} \quad f_2(w) = \frac{\psi(w)}{\phi(w)}$$

we get the simplest form of the Cauchy expression:

$$f(u) \times f_1(w) + f(v) = f_2(w) \qquad (20.26)$$

or

$$f(v) = f(u) \times f_1(w) + f_2(w) \qquad (20.27)$$

For example, an equation such as this

$$v = u \times w + w^2 \qquad (20.28)$$

corresponds to the simplest form of Cauchy formula and may be put into the nomographic form when the scales for v and u are regular, and the scale for w is curved (see Figure 20.31).

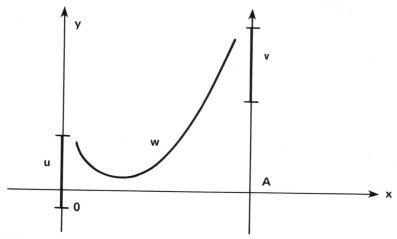

FIGURE 20.31. The Cauchy nomograph with the unfitted u, v, and w scales on a rectangular Cartesian plane.

Nomographs of this kind are relatively frequently used, although their construction is rather more complicated than that of the straight-line types, shown previously.

When the direction of both straight scales, u and v, is identical, the curved scale for w is outside these lines; when their directions are opposite, the curved scale is between these straight scales. An initial analysis of the equation given offers us an indication of which of the scales shall be a resulting one, and which one shall be placed as a central scale.

To construct a nomograph of this type, it is placed on the Cartesian plane in such a way that the left straight scale overlaps with the y axis; their zero point is situated at the intersecting point of the x and y axes.

When we denote as K the horizontal distance between the vertical, straight scales, and the moduli of scales for u and v are m_u and m_v, respectively, the equations for x and y coordinates describing the curved w scale are as follows:

$$X_w = \frac{K \times m_u \times f_1(w)}{m_v + m_u \times f_1(w)} \tag{20.29}$$

$$Y_w = \frac{m_u \times m_v \times f_2(w)}{m_v + m_u \times f_1(w)} \tag{20.30}$$

The sequence of steps in the construction of a nomograph of this type is as follows:

(a) We assume the range of variables u, v, w, and the length of scales u and v.
(b) We calculate the moduli m_u and m_v of the straight scales.
(c) We assume a suitable K value.
(d) We calculate a number of values of X_w and Y_w and we plot these values on the Cartesian plane.

In some cases, this nomograph is ready to use after these steps.

Cartesian Coordinates

When normal Cartesian coordinates with angles of $90°$ are used, it often occurs that the straight scales and the curved scale do not form a compact composition, and that the scales are not situated opposite one another. Such a situation is shown in Figure 20.32.

In this case rotation of the x axis around the zero point is necessary. The rotation is at such an angle that the appropriate ends of the u and v scales are situated exactly opposite one another. It looks graphically as in Figure 20.32.

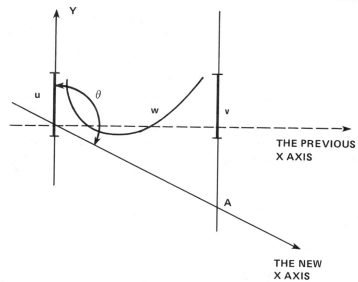

FIGURE 20.32. The Cauchy nomograph with fitting scales after the rotation of Cartesian plane.

When the appropriate angle θ is found, the new set of x_w and y_w coordinates has to be calculated, according to the following equations:

$$x'_w = x_w \times \sin \theta \qquad (20.31)$$

$$y'_w = y_w + x_w \times \cos \theta \qquad (20.32)$$

And finally, we calculate a sufficient number of points of the w scale according to these equations in order to ascertain the possibility of drawing a smooth curve. The long and tedious way described above is the "orthodox" method for drawing a nomograph of this type. However, it is not the only way to perform this task. By using a good millimeter graph paper with large dimensions, a precise ruler and a sharp pencil, it is possible to draw a very accurate nomograph by a simple geometrical method.

This method is especially accurate when the curved scale lies between the two straight lines; it is somewhat less accurate when this curved line lies outside both straight lines.

The Simplified Procedure

Before starting this procedure, all preparatory steps have to be fulfilled: the dimensions assumed, the directions of straight scales established, the moduli calculated and, afterward, the extreme points plotted on both straight scales, u and v.

The equation to be considered is

$$u = v \times w + w^2 \qquad (20.33)$$

with u being the independent variable, we assume the ranges of v and w as

$$0 < v < 10; \qquad 0 < w < 5$$

We then calculate the range of the variable u:

$$u_{min} = 0; \qquad u_{max} = 75; \qquad \Delta u = 75$$

We decide that the curved w scale shall be placed in the middle. Therefore we draw both straight side scales in opposite directions, and we plot the appropriate values on them according to the moduli found.

We then fix two values of v, each one very near to the opposite end of the v scale, for example, $v = 0$, $v_2 = 10$.

We then obtain two sets of auxiliary equations:

$$u = 0 \times w + w^2 \qquad (\text{i.e., } u = w^2)$$
$$u = 10 \times w + w^2$$

and we calculate a number of u values, corresponding to several integer w values.

We collect these values in a table:

w	0		1		2		3		4		5	
v	0	10	0	10	0	10	0	10	0	10	0	10
u	0	0	1	11	4	24	9	39	16	56	25	75

Now, we connect these values as follows:

(a) $v = 0$ to $u = 0$ and $v = 10$ to $u = 0$; the intersection point gives us the value 0 on the w scale.

(b) $v = 0$ to $u = 1$ and $v = 10$ to $u = 11$; the intersection point gives us the value 1 on the w scale.

(c) $v = 0$ to $u = 4$ and $v = 10$ to $u = 24$; this point gives us the value 2 on the w scale, and so on.

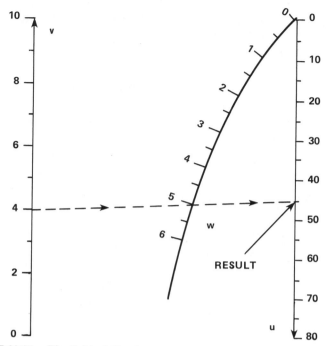

FIGURE 20.33. The finished Cauchy nomograph for the equation $u = v \times w + w^2$.

It is of course not enough to plot only 6 w values; the points for $w = 0.5$, 1.5, 2.5, and so on are found in the same way. When the smooth curve is finally drawn, the missing points on it may be added by manual interpolation.

If we assume the length of straight scales, $L_u = L_w = 10$ cm, and the distance $c = 8$ cm, then the final nomograph looks as in Figure 20.33.

NOTE. The same set of w values may be used to draw a nomograph for this equation when the curved w scale lies *outside* the straight scales. For this task, both straight scales have to be pointed in *the same direction*. This construction, however, demands considerably more skill and accuracy in order to be properly drawn.

20.5. NOMOGRAPHS WITH TWO STRAIGHT LINES AND ONE BINARY GRID

In some cases, a fourth variable may be inserted into the simplified Cauchy equation. When the fourth variable, t, is arranged in such a way that it is

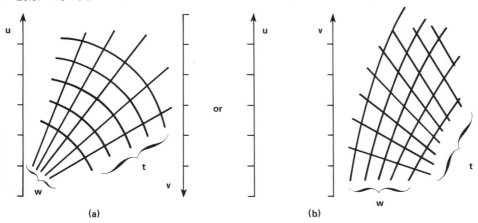

FIGURE 20.34. Possible nomographic forms of extended Cauchy expression.

paired with the variable w, the extended Cauchy equation looks as follows:

$$f(v) = f(u) \times f_1(w, t) + f_2(w, t) \qquad (20.34)$$

In such a case, this equation may be put into a nomographic form with two straight scales for $f(v)$ and $f(u)$, and with a binary grid consisting of two families of curves, representing the w and t variables. Schematically, such a nomograph looks as in Figure 20.34. The construction of such a nomograph is similar to that of a simple Cauchy nomograph (see Section 20.4); it requires, however, the application of a larger number of steps.

At first this equation is reduced to the ordinary Cauchy expression by assuming a certain, constant value of the variable t, $t = t_1 = $ const. Hence, the reduced equation looks as follows:

$$f(v) = f(u) \times f_1(w, t_1) + f_2(w, t_1) \qquad (20.35)$$

The nomograph is then built with straight scales for $f(v)$ and $f(u)$, and a curved scale for w, when $t = t_1 = $ const.

As a second step, another constant value is assumed for t, $t = t_2 = $ const, and the procedure is repeated. The scales for $f(v)$ and $f(u)$ *remain without change*; the only change is that another curve for w (at $t_2 = $ const) is drawn and *superimposed* on the drawing of the first nomograph. This procedure is repeated several times for various constant values for t: $t_1, t_2, t_3, \ldots, t_n$ until a dense network of t curves occurs in the body of the nomograph.

As a last step the identical values of w on the curves for various t values are connected with smooth curves, and the nomograph is ready for use (Figure 20.35 offers a graphical explanation).

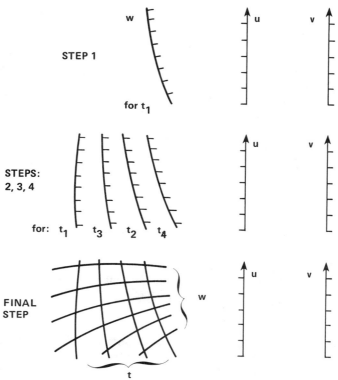

FIGURE 20.35. Consecutive steps in designing a nomograph for the extended Cauchy expression.

20.6. COMPOUND NOMOGRAPHS

Multivariable Equations

In all cases when the equation to be nomographed consists of more than three variables (i.e., four, five, six, etc.), and it does not correspond to the special case of the extended Cauchy expression, a so-called *compound nomograph* has to be designed.

A compound nomograph consists of at least two separate nomographs, where at least one scale is common, and belongs to both basic nomographs simultaneously.

As an example, we will take the simplest possible equation with four variables, u, v, w, and t, in the following form:

$$u + v + w = -t \tag{20.36}$$

or

$$u + v + w + t = 0 \tag{20.37}$$

First, we decide to assume an auxiliary variable, K. We then rewrite the above equation as a set of two equations, in one of six arbitrarily chosen ways:

$$
\begin{array}{lll}
\text{(a)} & K = u + v; & K + w + t = 0 \\
\text{(b)} & K = v + w; & K + u + t = 0 \\
\text{(c)} & K = w + t; & K + u + v = 0 \\
\text{(d)} & K = u + w; & K + v + t = 0 \\
\text{(e)} & K = u + t; & K + v + w = 0 \\
\text{(f)} & K = v + t; & K + u + w = 0
\end{array}
\qquad (20.38)
$$

We therefore get six possible variants; each variant consists of two equations, which are easily nomographable in the form of three parallel lines. The variable K is the common one in each set of two equations.

We take the variant (a), and rewrite it as follows:

$$
\text{(a1)} \quad K - u = v; \qquad \text{(a2)} \quad K + w = -t \qquad (20.39)
$$

Each of these nomographs may be drawn in a number of possible forms. We choose the forms of Figure 20.36.

According to the procedures previously presented, we calculate the moduli for each scale of both nomographs. The most important point is to design the *modulus of the K scale identical* for both basic nomographs, and to maintain the identical direction of the K scale for both of them.

Then, both nomographs may be superimposed one on the other, with the K scale common to each. The compound nomograph looks as in Figure 20.37.

When we assume that the variables u, v, and w are the given ones, and the t variable is the desired one, we proceed in the following way.

The nomograph is solved by *two* movements of a ruler. In the first movement, we connect two known variables u and v with a ruler, and the intermediate result, K, is obtained. The second movement of the ruler connects the just found K value with the known w value, and the intersection point of the ruler with the t scale gives us the desired result.

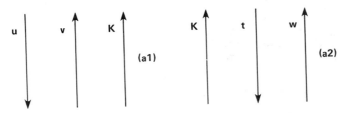

FIGURE 20.36. Configuration of basic nomographs.

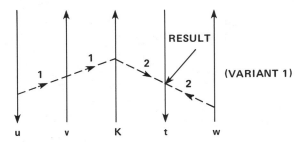

FIGURE 20.37.   First configuration of four-variable compound nomograph.

The intermediate result, K, is, as a rule, of no interest to the user of the nomograph. Therefore, also as a rule, the K scale remains *nongraduated*, and is called a *reference line* or a pivot line. An exception has been made only for the examples (see Figure 20.43, 20.46, 20.49), where the K scales are graduated in order to make these examples clearer.

Other Variants

Other variants of compound nomographs are possible. If we take once more the case (a) in equation (20.38) and rewrite these equations as follows,

$$(a1') \quad u + v = K; \qquad (a2') \quad K + w = -t$$

we get the basic forms as in Figure 20.38, and, observing the above mentioned rule of superimposing both basic nomographs, we get another form of the same compound nomograph as shown in Figure 20.39, with a different path of movements of the ruler to be used for its solution.

Further Adjustments

In some special cases, by carefully adjusting the distances between scales, the compound nomograph (variant 2) may be further compressed in such a way that the scales for v and t may be plotted on the same scale support. This possibility is very useful, especially for compound nomographs consisting of a relatively large number of variables, since it enables us to construct nomographs with a large number of variables with the least possible number of scales (Figure 20.40).

Example 20.11.   We construct a nomograph for the equation

$$u + v + w + t = 0 \quad \text{or} \quad u + v + w = -t$$

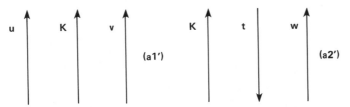

FIGURE 20.38. Another configuration of basic nomographs.

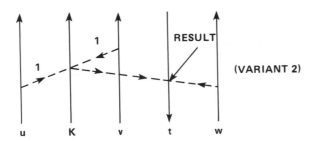

FIGURE 20.39. Second configuration of four-variable compound nomograph.

where u, v, and w are independent variables, and t is the dependent variable. The range of all three variables, u, v, and w varies from 0 to 10. The length of the scales is 10 cm.

Solution. We choose the set (a) of equation (20.38) and the variant 1 of the nomograph (Figure 20.37). Hence, the set of equations is

$$K = u + v; \qquad K + w + t = 0$$

The range of values for variable K equals

$$K_{\min} = v_{\min} + u_{\min} = 0 + 0 = 0$$
$$K_{\max} = v_{\max} + u_{\max} = 10 + 10 = 20$$

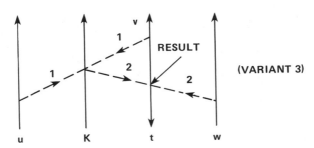

FIGURE 20.40. Third configuration of four-variable compound nomograph.

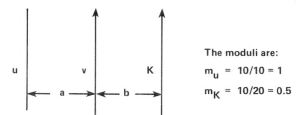

FIGURE 20.41. Configuration for the equation $K = u + v$.

The first nomograph, $K = u + v$, we write in the form $K - u = v$, and we choose the form shown in Figure 20.41.

The modulus for scale v equals

$$m_v = \frac{m_u \times m_K}{m_u + m_K} = \frac{1 \times 0.5}{1 + 0.5} = \frac{0.5}{1.5} = 0.3333$$

Let the distance between the u and K scales be 8 cm. Hence, the distance a equals

$$\frac{m_u}{a} = \frac{m_u + m_K}{8}; \qquad a = \frac{8 \times 1}{1 + 0.5} = \frac{8}{1.5} = 5.333 \text{ cm}$$

The second nomograph is

$$K + w + t = 0 \quad \text{or} \quad K + w = -t$$

The range for t is from -30 to 0, and the shape of the second nomograph is as in Figure 20.42. The moduli of the side scales are $m_K = 0.5$ (identical to the first nomograph), and the modulus for w is $m_w = \frac{10}{10} = 1$. Hence, the modulus for t equals

$$m_t = \frac{m_K \times m_w}{m_K + m_w} = \frac{0.5 \times 1}{0.5 + 1} = 0.3333$$

If the distance between K and w scales is 8 cm, the distance a also equals 5.333 cm. The finished nomograph is given in Figure 20.43.

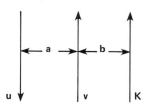

FIGURE 20.42. Configuration for the equation $K + w = -t$.

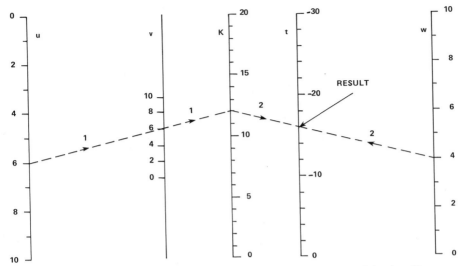

FIGURE 20.43. Nomograph for the equation $u + v + w + t = 0$ (variant 1).

Example 20.12. We construct a nomograph for the same equation in Example 20.11:

$$u + v + w + t = 0$$

in the form of variant 2 (Figure 20.39).

Solution. The range of all variables is identical, and the set (a) of equation (20.38) is chosen.

The first equation $u + v = K$ is put into the nomographic form of Figure 20.44.

The moduli for u and v scales are identical:

$$m_u = \tfrac{10}{10} = 1$$

$$m_v = \tfrac{10}{10} = 1$$

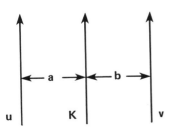

FIGURE 20.44. Nomograph for the equation $u + v = K$.

The modulus for K equals

$$m_K = \frac{m_u \times m_v}{m_u + m_v} = \frac{1 \times 1}{1 + 1} = 0.5$$

The second equation is identical to Example 20.11, namely,

$$K + w + t = 0 \quad \text{or} \quad K + w = -t$$

and its form is given in Figure 20.45. The modulus of the w scale equals $m_w = \frac{10}{10} = 1$.

The modulus for the t scale equals

$$m_t = \frac{m_K \times m_w}{m_K + m_w} = \frac{0.5 \times 1}{0.5 + 1} = \frac{0.5}{1.5} = 0.333$$

When we assume that the distance between scales u and v in the first nomograph is 6 cm, then the distance a equals

$$\frac{m_u}{a} = \frac{m_u + m_v}{6}; \qquad a = \frac{6 \times m_u}{m_u + m_v} = \frac{6 \times 1}{1 + 1} = 3$$

We assume that the distance between the scales K and w in the second nomograph is 12 cm, then the distance a' equals

$$\frac{m_K}{a'} = \frac{m_K + m_w}{12}; \qquad a' = \frac{12 \times m_K}{m_K + m_w} = \frac{12 \times 0.5}{0.5 + 1} = 4 \text{ cm}$$

Now, we may superimpose both nomographs, providing that the K scale is common for both.

The distances between scales are as follows:

$$a = 3 \text{ cm}; \qquad b = 3 \text{ cm}; \qquad b' = 12 - 4 = 8$$

and the final nomograph looks as in Figure 20.46.

Example 20.13. The problem and variables are identical to Examples

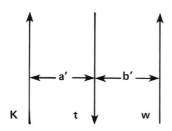

FIGURE 20.45. Nomograph for the equation $K + w = -t$.

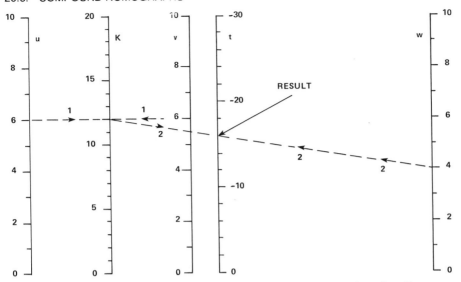

FIGURE 20.46. Nomograph for the equation $u + v + w + t = 0$ (variant 2).

20.11 and 20.12; however, the scope is to design a compressed nomograph with a common scale for v and t (variant 3, Figure 20.40).

Solution. After all preliminary calculations are made, the main aim is to establish distances b' and a' (see the previous example) as being identical.

For this purpose, the distance b' shall be fixed, and the distance a' shall be adjusted to be numerically equal to b' (Figure 20.47).

In the previous example, the b distance equaled 3 cm; hence for the revised compressed nomograph, the new a' distance has to be obviously *the same*.

So, the following equation has to be solved:

$$\frac{m_K}{a'} = \frac{m_K + m_v}{X}$$

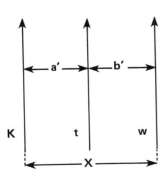

FIGURE 20.47. Configuration for a compressed nomograph with a common scale for v and t.

where all the above mentioned variables are known, and by X we denote the *unknown* distance between K and w scales.

Solving the equation for X, we get

$$X = \frac{a' \times (m_K + m_v)}{m_K}$$

$$X = \frac{4 \times (0.5 + 1)}{0.5} = \frac{6}{0.5} = 12$$

We have calculated that, if the distance between the K and w scales is taken as 12 cm, after superimposing both nomographs, the scales for K are identical, and the scales for v and t overlap, that is, they are plotted on *the same scale support*.

In this way we have succeeded in putting the equation on a compound nomograph consisting of four lines instead of five lines (as has been shown in both previous examples). The final nomograph is presented in Figure 20.48.

Practical Hints

In compound nomographs, the sequence of grouping the variables and choosing the intermediate K value cannot be performed in a random

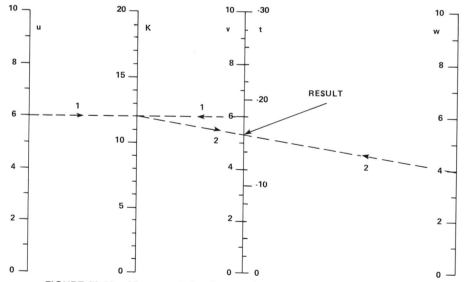

FIGURE 20.48. Nomograph for the equation $u + v + w + t = 0$ (variant 3).

manner. This order and sequence is of great importance and must be carefully designed. The final aim is to construct a nomograph with the resulting scale as large as possible—with the largest possible modulus, and with the distances between scales as uniform as possible.

This purpose may be achieved by grouping the variables *in the order of descending moduli*, and by choosing the intermediate variables K in the same way.

Example 20.14. An equation is given with five variables:

$$u + v + w + t = z \qquad (20.40)$$

We desire to put this equation into a nomograph with the largest possible modulus for the resulting z scale.

Solution. First, we examine the ranges of all variables, and we calculate the appropriate moduli.

The length of scales is chosen to be $L = 10$ cm, and the ranges of variables are as follows:

$$0 < u < 10; \quad 0 < v < 5; \quad -10 < w < -2; \quad 0 < t < 2$$

Hence, the moduli are

$$m_u = \tfrac{10}{10} = 1; \quad m_v = \tfrac{10}{5} = 2; \quad m_w = \tfrac{10}{8} = 1.25; \quad m_t = \tfrac{10}{2} = 5$$

The proper sequence of independent variables is selected:

$$t, v, w, u$$

and the three basic nomographs built according to the following auxiliary equations:

$$(1) \quad t + v = K_1; \qquad (2) \quad K_1 + w = K_2; \qquad (3) \quad K_2 + u = z \qquad (20.41)$$

Now, independent of the variant chosen (1, 2, or 3), the final nomograph should have the resulting z scale with the largest possible modulus and with the greatest accuracy, and the distance between separate scales should be as uniform as possible.

If this condition is not observed, the finished nomograph may be very inconvenient to use, since some of the scales may be placed extremely close to one another, while others may be found in relatively remote distances.

20.7. THE PROJECTIVE TRANSFORMATION OF NOMOGRAPHS

Ordinary Projective Transformation

There are frequent cases in which the finished nomograph, constructed according to one of the above mentioned methods, does not prove satisfactory in use. This may be due to numerous reasons; some of these are listed below.

(a) The values of variables of the nomograph scales are too convergent or too divergent in the vicinity of either of the ends.

(b) The ruler intersects the resulting scale in too acute an angle.

(c) One of the scales (especially a curved one) "escapes" too far from the rest of the scales, and so on.

Some of these disadvantages may be overcome by the application of so-called *projective transformation*.

Without entering deeply into the theory of projective geometry, it may be said that the projected scale is a new scale, which is obtained from a given one by projecting its points from a certain focus, S, on an oblique line, having a common zero point with the initial scale. This device is indicated in Figure 20.49.

By the appropriate selection of the L and C parameters we may obtain projected scales with the desired properties, either similar to the regular scales, or to the logarithmic ones in certain, selected ranges. The main

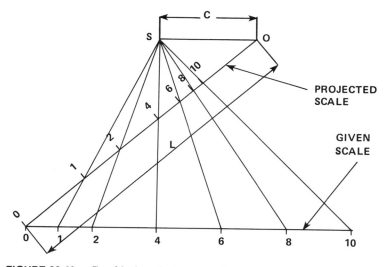

FIGURE 20.49.   Graphical method for drawing a projective transformed scale.

property of a projected scale is the presence of an infinite point, ∞, on it, a property that no other scale possesses.

Projective transformation may be performed not only on a separate scale, but on the total plane as well; due to this possibility the entire nomograph may be also transformed by projection. The main property of the projective transformed plane is that any three points, which have lain on a common straight line on the plane *before transformation*, will lie on a common straight line on the plane *after transformation* as well; this, of course, enables us to take advantage of this feature to transform the entire nomograph.

It must be remembered that, after the projective transformation, the nomograph built on three parallel lines is transformed into a nomograph with three straight lines which intersect in a common point which corresponds to the infinite value of all the scales in question.

It is also important to remember that, before the projective transformation, the lower values of the scales must be placed on the common horizontal line; if this is not observed, serious distortions in the lower range of scales may occur.

When the equation of a given scale for u is

$$Y_u = m_u \times f(u) \tag{20.42}$$

the equation of the projective transformed scale is

$$\overline{Y}(u) = L \times \frac{m_u \times f(u)}{C + m_u \times f(u)} \tag{20.43}$$

where the meaning of C and L values is shown in Figure 20.49.

When a nomograph based on three parallel lines has to be transformed, the set of transforming equations looks as follows:

$$Y_u = L_u \times \frac{m_u \times f(u)}{C + m_u \times f(u)} \tag{20.44}$$

$$Y_w = L_w \times \frac{m_w \times f(w)}{C + m_w \times f(w)} \tag{20.45}$$

$$Y_v = L_v \times \frac{m_v \times f(v)}{C + m_v \times f(v)} \tag{20.46}$$

The nomographs before and after the transformation look as in Figure 20.50.

The values of L_u, L_v, and L_w may be calculated according to the Pythagorean theorem from the initial data. In the simplest case, when the central w scale overlaps the Y axis, the length of L_w equals the length of the transformed Y axis, from the zero to the infinity points.

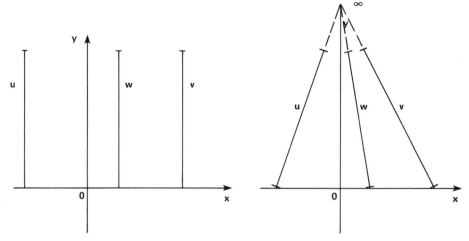

FIGURE 20.50. Schematic of projected transformation nomograph from three parallel scales.

The selection of proper L and C values may be performed in several ways. As a rule, the projective transformation is adjusted to the resulting scale. In most cases, we try to transform the resulting scale in such a way that it will be similar, as far as possible, to either a regular scale, or to a logarithmic scale.

The main property of a regular scale is that the value of the middle of the scale is an arithmetical average of its two extreme end values. Similarly, the logarithm of the value of the midpoint of the logarithmic scale is an arithmetical average of the logarithms of the extreme scale values. These properties may be used in the projective transformation, as shown in the following examples.

Example 20.15. We are given a function $f(u) = u^2$. The range of variables is

$$1 < u < 8$$

Hence

$$1 < u^2 < 64; \qquad \Delta u^2 = 63$$

When the length of the scale is assumed as $L = 10$ cm, its modulus is $m_u = \frac{10}{63} = 0.1587$, or $m_u = 0.16$. The final scale looks as in Figure 20.51.

As we observe, the scale is very dense in the low value range, and excessively accurate in the high value range; hence it is not convenient for nomographical use.

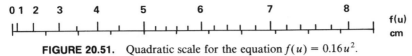

FIGURE 20.51. Quadratic scale for the equation $f(u) = 0.16u^2$.

This scale has to be transformed twice, to the forms resembling

(a) The regular scale.
(b) The logarithmic scale.

Solution. *Part (a).* The extreme values on the scale given are $u_1 = 1$, $u_2 = 8$, the average value being $u_{av} = (1 + 8)/2 = 4.5$. Hence, the scale, similar to the regular one, has to have the value of 4.5 in the middle of the transformed scale.

The equation of the original scale is $f(u) = 0.16u^2$. The values of the function for the extreme points, and the arithmetical averages are

$$f(1) = 0.16 \times 1^2 = 0.16$$
$$f(4.5) = 0.16 \times 4.5^2 = 3.24$$
$$f(8) = 0.16 \times 8^2 = 10.24$$

The length of the scale is $10.24 - 0.16 = 10.08$ cm and the dimension of the middle of the scale is

$$\frac{10.08}{2} + 0.16 = 5.04 + 0.16 = 5.2$$

We intend to retain the approximate length of the original scale, and similar dimensions of its extreme points, while the dimension of the $u = 4.5$ value is maintained approximately in the middle of the transformed scale.

By using equation (20.43),

$$\overline{Y}_u = L \times \frac{m_u \times f(u)}{C + m_u \times f(u)}$$

we construct the set of two equations (in which the values of the middle and the end of the new scale are slightly rounded).

$$\overline{Y}_{(4.5)} = L \times \frac{0.16 \times (4.5)^2}{C + 0.16 \times (4.5)^2} = 5$$

$$\overline{Y}_{(8)} = L \times \frac{0.16 \times 8^2}{C + 0.16 \times 8^2} = 10$$

These two equations are solved for the L and C values.

The results obtained are $C = 8.83$ and $L = 18.62$. After rounding, we assume $C = 8.8$ and $L = 18.6$, and the equation of the transformed scale looks as follows:

$$\overline{Y}_u = 18.6 \times \frac{0.16 \times u^2}{8.8 + 0.16 \times u^2}$$

or

$$\overline{Y}_u = \frac{2.98 \times u^2}{8.8 + 0.16 \times u^2} \tag{20.47}$$

We now prepare the table of values:

u	1	2	3	4	5	6	7	8
$\overline{Y}_U$	0.33	1.26	2.62	4.20	5.82	7.37	8.78	10.02

and the final, transformed scale looks as in Figure 20.52.

NOTE. The total length of the new scale is now $10.02 - 0.33 = 9.69$ cm instead of 10.08 cm as before. We may let it stay as such, or we may enlarge it to the previous dimension by multiplying the basic equation by the ratio of previous-to-present length ratio.

This factor, α, is $\alpha = L_1/L_2 = 10.08/9.69 = 1.04$. In this case, the final equation of the transformed scale is

$$\overline{\overline{Y}} = \alpha \times \overline{Y} \qquad \overline{\overline{Y}} = 1.04 \times \frac{2.98 \times u^2}{8.8 + 0.16 \times u^2}$$

$$\overline{\overline{Y}} = \frac{3.1 \times u^2}{8.8 + 0.16 \times u^2} \tag{20.48}$$

and we repeat the calculation of the scale values according to the corrected equation for $\overline{\overline{Y}}$.

Solution. Part (b). The same function in question has to be transformed into a scale resembling the logarithmical scale.

For this case, the value designed to be in the middle of the scale is calculated as follows:

$$\log(M) = \frac{\log(1) + \log(8)}{2}$$

$$\log(M) = \frac{0 + 0.9031}{2} = 0.4516$$

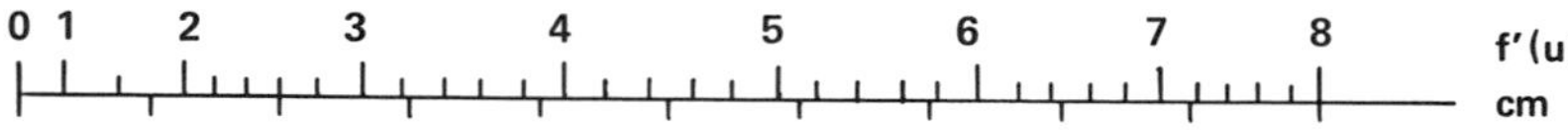

FIGURE 20.52. Quadratic scale transformed to "regular" scale.

hence,

$$M = 10^{0.4516}; \qquad M = 2.83$$

We repeat the calculation exactly, as in the previous example, only using other numerical values:

$$\overline{Y}_{(2.83)} = L \times \frac{0.16 \times (2.83)^2}{C + 0.16 \times (2.83)^2} = 5$$

and

$$\overline{Y}_{(8)} = L \times \frac{0.16 \times 8^2}{C + 0.16 \times 8^2} = 10$$

Solving these equations for L and C, we get the following values:

$$C = 1.707; \qquad L = 11.67$$

After rounding, we assume $C = 1.7$, $L = 11.7$, and the transformed scale has the following equation:

$$\overline{Y} = 11.7 \times \frac{0.16 \times u^2}{1.7 + 0.16 \times u^2}$$

or

$$\overline{Y} = \frac{1.87 \times u^2}{1.7 + 0.16u^2} \tag{20.49}$$

We prepare, as usual, the table of values for this scale:

u	1	2	3	4	5	6	7	8
$\overline{Y}_u$	1.005	3.2	5.36	7.02	8.20	9.02	9.60	10.02

The final scale looks as in Figure 20.53.

If the transformed scale is too short (or too long), we may multiply it by an appropriate factor. By this operation we change the L value, while the C value remains unchanged.

Example 20.16. In this example we shall perform the projective transformation of an entire nomograph.

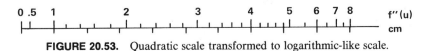

FIGURE 20.53. Quadratic scale transformed to logarithmic-like scale.

The Pythagorean theorem states that the square of the hypotenuse equals the sum of squares of the sides of a right triangle. This is expressed mathematically as follows:

$$u^2 + v^2 = w^2$$

This equation, as we already know, may be put into a nomograph with three parallel scales.

Assuming the range of variables $1 < (u = v) < 8$, the length of scales $L_u = L_v = L_w = 10$ cm, the distance between scales u and $v = 10$, and calculating the moduli (after rounding), we get the following results:

$$f(u) = 0.16 \times u^2$$
$$f(v) = 0.16 \times u^2$$
$$f(w) = 0.08 \times w^2$$

The finished nomograph looks as in Figure 20.54.

As already discussed, this nomograph is inconvenient for use because of its extremely dense marking in the low range of values. Therefore, we shall perform the projective transformation of this nomograph.

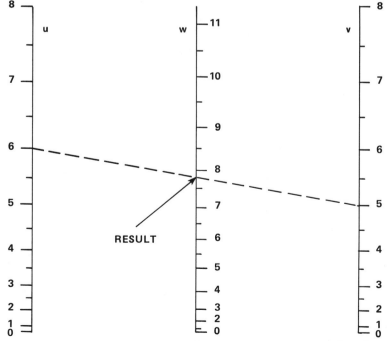

FIGURE 20.54.   Nomograph on three parallel scales for the equation $u^2 + v^2 = w^2$.

The best way of performing this transformation will be the conversion of the resulting w scale into a logarithmic-like one.

As before, we find the required value to be placed in the middle of the transformed w scale. We choose as the extreme values the points of $w_1 = \sqrt{2}$ and $w_2 = 11$. Hence, the value in the middle of the transformed scale is

$$\log(M) = \frac{\log(\sqrt{2}) + \log(11)}{2} = \frac{0.05 + 1.0414}{2}$$

$$\log(M) = 0.596; \qquad M = 3.95 \qquad (\text{we assume } M = 4)$$

We assume that the resulting w scale overlaps the Y axis on the Cartesian plane, having its zero point in the center of the system.

Hence, the new equation for the w scale is

$$\overline{Y} = L \times \frac{m_w \times f(w)}{C + m_w \times f(w)}$$

Taking as the two fixed points of the w scale the top value as $w = 11$ and the middle value as $w = 4$, we get the following two equations:

$$\overline{Y}_{(4)} = L \times \frac{0.08 \times 4^2}{C + 0.08 \times 4^2} = 5$$

$$\overline{Y}_{(11)} = L \times \frac{0.08 \times 11^2}{C + 0.08 \times 11^2} = 10$$

After solving for L and C we get the following results:

$$C = 1.74; \qquad L = 11.7$$

After rounding, we assume that $C = 1.7$ and $L = 12$, and we get the following equation for the transformed w scale.

$$\overline{Y} = 12 \times \frac{0.08 \times w^2}{1.7 + 0.08 \times w^2}$$

or

$$\overline{Y} = \frac{0.96 \times w^2}{1.7 + 0.08 \times w^2} \tag{20.50}$$

Afterward, we prepare the table of new values for the w scale, taking in account the value of $w = 0$ as well.

We then draw the Cartesian coordinates, marking the points of $x = -5$ and $x = +5$, the point of $Y = L = 12$, and we draw the triangle as in Figure 20.55.

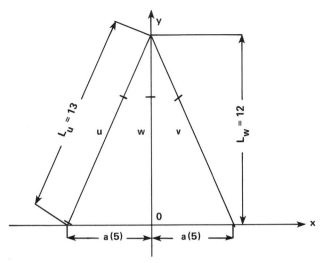

FIGURE 20.55. Geometrical calculation of parameters for the projective transformation.

The line connecting the values of a point $x = 0$ and $Y = 0$ with the point $Y = 12$ and $x = 0$ will be the scale support for the w scale; the line connecting points of $x = -5$ and $Y = 12$ will be the support for scale u, and the line connecting points of $x = +5$ and $Y = 12$ will be the support for scale v.

Since the triangle is isosceles, the parameters L_u and L_v are identical, and may be calculated as follows:

$$L_u = L_v = \sqrt{a^2 + L_w^2}\,; \qquad L_u = L_v = \sqrt{5^2 + 12^2}\,; \qquad L_u = L_v = 13$$

Hence, the equations for scales u and v after projective transformation are

$$\overline{Y}_u = 13 \times \frac{0.16 \times u^2}{1.7 + 0.16 \times u^2}$$

or

$$\overline{Y}_u = \frac{2.08 \times u^2}{1.7 + 0.16 \times u^2} \tag{20.51}$$

and for the v scale,

$$\overline{Y}_v = \frac{2.08 \times v^2}{1.7 + 0.16 \times v^2} \tag{20.52}$$

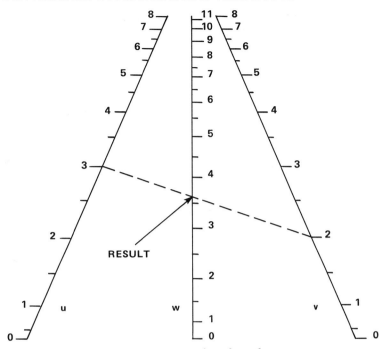

FIGURE 20.56. Nomograph for the equation $u^2 + v^2 = w^2$ after projective transformation.

The table of values is prepared according to these equations and the values are plotted on the already drawn u, v, and w scale supports. The finished nomograph looks as in Figure 20.56.

As may be seen, the accuracy in the low-value region has increased considerably, while the higher value region remains sufficiently accurate as well.

Nonprojective Transformation

It should be emphasized that, besides the projective transformation of nomographs, there exist a large number of *nonprojective* transformations which are also sometimes used in the nomographic techniques. One of the most frequently used is a transformation of a parallel line or a Z nomograph into a circle or an ellipse. This transformation is especially useful when one or more of the nomograph scales is stretched into infinity for certain values of the variables. The description of these methods is, however, beyond the scope of this chapter.

20.8. ATYPICAL NOMOGRAPHIC PRACTICES

This section is intended as an abstract only. Certain procedures, practices, and hints are only listed, since their detailed description exceeds the scope of this chapter.

For detailed description of these practices, the reader is referred to the more specific and advanced nomographic textbooks, as listed in the Bibliography.

How to Construct a Nomograph from a Set of Tabulated Data

The data are plotted on a Cartesian plane in such a way that the variables u and v are plotted along the x and y axis, and the variable w is represented as a family of curves (Figure 20.57).

If the family of w values constitutes a family of *straight lines* (either parallel, or not), the graph may be converted into a colinearic nomograph, where the scales for u and v are straight, and the scale for w is (as a rule) curved.

How to Construct a Nomograph from a Graph, Consisting of a Family of Curved Lines

If the lines are only moderately curved and their shape is not too sophisticated, they may be graphically straightened by the change of a characteristic of either the u or v scales (single anamorphosis), or both u and v scales (double anamorphosis).

After straightening all the w curves (or a great part of them), an *approximate* colinearic nomograph may be constructed.

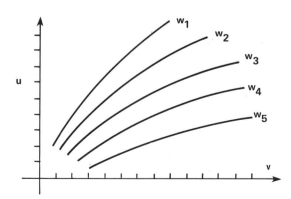

FIGURE 20.57.   Graphical plot from tabulated data on Cartesian plane.

How to Construct a Nomograph from Nonnomographable Equations

Some of the possibilities are listed below:

(a)   When an equation in question contains one specific variable, which changes in a relatively narrow range, and its influence on the result is negligible, this variable may be replaced by an average *constant* value. This operation frequently leads to a nomographable form of the equation.

(b)   Sometimes it is possible to develop the given equation into a Taylor or McLauren infinite series. Taking the first two members of the series, we get a simplified and approximate expression, which very frequently is nomographable. It has to be emphasized that such a nomograph gives *approximate* results only.

(c)   Sometimes the equation in question may be expressed as a sum of two nomographable equations. In this event, we proceed similarly by designing a compound nomograph, taking one of the variables twice in account. On the nomograph, they occur afterward as two scales: a primary and a secondary scale for the same variable.

(d)   It is always possible to put a *non-nomographable* equation with three variables into a nomograph with a *binary grid* (as the case of extended Cauchy expression) when one of the variables occurs twice: once as a linear scale, and a second time on a grid, in the form of a family of curves.

(e)   When an equation consists of four or more variables, and when a part of them forms a nomographable pattern, and part is evidently non-nomographable, it is possible to design an alignment chart which includes an ordinary graph *and* a nomograph, as a whole entity.

20.9. WORKING NOMOGRAPHS

Table 20.1 contains a listing of 21 nomographs of particular orientation to the coatings industry. These have been published over the years by various authors, at various times, and in various technical journals. They are arranged by an approximate order of gradually increasing difficulty. In some cases, the basis equation is not listed due to the length and/or complexity of the original form. For these, and for the original nomograph, the interested reader should consult the original references.

These nomographs have been carefully selected and all of them serve a dual purpose: (a) They are of interest to people working in the coatings industry, since all of them are directly, or partially connected with coating practice and related techniques and technologies, (b) They serve as practical illustrations of nomographic techniques described and presented in this chapter.

TABLE 20.1. Characteristic Types of Paint Industry Related Nomographs Arranged by Order of Difficulty

No.	Title	Formula	Source[a]
1	Nomograph for determining the viscosity of a liquid at some required temperature given the viscosity of the liquid at some other temperature	$\dfrac{d\eta}{dT} = f(\eta)$	(A)
2	Nomogram relating ball milling time to mill base volume	$\log t = a + 0.018B$	(B)
3	Nomogram relating ball diameter, difference in ball and mill base density, and optimum mill base viscosity	$ku = D^{0.75}(93 + 11\,\Delta p)$	(C)
4	Nomograph for the determination of the viscosity of liquid–solid suspensions	$\dfrac{\mu_s}{\mu_l} = \dfrac{1 + 0.5\,\phi_{\text{sol}}}{(1 - \phi_{\text{sol}})^4}$	(D)
5	Calculating the optimum and critical speeds for ball mills	$\cos\alpha = 4\pi^2\eta^2(R - r)/g$ $\eta_c = 54.2\left(\dfrac{1}{R - r}\right)^{1/2}$	(E)
6	Nomograph relating solubility parameter to molar volume, molar heat of vaporization, normal boiling point, and surface tension	$\delta = \left(\dfrac{\Delta E}{V}\right)^{1/2}$ $\delta = 4.1\left(\dfrac{\sigma}{V^{1/3}}\right)^{0.43}$ $\Delta E = \Delta H - RT$	(F)
7	Resin prices on a volume basis	$V = 3.61WD$	(G)

No.	Description	Equation	
8	Calculating intrinsic viscosity from a single viscosity ratio at a known polymer concentration	$[\eta] = \dfrac{(\eta_r - 1)}{4C} + \dfrac{3\ln\eta_1}{4C}$	(H)
9	Viscosity conversion for simple liquids	$ku = 17.2(\log W)^2 - 3.9\log W$ $\eta = 0.051(W - 34)$ $\eta = \rho v$	(I)
10	Nomogram for formulating glyceryl phthalate alkyds of varying oil length from blends of soya and tung oil	—	(J)
11	Nomogram relating volume percent solids, volume thinning ratio, coverage, and dry film thickness	—	(K)
12	Nomograph for calculating plating time	$t = \dfrac{T \times d}{1.667 \cdot E \cdot i_k \cdot f}$	(L)
13	Sizing balls and rods for use in grinding mills	$M = \sqrt{\dfrac{FW_i}{KC_s}}\sqrt{\dfrac{S}{\sqrt{D}}}$	(M)
14	Size distribution of ground products	$P = 80\left(\dfrac{d}{d_{80}}\right)^{a}$ $a = \dfrac{6.65\log(d_{80}) + 2.25}{38 - 3(C)^{0.4}}$	(N)

(continued)

TABLE 20.1. *(continued)*

15	Prediction of vapor pressures	Riedel's correlation	(O)
16	Nomograph for emulsion viscosity	$\mu = \dfrac{\mu_c}{\delta_c}\left(1 + \dfrac{1.5\mu_D \delta_D}{\mu_D + \mu_c}\right)$	(P)
17	Nomograph relating the evaporation and thickening of dilute solutions	$w = 100 \times \left(1 - \dfrac{y}{s}\right)$	(Q)
18	Temperatures for doubled reaction rates	$\dfrac{1}{T_2} = \dfrac{1}{T_1} - \dfrac{1.3769}{E}$	(R)
19	Densities of molten fatty acids and lower oleic alkyl esters	—	(S)
20	How to determine a "comparable cost" for paints	$N = \dfrac{w}{16.04V}$ $v = 100 - \dfrac{A}{C}(100 - B)$	(T)
21	Nomograms for solubility parameter	$\delta = \left(\dfrac{\Delta E}{V}\right)^{1/2}$	(U)

[a]Nomograph References:
(A) Patton, T. C., *Paint Flow and Pigment Dispersion*, 1st ed., Wiley, New York, 1964, p. 79.
(B) Patton, T. C., ibid (A), p. 282.
(C) Patton, T. C., ibid (A), pp. 275, 276
(D) Zanker, A., *The Chemical Engineer*, pp. 76, 77, February 1972.
(E) Zanker, A., *Chemical Engineering*, p. 111, August 2, 1976.
(F) Patton, T. C., ibid (A), pp. 352, 353.
(G) Grail, T. J., *Chemical Processing*, p. 377, February 1966.
(H) Patton, T. C., ibid (A), pp. 103, 104.
(I) Patton, T. C., ibid (A), p. 74.
(J) Patton, T. C., *Alkyd Resin Technology*, Wiley–Interscience, New York, 1962, pp. 141, 142.
(K) Avery, R. M., Jr., *Products Finishing*, pp. 55–57, January 1965.
(L) Zanker, A., *Plating*, pp. 700, 701, July 1971.

(M) Zanker, A., *Chemical Engineering*, p. 86, December 20, 1976.

(N) Zanker, A., *Chemical Engineering*, p. 113, July 31, 1978.

(O) Zanker, A., *Process Engineering*, pp. 132, 133, November 1977.

(P) Zanker, A., *Hydrocarbon Processing*, Vol. 48, No. 1, p. 178, January 1969.

(Q) Zanker, A., *Food Engineering*, p. 104, July 1974.

(R) Noddings, C. R., and Mullet, G. M., *Chemical Processing*, p. 49, March 9, 1964.

(S) Mapstone, G., *Chemical Processing*, p. 61, February 11, 1963.

(T) Volkening, V. B., and Wilson, J. T., Jr., *Corrosion*, Vol. 13, No. 8, pp. 33, 34, 1957.

(U) Jayasri, A., and Yaseen, M., *J. Ctngs. Tech.*, Vol. 52, No. 667, 1980, p. 41.

BIBLIOGRAPHY

Davis, D. S., *Nomography & Empirical Equations*, Reinhold, New York, 1955.

Davis, D. S., and Kulwiec, R. A., *Chemical Processing Nomographs*, Chemical Publishing Company, New York, 1969.

Douglass, R. D., and Adams, P. D., *Elements of Nomography*, McGraw-Hill, New York, 1947.

Johnson, L. J., *Nomography and Empirical Equations*, Wiley, New York, 1952.

Konorski, B., and Krysicki, W., *Nomography and Graphical Methods of Calculation*, W.N.T., 1973, Warsaw.

Krasnodebski, R., *Simple Nomographs*, PZWS, 1960, Warsaw.

Nevsky, B. A., *Nomographical Handbook*, G.I.T.T.L., 1951, Moscow.

Otto, E., *Nomography*, P.W.T., 1956, Warsaw.

Perry, R. H., et al., Ed., *Chemical Engineers' Handbook*, 4th ed., McGraw-Hill, New York, 1963.

Schmid, W., et al., *Graphisches Rechnen und Nomographie*, Berg-akademie, Fernstudium 1957, Germany.

Zanker, A., Nomographs—Outdated by the pocket calculator?, *Process Engineering*, June 1978.

GLOSSARY OF TERMS AND SYMBOLS

X	Value of variable (e.g. time)	$\overline{X}$	Mean of n values of X
Y	Value of an effect (e.g. yield)	μ	Mean of N values of X
N	No. of observations (population)	$\overline{\overline{X}}$	Grand average of X's
n	No. of observations (sample)	x,y	$X - \overline{X}, Y - \overline{Y}$
i	Class interval	+	Plus or positive
f	Frequency of observations	−	Minus or negative
s	Standard deviation (sample)	±	Plus or minus
σ	Standard deviation (population)	x or ·	Multiplied by
$s_{\overline{x}}$	Standard deviation of x	÷	Divided by
s^2	Variance (sample)	=	Equals
σ^2	Variance (population)	≠	Is not equal to
P	Probability (of occurrence)	≅	Equals approximately
t	Student's t statistic	>	Greater than
D.F., k, ν	Degrees of Freedom	<	Less than
χ	Chi square	≯	Not greater than
F	F variance ratio	≮	Not less than
Z	Z transformation coefficient	≥	Greater than or = to
r_s	Spearman's rank correlation coefficient	≤	Less than or = to
r	Correlation coefficient	~	Similar to
r^2	Coefficient of determination	$\sqrt{}$	Square root
b_0, b_1	Coefficients of regression equation	$\sqrt[n]{}$	n^{th} root
Q	Quartile point		
H_0	Null hypothesis	a^n	n^{th} power of a
H_1	Alternative hypothesis	log	common logarithm
$\int$	Integral of	ln	natural logarithm
$\int_a^b$	Integral between the limits a and b	f(x), F(x) or $\Phi(x)$	Function of x
		∞	Infinity
n!	n! = 1·2·3··· n		
≐	Observed to be equal to (approaches)	$\frac{dy}{dx}$ or $f'(x)$	Derivative of y = f(x) with respect to x
⇿	Equivalent to	α	Varies as
$\Sigma(\Sigma X)$	Sum (Sum of X values)	$|X|$	Numerical value of X

Greek Alphabet

Name of letter	Capital	Lower case	Trans-literation	Name of letter	Capital	Lower case	Transliteration
alpha	A	α	a	nu	N	ν	n
beta	B	$\mathit{6}$ or β	b	xi	Ξ	ξ	x
gamma	Γ	γ	g	omicron	O	o	o short
delta	Δ	δ	d	pi	Π	π	p
epsilon	E	ϵ	e short	rho	P	ρ	r
zeta	Z	ζ	z	sigma	Σ	σ or ς	s
eta	H	η	e long	tau	T	τ	t
theta	Θ	θ	th	upsilon	Υ	υ	y
iota	I	ι	i	phi	Φ	ϕ or φ	f
kappa	K	κ	k, c	chi	X	χ	ch as in German echt
lambda	Λ	λ	l	psi	Ψ	ψ	ps
mu	M	μ	m	omega	Ω	ω	o long

TABLES

TABLE A.1. Normal Distribution Areas

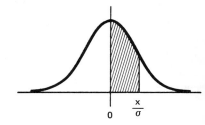

$\dfrac{x}{\sigma}$	0.00	0.01	0.02	0.03	0.04	0.05	0.06	0.07	0.08	0.09
0.0	0.0000	0.0040	0.0080	0.0120	0.0159	0.0199	0.0239	0.0279	0.0319	0.03
0.1	0.0398	0.0438	0.0478	0.0517	0.0557	0.0596	0.0636	0.0675	0.0714	0.07
0.2	0.0793	0.0832	0.0871	0.0910	0.0948	0.0987	0.1026	0.1064	0.1103	0.11
0.3	0.1179	0.1217	0.1255	0.1293	0.1331	0.1368	0.1406	0.1443	0.1480	0.15
0.4	0.1554	0.1591	0.1628	0.1664	0.1700	0.1736	0.1772	0.1808	0.1844	0.18
0.5	0.1915	0.1950	0.1985	0.2019	0.2054	0.2088	0.2123	0.2157	0.2190	0.22
0.6	0.2257	0.2291	0.2324	0.2357	0.2389	0.2422	0.2454	0.2486	0.2518	0.25
0.7	0.2580	0.2612	0.2642	0.2673	0.2704	0.2734	0.2764	0.2794	0.2823	0.28
0.8	0.2881	0.2910	0.2939	0.2967	0.2995	0.3023	0.3051	0.3078	0.3106	0.31
0.9	0.3159	0.3186	0.3212	0.3238	0.3264	0.3289	0.3315	0.3340	0.3365	0.33
1.0	0.3413	0.3438	0.3461	0.3485	0.3508	0.3531	0.3554	0.3577	0.3599	0.362
1.1	0.3643	0.3665	0.3686	0.3718	0.3729	0.3749	0.3770	0.3790	0.3810	0.383
1.2	0.3849	0.3869	0.3888	0.3907	0.3925	0.3944	0.3962	0.3980	0.3997	0.40
1.3	0.4032	0.4049	0.4066	0.4083	0.4099	0.4115	0.4131	0.4147	0.4162	0.41
1.4	0.4192	0.4207	0.4222	0.4236	0.4251	0.4265	0.4279	0.4292	0.4306	0.43
1.5	0.4332	0.4345	0.4357	0.4370	0.4382	0.4394	0.4406	0.4418	0.4430	0.44
1.6	0.4452	0.4463	0.4474	0.4485	0.4495	0.4505	0.4515	0.4525	0.4535	0.45
1.7	0.4554	0.4564	0.4573	0.4582	0.4591	0.4599	0.4608	0.4616	0.4625	0.46
1.8	0.4641	0.4649	0.4656	0.4664	0.4671	0.4678	0.4686	0.4693	0.4699	0.470
1.9	0.4713	0.4719	0.4726	0.4732	0.4738	0.4744	0.4750	0.4758	0.4762	0.476
2.0	0.4773	0.4778	0.4783	0.4788	0.4793	0.4798	0.4803	0.4808	0.4812	0.481
2.1	0.4821	0.4826	0.4830	0.4834	0.4838	0.4842	0.4846	0.4850	0.4854	0.485
2.2	0.4861	0.4865	0.4868	0.4871	0.4875	0.4878	0.4881	0.4884	0.4887	0.485
2.3	0.493	0.4896	0.4898	0.4901	0.4904	0.4906	0.4909	0.4911	0.4913	0.491
2.4	0.4918	0.4920	0.4922	0.4925	0.4927	0.4929	0.4931	0.4932	0.4934	0.493
2.5	0.4938	0.4940	0.4941	0.4943	0.4945	0.4946	0.4948	0.4949	0.4951	0.495
2.6	0.4953	0.4955	0.4956	0.4957	0.4959	0.4960	0.4961	0.4962	0.4963	0.496
2.7	0.4965	0.4966	0.4967	0.4968	0.4969	0.4970	0.4971	0.4972	0.4973	0.497
2.8	0.4974	0.4975	0.4976	0.4977	0.4977	0.4978	0.4979	0.4980	0.4980	0.498
2.9	0.4981	0.4982	0.4983	0.4984	0.4984	0.4984	0.4985	0.4985	0.4986	0.498
3.0	0.49865	0.4987	0.4987	0.4988	0.4988	0.4988	0.4989	0.4989	0.4989	0.499
3.1	0.49903	0.4991	0.4991	0.4991	0.4992	0.4992	0.4992	0.4992	0.4993	0.499
3.2	0.4993129									
3.3	0.4995166									
3.4	0.4996631									
3.5	0.4997674									
3.6	0.4998409									
3.7	0.4998922									
3.8	0.4999277									
3.9	0.4999519									
4.0	0.4999683									
4.5	0.4999966									
5.0	0.4999997133									

TABLE A.2. Normal Curve Areas

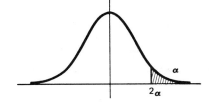

z	.00	.01	.02	.03	.04	.05	.06	.07	.08	.09
0.0	.5000	.4960	.4920	.4880	.4840	.4801	.4761	.4721	.4681	.4641
0.1	.4602	.4562	.4522	.4483	.4443	.4404	.4364	.4325	.4286	.4247
0.2	.4207	.4168	.4129	.4090	.4052	.4013	.3974	.3936	.3897	.3859
0.3	.3821	.3783	.3745	.3707	.3669	.3632	.3594	.3557	.3520	.3483
0.4	.3446	.3409	.3372	.3336	.3300	.3264	.3228	.3192		.3121
0.5	.3085	.3050	.3015	.2981	.2946	.2912	.2877	.2823	.2810	.2776
0.6	.2743	.2709	.2676	.2643	.2611	.2578	.2546	.2514	.2483	.2451
0.7	.2420	.2389	.2358	.2327	.2296	.2266	.2236	.2206	.2177	.2148
0.8	.2119	.2090	.2061	.2033	.2005	.1977	.1949	.1922	.1894	.1867
0.9	.1841	.1814	.1788	.1762	.1736	.1711	.1685	.1660	.1635	.1611
1.0	.1587	.1562	.1539	.1515	.1492	.1469	.1446	.1423	.1401	.1379
1.1	.1357	.1335	.1314	.1292	.1271	.1251	.1230	.1210	.1190	.1170
1.2	.1151	.1131	.1112	.1093	.1075	.1056	.1038	.1020	.1003	.0985
1.3	.0968	.0951	.0934	.0918	.0901	.0885	.0869	.0853	.0838	.0823
1.4	.0808	.0793	.0778	.0764	.0749	.0735	.0721	.0708	.0694	.0681
1.5	.0668	.0655	.0643	.0630	.0618	.0606	.0594	.0582	.0571	.0559
1.6	.0548	.0537	.0526	.0516	.0505	.0495	.0485	.0475	.0465	.0455
1.7	.0446	.0436	.0427	.0418	.0409	.0401	.0392	.0384	.0375	.0367
1.8	.0359	.0351	.0344	.0336	.0329	.0322	.0314	.0307	.0301	.0294
1.9	.0287	.0281	.0274	.0268	.0262	.0256		.0244	.0239	.0233
2.0	.0228	.0222	.0217	.0212	.0207	.0202	.0197	.0192	.0188	.0183
2.1	.0179	.0174	.0170	.0166	.0162		.0154	.0150	.0146	.0143
2.2	.0139	.0136	.0132	.0129	.0125	.0122	.0119	.0116	.0113	.0110
2.3	.0107	.0104	.0102	.00990	.00964	.00939	.00914	.00889	.00868	.00842
2.4	.00820	.00798	.00776	.00755	.00734	.00714	.00695	.00676	.00657	.00639
2.5	.00621	.00604	.00587	.00570	.00554	.00539	.00523	.00508	.00494	.00480
2.6	.00466	.00453	.00440	.00427	.00415	.00402	.00379	.00391	.00368	.00357
2.7	.00347	.00336	.00326	.00317	.00307	.00298	.00289	.00280	.00272	.00264
2.8	.00256	.00248	.00240	.00233	.00226	.00219	.00212	.00205	.00199	.00193
2.9	.00187	.00181	.00175	.00169	.00164	.00159	.00154	.00149	.00144	.00139

TABLE A.3. Critical Values of Student's *t*-Distribution

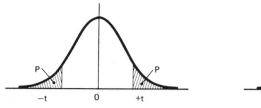

Two-tail   One-tail

DF	Probability (Two-tail)							
	0.80	0.40	0.20	0.10	0.05	0.02	0.01	0.001
1	0.325	1.376	3.078	6.314	12.706	31.821	63.657	636.619
2	0.289	1.061	1.886	2.920	4.303	6.965	9.925	31.598
3	0.277	0.978	1.638	2.353	3.182	4.541	5.841	12.941
4	0.271	0.941	1.533	2.132	2.776	3.747	4.604	8.610
5	0.267	0.920	1.476	2.015	2.571	3.365	4.032	6.859
6	0.265	0.906	1.440	1.943	2.447	3.143	3.707	5.959
7	0.263	0.896	1.415	1.895	2.365	2.998	3.499	5.405
8	0.262	0.889	1.397	1.860	2.306	2.896	3.355	5.041
9	0.261	0.883	1.383	1.833	2.262	2.821	3.250	4.781
10	0.260	0.879	1.372	1.812	2.228	2.764	3.169	4.587
11	0.260	0.876	1.363	1.796	2.201	2.718	3.106	4.437
12	0.259	0.873	1.356	1.782	2.179	2.681	3.055	4.318
13	0.259	0.870	1.350	1.771	2.160	2.650	3.012	4.221
14	0.258	0.868	1.345	1.761	2.145	2.624	2.977	4.140
15	0.258	0.866	1.341	1.753	2.131	2.602	2.947	4.073
16	0.258	0.865	1.337	1.746	2.120	2.583	2.921	4.015
17	0.257	0.863	1.333	1.740	2.110	2.567	2.898	3.965
18	0.257	0.862	1.330	1.734	2.101	2.552	2.878	3.922
19	0.257	0.861	1.328	1.729	2.093	2.539	2.861	3.883
20	0.257	0.860	1.325	1.725	2.086	2.528	2.845	3.850
21	0.257	0.859	1.323	1.721	2.080	2.518	2.831	3.819
22	0.256	0.858	1.321	1.717	2.074	2.508	2.819	3.792
23	0.256	0.858	1.319	1.714	2.069	2.500	2.807	3.767
24	0.256	0.857	1.318	1.711	2.064	2.492	2.797	37.45
25	0.256	0.856	1.316	1.708	2.060	2.485	2.787	3.725
26	0.256	0.856	1.315	1.706	2.056	2.479	2.779	3.707
27	0.256	0.855	1.314	1.703	2.052	2.473	2.771	3.690
28	0.256	0.855	1.313	1.701	2.048	2.467	2.763	3.674
29	0.256	0.854	1.311	1.699	2.045	2.462	2.756	3.659
30	0.256	0.854	1.310	1.697	2.042	2.457	2.750	3.646
40	0.255	0.851	1.303	1.684	2.021	2.423	2.704	3.551
60	0.254	0.848	1.296	1.671	2.000	2.390	2.660	3.460
120	0.254	0.845	1.289	1.658	1.980	2.358	2.617	3.373
∞	0.253	0.842	1.282	1.645	1.960	2.326	2.576	3.291
	0.40	0.20	0.10	0.05	0.025	0.01	0.005	0.0005
	Probability (One-tail)							

TABLE A.4. Distribution of Chi Square

DF	0.995	0.99	0.98	.975	0.95	0.90	0.80	.75	0.20	0.10	0.05	0.02	0.01	0.001
1	$.0^4393$	$.0^3157$	$.0^3628$	$.0^3982$	.00393	0.0158	0.0642	.1015	1.642	2.706	3.841	5.412	6.635	10.827
2	.0100	.0201	.0404	.0506	.103	.211	.446	.575	3.219	4.605	5.991	7.824	9.210	13.815
3	.0717	.115	.185	.216	.352	.584	1.005	1.213	4.642	6.251	7.815	9.837	11.345	16.268
4	.207	.297	.429	.484	.711	1.064	1.649	1.923	5.989	7.779	9.488	11.668	13.277	18.465
5	.412	.554	.752	.831	1.145	1.610	2.343	2.675	7.289	9.236	11.070	13.388	15.086	20.517
6	.676	.872	1.134	1.237	1.635	2.204	3.070	3.455	8.558	10.645	12.592	15.033	16.812	22.457
7	.989	1.239	1.564	1.690	2.167	2.833	3.822	4.255	9.803	12.017	14.067	16.622	18.475	24.322
8	1.344	1.646	2.032	2.180	2.733	3.490	4.594	5.071	11.030	13.362	15.507	18.168	20.090	26.125
9	1.735	2.088	2.532	2.700	3.325	4.168	5.380	5.899	12.242	14.684	16.919	19.679	21.666	27.877
10	2.156	2.558	3.059	3.247	3.940	4.865	6.179	6.737	13.442	15.987	18.307	21.161	23.209	29.588
11	2.603	3.053	3.609	3.816	4.575	5.578	6.989	7.584	14.631	17.275	19.675	22.618	24.725	31.264
12	3.074	3.571	4.178	4.404	5.226	6.304	7.807	8.438	15.812	18.549	21.026	24.054	26.217	32.909
13	3.565	4.107	4.765	5.009	5.892	7.042	8.634	9.299	16.985	19.812	22.362	25.472	27.688	34.528
14	4.075	4.660	5.368	5.629	6.571	7.790	9.467	10.165	18.151	21.064	23.685	26.873	29.141	36.123
15	4.601	5.229	5.985	6.262	7.261	8.547	10.307	11.037	19.311	22.307	24.996	28.259	30.578	37.697
16	5.142	5.812	6.614	6.908	7.962	9.312	11.152	11.912	20.465	23.542	26.296	29.633	32.000	39.252
17	5.697	6.408	7.255	7.564	8.672	10.085	12.002	12.792	21.615	24.769	27.587	30.995	33.409	40.790
18	6.265	7.015	7.906	8.231	9.390	10.865	12.857	13.675	22.760	25.989	28.869	32.346	34.805	42.312
19	6.844	7.633	8.567	8.907	10.117	11.651	13.716	14.562	23.900	27.204	30.144	33.687	36.191	43.820
20	7.434	8.260	9.237	9.591	10.851	12.443	14.578	15.452	25.038	28.412	31.410	35.020	37.566	45.315
21	8.034	8.897	9.915	10.283	11.591	13.240	15.445	16.344	26.171	29.615	32.671	36.343	38.932	46.797
22	8.643	9.542	10.600	10.982	12.338	14.041	16.314	17.240	27.301	30.813	33.924	37.659	40.289	48.268
23	9.260	10.196	11.293	11.689	13.091	14.848	17.187	18.137	28.429	32.007	35.172	38.968	41.638	49.728
24	9.886	10.856	11.992	12.401	13.848	15.659	18.062	19.037	29.553	33.196	36.415	40.270	42.980	51.179
25	10.520	11.524	12.697	13.120	14.611	16.473	18.940	19.939	30.675	34.382	37.652	41.566	44.314	52.620
26	11.160	12.198	13.409	13.844	15.379	17.292	19.820	20.843	31.795	35.563	38.885	42.856	45.642	54.052
27	11.808	12.879	14.125	14.573	16.151	18.114	20.703	21.749	32.912	36.741	40.113	44.140	46.963	55.476
28	12.461	13.565	14.847	15.308	16.928	18.939	21.588	22.657	34.027	37.916	41.337	45.419	48.278	56.893
29	13.121	14.256	15.574	16.047	17.708	19.768	22.475	23.567	35.139	39.087	42.557	46.693	49.588	58.302
30	13.787	14.953	16.306	16.791	18.493	20.599	23.364	24.478	36.250	40.256	43.773	47.962	50.892	59.703

TABLE A.5. Distribution of Variance Ratio

F Distribution, Upper 5% Points (F.95)

f₁ degrees of freedom in numerator
f₂ degrees of freedom in denominator

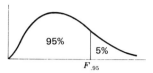

95%
5%
$F_{.95}$

f₂\f₁	1	2	3	4	5	6	7	8	9	10	12	15	20	24	30	40	60	120	∞
1	161	200	216	225	230	234	237	239	241	242	244	246	248	249	250	251	252	253	254
2	18.5	19.0	19.2	19.2	19.3	19.3	19.4	19.4	19.4	19.4	19.4	19.4	19.4	19.5	19.5	19.5	19.5	19.5	19.5
3	10.1	9.55	9.28	9.12	9.01	8.94	8.89	8.85	8.81	8.79	8.74	8.70	8.66	8.64	8.62	8.59	8.57	8.55	8.53
4	7.71	6.94	6.59	6.39	6.26	6.16	6.09	6.04	6.00	5.96	5.91	5.86	5.80	5.77	5.75	5.72	5.69	5.66	5.63
5	6.61	5.79	5.41	5.19	5.05	4.95	4.88	4.82	4.77	4.74	4.68	4.62	4.56	4.53	4.50	4.46	4.43	4.40	4.37
6	5.99	5.14	4.76	4.53	4.39	4.28	4.21	4.15	4.10	4.06	4.00	3.94	3.87	3.84	3.81	3.77	3.74	3.70	3.67
7	5.59	4.74	4.35	4.12	3.97	3.87	3.79	3.73	3.68	3.64	3.57	3.51	3.44	3.41	3.38	3.34	3.30	3.27	3.23
8	5.32	4.46	4.07	3.84	3.69	3.58	3.50	3.44	3.39	3.35	3.28	3.22	3.15	3.12	3.08	3.04	3.01	2.97	2.93
9	5.12	4.26	3.86	3.63	3.48	3.37	3.29	3.23	3.18	3.14	3.07	3.01	2.94	2.90	2.86	2.83	2.79	2.75	2.71
10	4.96	4.10	3.71	3.48	3.33	3.22	3.14	3.07	3.02	2.98	2.91	2.85	2.77	2.74	2.70	2.66	2.62	2.58	2.54
11	4.84	3.98	3.59	3.36	3.20	3.09	3.01	2.95	2.90	2.85	2.79	2.72	2.65	2.61	2.57	2.53	2.49	2.45	2.40
12	4.75	3.89	3.49	3.26	3.11	3.00	2.91	2.85	2.80	2.75	2.69	2.62	2.54	2.51	2.47	2.43	2.38	2.34	2.30
13	4.67	3.81	3.41	3.18	3.03	2.92	2.83	2.77	2.71	2.67	2.60	2.53	2.46	2.42	2.38	2.34	2.30	2.25	2.21
14	4.60	3.74	3.34	3.11	2.96	2.85	2.76	2.70	2.65	2.60	2.53	2.46	2.39	2.35	2.31	2.27	2.22	2.18	2.13
15	4.54	3.68	3.29	3.06	2.90	2.79	2.71	2.64	2.59	2.54	2.48	2.40	2.33	2.29	2.25	2.20	2.16	2.11	2.07
16	4.49	3.63	3.24	3.01	2.85	2.74	2.66	2.59	2.54	2.49	2.42	2.35	2.28	2.24	2.19	2.15	2.11	2.06	2.01
17	4.45	3.59	3.20	2.96	2.81	2.70	2.61	2.55	2.49	2.45	2.38	2.31	2.23	2.19	2.15	2.10	2.06	2.01	1.96
18	4.41	3.55	3.16	2.93	2.77	2.66	2.58	2.51	2.46	2.41	2.34	2.27	2.19	2.15	2.11	2.06	2.02	1.97	1.92
19	4.38	3.52	3.13	2.90	2.74	2.63	2.54	2.48	2.42	2.38	2.31	2.23	2.16	2.11	2.07	2.03	1.98	1.93	1.88
20	4.35	3.49	3.10	2.87	2.71	2.60	2.51	2.45	2.39	2.35	2.28	2.20	2.12	2.08	2.04	1.99	1.95	1.90	1.84
21	4.32	3.47	3.07	2.84	2.68	2.57	2.49	2.42	2.37	2.32	2.25	2.18	2.10	2.05	2.01	1.96	1.92	1.87	1.81
22	4.30	3.44	3.05	2.82	2.66	2.55	2.46	2.40	2.34	2.30	2.23	2.15	2.07	2.03	1.98	1.94	1.89	1.84	1.78
23	4.28	3.42	3.03	2.80	2.64	2.53	2.44	2.37	2.32	2.27	2.20	2.13	2.05	2.01	1.96	1.91	1.86	1.81	1.76
24	4.26	3.40	3.01	2.78	2.62	2.51	2.42	2.36	2.30	2.25	2.18	2.11	2.03	1.98	1.94	1.89	1.84	1.79	1.73
25	4.24	3.39	2.99	2.76	2.60	2.49	2.40	2.34	2.28	2.24	2.16	2.09	2.01	1.96	1.92	1.87	1.82	1.77	1.71
26	4.23	3.37	2.98	2.74	2.59	2.47	2.39	2.32	2.27	2.22	2.15	2.07	1.99	1.95	1.90	1.85	1.80	1.75	1.69
27	4.21	3.35	2.96	2.73	2.57	2.46	2.37	2.31	2.25	2.20	2.13	2.06	1.97	1.93	1.88	1.84	1.79	1.73	1.67
28	4.20	3.34	2.95	2.71	2.56	2.45	2.36	2.29	2.24	2.19	2.12	2.04	1.96	1.91	1.87	1.82	1.77	1.71	1.65
29	4.18	3.33	2.93	2.70	2.55	2.43	2.35	2.28	2.22	2.18	2.10	2.03	1.94	1.90	1.85	1.81	1.75	1.70	1.64
30	4.17	3.32	2.92	2.69	2.53	2.42	2.33	2.27	2.21	2.16	2.09	2.01	1.93	1.89	1.84	1.79	1.74	1.68	1.62
40	4.08	3.23	2.84	2.61	2.45	2.34	2.25	2.18	2.12	2.08	2.00	1.92	1.84	1.79	1.74	1.69	1.64	1.58	1.51
60	4.00	3.15	2.76	2.53	2.37	2.25	2.17	2.10	2.04	1.99	1.92	1.84	1.75	1.70	1.65	1.59	1.53	1.47	1.39
120	3.92	3.07	2.68	2.45	2.29	2.18	2.09	2.02	1.96	1.91	1.83	1.75	1.66	1.61	1.55	1.50	1.43	1.35	1.25
∞	3.84	3.00	2.60	2.37	2.21	2.10	2.01	1.94	1.88	1.83	1.75	1.67	1.57	1.52	1.46	1.39	1.32	1.22	1.00

(a)

TABLE A.5. Distribution of Variance Ratio (*continued*)

Distribution, Upper 1% Points (F$_{.99}$)

f_1 degrees of freedom in numerator
f_2 degrees of freedom in denominator

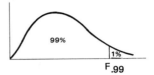

F$_{.99}$

f_2 \ f_1	1	2	3	4	5	6	7	8	9	10	12	15	20	24	30	40	60	120	∞
1	4052	5000	5403	5625	5764	5859	5928	5981	6023	6056	6106	6157	6209	6235	6261	6287	6313	6339	6366
2	98.5	99.0	99.2	99.2	99.3	99.3	99.4	99.4	99.4	99.4	99.4	99.4	99.4	99.5	99.5	99.5	99.5	99.5	99.5
3	34.1	30.8	29.5	28.7	28.2	27.9	27.7	27.5	27.3	27.2	27.1	26.9	26.7	26.6	26.5	26.4	26.3	26.2	26.1
4	21.2	18.0	16.7	16.0	15.5	15.2	15.0	14.8	14.7	14.5	14.4	14.2	14.0	13.9	13.8	13.7	13.7	13.6	13.5
5	16.3	13.3	12.1	11.4	11.0	10.7	10.5	10.3	10.2	10.1	9.89	9.72	9.55	9.47	9.38	9.29	9.20	9.11	9.02
6	13.7	10.9	9.78	9.15	8.75	8.47	8.26	8.10	7.98	7.87	7.72	7.56	7.40	7.31	7.23	7.14	7.06	6.97	6.88
7	12.2	9.55	8.45	7.85	7.46	7.19	6.99	6.84	6.72	6.62	6.47	6.31	6.16	6.07	5.99	5.91	5.82	5.74	5.65
8	11.3	8.65	7.59	7.01	6.63	6.37	6.18	6.03	5.91	5.81	5.67	5.52	5.36	5.28	5.20	5.12	5.03	4.95	4.86
9	10.6	8.02	6.99	6.42	6.06	5.80	5.61	5.47	5.35	5.26	5.11	4.96	4.81	4.73	4.65	4.57	4.48	4.40	4.31
10	10.0	7.56	6.55	5.99	5.64	5.39	5.20	5.06	4.94	4.85	4.71	4.56	4.41	4.33	4.25	4.17	4.08	4.00	3.91
11	9.65	7.21	6.22	5.67	5.32	5.07	4.89	4.74	4.63	4.54	4.40	4.25	4.10	4.02	3.94	3.86	3.78	3.69	3.60
12	9.33	6.93	5.95	5.41	5.06	4.82	4.64	4.50	4.39	4.30	4.16	4.01	3.86	3.78	3.70	3.62	3.54	3.45	3.36
13	9.07	6.70	5.74	5.21	4.86	4.62	4.44	4.30	4.19	4.10	3.96	3.82	3.66	3.59	3.51	3.43	3.34	3.25	3.17
14	8.86	6.51	5.56	5.04	4.70	4.46	4.28	4.14	4.03	3.94	3.80	3.66	3.51	3.43	3.35	3.27	3.18	3.09	3.00
15	8.68	6.36	5.42	4.89	4.56	4.32	4.14	4.00	3.89	3.80	3.67	3.52	3.37	3.29	3.21	3.13	3.05	2.96	2.87
16	8.53	6.23	5.29	4.77	4.44	4.20	4.03	3.89	3.78	3.69	3.55	3.41	3.26	3.18	3.10	3.02	2.93	2.84	2.75
17	8.40	6.11	5.19	4.67	4.34	4.10	3.93	3.79	3.68	3.59	3.46	3.31	3.16	3.08	3.00	2.92	2.83	2.75	2.65
18	8.29	6.01	5.09	4.58	4.25	4.01	3.84	3.71	3.60	3.51	3.37	3.23	3.08	3.00	2.92	2.84	2.75	2.66	2.57
19	8.18	5.93	5.01	4.50	4.17	3.94	3.77	3.63	3.52	3.43	3.30	3.15	3.00	2.92	2.84	2.76	2.67	2.58	2.49
20	8.10	5.85	4.94	4.43	4.10	3.87	3.70	3.56	3.46	3.37	3.23	3.09	2.94	2.86	2.78	2.69	2.61	2.52	2.42
21	8.02	5.78	4.87	4.37	4.04	3.81	3.64	3.51	3.40	3.31	3.17	3.03	2.88	2.80	2.72	2.64	2.55	2.46	2.36
22	7.95	5.72	4.82	4.31	3.99	3.76	3.59	3.45	3.35	3.26	3.12	2.98	2.83	2.75	2.67	2.58	2.50	2.40	2.31
23	7.88	5.66	4.76	4.26	3.94	3.71	3.54	3.41	3.30	3.21	3.07	2.93	2.78	2.70	2.62	2.54	2.45	2.35	2.26
24	7.82	5.61	4.72	4.22	3.90	3.67	3.50	3.36	3.26	3.17	3.03	2.89	2.74	2.66	2.58	2.49	2.40	2.31	2.21
25	7.77	5.57	4.68	4.18	3.86	3.63	3.46	3.32	3.22	3.13	2.99	2.85	2.70	2.62	2.54	2.45	2.36	2.27	2.17
26	7.72	5.53	4.64	4.14	3.82	3.59	3.42	3.29	3.18	3.09	2.96	2.82	2.66	2.58	2.50	2.42	2.33	2.23	2.13
27	7.68	5.49	4.60	4.11	3.78	3.56	3.39	3.26	3.15	3.06	2.93	2.78	2.63	2.55	2.47	2.38	2.29	2.20	2.10
28	7.64	5.45	4.57	4.07	3.75	3.53	3.36	3.23	3.12	3.03	2.90	2.75	2.60	2.52	2.44	2.35	2.26	2.17	2.06
29	7.60	5.42	4.54	4.04	3.73	3.50	3.33	3.20	3.09	3.00	2.87	2.73	2.57	2.49	2.41	2.33	2.23	2.14	2.03
30	7.56	5.39	4.51	4.02	3.70	3.47	3.30	3.17	3.07	2.98	2.84	2.70	2.55	2.47	2.39	2.30	2.21	2.11	2.01
40	7.31	5.18	4.31	3.83	3.51	3.29	3.12	2.99	2.89	2.80	2.66	2.52	2.37	2.29	2.20	2.11	2.02	1.92	1.80
60	7.08	4.98	4.13	3.65	3.34	3.12	2.95	2.82	2.72	2.63	2.50	2.35	2.20	2.12	2.03	1.94	1.84	1.73	1.60
120	6.85	4.79	3.95	3.48	3.17	2.96	2.79	2.66	2.56	2.47	2.34	2.19	2.03	1.95	1.86	1.76	1.66	1.53	1.38
∞	6.63	4.61	3.78	3.32	3.02	2.80	2.64	2.51	2.41	2.32	2.18	2.04	1.88	1.79	1.70	1.59	1.47	1.32	1.00

(b)

TABLE A.6. Values of the Correlation Coefficient for Different Levels of Significance

ν	$a = 0.10$	0.05	0.02	0.01
1	0.98769	0.996917	0.9995066	0.998766
2	0.90000	0.95000	0.98000	0.990000
3	0.8054	0.8783	0.93433	0.95873
4	0.7293	0.8114	0.8822	0.91720
5	0.6694	0.7545	0.8329	0.8745
6	0.6215	0.7067	0.7887	0.8343
7	0.5822	0.6664	0.7498	0.7977
8	0.5494	0.6319	0.7155	0.7646
9	0.5214	0.6021	0.6851	0.7348
10	0.4973	0.5760	0.6581	0.7079
11	0.4762	0.5529	0.6339	0.6835
12	0.4575	0.5324	0.6120	0.6614
13	0.4409	0.5139	0.5923	0.6411
14	0.4259	0.4973	0.5742	0.6226
15	0.4124	0.4821	0.5577	0.6055
16	0.4000	0.4683	0.5425	0.5897
17	0.3887	0.4555	0.5285	0.5751
18	0.3783	0.4438	0.5155	0.5614
19	0.3687	0.4329	0.5034	0.5487
20	0.3598	0.4227	0.4921	0.5368
25	0.3233	0.3809	0.4451	0.4869
30	0.2960	0.3494	0.4093	0.4487
35	0.2746	0.3246	0.3810	0.4182
40	0.2573	0.3044	0.3578	0.3932
45	0.2428	0.2875	0.3384	0.3721
50	0.2306	0.2732	0.3218	0.3541
60	0.2108	0.2500	0.2948	0.3248
70	0.1954	0.2319	0.2737	0.3017
80	0.1829	0.2172	0.2565	0.2830
90	0.1726	0.2050	0.2422	0.2673
100	0.1638	0.1946	0.2301	0.2540

For a total correlation, ν is 2 less than the number of pairs in the sample; for a partial correlation, the number of eliminated variates also should be subtracted.

TABLE A.7. Critical Values for Testing Outliers

(x_1 is the extreme value)

Statistic	Number of Means, n	Critical Values † $a = 0.05$	$a = 0.01$
For a single outlier x_1	3	0.941	0.988
	4	0.765	0.889
$r_{10} = \dfrac{x_2 - x_1}{x_n - x_1}$	5	0.642	0.780
	6	0.560	0.698
	7	0.507	0.637
	8	0.468	0.590
	9	0.437	0.555
	10	0.412	0.527
For an outlier x_1	8	0.554	0.683
when x_n is suspect	9	0.512	0.635
	10	0.477	0.597
$r_{11} = \dfrac{x_2 - x_1}{x_{n-1} - x_1}$			
For an outlier x_1	10	0.612	0.726
when x_2 and x_n are both suspect;	11	0.576	0.679
	12	0.546	0.642
$r_{21} = \dfrac{x_3 - x_1}{x_{n-1} - x_1}$	13	0.521	0.615
For an outlier x_1,	14	0.546	0.641
when x_2, x_{n-1}, and x_n	15	0.525	0.616
are all suspect	16	0.507	0.595
	17	0.490	0.577
	18	0.475	0.561
$r_{22} = \dfrac{x_3 - x_1}{x_{n-2} - x_1}$	19	0.462	0.547
	20	0.450	0.535
	21	0.440	0.524
	22	0.430	0.514
	23	0.421	0.505
	24	0.413	0.497
	25	0.406	0.489
	26	0.399	0.486
	27	0.393	0.475
	28	0.387	0.469
	29	0.381	0.463
	30	0.376	0.457

† These are for a "one-sided" test. For a "two-sided" test, $a = 0.10$ and 0.02. $P_r(r_{ij} > R) = a$.

TABLE A.8. Critical Values of the Spearman Rank Order Correlation Coefficient r_{rho}

	Level of significance for one-tailed test			
	0.05	0.025	0.01	0.005
	Level of significance for two-tailed test			
n	0.10	0.05	0.02	0.01
5	0.900	1.000	1.000	--
6	0.829	0.886	0.943	1.000
7	0.714	0.786	0.893	0.929
8	0.643	0.738	0.833	0.881
9	0.600	0.683	0.783	0.833
10	0.564	0.648	0.746	0.794
12	0.506	0.591	0.712	0.777
14	0.456	0.544	0.645	0.715
16	0.425	0.506	0.601	0.665
18	0.399	0.475	0.564	0.625
20	0.377	0.450	0.534	0.591
22	0.359	0.428	0.508	0.562
24	0.343	0.409	0.485	0.537
26	0.329	0.392	0.465	0.515
28	0.317	0.377	0.448	0.496
30	0.306	0.364	0.432	0.478

APPENDIX 3

ACKNOWLEDGMENTS

Table A.1. Normal Distribution Areas. Source: Eugene L. Grant, *Statistical Quality Control*, 3d ed., McGraw-Hill, New York, 1964. With permission.

Table A.2. Normal Curve Areas. Reprinted by permission from Frederick E. Croxton, *Elementary Statistics with Applications in Medicine*, Prentice-Hall, Englewood Cliffs, New Jersey, 1953, p. 323.

Table A.3. Critical Values of Student's *t* Distribution. Reproduced with permission from E. S. Pearson, Tables of the percentage points of the incomplete beta function, *Biometrika*, Vol. 32, 1941, pp. 168–181, with the kind permission of the Biometrika Trustees, and from C. A. Bennett and N. L. Franklin, *Statistical Analysis in Chemistry and the Chemical Industry*, p. 696, John Wiley & Sons, New York, 1954.

Table A.4. Distribution of Chi Square. Reproduced with permission from E. S. Pearson, Tables of the percentage points of the chi square distribution, *Biometrika*, Vol. 32, 1941, pp. 188–189, with the kind permission of the Biometrika Trustees, and from C. A. Bennett and N. L. Franklin, *Statistical Analysis in Chemistry and the Chemical Industry*, pp. 694, 695, John Wiley & Sons, New York, 1954.

Table A.5. Distribution of Variance Ratio. Reproduced with permission from E. S. Pearson, Tables of percentage points of the inverted beta (*F*) distribution, *Biometrika*, Vol. 32, 1943, pp. 73–88, with the kind permission of the Biometrika Trustees, and from C. A. Bennett and N. L. Franklin, *Statistical Analysis in Chemistry and the Chemical Industry*, pp. 702–711, John Wiley & Sons, New York, 1954.

Table A.6. Values of the Correlation Coefficient for Different Levels of Significance. Source: R. A. Fisher, Table V.A (p. 211), *Statistical Methods for Research Workers*, 14th ed., Copyright 1970, University of Adelaide. With permission, Macmillan, New York.

Table A.7. Critical Values for Testing Outliers. Source: Entries are extracted from W. J. Dixon, Processing data for outliers, *Biometrics*, Vol. 9, 1953, p. 74, with the kind permission of the Biometrika Trustees.

Table A.8. Critical Values of the Spearman Rank Order Correlation Coefficient r_{rho}. Source: F. N. David, *Tables of the Correlation Coefficient*, Biometrika, Cambridge, The University Press, 1938. With the kind permission of the Biometrika Trustees.

Table 4.3. Number of Observations for *t* Test of Mean. Source: O. L. Davies, *Design and Analysis of Industrial Experiments*, Oliver & Boyd, Edinburgh and London, 1956.

Table 4.4. Number of Observations for *t* Test of Difference Between Two Means. Source: O. L. Davies, *Design and Analysis of Industrial Experiments*, Oliver & Boyd, Edinburgh and London, 1956.

Table 4.5. Number of Observations Required for the Comparison of a Population Variance with a Standard Value Using the Chi Square Test. Source: *Selected Techniques of Statistical Analysis* by C. Eisenhart, M. W. Hastay, and W. A. Wallis, Eds., Copyright 1947 by the Statistical Research Group, Columbia University. Used with the permission of McGraw-Hill, New York.

659

Table 4.6. Number of Observations Required for the Comparison of Two Population Variances Using the *F*-Test. Source: *Selected Techniques of Statistical Analysis* by C. Eisenhart, M. W. Hastay, and W. A. Wallis, Eds., Copyright 1947 by the Statistical Research Group, Columbia University. Used with the permission of McGraw-Hill, New York.

Table 8.3. Sample Size for Acceptance Sampling. Reprinted with permission from H. Burstein, *Attribute Sampling: Tables and Explanations*, McGraw-Hill, New York, 1971.

Table 8.7. Factors to Calculate AOQL Using Values of *y* for Given Values of *c*. Abridged and reproduced with permission from H. F. Dodge and H. G. Romig, *Sampling Inspection Tables—Single & Double Sampling*, 2nd ed., John Wiley & Sons, New York, 1959, and with the specific permission of Bell Laboratories, copyright 1944.

Table 8.10. Values for Use in Designing a Sampling Plan. Source: A. J. Duncan, *Quality Control and Industrial Statistics*, 4th ed., Irwin, Homewood, Illinois, 1974 and Chemical Corps. Engineering Agency, Manual No. 2.

Table 8.11. Values for Use in Designing a Sampling Plan. Adapted and reproduced with permission from A. J. Duncan, *Quality Control and Industrial Statistics*, 4th ed., Irwin, Homewood, Illinois, 1974.

Table 8.16. Values for Use in Designing a Multiple Sample Plan. Source: A. J. Duncan, *Quality Control and Industrial Statistics*, 4th ed., Irwin, Homewood, Illinois 1974 and Chemical Corps. Engineering Agency, Manual No. 2.

Table 8.24. Dodge–Romig Single Sampling Table for Lot Tolerance Percent Defective (LTPD) = 5.0%. Abridged and reproduced with permission from H. F. Dodge and H. G. Romig, *Sampling Inspection Tables—Single and Double Sampling*, 2nd ed., John Wiley & Sons, New York, 1959, and with specific permission of Bell Laboratories, copyright 1944.

Table 8.25. Dodge–Romig Double Sampling Table for Lot Tolerance Percent Defective (LTPD) = 5.0%. Abridged and reproduced with permission from H. F. Dodge and H. G. Romig, *Sampling Inspection Tables—Single and Double Sampling*, 2nd ed., John Wiley & Sons, New York, 1959, and with the specific permission of Bell Laboratories, copyright 1944.

Table 8.26. Dodge–Romig Single Sampling Table for Average Outgoing Quality Limit (AOQL). Abridged and reproduced with permission from H. F. Dodge and H. G. Romig, *Sampling Inspection Tables—Single and Double Sampling*, 2nd ed., John Wiley & Sons, New York, 1959, and with the specific permission of Bell Laboratories, copyright 1944.

Table 8.27. Dodge–Romig Double Sampling Table for Average Outgoing Quality Limit (AOQL). Abridged and reproduced with permission from H. F. Dodge and H. G. Romig, *Sampling Inspection Tables—Single and Double Sampling*, 2nd ed., John Wiley & Sons, New York, 1959, and with the specific permission of Bell Laboratories, copyright 1944.

Table 9.1. Percentiles of Hartley's *H* Statistic Distribution. Source: Reprinted with permission from H. A. David, Upper 5 and 1% points of the maximum *F*-ratio, *Biometrika*, Vol. 39, 1952, pp. 422–424. With the kind permission of the Biometrika trustees.

Table 9.2. Cochran's Table. Source: *Selected Techniques of Statistical Analysis* by C. Eisenhart, M. W. Hastay, and W. A. Wallis, Eds., copyright 1947, by the Statistical Research Group, Columbia University. Used with permission of McGraw-Hill, New York.

Figure 9.4. Pearson and Hartley Charts for the Power of the *F*-Test. Source: E. S. Pearson and H. O. Hartley in *Biometrika*, Vol. 38, 1951, pp. 115, 122, for $v_1 = 3$. Used with the kind permission of the Biometrika Trustees.

Table 11.3. Lower-Tail Critical Values for Wilcoxon's Signed-Rank Statistics. Reprinted with permission from the *CRC Handbook of Tables for Probability and Statistics*, 2nd ed., 1968, W. H. Beyer, Ed., Table X.2, pp. 399–400, copyright, The CRC Press, Boca Raton, Florida.

Table 11.4. Lower-Tail Critical Values for the Mann–Whitney *U*-Statistic. Reprinted with permission from the *CRC Handbook of Tables for Probability and Statistics*, 2nd ed., 1968, W. H. Beyer, Ed., Table X.4, pp. 405–408, copyright, The CRC Press, Boca Raton, Florida.

Table 11.5. Critical Values for Friedman's Statistic. Source: Donald B. Owen, *Handbook of Statistical Tables*, copyright 1962. U.S. Department of Energy. Table 14.1. Published by Addison-Wesley, Reading, Massachusetts. Reprinted with permission of the publisher.

Table 11.6. Upper Tail Critical Values for Spearman's Rank Correlation Coefficient. Source: R. A. Fisher, Table V.A (p. 211), *Statistical Methods for Research Workers*, 14th ed., copyright 1970, University of Adelaide. With permission of Macmillan, New York.

Table 11.7. Critical Values for the Total Number of Runs (R). Reprinted, with permission, from the *CRC Handbook of Tables for Probability and Statistics*, 2nd ed., 1968, W. H. Beyer, Ed., Table X.6, pp. 414–424, copyright. The CRC Press, Boca Raton, Florida.

Table 16.4. Factors for Computing Control Chart Lines. Copyright American Society for Testing and Materials, Philadelphia, Pennsylvania. Adapted, with permission; and from *Statistical Quality Control*, 3rd ed., by E. L. Grant, used with the permission of McGraw-Hill, New York.

Portions of Chapter 7 appeared in *Industrial Engineering*, Published by the American Institute of Industrial Engineers, and are used with the kind permission of *Industrial Engineering* magazine.

Portions of Chapter 12 appeared in the *Journal of Coatings Technology* and are used with the kind permission of the journal and its publisher, the Federation of Societies for Coatings Technology.

Portions of the tables and figures for Chapter 8 were based on tables from D.O.D. Military Standard MIL-STD-105D, April 29, 1963, and Military Handbook Guide for Attribute Lot Sampling Inspection, MIL-HDBK-53-1A, February 1, 1982, as follows: **II-A.** Single Sampling Plans for Normal Inspection; **II-B.** Single Sampling Plans for Tightened Inspection; **II-C.** Single Sampling Plans for Reduced Inspection; **III-A.** Double Sampling Plans for Normal Inspection; **III-B.** Double Sampling Plans for Tightened Inspection; **III-C.** Double Sampling Plans for Reduced Inspection; **IV-A.** Multiple Sampling Plans for Normal Inspection; **IV-B.** Multiple Sampling Plans for Tightened Inspection; **IV-C.** Multiple Sampling Plans for Reduced Inspection; **V-A.** Average Outgoing Quality Limit Factors for *Normal* Inspection (Single Sampling); **V-B.** Average Outgoing Quality Limit Factors for *Tightened* Inspection (Single Sampling); **VI-A.** Limiting Quality for Pa = 10 Percent (defects per 100 units); **VI-B.** Limiting Quality for Pa = 10 Percent (defects per 100 units); **VII-A.** Limiting Quality for Pa = 5 Percent (defects per 100 units); **VII-B.** Limiting Quality for Pa = 5 Percent (defects per 100 units); **VIII.** Limit Numbers for Reduced Inspection; **X A-R.** Tables for Sample Size Code Letter A-R. Charts A-R. OC Curves for Single Sampling Plans; MIL-HDBK-53-1A February 1, 1982. Military Handbook Guide for Attribute Lot Sampling. Inspection and MIL-STD-105 DOD; · **C A-R.** Table and Chart C A-R for Sample Size Code Letter A-R. Charts C A-R. OC Curves for Normal-Tightened Single Sampling Schemes.

INDEX